Process Intensification for Green Chemistry

Process Intensification for Green Chemistry

Engineering Solutions for Sustainable Chemical Processing

Edited by

KAMELIA BOODHOO and ADAM HARVEY

School of Chemical Engineering & Advanced Materials
Newcastle University, UK

WILEY

A John Wiley & Sons, Ltd., Publication

This edition first published 2013

Registered office
John Wiley & Sons Ltd., The Atrium, Southern Gate, Chichester, West Sussex, PO19 8SQ, United Kingdom

For details of our global editorial offices, for customer services and for information about how to apply for permission to reuse the copyright material in this book please see our website at www.wiley.com.

Library of Congress Cataloging-in-Publication Data applied for.

A catalogue record for this book is available from the British Library.

ISBN: 9780470972670
Set in 10/12 pt Times by Thomson Digital, Noida, India

Contents

List of Contributors

Kamelia Boodhoo School of Chemical Engineering & Advanced Materials, Newcastle University, UK

Svetlana Borukhova Department of Chemical Engineering and Chemistry, Micro Flow Chemistry & Process Technology, Eindhoven University of Technology, Eindhoven, The Netherlands

Vitaly Budarin Green Chemistry Centre of Excellence, University of York, York, UK

James Clark Green Chemistry Centre of Excellence, University of York, York, UK

Enrico Drioli Institute on Membrane Technology, CNR-ITM, University of Calabria, Rende, Calabria, Italy

Dag Eimer D-IDE AS, Teknologisenteret, Porsgrunn, Norway

Niels Eldrup Sivilingeniør Eldrup AS, Teknologisenteret, Porsgrunn, Norway

Dena Ghiasy School of Chemical Engineering & Advanced Materials, Newcastle University, UK

Lidietta Giorno Institute on Membrane Technology, CNR-ITM, University of Calabria, Rende, Calabria, Italy

Parag Gogate Chemical Engineering Department, Institute of Chemical Technology, Matunga, Mumbai, India

Mark Gronnow Green Chemistry Centre of Excellence, University of York, York, UK

Jan Harmsen Harmsen Consultancy BV, Nieuwerkerk aan den Ijssel, The Netherlands

Adam Harvey School of Chemical Engineering & Advanced Materials, Newcastle University, UK

Volker Hessel Department of Chemical Engineering and Chemistry, Micro Flow Chemistry & Process Technology, Eindhoven University of Technology, Eindhoven, The Netherlands

Anton A. Kiss Arnhem, The Netherlands

Keat T. Lee School of Chemical Engineering, Universiti Sains Malaysia, Engineering Campus, Pulau Pinang, Malaysia

Steven Lim School of Chemical Engineering, Universiti Sains Malaysia, Engineering Campus, Pulau Pinang, Malaysia

Duncan Macquarrie Green Chemistry Centre of Excellence, University of York, York, UK

Rosalinda Mazzei Institute on Membrane Technology, CNR-ITM, University of Calabria, Rende, Calabria, Italy

Vijayanand Moholkar Chemical Engineering Department, Indian Institute of Technology, Guwahati, Assam, India

Aniruddha Pandit Chemical Engineering Department, Institute of Chemical Technology, Matunga, Mumbai, India

Emma Piacentini Institute on Membrane Technology, CNR-ITM, University of Calabria, Rende, Calabria, Italy

Joseph Wood School of Chemical Engineering, University of Birmingham, Birmingham, UK

Preface

Of late, a tremendous effort has been made to implement more sustainable and environmentally friendly processes in the chemical industry. Increased legislation on emissions and waste disposal and the need for businesses to remain highly competitive and to demonstrate their social responsibility are just some of the reasons for this drive towards greener processing. The successful implementation of greener chemical processes relies not only on the development of more efficient catalysts for synthetic chemistry but also, and as importantly, on the development of reactor and separation technologies that can deliver enhanced processing performance in a safe, cost-effective and energy-efficient manner. In some sectors, particularly those related to pharmaceuticals and fine chemicals processing, separations is often the stage at which the most waste is generated, through large amounts of solvents for purification, and this must therefore be addressed at the outset when novel green reactions are explored. The ideal process is one in which byproducts are reduced or eliminated altogether at the reaction stage, rather than removed *after* they are formed – a concept referred to as *waste minimization at source*.

Process intensification (PI) has emerged as a promising field that can effectively tackle these process challenges while offering at the same time the potential for 'clean' or 'green' processing in order to diminish the environmental impact presented by the chemical industry. One of the ways this is made possible is by minimizing the scale of reactors operating ideally in continuous mode so that more rapid heat/mass-transfer/mixing rates and plug flow behaviour can be achieved for high selectivity in optimized reaction processes.

This book covers the latest developments in a number of intensified technologies, with particular emphasis on their application to green chemical processes. The focus is on intensified *reactor* technologies, such as spinning disc reactors, microreactors, monolith reactors, oscillatory flow reactors and so on, and a number of combined or *hybrid* reactor/separator systems, the most well known and widely used in industry being reactive distillation (RD). PI is about not only the implementation of novel designs of reaction/separation units but also the use of *novel processing methods* such as alternative forms of energy input to promote reactions. A notable example here is ultrasonic energy, applications for which are also highlighted in this book. Each chapter presents relevant case studies examining the green processing aspect of these technologies. Towards the end of the book, we have included four chapters to emphasize the industry relevance of PI, with particular focus on the general business context within which intensification technology development and application takes place; on process economics and environmental impact; on the energy-saving potential of intensification technologies; and on practical considerations for industrial implementation of PI.

The book is intended to be a useful resource for practising engineers and chemists alike who are interested in applying intensified reactor and/or separator systems in a range of industries, such as petrochemicals, fine/specialty chemicals, pharmaceuticals and so on. Not only will it provide a basic knowledge of chemical engineering principles and PI for chemists and engineers who may be unfamiliar with these concepts, but it will be a valuable tool for chemical engineers who wish to fully apply their background in reaction and separation engineering to the design and implementation of green processing technologies based on PI principles. Students on undergraduate and post-graduate degree programmes which cover topics on advanced reactor designs, PI, clean technology and green chemistry will also have at their disposal a vast array of material to help them gain a better understanding of the practical applications of these different areas.

We would like to thank all contributors to this book for their commitment in producing their high-quality manuscripts. Our heartfelt gratitude goes to Sarah Hall, Sarah Tilley and Rebecca Ralf at Wiley-Blackwell, whose support and encouragement throughout this project made it all possible.

Kamelia Boodhoo
Adam Harvey
August 2012

1

Process Intensification: An Overview of Principles and Practice

Kamelia Boodhoo and Adam Harvey

School of Chemical Engineering & Advanced Materials, Newcastle University, UK

1.1 Introduction

The beginning of the 21st century has been markedly characterized by increased environmental awareness and pressure from legislators to curb emissions and improve energy efficiency by adopting 'greener technologies'. In this context, the need for the chemical industry to develop processes which are more sustainable or eco-efficient has never been so vital. The successful delivery of green, sustainable chemical technologies at industrial scale will inevitably require the development of innovative processing and engineering technologies that can transform industrial processes in a fundamental and radical fashion. In bioprocessing, for example, genetic engineering of microorganisms will obviously play a major part in the efficient use of biomass, but development of novel reactor and separation technologies giving high reactor productivity and ultimately high-purity products will be equally important for commercial success. Process intensification (PI) can provide such sought-after innovation of equipment design and processing to enhance process efficiency.

Process Intensification for Green Chemistry: Engineering Solutions for Sustainable Chemical Processing, First Edition. Edited by Kamelia Boodhoo and Adam Harvey.

1.2 Process Intensification: Definition and Concept

PI aims to make dramatic reductions in plant volume, ideally between 100- and 1000-fold, by replacing the traditional unit operations with novel, usually very compact designs, often by combining two or more traditional operations in one hybrid unit. The PI concept was first established at Imperial Chemical Industries (ICI) during the late 1970s, when the primary goal was to reduce the capital cost of a production system. Although cost reduction was the original target, it quickly became apparent that there were other important benefits to be gained from PI, particularly in respect of improved intrinsic safety and reduced environmental impact and energy consumption, as will be discussed later in this chapter.

Over the last 2 decades, the definition of PI has thus evolved from the simplistic statement of '*the physical miniaturisation of process equipment while retaining throughput and performance*' [1] to the all-encompassing definition '*the development of innovative apparatus and techniques that offer drastic improvements in chemical manufacturing and processing, substantially decreasing equipment volume, energy consumption, or waste formation, and ultimately leading to cheaper, safer, sustainable technologies*' [2]. Several other definitions with slight variations on the generic theme of innovative technologies for greater efficiency have since emerged [3].

The reduction in scale implied by intensification has many desirable consequences for chemical engineering operations. First, the lower mass- and heat-transfer resistances enabled by the reduced path lengths of the diffusion/conduction interfaces, coupled with more intense fluid dynamics in active enhancement equipment, allow reactions to proceed at their inherent rates. By the same token, the more rapid mixing environment afforded by the low reaction volumes should enable conversion and selectivity to be

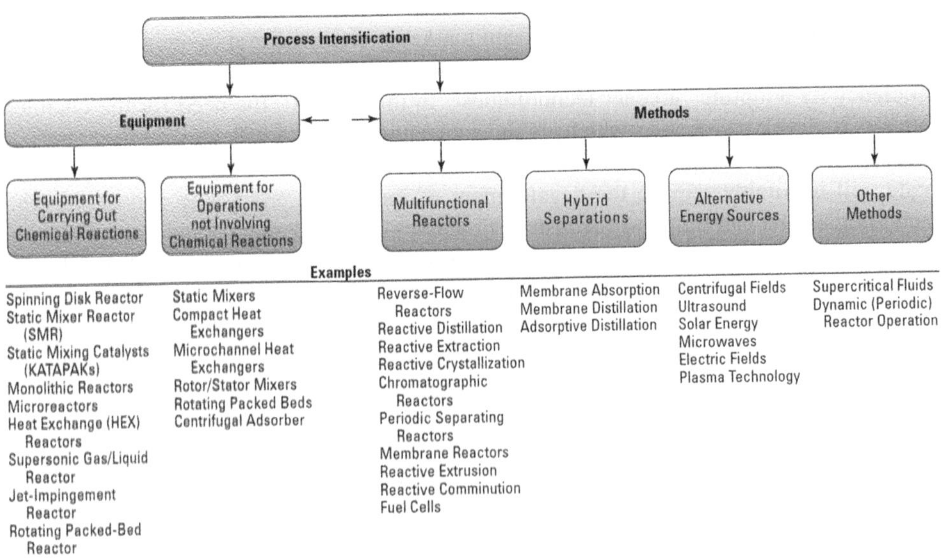

Figure 1.1 *Classification of PI equipment and methods. Reproduced from [ref 2] with permission of American Institute of Chemical Engineers copyright (2000).*

maximized. Residence times of the order of minutes and seconds may be substituted for the hour-scale processing times associated with large conventional batch operations, with beneficial consequences for energy consumption and process safety.

PI covers a wide range of processing equipment types and methodologies, as aptly illustrated in Figure 1.1 [2]. Many of the equipment types classed as 'intensified technologies' have long been implemented in the chemical industry, such as compact heat exchangers, structured packed columns and static mixers. More recent developments include the spinning disc reactor (SDR), oscillatory baffled reactor, loop reactor, spinning tube-in-tube reactor, heat-exchange reactor, microchannel reactor and so on. Lately, it has become increasingly important for the chemical processing industries not only to remain cost competitive but to do so in an environmentally friendly or 'green' manner. It is fitting, therefore, that many of the processes based on the PI philosophy also enable *clean technology* to be practised. For instance, high selectivity operations in intensified reactors will on their own reduce or ideally eliminate the formation of unwanted byproducts. Combining such intensified reactors with renewable energy sources such as solar energy would give even greater impetus to achieving these green processing targets.

1.3 Fundamentals of Chemical Engineering Operations

1.3.1 Reaction Engineering

Reactor engineering starts with the simple mass balance:

$$\text{In} + \text{Made} = \text{Out} + \text{Accumulated} \tag{1.1}$$

Where 'Made' is the rate at which a species is created or lost by reaction. The rate of this reaction in a well-mixed system is governed by the reaction kinetics, which depend only upon the concentrations of species and temperature. However, not all systems are well mixed, particularly at larger scales, and mixing can be rate-determining. The different degrees and types of mixing are introduced in Section 1.3.2. The 'Accumulated' term will be zero for continuous reactors running in steady state, but will be of interest during start-up or shut-down. Determining the rate at which species are created or destroyed in a reactor requires knowledge of mixing, reaction kinetics and heat transfer. Once these are known they can be input into a reactor model. An important part of this model for continuous reactors (as most intensified reactors are) is the residence time distribution (RTD), which is the probability distribution for the length of time elements of fluid will spend in a given reactor design. It can be envisaged as the response to the input of an infinitely narrow pulse of a tracer. All real reactors fall between two extreme cases: the plug flow reactor (PFR) and the continuously stirred tank reactor (CSTR).

1.3.1.1 Plug Flow Reactor

'Plug flow' refers to fluid flowing in discrete 'plugs'; that is, without interaction between the elements. The RTD of a perfect PFR is infinitely thin. Any input tracer pulse to the reactor will remain unchanged, as shown in Figure 1.2.

Real PFRs have symmetrical Gaussian RTDs centred on the mean residence time, the breadth of the RTD decreasing with increasing proximity to ideal plug flow. In practice, this

Figure 1.2 A perfect PFR, showing the response to a perfect input pulse.

is usually achieved by ensuring a high level of turbulence in the flow, as this produces a flat velocity profile. The most conventional form of PFR is a tubular reactor in very turbulent flow. However, there are many variations on this basic form, and other ways of achieving plug flow. Chapters 3 and 5 cover examples of unconventional, intensified PFRs.

1.3.1.2 Continuously Stirred Tank Reactor

The CSTR is, at its simplest, a batch-stirred tank to which an inflow and outflow have been added (of equal flow rate, when at steady state). To determine the RTD of such a reactor, we must picture a pulse of fluid entering it. A 'perfect' CSTR is perfectly mixed, meaning that fluid is uniformly dispersed the instant it enters the reactor. The outflow is at the same concentration of tracer as the bulk of the reactor. Tracer will initially flow out at this concentration, while being replaced with fluid containing no tracer; that is, the tank gradually becomes diluted, and the concentration in the outflow decreases. This leads to a monotonic decrease in concentration, which can be shown to follow an exponential decay (Figure 1.3).

1.3.1.3 The Plug-Flow Advantage

A CSTR's RTD is generally not desirable, as, for a given desired mean average residence time:

- Much of the material in the reactor will spend too long in the reactor (due to the long tail in the RTD) and will consequently be 'overcooked'. The main problem with this is that it allows competing reactions to become more significant.
- Much of the material will be in the reactor for less than the desired residence time. It will therefore not reach the desired level of conversion.

The CSTR can thus lead to increased by-product formation and unsatisfactory conversion. In contrast, plug flow means that each element of fluid experiences the same processing history: each spends exactly the same amount of time in the reactor as every other, and is subject to exactly the same sequence of conditions. This reduces by-product formation and

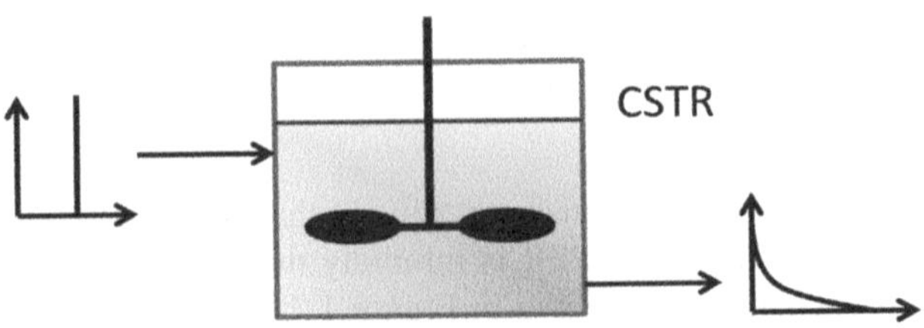

Figure 1.3 RTD for an ideal CSTR.

ensures that the desired conversion is achieved. Furthermore, in practice a PFR will have a smaller volume than an equivalent CSTR, for the following reasons:

- The reactor will be the correct size. CSTRs are usually oversized to compensate for the poor RTD.
- No headspace is required, as is the case in any tank reactor.
- For most reaction kinetics (the most notable exception perhaps being autocatalytic reactions), simply following the design equations will lead to a PFR design that is smaller than a CSTR. For an explanation of this, the reader is advised to consult Sections 5.2 and 5.3 in reference [4].
- Stirred tanks do not scale up in a predictable manner. Uniform mixing becomes difficult to achieve, which can reduce the rate of reaction, necessitating a larger reactor. This is less of an issue with tubular reactors.

For these reasons, PFRs are often preferred in principle. In practice they are difficult to use at long residence times (above a few minutes) and multiphase reactions can be difficult.

1.3.2 Mixing Principles

Mixing is the process of bringing separated fluid elements into close proximity, in a system which, in the simplest case, aims to reduce non-uniformity in a particular property, such as concentration, viscosity or temperature. Most mixing processes occur alongside heat- and/or mass-transfer operations and chemical reactions.

1.3.2.1 Influence of Mixing on Reactions

Mixing is a particularly important process in reactor design, especially in continuous-flow reactors. Designing the mixing process to yield a much shorter mixing time in comparison to the mean residence time of the reactants in the reaction vessel is of paramount importance for good operation of the reactor. If mixing is slow, large and varying concentration gradients of reactant species will exist in different parts of the reactor, resulting in wide variations in product concentrations and properties, which may be deemed off-spec in many applications. In fact, the rate of mixing often determines the rate of these processes and may have a significant impact on the product distribution obtained, especially if many competing reaction steps are involved.

1.3.2.2 Turbulent Mixing: Mixing Scales, Mechanisms and Mixing Times

In a single-phase turbulent flow system, there are three distinct mixing scales that influence a chemical process: macromixing, mesomixing and micromixing [5,6]. These are defined on the basis of their characteristic length scale, as depicted in Figure 1.4, and are directly correlated with the turbulent energy dissipation rate, ε.

The intensity of mixing at each of these scales is significantly influenced by the mechanical energy input into the system by the mixing device. It is generally assumed that higher energy input translates into a higher energy dissipation rate for better mixing – but this is not always the case, as energy may be wasted, for example, in vortex formation at a higher agitation rate in an unbaffled vessel. The energy input causes the fluid to undergo motion across the cascade of length scales described in this section, so that any

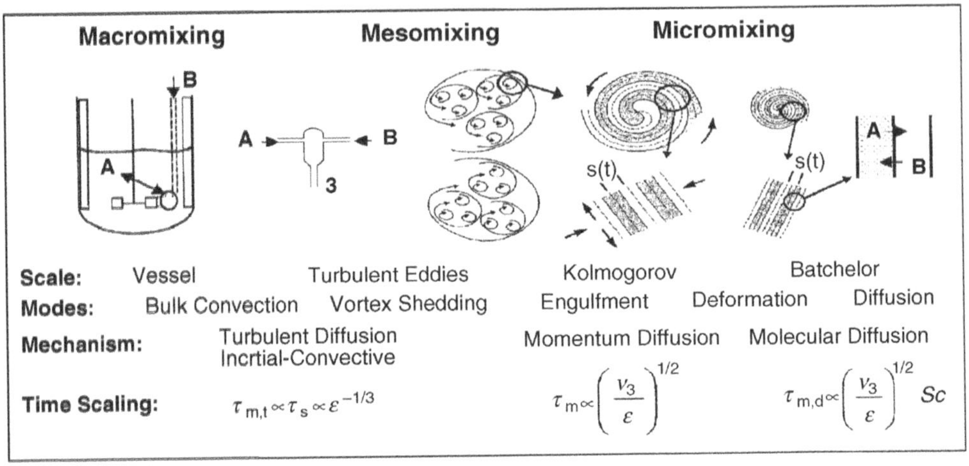

Figure 1.4 Turbulent mixing mechanisms across various length scales. Reproduced from [ref 7] by permission of John Wiley & Sons. © 2003.

concentration inhomogeneities are gradually reduced and eliminated. The kinetic energy thus imparted to the fluid is ultimately dissipated as internal energy, which occurs at the smallest length scales of turbulence; that is, at the Kolgomorov scale.

Various mixers/reactors have been characterized in terms of their energy dissipation rates, as shown in Table 1.1. This illustrates the potential capability of intensified systems such as static mixers, rotor-stator mixers and the SDR, among others, to provide a higher level of mixing intensity than the conventional stirred tank reactor. It is important to remember, however, that higher energy input will be a penalty incurred in terms of energy consumption, and the benefits from the mixing process under these conditions have to demonstrate significant process improvement.

Macromixing. Macromixing involves mixing on the macroscopic scale, which refers to the scale of the vessel or reactor. The process is often referred to as 'distributive mixing' [6,14], which is achieved by bulk motion or convective transport of the liquid at the macroscopic scale, resulting in uniform spatial distribution of fluid elements within the

Table 1.1 *Comparison of energy dissipation rates in a range of mixers/reactors.*

Reactor/mixer type	Energy dissipation rate (W/kg)	References
Stirred tank reactor	0.1–100	[8,9]
Static mixers	1–1000	[9,10]
Impinging jet reactor	20–6800	[11]
Rotor-stator spinning disc reactor (27 cm disc diameter, 240–2000 rpm)	≤6000	[12]
Thin-film spinning disc reactor (10 cm disc diameter, range of disc speeds 200–2400 rpm)	≤2000	[13]

reactor volume. In a continuous flow reactor, the macromixing process directly influences the RTD of a feed stream introduced into the contents of the vessel.

The macromixing time in a mechanically stirred, baffled tank, τ_{mac}, is a function of the mean circulation time, τ_c, in the vessel. In a vessel configured for optimized mixing, $\tau_{mac} = 3\tau_c$, while in a non-optimized system, $\tau_{mac} = 5\tau_c$ [6].

The mean circulation time, τ_c, is generally expressed in terms of the impeller pumping capacity, Q_c [14]:

$$\tau_c = \frac{V}{Q_c} \tag{1.2}$$

or

$$\tau_c = \frac{V}{C_D N D_i^3} \tag{1.3}$$

where C_D, the discharge coefficient of the impeller, is a constant, which typically varies between 0.7 and 1.0, depending upon the impeller used [14].

Mesomixing. Mesomixing refers to coarse-scale, dispersive mixing via turbulent eddies. It is typically characterized by two different mechanisms [5,15]: (1) turbulent dispersion of a fresh feed introduced to a vessel which mixes with its local surroundings; and (2) inertial-convective break-up of large eddies that are larger than the Kolgomorov length scale.

The characteristic timescale associated with turbulent dispersion, τ_D, can be defined by either equation 1.4 or equation 1.5, depending on the radius of the feed pipe, r_{pipe}, with respect to the characteristic length scale for dispersion, L_D [5,15]:

$$\tau_D = \frac{Q_f}{u D_{turb}} \quad (\text{if } r_{pipe} << L_D) \tag{1.4}$$

$$\tau_D = \frac{{r_{pipe}}^2}{D_{turb}} \quad (\text{if } r_{pipe} \approx L_D \text{ or } r_{pipe} > L_D) \tag{1.5}$$

where $D_{turb} = 0.12\varepsilon^{1/3}L_D^{4/3}$ [5,16].

Baldyga *et al.* [15] have expressed the inertial-convective mesomixing timescale, τ_s, as:

$$\tau_s = A\left(\frac{L_C^2}{\varepsilon}\right)^{1/3} \tag{1.6}$$

where A is a constant having a value between 1 and 2, depending on the turbulence level in the system.

Micromixing. Micromixing represents the final stage of the turbulent mixing process, which proceeds at much finer length scales than macro- and mesomixing, referred to as the

Kolgomorov or Batchelor length scale. At the microscale level, the Kolgomorov length scale, η_K (representing smallest scales of *turbulence* before viscosity effects dominate), and Batchelor length scale, η_B (representing smallest scales of fluctuations prior to *molecular diffusion*), are defined as [17–19]:

$$\eta_K = \left(\frac{\nu^2}{\varepsilon}\right)^{1/4} \tag{1.7}$$

$$\eta_B = \left(\frac{\nu D_l^2}{\varepsilon}\right)^{1/4} = \frac{\eta_K}{\sqrt{Sc}} \tag{1.8}$$

where the Schmidt number, Sc $\left(=\frac{\nu}{D_l}\right)$, for liquids is typically of the order of 10^3, so that $\eta_B << \eta_K$. For aqueous solutions in turbulent regimes, η_K is of the order of 10–30 μm.

The physical phenomena of the micromixing process include engulfment, deformation by shear and diffusion of the fine-scale fluid elements. The relevant mixing times associated with these processes are [5]:

$$\text{Engulfment}: \tau_e = 17.2\left(\frac{\nu}{\varepsilon}\right)^{0.5} \tag{1.9}$$

$$\text{Shear deformation and diffusion}: \tau_{Ds} \approx 2\left(\frac{\nu}{\varepsilon}\right)^{0.5} arc\sinh\,(0.05\,Sc) \tag{1.10}$$

More often than not, $\tau_{D_S} << \tau_e$, resulting in the overall micromixing process being dictated by the progression of the engulfment phenomenon taking place at the Kolgomorov length scale.

Although the actual molecular mass transfer process before the reaction is ultimately achieved by molecular diffusion, enhancing the rates of macro- and mesomixing through turbulent hydrodynamic conditions enables faster attainment of the fluid state, where micromixing and therefore molecular diffusion prevail.

1.3.3 Transport Processes

Understanding transport processes is at the heart of PI, as the subject can be defined as a search for new ways of enhancing or achieving transport of mass, heat or momentum.

Transport processes – heat, mass and momentum transfer – are generally governed by equations of the same form. They are all flows in response to a 'driving force' – a temperature difference, a concentration difference and a pressure difference, respectively – opposed by their respective resistances. Brief overviews of the intensification of mass, heat and momentum transfer follow.

1.3.3.1 Heat Transfer

Heat transfer – the transport of energy from one region to another, driven by a temperature difference between the two – is a key consideration in the design of all unit operations. Unit operations have defined operating temperatures, so the heat flows in and out must be understood in order to maintain the temperature within a desired range. Reactors, for instance, must be supplied with heat or must have it removed at a rate that depends upon the exo/endothermicity of the reaction, the heat-transfer characteristics of the reactor and the heat flows in and out, in order to ensure that the reaction takes place at the correct temperature and therefore the correct rate.

Furthermore, the streams into and out of unit operations must be maintained at the correct temperatures. This is usually achieved using heat exchangers: devices for transferring heat between fluid streams without the streams mixing. It was always been a given in heat exchanger design that they must operate in turbulent flow wherever possible, as turbulent flow results in considerably higher heat-transfer coefficients than laminar. Hence, heat exchangers were not designed with narrow channels, as the achievement of turbulence depends upon exceeding a certain Reynolds number, which is directly proportional to the diameter of the channel:

$$\mathrm{Re} = \rho \mathrm{v} \mathrm{D}/\mu \tag{1.11}$$

Reassessing such assumptions about heat and mass transfer is at the heart of PI, and has led to the development of 'compact heat exchangers', which have extremely narrow channels.

This only makes sense if the heat transfer itself rather than just the heat-transfer coefficient is considered. The rate of heat transfer in a heat exchanger is not only a function of the heat-transfer coefficient, as can be observed in the 'heat exchanger design equation':

$$\mathrm{q} = \mathrm{U}\mathrm{A}_\mathrm{s}\Delta\mathrm{T}_\mathrm{lm} \tag{1.12}$$

It is also clearly a function of the heat-transfer surface area A_s. Compact heat exchangers have very narrow channels (sub-mm), so the flow is laminar (as Re depends upon channel width, D) and therefore has a significantly lower heat-transfer coefficient than a turbulent flow. However, this is more than compensated for by the increase in heat-transfer surface area per unit volume, giving a higher heat-transfer rate per unit volume than conventional heat exchanger designs (such as 'shell-and-tube'). A concise overview of compact heat exchangers is given by Reay *et al.* [20].

There are also a range of devices ('turbulence promoters') that are designed to perturb flow in order to bring about the onset of turbulence at lower Re. These promoters allow the higher heat- and mass-transfer coefficients associated with turbulence to be accessed at lower velocities, thereby reducing the associated pumping duties. They can also be classified as intensified devices, although the degree of intensification is nowhere near as great as that in the compact heat exchanger. They suffer less from fouling, however, which is one of the main drawbacks of compact heat exchangers: their applications are limited to 'clean' fluids, as they are very easily blocked by fouling. As with most technologies, the strengths and weaknesses of intensified technologies must be assessed so as to define a 'niche' or parameter space within which they are the best-performing.

1.3.3.2 Mass Transfer

An appreciation of mass transfer is required for the intensification of separation processes. Common separation unit operations are distillation, crystallization, ad/absorption and drying.

In many processes, the heat and mass transfer are interrelated. Generally, what enhances one enhances the other. Indeed, the mechanisms for transfer are often the same or are closely related. Experiments in heat transfer have often been used to draw conclusions about mass transfer (and vice versa) through analogies. Various equations describing one or the other are based upon analogy. Compare for instance the Dittus–Boelter equations for heat and mass transfer:

$$\text{Heat: Nu} = C_1.\text{Re}^{0.8}.\text{Pr}^{0.33} \quad (1.13)$$

$$\text{Mass: Sh} = C_2.\text{Re}^{0.8}.\text{Sc}^{0.33} \quad (1.14)$$

An example of an intensified mass-transfer device is the rotating liquid–liquid extractor. The conventional design of liquid–liquid extractors was based on using the density difference between the liquids to drive a countercurrent flow, by inputting the denser fluid at the top of the column and the lighter at the bottom. One of the variables, although it may not appear to be a variable initially, is g, the acceleration due to gravity. This can of course be increased by applying a centrifugal field, in which case the lighter fluid is introduced from the outside and travels inward countercurrent to the denser fluid. The first example of this kind of device was the Podbielniak liquid–liquid contactor, originally developed in the 1940s for penicillin extraction. There are currently hundreds of Podbielniak contactors in use worldwide for a range of applications, including antibiotic extraction, vitamin refining, uranium extraction, removal of aromatics, ion exchange, soap manufacture and extraction of various organics [21]. This illustrates that there are many successful examples of PI in industry today, although they are not viewed as such, as they are not a new technology (and the term 'process intensification' did not exist when they were invented). Indeed, any continuous process is an example of an intensified process.

1.3.3.3 Momentum Transfer

Momentum transfer occurs due to velocity gradients within fluids. Many of the technologies listed above to enhance mass and heat transfer, also involve enhanced momentum transfer. Again, as illustrated by the equations in section 1.3.3.2 (between heat and mass transfer), there are analogies between this transfer process and others that lead to meaningful quantitative relationships. Theories such as the Reynolds analogy (see Ref [22] for a concise explanation), and its more sophisticated and accurate descendants, are based on heat, mass and momentum transfer processes having the same mechanism: in this particular analogy, the mechanism for all is the transport of turbulent eddies from a bulk medium to a surface.

Essentially, any technology that enhances the flow increases momentum transfer. The rotational fields applied to flows in section 'Centrifugal Fields' (see section 1.4.1.1) and the turbulence promoters mentioned in 1.3.3.1 are just two examples of enhanced momentum transfer (along with enhancement of other transfer properties). It should be noted that enhancement of momentum transfer is often not performed for its own sake, but rather to promote other transfer properties.

1.4 Intensification Techniques

Intensification of a process may be achieved through a variety of means, including enhancing mixing and heat/mass transfer by additional energy input via external force fields or via enhanced surface configurations, transforming processes from batch to continuous mode in order to achieve smaller process volume and integrating process steps in hybrid technologies. Each of these will be discussed briefly in this section.

1.4.1 Enhanced Transport Processes

Heat and mass transport rates are largely influenced by the fluid dynamics, which directly affect the heat/mass-transfer coefficients and the available area on which the transfer of energy/mass can occur. Mixing rates are similarly affected by these parameters. Therefore, any attempt at intensifying these processes should focus on enhancing the turbulence in the system and/or increasing the transfer surface area. One way of achieving this is by subjecting the reaction environment to external force fields, such as centrifugal, electric and ultrasonic fields.

1.4.1.1 Enhanced Force Fields

Centrifugal Fields. Surface rotation as a technique for intensification has stimulated keen interest from academic workers for many years. As early as the 1950s, Hickman's research efforts into two-phase heat transfer on spinning disc surfaces culminated in the development of the first successful centrifugal evaporator used in sea water desalination [23].

The benefits that can be extracted from the exploitation of high centrifugal fields created by rotation are as follows:

- The rotational speed of the spinning surface provides an additional degree of freedom, which can be readily manipulated for optimum equipment performance.
- The extremely high gravity fields thus generated are capable of producing very thin films, in which heat transfer, mass transfer and mixing rates are greatly intensified. The short path lengths and the high surface area per unit volume provide the opportunity for rapid molecular diffusion and enhanced heat transfer, even on scale-up (Figure 1.5). The

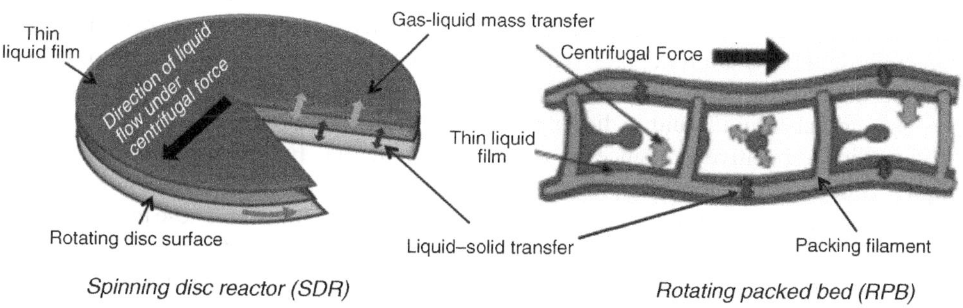

Figure 1.5 *Thin-film processing in an SDR and a RPB, illustrating the short diffusion/conduction path lengths and high surface area for enhanced heat and mass transfer.*

performance of multiple phase processes in particular stands more chance of being enhanced under the influence of high gravitational forces as a result of increased interphase buoyancy and slip velocity [24].

- Applications in which the solid content of a process fluid poses problems with regard to fouling in conventional devices can, in principle, be handled by the rotating equipment. The rotating action in itself provides a scraping or 'self-cleaning' mechanism strong enough to shift most solid deposits away from the surface of revolution, thereby ensuring maximum exposed area at all times during operation.
- The very short and controllable residence times achieved under the centrifugal action enable heat-sensitive materials to be processed with minimal risk of degradation.

Several unit operations have been identified in which the centrifugal acceleration generated on the surfaces of revolution presents remarkable potential for intensification. Typical operations include distillation, extraction, boiling, condensation, crystallization, precipitation and gas–liquid reactions.

The SDR and the rotating packed bed (RPB) are two well-known examples of centrifugal field processing equipment. The SDR will be treated in more detail in Chapter 3.

Alternative Force Fields. Alternative force fields commonly employed to intensify processes include ultrasound, electric fields and energy of electromagnetic radiation, whose applications to chemical and biochemical processes in the context of PI have been reviewed by Stankiewicz [25].

Ultrasonic Fields. 'Ultrasound' refers to sound waves beyond the audible range of the human ear, with a frequency of approximately 20 kHz to 500 MHz. The frequency typically applied to chemical processing is generally no higher than 2 MHz [26]. Ultrasound is propagated through a liquid medium in alternating cycles of compression and stretching, or rarefaction. These induce an effect known as cavitation, whereby microbubbles are generated, expand and are subsequently destroyed in successive compression cycles, releasing a large amount of heat and pressure energy in the local environment of the bubbles (Figure 1.6). Local temperatures and pressures after the collapse of microbubbles can reach as high as 5000 °C and 2000 atmospheres, respectively, depending on the power input [26]. Mechanical or chemical effects can arise from such extreme conditions in the system, as discussed in many review articles on the subject [27,28]. Thus, for instance, the mechanical effects are characterized by the pressure waves or shock waves resulting from the collapse of cavitation bubbles. These waves generate intense mixing conditions and enhanced transport rates throughout the bulk of the liquid medium in homogeneous systems and at liquid/liquid or liquid/solid interfaces in heterogeneous systems, which have a direct, positive influence on a chemical reaction. Furthermore, in immiscible liquid systems much finer droplets can be formed under ultrasound exposure than by mechanical agitation, creating a greater surface area for mass transfer. These mechanical or physical effects are generally thought to be responsible for the rate enhancements and improved product properties observed in many chemical processes subjected to ultrasonic irradiation [28,29].

Chemical effects due to ultrasound arise if the chemical compounds in the processing medium can fracture into reactive intermediates such as free radicals (a process often

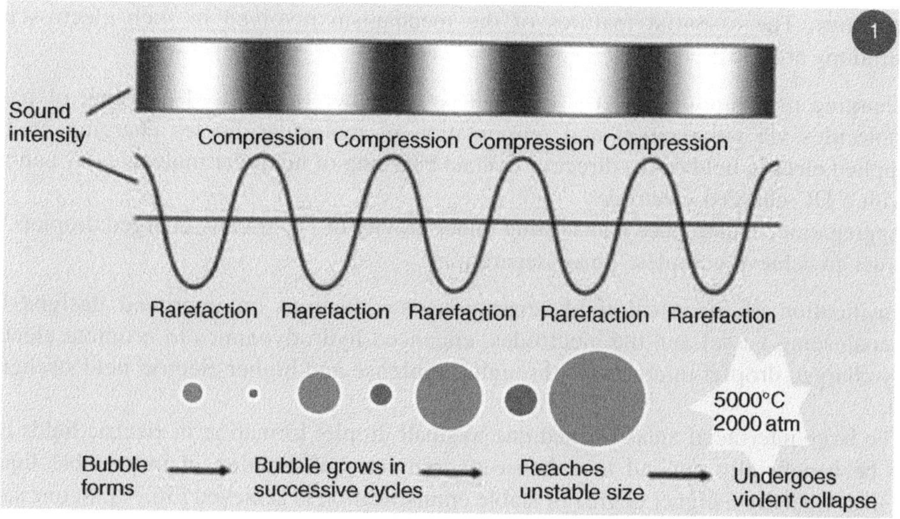

Figure 1.6 *Generation and collapse of an acoustic cavitation bubble. Reprinted from [ref 29] © 2010, with permission from Elsevier.*

referred to as 'sonolysis') under the very intense local temperatures attained in the bubble cavities. The literature has many documented examples of sonolysis [26,30,31]. Such chemical effects are usually reflected in changes in mechanisms and product distributions [32]. The well-documented influence of ultrasound on crystallization processes illustrates the latter effect most appropriately [33].

In summary, ultrasonic processing for intensification of chemical processes is associated with the following benefits, all of which have wider implications for greener processing [26,29]:

- Increases in both reaction speed and yield in an extensive range of heterogeneous and homogeneous systems, as highlighted by Thompson and Doirasawmy [28].
- Waste minimization through increased selectivity.
- Changes in and simplification of reaction pathways, which can lead to milder processing conditions (e.g. ambient temperatures and pressures, reduced solvent use) and higher energy efficiency.
- The use of environmentally benign reactants and solvents while retaining or even enhancing the reaction rate under ultrasonication.

Chapter 7 explores these concepts in more detail.

Electric Fields. High-intensity electrical fields have long been known to have a destabilizing effect on dispersed systems containing polar molecules, such as water, and to enhance mass-transfer processes via promoted coalescence of the dispersed phase. The removal of dust from air in the Cottrell precipitator [34] and the dehydration of crude oil emulsion in oilfields were among the first industrial processes developed almost a century ago to harness the beneficial effects of electric fields in such phase separations. The essential features of the mechanism involved in such electrostatic separations are [35]:

- Charging of a liquid droplet – this can be by either (a) induced charging of polar molecules via polarization and reorientation of molecular-dipoles charging by the applied electric field or (b) direct or contact charging of nonpolar molecules by contact with a DC-charged electrode
- Aggregation, coalescence and settling under gravity of electrically charged droplets, in order to achieve complete phase separation

Intensification of electrical field separations has focused on improved designs for the coalescing vessel and the electrodes, enhanced hydrodynamics to promote electrically charged droplet interactions through turbulence and higher electric field strengths [35].

The large interfacial areas formed due to small droplet formation in electric fields can also be beneficially applied to enhance overall rates of reaction in immiscible liquid systems, whereby a higher degree of stable emulsification is achieved [36,37]. In one such process involving the enzymatic hydrolysis of triglyceride esters to yield free fatty acids and glycerol, studied by Weatherley and Rooney [38], electrostatic fields were used to intensify the dispersion of the aqueous phase into the oil substrate by creating large interfacial areas between the reacting species and thus enhancing the overall rate of reaction at relatively modest temperatures and pressures.

Electric field effects on intensification of heat-transfer processes in general and of boiling in particular are also well documented [39–42], and the mechanisms are generally well understood [43–45]. In nucleate boiling, for instance, not only are more bubbles released from the surface when an electric field is applied but also they are smaller than in its absence [44].

The combined effects of electric and centrifugal fields on hydrodynamics in thin-film flow in an SDR have been analysed via numerical simulations by Matar and Lawrence [46]. The applied electric field was shown to induce turbulence on the film surface through the formation of an increased intensity of large-amplitude waves. These simulation results suggest that electric fields have the potential to further enhance heat and mass transfer and reaction rates in thin-film processing.

Electromagnetic Fields. The electromagnetic spectrum covers a wide range of energy fields, such as microwaves, light, X-rays and γ-rays. Here only microwaves and light will be considered with respect to the intensification of chemical processes.

Microwaves. Microwaves are a form of electromagnetic energy, with frequencies in the range of 300 MHz to 300 GHz. The commonly used frequency for microwave heating and chemical processing is 2.45 GHz. Microwave heating of materials is quite distinct from conventional heating, where conduction and convection are the main mechanisms for the transfer of heat. Under microwave exposure, on the other hand, either dipole interactions or ionic conduction come into play, depending on the chemical species involved [47,48]. Dipole interactions occur with polar molecules that have high dielectric constants, such as water and alcohols, while migration of dissolved ions in the electric field takes place in ionic conduction. Both mechanisms require effective coupling between components of the target material and the rapidly oscillating electrical field of the microwaves. Heat is generated by molecular collision and friction.

Microwave-assisted processing is associated with many features that have positive implications for intensification of processes. In particular, the more rapid, controlled and uniform heating rates afforded by microwave exposure result not only in higher rates of reaction than conventional heating methods but also in better product quality, through improved selectivity. Recent publications by Toukoniitty *et al.* [27] and Leonelli and Mason [29] provide good reviews of these aspects, with the former focusing on heterogeneous catalytic systems.

There are a number of examples of the industrial use of microwave heating in the food, rubber and wood industries, as highlighted by Leonelli and Mason [29]. Although small-scale batch and continuous reactor systems with microwave irradiation capabilities are commercially available [47], there is still much scope for further development of industrial-scale microwave reactors, with continuous flow systems that have microwave irradiation or indeed coupled microwave and ultrasound irradiation capabilities being ideally suited for this purpose [25,29].

Light Energy. The observation that certain compounds could be affected by sunlight to give materials with the same chemical composition but very different physical properties was first made in 1845, when Blyth and Hofmann [49] noted that styrene was converted from a liquid to a glassy solid when exposed to sunlight. Since then, the industrial potential of photochemistry has been widely demonstrated in a number of reactions, such as the production of caprolactam, used in the manufacture of nylon 6, and the formation of vitamins D_2 and D_3 [50].

Photoinitiation is an attractive alternative to thermal activation of reactions for a number of reasons. It is an inherently clean process, requiring only the reacting molecules in order for the reaction to be activated. Additional and often expensive and environmentally unfriendly reagents and catalysts can be minimized. The irradiation by a specific wavelength, and therefore a well-defined energy input in the form of photons, allows only certain reactions to be targeted, thereby reducing byproduct formation and costly downstream separation processes. Since activation is by light, ambient operating temperatures can be utilized, which not only reduce thermal energy consumption but can also result in better control of the process, especially if side reactions can be minimized at lower temperatures. This is particularly relevant for polymer processing, where low temperatures generally result in (1) better tacticity control of the polymer, (2) reduced transfer effects, which can be a cause of excessively branched macromolecules, and (3) minimal thermal degradation risks for the polymer formed.

In spite of the clear advantages offered by photochemistry as a reaction-initiation technique, its use in industry is surprisingly rather limited. This is mostly because of the technical problems associated with the uniform irradiation of large reaction volumes, such as those encountered in conventional batch set-ups. Batch reactors, especially those at commercial scale, possess particularly low surface area to volume ratios. This poses enormous processing challenges in that the photons emanating from the light source have extremely limited penetration depths into the fluid (a few centimetres at most), resulting in quite ineffective and non-uniform initiation of reactions in conventional stirred tank reactor configurations. These issues have been addressed to a certain extent by the development of the falling film reactor, but other reactor configurations such as the thin-film SDR or microreactor might potentially be applied for more efficient processing of photochemical processes. This will be dealt with in more detail in Chapter 3, where the SDR is considered.

1.4.1.2 Enhanced Surface Configurations

The influence of well-defined structures or enhanced surface configurations on the increased efficiency of chemical processes, in terms of heat/mass transfer and mixing rates, is well established. Such structures can be classified according to the time and length scales at which they function. For example, molecular-scale structures such as zeolite supports, which are termed 'molecular reactors' [6], have been used to improve chemical transformations at molecular scales. A range of examples of such molecular reactors applied in chemical reactions is provided in a recent paper by Van Gerven and Stankiewicz [6]. At the meso- and macroscales, structures such as channel reactors (including milli- and microchannels), monoliths, foams and static mixers have all been used to improve process performances such as yield and selectivity. A brief overview of these areas is given in this subsection; a selection of them will form the subject of later chapters in the book.

Micro/millichannel Reactors. The use of micrometre- or millimetre-scale (typical dimensions of 10–100 μm and 0.5–2.0 mm, respectively) reaction spaces in the form of channels of various shapes (Figure 1.7) allows for much more precise control of diffusion, heat exchange, retention/residence times and flow patterns in chemical reactions [51]. Although flow in these types of reactor systems tends to be mainly laminar (Re $<$ 100, typically), effective mass transfer can be achieved via the very short diffusion path lengths, determined by the diameter of the channels. Thus, given that the characteristic diffusion time, t_D, is expressed as:

$$t_D = \frac{l^2}{D_l} \tag{1.15}$$

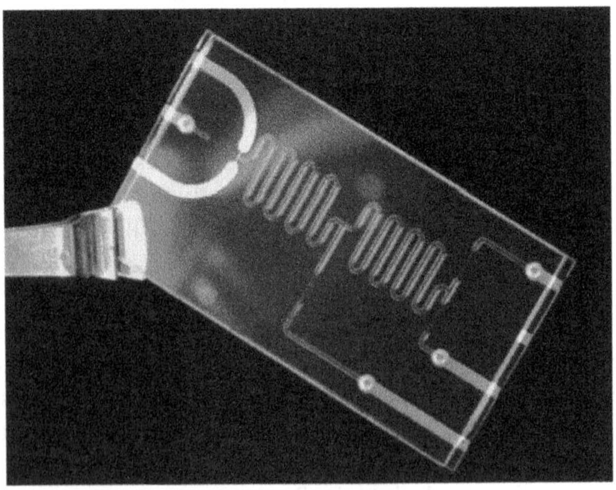

Figure 1.7 A glass microreactor. Reproduced from [ref 52] by permission of John Wiley & Sons. © 2005.

the diffusion time across a macroscale channel (of the order of 1 cm diameter) versus a microchannel of 100 μm diameter can vary dramatically from 10^5 to 10 s, respectively, for solvated molecules of typical diffusivity 10^{-9} m^2/s. Although the diffusion timescale is greatly reduced in the microchannel, it is nevertheless still a slow process for efficient micromixing in fast reactions where micromixing times of the order of milliseconds are required. In such cases, it is therefore important to also provide very large interfacial areas for efficient contact between reacting streams in the channel, which can be achieved through a number of 'passive' techniques, as described by Hessel *et al.* [53] – for example, the use of miniaturized static inserts, which continuously split and recombine the flow streams, or of turbulent collision of injected streams, among many other techniques – albeit at the expense of greater pressure drop. Moreover, when more than one phase is present, vortex motions generated by the shearing motion within the slug flow are also known to enhance the mixing within the slugs and improve mass-transfer rates across the interface [54,55]. Near-plug flow behaviour that gives control of RTDs in micro-/millichannel reactors can be achieved by introducing segmented flow via a second phase in the flowing system [53]. Very good control of highly exothermic reactions in small-diameter channel reactors is also possible, due to the extremely high surface area to volume ratios, of up to 50 000 m^2/m^3, when compared with conventional stirred tank reactor geometries.

As a result of these admirable traits, much attention has been given to micro- and mesoscale channel reactors as alternative systems for a host of reactions that are typically limited in some way (mass transfer, heat transfer or mixing) when conducted in conventional batch reactors. Mason *et al.* [56] offers an extensive review of the current literature in this area, but specific examples worthy of note include hydrogenations reactions [57], a nitration reaction [58] and a simple acid–base reaction [54], all of which require rapid mass transfer in order to keep pace with the inherently fast reaction kinetics.

The characteristics of microreactor technologies and their processing advantages will be covered in greater depth in Chapter 4.

Monolithic Structures. Monoliths represent a type of structured packing unit consisting of a large number of parallel, straight capillary channels, separated from each other by walls. The 'honeycomb' channels come in various shapes and sizes, with rectangular channels being most commonly used. They are usually made from ceramic or metallic materials and they have traditionally been used as catalyst supports in the treatment of NOx and CO emissions in the automotive industry [59]. Monolithic catalyst structures are characterized by several features that make them highly attractive for use in multiphase catalytic reactors [59,60]. In brief, flow in the channels is uniformly distributed, with relatively low pressure drops. The large catalyst-coated surface areas presented by the monolith arrangement, which can be further augmented by the provision of fins in the channel walls, are conducive to extremely good contact between the gas/liquid phase and the catalyst. Combined with the short diffusion path length offered by the thin catalyst coating, the large contact areas between the catalyst and the reacting phases give rise to intense mass transfer to and from the catalyst.

Other structures include porous meshes [61,62] and metal foams [63]. These have been shown to enhance gas–liquid mass transfer and uniformity of mixing, respectively.

Monolith catalytic reactors will be covered in more detail in Chapter 6.

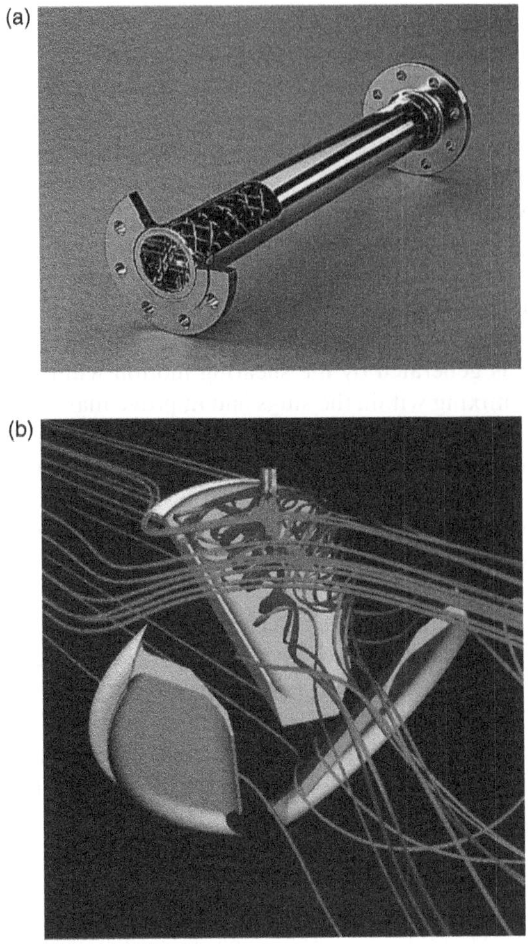

Figure 1.8 (a) SMX static mixer for laminar mixing; (b) CompaX static mixer for turbulent mixing. Reproduced with permission from Sulzer ChemTech © 2012.

Static Mixers. Static mixers are motionless pipeline inserts designed to promote mixing. Various static mixer designs are available for a range of applications, including blending of viscous materials and laminar or turbulent mixing in single-phase liquids or multiphase systems. For laminar mixing, the mixer elements such as those of the SMX configuration (Figure 1.8a) are designed to repeatedly split the incoming flow into many layers, which are redistributed around the mixer structure in the transverse direction relative to the net flow. This gives rise to greater uniformity of component distribution across the flow cross-section of the pipe and increased interfacial area for enhanced diffusion. On the other hand, static mixers designed for turbulent mixing seek to enhance the formation of turbulent eddies in the flow stream (Figure 1.8b).

Several benefits of static mixers can be exploited, including homogenous mixing, narrow RTD, low cost and low maintenance due to lack of moving parts. One of their limitations is

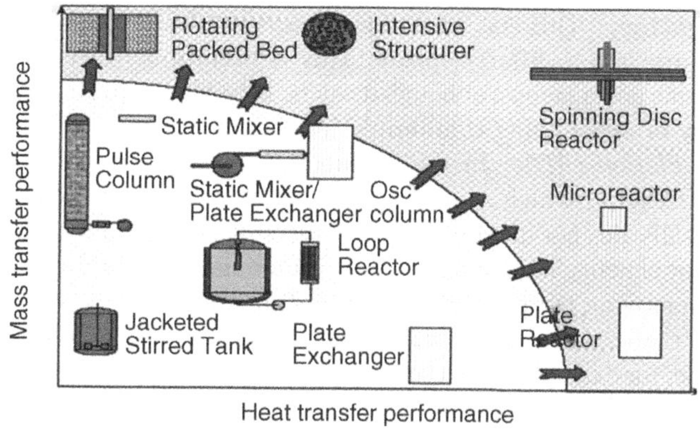

Figure 1.9 *Heat- and mass-transfer performance of several intensified units compared to the jacketed stirred tank. Reproduced from [ref 69] with permission from Britest © 2000.*

related to the handling of very viscous materials, which may result in extremely high pressure drops, although new designs such as the SMX plus mixer have been introduced to alleviate such processing difficulties [64]. Moreover, unless the structures are very open, static mixers can be prone to blockages by suspended solids (e.g. catalysts).

Static mixers have found wide applications in many industrial mixing operations involving gas–liquid dispersion and liquid–liquid extraction for enhanced mass transfer [65], and they have long been regarded as one of the successful applications of PI in industry. They can also be employed as an in-line mixer in multifunctional heat-exchanger reactors, where they not only facilitate rapid mixing between reacting species but also provide additional surfaces for efficient heat transfer [66–68].

The heat- and mass-transfer intensification abilities of the various technologies described in this section are neatly summed up in Figure 1.9. The comparatively poor performance of the workhorse of the chemical industry that is the jacketed stirred tank reactor sits in stark contrast with the intensified systems.

1.4.2 Integrating Process Steps

One of the most important intensification strategies is combining process steps, thereby reducing the overall number in a process. This can have a range of benefits:

- Reduction in capital cost
- Reduction in running cost
- Reduction in duty of downstream unit operations

One of the most familiar integrations is between reaction and separation. Examples include:

- *Membrane Reactors:* Membranes that are permeable to only one of the products are used to remove this product in situ, thereby overcoming equilibrium constraints. In this way, membrane reactors have often been used to bring equilibrium reactions to completion.

- *Reactive Extraction:* This is a separation process in which reaction is used to cause a product to move into a different phase. This is usually between two liquid phases, normally organic and aqueous, although it can be between solids and liquids. Again, this is often a way of overcoming equilibrium limitations.
- *Reactive Distillation (RD):* In this method, the distillation column has a second function as a reactor, which removes the reactor from the flow-sheet. The greatest advantage usually lies not in reducing the number of unit operations (although this can be a significant benefit) in the plant, but in overcoming the equilibrium limitations of a reaction, by removing the product through distillation.
- *Supercritical Operation:* A key advantage of performing reactions in supercritical carbon dioxide, for example, is that separation of the solvent (the CO_2) from the reaction mixture becomes facile, and does not require a separate unit operation. Instead, all that is required is that the pressure be released to the necessary degree for the reaction mixture to come out of solution.

There are many other examples of combined reaction and separation: these are simply a selection to illustrate the range and the possibilities for process improvement.

To focus on one example, RD, in the 200 kte[1]/yr Eastman process for methyl acetate production, the following reaction takes place:

$$\text{acetic acid} + \text{methanol} \leftrightarrow \text{methyl acetate} + \text{water} \tag{1.16}$$

Eastman successfully used RD to replace 11 distillation vessels (with associated condensers and reboilers) with just 3 RD vessels [70], resulting in significant running and capital cost savings (as distillation columns are among the most energy-intensive unit operations in any process). This intensification was shown to be successful because the reaction fitted the requirements of RD, the most important of which was that one of the products (the methyl acetate itself) had a significantly lower boiling point (57 °C) than any other species present.

When integrating process steps, as with any other intensification, the effects on the whole process should be assessed. It may be that the effects on the process as a whole are minimal or even undesirable, or that the economics are not significantly affected by the new design. However, there is a wide range of process steps that can be combined, and with any process it is worth taking the time to assess the scope for intensification by integrating steps.

1.4.3 Moving from Batch to Continuous Processing

Generally, continuous processing is more efficient than batch processing. Continuous reactors are usually smaller than equivalent batch reactors for two reasons:

1. *Greater Occupancy:* Batch vessels are tied to the 'batch cycle', meaning that much of their time is spent on processes such as filling, emptying, cleaning, heating and cooling, all of which reduce their space-time yield, as during these processes the reactor is not generating product.

[1]Kilotonnes

2. *More Effective Mixing:* The more efficient mixing observed in continuous reactors often leads to higher rates of reaction, which lead to smaller reactors (for the same production rate). This is only the case when the reaction is 'mixing-limited', but it often occurs at larger scales, because good mixing is difficult to achieve at scale, which limits the reaction rate. Batch-stirred vessels 'do not scale up well', which essentially means that their performance at laboratory scale/pilot scale may not be a reliable guide to their performance at industrial scale. A particular problem is that energy densities that are easily achieved at laboratory scale are practically unattainable at the scale of 10s of m^3, as the power scales up on the diameter of the impeller to the power 5. Continuous reactors often have better mixing simply because the length scales are smaller, the diameter of a tubular reactor being considerably less than that of an equivalent tank. Beyond this, there may well be different mixing patterns in continuous reactors that are more effective than those of typical batch agitation.

The reduced scale and improved mixing patterns lead to a range of benefits of continuous operation, such as:

- Improved control
- Better product quality
- Less hazardous operation, due to the reduced volumes of hazardous material
- Reduced capital costs: this is due to the smaller size of the equipment, although in some instances this can be somewhat counteracted by costs associated with the mixing mechanisms or pumps
- Reduced running costs: better control of the temperatures in a reactor usually means that less energy is required to run the process, as the reaction mixture will not need to be overheated to allow for inhomogeneities. When reaction times are reduced by conversion from batch to continuous, the reaction mixture has to be held at the correct temperature for a shorter period of time, which reduces the heating duty overall

It should be noted that these are generalities, so are not all true for all cases of batch–continuous conversion.

In practice, however, batch processing is often used, and there are a range of reasons for this:

- *Flexibility:* Batch vessels can be used for a range of different reactions, and indeed different process steps, whereas continuous equipment needs to be purpose-designed.
- *Batch Integrity:* In pharmaceutical processing, every operation undergone by a specific batch of reactants must be clearly defined, and this is achieved much more easily when there is a well-defined batch of material.
- *Small-Scale Operation:* At very small scales of production, the benefits of moving to continuous operation can be minimal. Continuous processing is usually one of the main economies of scale in chemicals manufacture.
- *High-Added-Value Products:* When the value of a product is high, the processing costs can be immaterial, and other process qualities will become more important. Often the time-to-market will be critical, and in such cases it is often perceived as being less risky to rely on known batch processes, or to scale up in batch, as the processes will have been developed in that way in the laboratory.

- *Long Residence Times:* When reaction times are more than a few minutes, conventional reactors start to become increasingly impractical or expensive if tight control of product specification (achieved by 'plug flow') is to be realized. Very few reactions that take hours are performed continuously. This is the main 'niche' of the oscillatory baffled reactor (OBR): converting 'long' batch processes to more efficient continuous processing.

1.5 Merits of PI Technologies

PI is associated with several business-, process- and environment-related benefits, as summarized in Figure 1.10.

1.5.1 Business

Cutting down the size of the main plant items such as reactors, heat exchangers and separators in a process entails not only a reduction in individual equipment cost but also – and to a more appreciable extent – huge savings in installation costs as a result of reduced pipework, civil engineering and support structure, to name but a few factors. Furthermore, it is envisaged that smaller-sized plants will become more 'mobile' and hence be capable of being transported to the customer or to resources such as an oil field, for instance, for processing of flare gas into methanol. In this distributed manufacturing scenario, the logistics associated with such small, portable plants will be greatly simplified. Another benefit, especially relevant to the pharmaceutical industry, is the ability to introduce new products to the market more rapidly than is currently possible. With PI, a lab-scale continuous reactor can become the manufacturing unit, if the throughput matches the

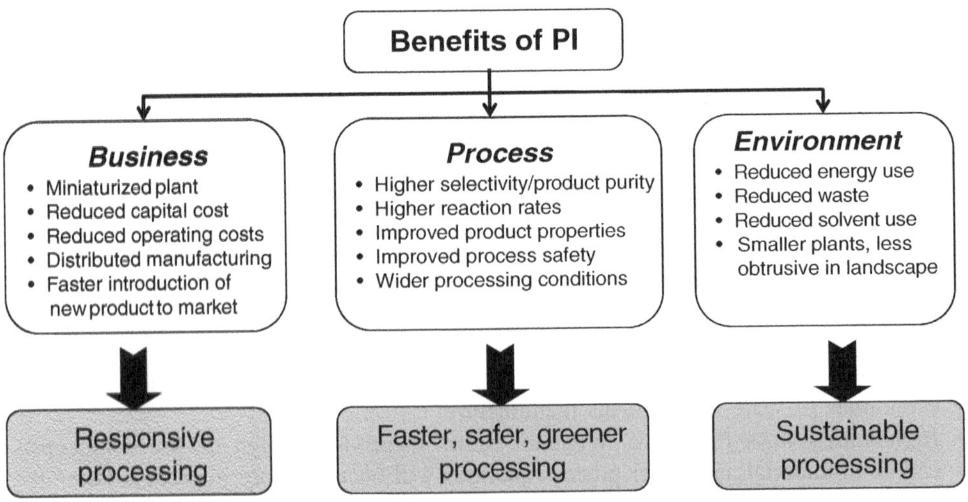

Figure 1.10 *Business, process and environmental benefits of PI.*

desired production rate. This concept has been demonstrated in a feasibility study undertaken by SmithKline Beecham in the SDR [71], where a lab-scale reactor could match the production capacity of a batch process giving 8 tonnes of product per annum. Cutting down on the process development stages can avoid lengthy delays, as these stages alone can take up to 3 years when scaling up batch processes, which is a considerable amount of time in a product lifespan of 20–25 years before patent expiry.

1.5.2 Process

The process benefits of higher selectivity, faster reaction rates and better product properties were explored in great detail under the various technologies considered in Section 1.4.

Safety in the chemical industry is one of the highest priorities for health-and-safety regulatory bodies following a number of fatal incidents involving large inventories of hazardous materials and exothermic runaways in batch reactors, such as the Bhopal disaster in 1984 [72]. Adopting the PI approach can substantially improve the intrinsic safety of a process as there will be a significantly reduced volume of potentially hazardous chemical at any one time, in a smaller intensified unit. In addition, one of the objectives of PI is to move away from batch processing to small continuous reactors, which have a more efficient overall operation, especially in the case of hugely exothermic reactions, in which the heat can be removed continuously as it is being released. It is to be borne in mind, however, that when intensifying a process, for example by integrating process steps, the process will be prone to becoming more complex, and overall safety considerations on the basis of a number of parameters need to be properly assessed, as highlighted in a recent study [73].

Another major advantage of PI technologies is that by virtue of their greatly improved heat and mass transfer, they allow process conditions to be employed which would previously have been impossible with conventional technologies. These are what Hessel refers to as the 'novel process windows' offered by PI [74]. Several applications, particularly related to microreactors [74–76] and SDRs [77,78], in which these more intense operating conditions – such as highly concentrated reaction systems, higher temperatures and higher pressures – have been investigated, are documented in the literature.

1.5.3 Environment

PI technologies have tremendously appealing environmental implications, whereby a small, compact, highly intensified plant is more likely to be below the tree line, making it far less of an eyesore for the general public than the unsightly and massive steel works characterizing our present chemical plants. Furthermore, novel reactor designs based on the PI concept will enable clean technology to be practised by enabling waste minimization at source. In other words, high-selectivity operation in intensified reactors will reduce or eliminate altogether the formation of unwanted byproducts, which, if not removed from the effluent before discharge, can cause irreversible damage to the environment. High-purity product – which is hence of improved quality – will thus be obtained without incurring enormous downstream purification costs.

The improved energy efficiency foreseeable in intensified unit operations constitutes yet another highly attractive benefit of PI in a world where there is overwhelming concern about the ever-growing demand on non-renewable energy resources. In order to address this concern, there is a great and urgent need for new process technologies that can utilize energy in an efficient manner. In this respect, PI is a positive step for the chemical industry. Great enhancements in heat and mass transfer, two of the most fundamental and frequently encountered operations in chemical engineering processes, can be achieved in intensified units, as has been described in this chapter. Such improvements mean that process times and therefore energy consumption can be dramatically reduced for a given operation. Furthermore, alternative energy sources such as microwave and light energy, which can be targeted to desired process chemistries, can result in less energy waste than is usually encountered with conventional thermal energy sources. A case study highlighting the energy-saving potential of a PI technology is presented in Chapter 15.

1.6 Challenges to Implementation of PI

PI was first developed as a coherent philosophy by Ramshaw *et al.* in the 1970s, so it might be surprising that it has not had more impact by now. There are manifold reasons for the lack of implementation in industry. Some are:

- *Perception of Risk:* Most chemical/pharmaceutical companies are (understandably) risk-averse, and for new technologies the risk is difficult to assess. There is typically a lack of data and a lack of previous case studies on which to base designs and operation, and therefore economic calculations. This results in a so-called 'rush-to-be-second' mentality. The lack of risk and cost data makes it difficult to assess the risk–reward balance and perform a reliable cost–benefit analysis for the process change.
- *Lack of 'Champions' within Industry:* For any given change in an organization, there is an associated individual with responsibility for it. When this change is substantial and technical, such as the change to a novel unit operation in a chemical plant, the 'champion' will have a lot of inertia to overcome, and must be dedicated and well-informed.The champion has to fully understand the technology and its benefits. It is the responsibility of technology developers to inform possible users of the benefits of their technology, but changes cannot be effected without a substantial 'pull' from the users.
- *Control/Monitoring:* The greatly enhanced rate of some intensified processes leads to problems in monitoring, as the response times of the instruments become significant in comparison to the process time. This can be an important added technical development step, and as such an obstacle to commercialization. To address these issues, a large body of research work has been undertaken on control of more established PI technologies, such as RD and dividing wall columns, as highlighted in a recent review of control aspects in PI [79]. Work is currently ongoing to develop appropriate control strategies for newer developments such as the SDR [80].
- *PI's Limitations:* Sometimes PI is not the solution. Any technology has a specific problem or set of problems that it can solve, and these need to be clearly identified and explained. The greatest danger of not doing so is that feasibility studies/pilot-scale trials

are performed that fail because the premise was incorrect (i.e. it was the wrong technology from the start), rather than because of any technical failing, and the technology itself is then labelled as inadequate. A number of technologies developed have been characterized as 'solutions seeking problems', and these problems will sometimes be found. However, sometimes the question that is asked is, 'Can this technology be used for this application?', when it should be, 'Does this technology have realizable economic benefits for the process overall?'

Many of these challenges are familiar to anyone who has had a new technology of any sort to promote. The key issue is simply that the new process/technology must have a *provable and significant* economic advantage. It must be borne in mind that an economic advantage is not only achieved by reducing the size of a piece of process equipment (or, preferably, the entire plant), but can be brought about in a variety of ways:

- *Safer Operation:* Accidents are a cost. Some function of the risk and consequent cost of an accident should be factored into the economic evaluation of any project. Intensified technologies in general will significantly reduce the hazard, and therefore cost, of any incident. More effective monitoring, which is more likely with smaller intensified equipment, should help to reduce the rate of incidence of accidents.
- *Product Quality:* Improved product quality reduces the load on downstream separations, perhaps allowing a reduction in size, or alternative technologies to be used, or process steps to be removed altogether, thereby reducing capital and running costs in other areas of the plant. The higher product quality produced in intensified processes due to the enhanced effectiveness of heat and mass transfer reduces the amount of waste produced, thereby lessening the amount that has to be disposed of. In some industries, where the wastes are highly toxic, for instance, disposal and/or treatment is one of the largest running costs. This will become more of a driver in the future, as environmental legislation becomes more stringent worldwide.
- *Reduced Running Costs:* The greater efficiency of intensified plants reduces energy costs. In a fossil fuel-based energy supply system, as most currently are, this equates directly to reduced CO_2 emissions. Such issues should become increasingly important as carbon trading becomes more standard.

1.7 Conclusion

As a revolutionary approach to chemical processing, PI has gained remarkable momentum in the last 2 decades. It was described more than 10 years ago as 'the key to survival of the fittest in international competition' [81]. It is not surprising, therefore, given the intensely competitive environment in which businesses in the chemical industry operate, that great strides have been made in the research and development of PI technologies over the last few years.

With greater emphasis being placed on sustainable development nowadays, PI can be an important element in making future chemical and pharmaceutical industries greener. This is because it can be used to greatly reduce the size of many unit operations, and in doing so reduce not only the energy required to run them but also the surrounding infrastructure and the energy and materials used in their manufacture. In an intensified environment, higher selectivities and yields can result in improved atom efficiency and waste minimization at source, leading to reduced downstream processing.

There is no doubt that PI is here to stay and that it will continue to play a major role in the development and growth of the chemical and processing industry in the years to come. Whether or not large plants should be considered 'dinosaurs' that will be extinct within the next 50 years remains to be seen, but it is a vision that is not altogether unrealistic. There are still many challenges ahead, which will require a concerted effort from both the academic community and industry partners before PI is more widely implemented and practised. It is also worthy of note here that how successful PI is in its application to a given process is very much dependent on the characteristics of that process. An inherently slow process will not be made any faster by intensifying heat- and mass-transfer characteristics. Thus, it is important to clearly understand the fundamentals of a process before any attempt at intensifying it is made. There are a few other process-related limitations to the general application of PI, such as:

- Some processes operate at too small a scale for PI to be economically worthwhile.
- Various plants require a flexibility of operation that is difficult to achieve in a continuous plant, which is usually purpose-designed.

However, in spite of these application limitations and challenges, the potential for PI remains largely untapped, but the economic rewards for those companies that do introduce PI are likely to be substantial – and this will have environmental benefits for a more sustainable future for generations to come.

Nomenclature

A	Constant
A_s	Heat-transfer surface area (m^2)
C_1, C_2	Constants in Nusselt number and Sherwood number correlations, respectively
C_D	Coefficient of discharge of impeller (–)
C_p	Specific heat capacity (J/kg K)
D	Pipe or channel diameter (m)
D	Diffusivity (m^2/s)
D_i	Impeller diameter (m)
D_l	Liquid diffusivity (m^2/s)
D_{turb}	Turbulent diffusivity (m^2/s)
h	Heat-transfer coefficient (W/m^2 K)
k	Thermal conductivity (W/m K)
k_m	Mass-transfer coefficient (m/s)
l	Characteristic length for diffusion (m)
L	Characteristic length scale of large eddies (at macroscale) (m)
L_c	Integral length scale of turbulence (m)
L_D	Characteristic length scale for turbulent dispersion (m)
N	Impeller rotational speed (Hz)
Nu	Nusselt number (= hD/k)
Q	Heat-transfer rate (W)

Q_c	Impeller pumping capacity (m^3/s)
Q_f	Flow rate of injected fluid (m^3/s)
r_{pipe}	Radius of pipe for feed introduction (m)
Re	Reynolds number (= $\rho uD/\mu$)
Sc	Schmidt number (= $\nu/\mathbf{D}$ or ν/D_l)
Sh	Sherwood number (= $kml/\mathbf{D}$)
t_D	Characteristic diffusion time (s)
ΔT_{lm}	Logarithmic mean temperature difference (K)
u	Average velocity (m/s)
U	Overall heat transfer (W/m^2 K)
V	Vessel volume (m^3)

Greek Letters

ε	Turbulent energy dissipation rate (W/kg)
η_K	Kolgomorov length scale (m)
η_B	Batchelor length scale (m)
ν	Kinematic viscosity of fluid (m^2/s)
τ_c	Circulation time (s)
τ_D	Characteristic mesomixing timescale for turbulent dispersion (s)
τ_{DS}	Characteristic micromixing time for shear deformation and diffusion at Batchelor scale (s)
τ_e	Characteristic micromixing time for engulfment (s)
τ_{mac}	Macromixing time (s)
τ_s	Inertial-convective mesomixing time (s)
μ	Viscosity (Pa.s)
ρ	Density (kg/m^3)

References

(1) W. T. Cross and C. Ramshaw, Process intensification – laminar flow heat transfer. Chem. Eng. Res. Des., 64(4), 293–301 (1986).

(2) A. I. Stankiewicz and J. A. Moulijn, Process intensification: transforming chemical engineering. Chem. Eng. Prog., 96(1), 22–34 (2000).

(3) T. Van Gerven and A. Stankiewicz, Structure, energy, synergy, time – the fundamentals of process intensification. Ind. Eng. Chem. Res., 48(5), 2465–2474 (2009).

(4) O. Levenspiel, Chemical Reaction Engineering, 3rd edition. John Wiley & Sons: New York (1999).

(5) J. Baldyga and R. Pohorecki, Turbulent micromixing in chemical reactors – a review. Chem. Eng. J. Bioch. Eng., 58(2), 183–195 (1995).

(6) J. R. Bourne, Mixing in single-phase chemical reactors, in Mixing in the Process Industries, A. W. Nienow, M. F. Edwards and N. Harnby (Eds.), Butterworth-Heinemann: Burlington (1997).

(7) B. K. Johnson and R. K. Prud'homme, Chemical processing and micromixing in confined impinging jets. AIChE J., 49(9), 2264–2282 (2003).

(8) L. H. Nouri, J. Legrand, N. Benmalek, F. Imerzoukene, A.-R. Yeddou and F. Halet, Characterisation and comparison of the micromixing efficiency in torus and batch stirred reactors. Chem. Eng. J., 142(1), 78–86 (2008).
(9) M. D. Wood, A methodical approach to process intensification. PhD Thesis, Cranfield University (2000).
(10) S. J. Hearn, Turbulent mixing mechanisms in motionless mixers. PhD Thesis, University of Birmingham (1995).
(11) S. W. Siddiqui, Y. Zhao, A. Kukukova and S. M. Kresta, Characteristics of a confined impinging jet reactor: energy dissipation, homogeneous and heterogeneous reaction products, and effect of unequal flow. Ind. Eng. Chem. Res., 48(17), 7945–7958 (2009).
(12) M. Meeuwse, E. Hamming, J.van der Schaaf and J. C. Schouten, Effect of rotor-stator distance and rotor radius on the rate of gas-liquid mass transfer in a rotor-stator spinning disc reactor. Chemical Engineering and Processing, 50(10), 1095–1107 (2011).
(13) K. V. K. Boodhoo and S. Al-Hengari, Micromixing characteristics in a small scale spinning disk reactor. Chem. Eng. Technol., 35(7), 1229–1237 (2012).
(14) G. B. Tatterson, Fluid Mixing and Gas Dispersion in Agitated Tanks, McGraw-Hill Inc.: New York (1991).
(15) J. Baldyga, J. R. Bourne and S. J. Hearn, Interaction between chemical reactions and mixing on various scales. Chem. Eng. Sci., 52(4), 457–466 (1997).
(16) J. Baldyga and J. R. Bourne, Interactions between mixing on various scales in stirred tank reactors. Chem. Eng. Sci., 47(8), 1839–1848 (1992).
(17) A. N. Kolgomorov, Dokl. Akad. Nauk SSSR, 30, 301 (1941).
(18) S. B. Pope, Turbulent Flows, Cambridge University Press (2000).
(19) G. K. Batchelor, Small-scale variation of convected quantities like temperature in turbulent fluid. Part 1. General discussion and the case of small conductivity. J. Fluid Mech., 5, 113–133 (1959).
(20) D. A. Reay, C. Ramshaw and A. P. Harvey, Process Intensification: Engineering for Efficiency, Sustainability and Flexibility, Chapter 4, Elsevier (2008).
(21) G. Baier, Liquid-liquid extraction based on a new flow pattern: two-fluid Taylor Couette flow, PhD Thesis, University of Wisconsin-Madison (1999).
(22) J. M. Coulson, J. F. Richardson, J. R. Backhurst and J. H. Harker, "Chemical Engineering Volume 1", 6th ed. pp 720–735, Butterworth Heinemann (1999).
(23) K. C. D. Hickman, Centrifugal boiler compression still. Ind. Eng. Chem., 49(5), 786–789 (1957).
(24) C. Ramshaw, The opportunities for exploiting centrifugal fields. Heat Recov. Syst. CHP, 13(6), 493–513 (1992).
(25) A. Stankiewicz, Energy matters – alternative sources and forms of energy for intensification of chemical and biochemical processes. Chem. Eng. Res. Des., 84(A7) 511–521 (2006).
(26) T. J. Mason and P. Cintas, Sonochemistry, in Handbook of Green Chemistry and Technology, J. H. Clark and D. J. Macquarrie (Eds.), Blackwell: Oxford (2002).
(27) B. Toukoniitty, J. P. Mikkola, D. Y. Murzin and T. Salmi, Utilization of electromagnetic and acoustic irradiation in enhancing heterogeneous catalytic reactions. Appl. Catal. A-Gen., 279 (1–2), 1–22 (2005).
(28) L. H. Thompson and L. K. Doraiswamy, Sonochemistry: science and engineering. Ind. Eng. Chem. Res., 38(4), 1215–1249 (1999).
(29) C. Leonelli and T. J. Mason, Microwave and ultrasonic processing: now a realistic option for industry. Chem. Eng. Process., 49(9), 885–900 (2010).
(30) C. Petrier, A. Jeunet, J. L. Luche and G. Reverdy, Unexpected frequency effects on the rate of oxidative processes induced by ultrasound. J. Am. Chem. Soc., 114(8), 3148–3150 (1992).

(31) G. Mark, A. Tauber, R. Laupert, H.-P. Schuchmann, D. Schulz, A. Mues and C. Sonntag, OH-radical formation by ultrasound in aqueous solution. Part II: Terephthalate and Fricke dosimetry and the influence of various conditions on the sonolytic yield. Ultrason. Sonochem., 5(2), 41–52 (1998).
(32) J.-L. Luche and P. Cintas, in Advances in Sonochemistry, T. J. Mason (Ed.), Volume 5 JAI Press (1998).
(33) L. McCausland and P. Cains, Ultrasound to make crystals. Chem. Ind. (9) 15 (2003).
(34) W. G. Eddy and H. C. Eddy, Discussion on electrical dehydration of crude oil. J. Ind. Eng. Chem., 13, 1016 (1921).
(35) S. E. Taylor, Theory and practice of electrically-enhanced phase separation of water-in-oil emulsions. Chem. Eng. Res. Des., 74(A5) 526–540 (1996).
(36) J. F. Hughes and I. D. Pavey, Electrostatic emulsification. J. Electrostat., 10, 45–55 (1981).
(37) T. C. Scott, Surface area generation and droplet size control using pulsed electric fields. AIChE J., 33, 1557–1559 (1987).
(38) L. R. Weatherley and D. Rooney, Enzymatic catalysis and electrostatic process intensification for processing of natural oils. Chem. Eng. J., 135(1–2), 25–32 (2008).
(39) T. B. Jones, Electrohydrodynamically enhanced heat transfer in liquids: a review. Adv. Heat Transfer, 14, 107–148 (1978).
(40) H. Kawahira, Y. Kubo, T. Yokoyama and J. Ogata, The effect of an electric field on boiling heat transfer of Refrigerant-11 – boiling on a single tube. IEEE T. Ind. Appl., 26(2), 359–365 (1990).
(41) A. Wolny, Intensification of the evaporative process by electric field. Chem. Eng. Sci., 47(3), 551–554 (1992).
(42) J. Ogata and A. Yabe, Augmentation of boiling heat transfer by utilizing the EHD effect: EHD behaviour of boiling bubbles and heat transfer characteristics. Int. J. Heat Mass Tran., 36, 783–791 (1993).
(43) E. Bonjour, J. Verdier and L. Weil, Electroconvection effects on heat transfer. Chem. Eng. Prog., 58, 63 (1962).
(44) R. F. Lovenguth and D. Hanesian, Boiling heat transfer in the presence of non-uniform direct current electric fields. Ind. Eng. Chem. Fundam., 10, 570 (1971).
(45) M. Cipriani, P. Di Marco and W. Grassi, Effect of an externally applied electric field on pool film boiling of fc-72. Heat Transfer Eng., 25, 3–13 (2004).
(46) O. K. Matar and C. J. Lawrence, The flow of a thin conducting film over a spinning disc in the presence of an electric field. Chem. Eng. Sci., 61(12), 3838–3849 (2006).
(47) C. R. Strauss, Applications of microwaves for environmentally benign organic chemistry, in Handbook of Green Chemistry and Technology, J. H. Clark and D. J. Macquarrie (Eds.), Blackwell: Oxford (2002).
(48) E. D. Neas and M. J. Collins, Introduction to Microwave Sample Preparation, H. M. Kingston and L. B. Jassie (Eds.), ACS: Washington, DC (1988).
(49) J. Blyth and A. W. Hoffmann. Ann., 53, 292 (1845).
(50) I. Dunkin, Photochemistry, in Handbook of Green Chemistry and Technology, J. H. Clark and D. J. Macquarrie (Eds.), Blackwell: Oxford (2002).
(51) S. Kataoka, A. Endo, A. Harada, Y. Inagi and T. Ohmori, Characterization of mesoporous catalyst supports on microreactor walls. Appl. Catal. A-Gen., 342(1–2), 107–112 (2008).
(52) P. Watts and S. J. Haswell, The application of microreactors for small scale organic synthesis. Chem. Eng. Technol., 28(3), 290–301 (2005).
(53) V. Hessel, H. Lowe and F. Schonfeld, Micromixers – a review on passive and active mixing principles. Chem. Eng. Sci., 60(8–9), 2479–2501 (2005).
(54) J. R. Burns and C. Ramshaw, The intensification of rapid reactions in multiphase systems using slug flow in capillaries. Lab on a Chip, 1(1), 10–15 (2001).

(55) J. Aubin, M. Ferrando and V. Jiricny, Current methods for characterising mixing and flow in microchannels. Chem. Eng. Sci., 65(6), 2065–2093 (2010)

(56) B. P. Mason, K. E. Price, J. L. Steinbacher, A. R. Bogdan and D. T. McQuade, Greener approaches to organic synthesis using microreactor technology. Chem. Rev., 107(6), 2300–2318 (2007).

(57) S. Tadepalli, D. Y. Qian and A. Lawal, Comparison of performance of microreactor and semi-batch reactor for catalytic hydrogenation of o-nitroanisole. Catal. Today, 125(1–2), 64–73 (2007).

(58) J. R. Burns and C. Ramshaw, Development of a microreactor for chemical production. Chem. Eng. Res. Des., 77(A3) 206–211 (1999).

(59) S. Roy, T. Bauer, M. Al-Dahhan, P. Lehner and T. Turek, Monoliths as multiphase reactors: a review. AIChE J., 50(11), 2918–2938 (2004)

(60) A. Stankiewicz, Process intensification in in-line monolithic reactor. Chem. Eng. Sci., 56(2), 359–364 (2001).

(61) K. V. K. Boodhoo, M. Vicevic, C. D. Cartwright, T. Ndlovu and E. Toogood, Intensification of gas–liquid mass transfer using a rotating bed of porous packings for application to an E. coli batch fermentation process. Chem. Eng J., 135, 141–150 (2008).

(62) H. Nikakhtari and G. A. Hill, Enhanced oxygen mass transfer in an external loop airlift bioreactor using a packed bed. Ind. Eng. Chem. Res., 44(4), 1067–1072 (2005)

(63) C. Hutter, C. Allemann, S. Kuhn and P. R.von Rohr, Scalar transport in a milli-scale metal foam reactor. Chem. Eng. Sci., 65(10), 3169–3178 (2010).

(64) S. Hirschberg, R. Koubek, F. Moser and J. Schoeck, An improvement of the Sulzer SMX (TM) static mixer significantly reducing the pressure drop. Chem. Eng. Res. Des., 87(4A), 524–532 (2009).

(65) R. K. Thakur, C. Vial, K. D. P. Nigam, E. B. Nauman and G. Djelveh, Static mixers in the process industries – a review. Chem. Eng. Res. Des., 81(A7) 787–826 (2003).

(66) T. Lemenand, C. Durandal, D. Della Valle and H. Peerhossaini, Turbulent direct-contact heat transfer between two immiscible fluids. Int. J. Therm. Sci., 49(10), 1886–1898 (2010)

(67) C. Habchi, T. Lemenand, D. Della Valle and H. Peerhossaini, Turbulent mixing and residence time distribution in novel multifunctional heat exchangers-reactors. Chem. Eng. Process., 49 (10), 1066–1075 (2010)

(68) X. Fan, M. G. Manchon, K. Wilson, S. Tennison, A. Kozynchenko, A. A. Lapkin and P. K. Plucinski, Coupling of Heck and hydrogenation reactions in a continuous compact reactor. J. Catal., 267(2), 114–120 (2009).

(69) K. Wall, P. N. Sharratt and N. Sadr-kazemi, Development of separation operations in low-tonnage chemical process design, in IChemE Separation Symposium, Manchester, 13 September 2000.

(70) V. H. Agreda, L. R. Partin and W. H. Heise, High-purity methyl acetate via reactive distillation. Chem. Eng. Prog., 86(2), 40–46 (1990).

(71) P. Oxley, C. Brechtelsbauer, F. Ricard, N. Lewis and C. Ramshaw, Evaluation of spinning disk reactor technology for the manufacture of pharmaceuticals. Ind. Eng. Chem. Res., 39(7), 2175–2182 (2000).

(72) J. C. Etchells, Process intensification – safety pros and cons. Process Saf. Environ., 83(B2) 85–89 (2005).

(73) F. Ebrahimi, T. Virkki-Hatakka and I. Turunen, Safety analysis of intensified processes. Chem. Eng. Process., 52, 28–33 (2012).

(74) V. Hessel, Novel process windows – gate to maximizing process intensification via flow chemistry. Chem. Eng. Technol., 32(11), 1655–1681 (2009).

(75) N. Kockmann and D. M. Roberge, Harsh reaction conditions in continuous-flow microreactors for pharmaceutical production. Chem. Eng. Technol., 32(11), 1682–1694 (2009).

(76) T. Razzaq, T. N. Glasnov and C. O. Kappe, Accessing novel process windows in a high-temperature/pressure capillary flow reactor. Chem. Eng. Technol., 32(11), 1702–1716 (2009).

(77) K. V. K. Boodhoo, W. A. E. Dunk, M. Vicevic, R. J. Jachuck, V. Sage, D. J. Macquarrie and J. H. Clark, Classical cationic polymerization of styrene in a spinning disc reactor using silica-supported BF_3 catalyst. J. Appl. Polym. Sci., 101(1), 8–19 (2006).

(78) M. R. Moghbeli, S. Mohammadi and S. M. Alavi, Bulk free-radical polymerization of styrene on a spinning disc reactor. J. Appl. Polym. Sci., 113(2), 709–715 (2009).

(79) N. M. Nikacevic, A. E. M. Huesman, P. M. J.Van den Hof and A. I. Stankiewicz, Opportunities and challenges for process control in process intensification. Chem. Eng. Process., 52, 1–15 (2012).

(80) D. Ghiasy, K. V. K. Boodhoo and M. T. Tham, Control of intensified equipment: a simulation study for pH control in a spinning disc reactor. Chem. Eng. Process., 55, 1–7 (2012).

(81) J. Semel, Process intensification – today's challenge for the chemical industry, in2nd International Conference on Process Intensification in Practice – Applications and Opportunities, BHR Conference Series No. 28 (1997).

2

Green Chemistry Principles

James Clark, Duncan Macquarrie, Mark Gronnow and Vitaly Budarin

Green Chemistry Centre of Excellence, University of York, York, UK

2.1 Introduction

'Green chemistry' was coined as a term in the early 1990s by staff at the US Environmental Protection Agency (EPA) and subsequently by Anastas and Warner [1] as an attempt to bring together the work being undertaken worldwide on clean chemical technologies. However, attempts to think 'green' were already underway prior to this, due to the increase in stringent environmental protection laws and social pressures. A definition of 'green chemistry' is difficult to provide, as numerous, albeit similar, ones exist. It is defined by Warner and Anastas as '*the design of chemical products and processes that reduce or eliminate the use and generation of hazardous substances*' [1], whereas the EPA defines it thus: '*Green chemistry, also known as sustainable chemistry, is the design of chemical products and processes that reduce or eliminate the use or generation of hazardous substances. Green chemistry applies across the life cycle of a chemical product, including its design, manufacture, and use*' [2]. Embracing life cycle considerations is in keeping with the current way of thinking in research and commerce.

Green chemistry focuses on the design of chemical products and processes that are environmentally benign. In other words, if green chemistry is used in a process, its inputs and outputs should become safer, its energy consumption should be reduced and its costs should be reduced as yields increase, separations become simpler and less waste is generated. Yes – green chemistry can be cheaper! However, in many cases industrial

Process Intensification for Green Chemistry: Engineering Solutions for Sustainable Chemical Processing, First Edition. Edited by Kamelia Boodhoo and Adam Harvey.

inertia prevents uptake of new technology, since it may involve reassessing old procedures or purchasing new equipment. Furthermore, regulatory issues, particularly in the pharmaceutical industry, can lead to reluctance to incorporate clean chemical technologies into manufacturing processes.

Practically, green chemistry is a term of two parts: first, there is new technology and hardware, such as flow reactors and microwave reactors; and second, there is general good housekeeping and organic chemistry, which can be considered as 'green'. For instance, lower temperatures, shorter reaction times, the choice of route, increased yields and the use of fewer washings at work up all improve the 'cleanness' of a reaction by saving energy, saving time (and thereby energy) and reducing waste; all of which appear in the 'Twelve Principles of Green Chemistry' [1] (see Section 2.2).

Today there is emphasis on feedstocks (which should be renewable) and products (which should be safe for humans and the environment, and increasingly should be designed for the recovery of resources or for 'closed-loop manufacturing'). This means that green chemistry research and application encompass the use of biomass as a source of organic carbon (and potentially other elements) and the design of new greener products, typically to replace existing products that are unacceptable in the light of new legislation (e.g. REACH) or consumer perception.

In this chapter the philosophy and motivations of green chemistry are discussed, along with a number of the tools and techniques associated with this maturing field.

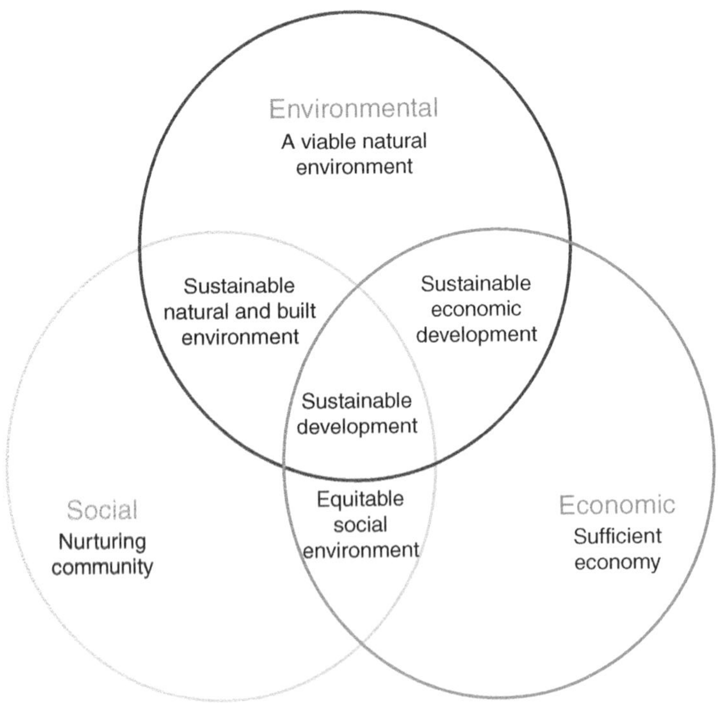

Figure 2.1 'Triple bottom line' of social, environmental and economic factors representing sustainable development.

2.1.1 Sustainable Development and Green Chemistry

Closely linked to green chemistry is the term 'sustainable development'. As with green chemistry, multiple definitions exist for this, and its interpretation and implementation are very different for different people, but the 1987 definition proposed by Gro Harlem Brundtland is generally accepted. Brundtland chaired a review panel that authored a report entitled 'Our Common Future', which defined sustainable development as '*development that meets the needs of the present without compromising the ability of future generations to meet their own needs*' [3].

Sustainable development lies at the overlap of social, environmental and economic development, as shown in Figure 2.1.

Green chemistry can be seen as a tool by which sustainable development can be achieved. The application of green chemistry is relevant to social, environmental and economic concerns.

The achievement of sustainable development will require action by the international community, national governments, commercial and noncommercial organizations and individuals from a wide variety of disciplines. Acknowledgement of sustainable development has been taken forward into policy by many governments, including European [4], Chinese [5] and American [6].

2.2 The Twelve Principles of Green Chemistry

Clean chemical technology is not a new field; rather, it is a thought process at the design stage of any synthesis using new and existing tools and knowledge. To better illustrate this, the pioneers of this field suggested a range of foundations for the design of more benign chemistry. The first of these is Anastas' 'Twelve Principles of Green Chemistry' [1]:

1. *Prevention*: It is better to prevent waste than to treat or clean up waste after it has been created.
2. *Atom Economy*: Synthetic methods should be designed to maximize the incorporation of all materials used in the process into the final product.
3. *Less Hazardous Chemical Synthesis*: Wherever practical, synthetic methods should be designed to use and generate substances that possess little or no toxicity to human health and the environment.
4. *Design for Safer Chemicals*: Chemical products should be designed to effect their desired function while minimizing their toxicity.
5. *Safer Solvents and Auxiliaries*: The use of auxiliary substances (e.g. solvents, separation agents, etc.) should be made unnecessary wherever possible and innocuous where necessary.
6. *Design for Energy Efficiency*: The energy requirements of chemical processes should be noted for their environmental and economic impacts and should be minimized. If possible, synthetic methods should be conducted at ambient temperature and pressure.
7. *Use of Renewable Feedstocks*: A raw material or feedstock should be renewable rather than depleting whenever technically and economically practical.

8. *Reduction of Derivatives*: Unnecessary derivatization (e.g. use of blocking groups, protection/deprotection, temporary modification of physical/chemical processes) should be minimized or avoided if possible, because such steps require additional reagents, energy use and generate waste.
9. *Catalysis*: Catalytic reagents (as selective as possible) are superior to stoichiometric reagents.
10. *Design for Degradation*: Chemical products should be designed so that at the end of their function they break down into innocuous degradation products that do not persist in the environment.
11. *Real-time Analysis for Pollution Prevention*: Analytical methodologies need to be further developed to allow for real-time in-process monitoring and control prior to the formation of hazardous substances.
12. *Inherently Safer Chemistry for Accident Prevention*: Substances and the forms of substances used in a chemical process should be chosen to minimize the potential for chemical accidents, including releases, explosions and fire.

These Twelve Principles have been built on and adapted by many other scientists in pursuit of the ideal clean synthesis. As already indicated, they are more process focused than modern green chemistry thinking, but nonetheless they remain a useful guide, especially for process chemists.

2.2.1 Ideals of Green Chemistry

In Figures 2.2 and 2.3 the ideals of green chemical synthetic design are shown. It is important to note that these points are most effectively dealt with and are easier to apply if they are considered at the design stage rather than retrospectively.

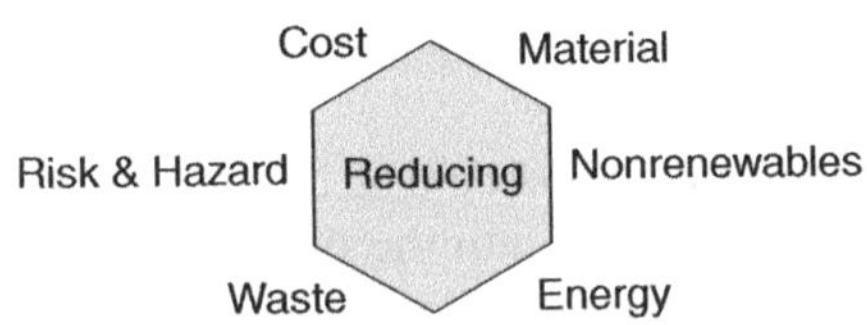

Figure 2.2 *Factors for reduction in syntheses.*

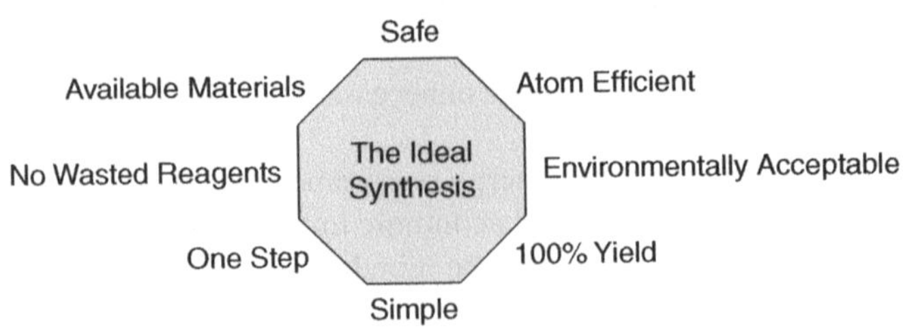

Figure 2.3 *The eight parts of an ideal synthesis.*

When considering the importance of focusing on green aspects at the design stage, comparisons can be made with the chemical engineering discipline. Traditionally, chemical plants concentrated on mechanical safety devices, estimation of the probability of accident and reduction of that probability. However, mechanical devices are not infallible and safety measures cannot completely prevent accidents from happening. Therefore, chemical engineering created the concept of 'Inherently Safer Design' (ISD), with the intention of eliminating rather than preventing hazards. Trevor Kletz, a leader in this field, states, '*What you don't have can't leak*' [7]. ISD means not holding inventories of hazardous chemicals or not using them in the first place.

This logic would have prevented the accident at Bhopal, India in 1984, in which many thousands of people were killed or seriously injured. One of the chemicals used in the process at the Union Carbide factory (methylisocyanate, MIC) was highly water sensitive and when a watertight holding tank was breached, the accident occurred, releasing the chemical into the air and affecting villages surrounding the factory. MIC was not essential for the chemistry and an ISD approach would have used an alternative, thus eliminating the risk of such an accident ever occurring. While this is an extreme example, it demonstrates the importance of considerations at the design stage.

In the area of green chemistry, research has led to the invention of a number of ingenious processing technologies that can save time, save energy or reduce waste production. These technologies by and large only exist in academia, however, and, with very few exceptions, industry has been slow to utilize them. Techniques have included photochemistry, microwave chemistry, sonochemistry and synthetic electrochemistry. All these replacements for conventional conductive heating can lead to improved yields, reduced reaction times and reduced byproduct formation. Engineered greener technologies also exist, including a number of replacements for the stirred tank batch reactor, such as continuous stirred tanks, fluidized bed reactors, microchannel reactors and spinning disc reactors (SDRs), all of which increase throughput while decreasing energy usage and waste. Unfortunately, despite these many new processes, industry is reluctant to use these hardware solutions. One of the reasons for this is the often massive financial expenditure involved in purchasing these items and the limited number of chemistries that have been demonstrated with them to date.

2.3 Metrics for Chemistry

Having made a 'greening' improvement to a chemical process, it is important to be able to quantify the change. By doing so, a tangible element or benefit of the new technology can be seen, which is likely to aid the communication of the work and may potentially facilitate transfer to industry. For a non-chemist, the most attractive way of quoting the improvement would be of the form 'a decrease of £X per kilo of compound Y'. This, however, is an oversimplification and does not allow a chemist to visualize the improvement made or to make allowance for toxicity/hazard. For yield improvements and selectivity increases, percentages are suitable, but this simplistic approach may not always be appropriate. For example, when a highly pyrophoric reagent is replaced by a benign one, a numerical value is difficult to assign, but the improvement is obvious, if all other factors are similar.

Numerous metrics have been formulated over time and their suitability has been discussed at great length [8–11]. The problem observed is that the more accurate and universally applicable the metric devised, the more complex and unemployable it becomes. A good metric must be clearly defined, simple, measurable, objective rather than subjective and must ultimately drive the desired behaviour [12]. In this section a number of established metrics for measuring greenimprovements are discussed.

2.3.1 Effective Mass Yield

Effective mass yield is defined as the percentage of the mass of the desired product relative to the mass of all nonbenign materials used in its synthesis. Hudlicky *et al.* [13] suggest the following equation:

$$Effective\,mass\,yield\,(\%) = \frac{\text{Mass of products}}{\text{Mass of nonbengin reagents}} \times 100 \qquad (2.1)$$

This metric requires further definition of what is meant by a 'benign substance'. Hudlicky defines this as *'those by-products, reagents or solvents that have no environmental risk associated with them, for example, water, low-concentration saline, dilute ethanol, autoclaved cell mass, etc.'* [13]. This definition leaves the metric open to criticism, as all substances have some degree of environmental risk or impact associated with them (which is a subjective term). The formula also fails to address the level of toxicity associated with a process. Until all toxicology data are available for all chemicals and a term dealing with these levels of 'nonbenign' reagents is written into the formula, the effective mass yield is not the best metric for chemistry.

2.3.2 Carbon Efficiency

Carbon efficiency is a simplified formula developed at GlaxoSmithKline (GSK).[11] Its mathematical representation is as follows:

$$Carbon\,Efficiency\,(\%) = \frac{\text{Amount of carbon in product}}{\text{Total carbon present inreagents}} \times 100 \qquad (2.2)$$

This metric is a good simplification for use in the pharmaceutical industry as it takes into account the stoichiometry of reactants and products. Furthermore, it is of interest to the pharmaceutical industry, where development of carbon skeletons is essential.

2.3.3 Atom Economy

Atom economy was designed in a different way to the other metrics; most were designed to measure the improvement made, while Barry Trost [14] designed atom economy as a method by which organic chemists would pursue 'greener' chemistry. The simple definition of atom economy is a calculation of how much of the reactants remain in the final product. For a generic multistage reaction, the atom economy is as follows:

$$A + B \rightarrow C \qquad (2.3)$$

$$C + D \rightarrow E$$

$$E + F \rightarrow G$$

$$Atom\,Economy\,(\%) = \frac{M.W._{G}}{M.W._{A} + M.W._{B} + M.W._{D} + M.W._{F}} \times 100$$

The drawback of atom economy is that assumptions have to be made. For example, inorganic reagents (such as potassium carbonate in a Williamson's ether formation) are ignored as they are not incorporated into the final product. Solvents are also ignored, as is the stoichiometry of the reagents.

The atom economy calculation is a very simple representation of the 'greenness' of a reaction as it can be carried out without the need for experimental results. However, it is useful because a low atom economy at the design stage of a reaction prior to entry into the laboratory can drive the formulation of a cleaner synthetic strategy.

2.3.4 Reaction Mass Efficiency

Again developed by GSK [11], the reaction mass efficiency takes into account atom economy, yield and stoichiometry. The formula can take one of the following two forms.

From a generic reaction where $A + B \rightarrow C$:

$$Reaction\,Mass\,Efficiency\,(\%) = \frac{M.W._{C}}{M.W._{A} + M.W._{B}\frac{M.W._{B}}{M.W._{A}}} \times 100 \qquad (2.4)$$

Or more simply:

$$Reaction\,Mass\,Efficiency\,(\%) = \frac{\text{Mass of product C}}{\text{Mass of A} + \text{Mass of B}} \times 100 \qquad (2.5)$$

Like carbon efficiency, this measure shows the 'cleanness' of a reaction but not of a process; for example, neither metric takes into account waste produced. These metrics might thus present a rearrangement as 'very green' but would fail to address any solvent, work up and energy issues arising.

2.3.5 Environmental (E) Factor

The first general metric for green chemistry remains one of the best. Roger Sheldon's [15,16] E-factor can be made as complex and thorough or as simple as required. Assumptions concerning solvent and other factors can be made, or a total analysis can be performed.

The E-factor calculation is defined by the ratio of the mass of waste per unit of product. Equation (2.6) shows this mathematically:

$$E\text{-}factor = \frac{\text{Total waste (kg)}}{\text{Product (kg)}} \qquad (2.6)$$

Table 2.1 E-factors across the chemical industry.

Industry sector	Annual production (tonnage)	E-factor	Waste produced (tonnage)
Oil refining	10^6–10^8	~0.1	10^6
Bulk chemicals	10^4–10^6	<1 to 5	10^5
Fine chemicals	10^2–10^4	5 to >50	10^4
Pharmaceuticals	10–10^3	25 to >100	10^3

The metric is very simple to understand and to use. It highlights the waste produced in the process as opposed to the reaction, thus helping those who try to fulfil one of the Twelve Principles of Green Chemistry to avoid waste production. E-factors ignore recyclable factors such as recycled solvents and reused catalysts, which obviously increases their accuracy but ignores the energy involved in the recovery (these are often included theoretically by assuming 90% solvent recovery). The main difficulty with E-factors is the need to define system boundaries before calculations can be made, and these vary from scientist to scientist. This limitation is the major drawback of all green chemistry metrics, with the exception of the extremely complex life cycle assessment [17].

Sheldon took his publications one stage further and produced Table 2.1. Crucially, this metric is simple to apply industrially, as a production facility can measure how much material enters the site and how much leaves as product and waste, giving a direct, accurate global E-factor for the site. Table 2.1 shows that oil companies produce a lot less waste than pharmaceuticals as a percentage of material processed. This reflects the fact that the profit margins in the oil industry require them to minimize waste and find uses for products that would normally be discarded as waste. By contrast, the pharmaceutical sector is more focused on molecule manufacture and quality. The (currently) high profit margins within the sector mean that there is less concern about the comparatively large amounts of waste that are produced (especially considering the volumes used), although it has to be noted that, despite the percentage waste and E-factor being high, the pharmaceutical section produces a much lower tonnage of waste than any other sector. Table 2.1 encouraged a number of large pharmaceutical companies to commence green chemistry programmes.

Through incorporating yield, stoichiometry and solvent usage, the E-factor is an excellent metric. Crucially, E-factors can be combined to assess multistep reactions step by step or in one calculation.

2.3.6 Comparison of Metrics

A group of scientists at GSK [18] attempted to compare the currently available metrics, with the goal of designing a metric set suitable for their business. Following lengthy comparisons of different metrics for 28 reaction types and a comparison of simple and complex costings for four different drug molecules, they came up with a number of conclusions. It was suggested that yield is still a very good metric, especially for high-value-added chemistries such as those used in pharmaceutical synthesis, but conceded that yield does not encourage sustainable practices. Atom economy has its uses in conjunction with other metrics but is not suitable as a standalone green chemistry metric. Reaction mass efficiency combines both process and chemistry features and therefore has potential to be used optimally by chemists, process chemists and chemical engineers. The reaction mass efficiency is more likely to encourage sustainable practices as it focuses attention away from the waste and towards the use of materials. However, in terms of measuring the process and ease of use, E-factors are the most effective calculation at the current time.

2.4 Catalysis and Green Chemistry

The use of catalysis can help fulfil some of the Twelve Principles. Clearly, the use of catalytic processes is desirable in the vast majority of cases, as it can reduce the time and energy requirements for a reaction, and often enhance the selectivity of the process, reducing the E-factor significantly. It is important to consider the entire process, not just the reaction, and separation of the product away from the catalyst is clearly an additional factor not encountered in noncatalytic routes. Therefore, heterogeneous catalysis is optimal for greener chemistry as its activity can be as high as that of homogeneous catalysis and its selectivity can be better than heterogeneous due to the adsorption properties of the support material and the ease of separation through simple filtration. Recovery and reuse of the catalyst is also a key factor, with many catalysts being complex to synthesize or requiring expensive, often scarce, metals. Here we must also be mindful of the diminishing availability of several elements – the focus on carbon has disguised the fact that there are other key elements that are being depleted and whose availability will change in the short to medium term: the so-called issue of elemental sustainability. A heterogeneous catalyst is generally a significant benefit for both of these factors, as it resides in a separate phase and can easily be separated by filtration or centrifugation (or even by magnetism, in the case of magnetic support materials) [19,20] from a batch process. It can be used in a range of continuous processes where reaction and separation occur simultaneously.

In terms of green chemistry, the synthesis of the support must also be taken into account, as this can potentially be a complex process involving several steps, and usually requires significant washing/conditioning in order to ensure that there is no loosely bound material on the catalyst which will detach on use, leading to the same difficulties encountered with homogeneous catalysts and obviating the benefits of the heterogenization. As the stability of a catalyst is best judged in the precise conditions of its use, one option when conditioning the catalyst is to run a reaction with it, and treat the recovered catalyst as being the conditioned catalyst, on the assumption that the first use will have washed off any loose material and that what is left is likely to be tightly bound and stable. It can often be seen that the first use of a heterogeneous catalyst gives faster rates than subsequent ones, which are relatively consistent with each other. Such a scenario suggests that the first run has a significant homogeneous component, and that subsequent runs reveal the true heterogeneous activity of the catalyst. (However, it should be noted that there may be other explanations for this behaviour and that tests for homogeneity should still be carried out [21].)

Making use of waste streams rather than virgin materials is an important consideration. Here, too, progress is being made in the preparation of inorganic materials, catalysts and catalyst supports from a range of waste streams, including the ash from biomass combustion (which is often rich in silica [22]).

The majority of the heterogeneous catalysts that have been described in the literature are based on the functionalization of silica with organic groups, almost exclusively via silane linkers $(RO)_3SiR'$, where R′ is the catalytic group or a precursor thereof [23,24]. These can be prepared by attachment of the silanes on to a preformed silica surface, or by a co-condensation route which forms the silica with the silanes incorporated into (and preferably at the surface of) the resultant material. Templated silicas with organic functionalities are known, in which the template directs the synthesis of a material with very well-defined regular pores of uniform size and shape. This can be important for diffusion properties, meaning that mass transport is better and rates are often higher. A recent publication highlights the need for and key directions towards environmentally responsible design and syntheses of such catalysts [25].

The second additional aspect of a heterogeneous catalyst which must be considered relative to its environmental impact on a process is whether reactivation is required and what the nature of the reactivation is. This might be thermal, physical or chemical (e.g. via washing off a build-up of impurities which block the surface or pores of the material).

Finally, consideration must be given to the end of life – what is the fate of the catalyst? Can any components (e.g. metals or supports) be recovered and recycled or used for an alternative purpose?

The overall impact of the synthesis, recovery/reactivation and fate at end of life is clearly minimized when the turnover number (TON) (moles product/moles catalyst) is maximized; there is therefore a clear need to maximize TONs. Blaser *et al.* recently published work which indicated that TONs of around 50 000 are likely to be required for most industrial applications [26]. While the actual TON required will vary from case to case, and may sometimes be significantly lower than that quoted by Blaser, the numbers are nonetheless considerably higher (often by orders of magnitude) than those published in the literature. It is often possible to substantially increase TONs from the initial studies by careful optimization and modification, but this is rarely done in academic studies, and acceptable TONs are generally a long way off.

The rate of reaction is often lower in a heterogeneous system than in the corresponding homogeneous system (in cases where such a comparison is valid). This is due to the additional transport phenomena associated with moving reactants and products from the solid catalyst surface to the solution phase, through narrow-pore networks in the catalyst (the vast majority of heterogeneous systems are porous, in order to provide a sufficiently high surface area for reaction to take place; a nonporous material has a tiny surface area, and therefore virtually nowhere for the reaction to take place). The first of these two barriers is the hydrodynamic barrier, which exists close to the surface of a particle in solution. Here, there is a thin layer of solution, which can be thought of as being essentially physisorbed to the surface; that is, it is stationary with respect to the particle. This layer coats the catalyst particle and moves with the particle. Reactants have to cross this boundary in order to react and products must do likewise to escape into the bulk solution. Once inside, the reactants and products have to diffuse through the pore system of the catalyst, which is often narrow and irregular. Both these phenomena slow the reaction down in comparison to a homogeneous system, meaning that most heterogeneous catalysts are slower than their homogenous equivalents.

The correct choice of solvent can be particularly critical, as the hydrodynamic boundary is dependent on solvent viscosity, and solvents with low viscosity can reduce the

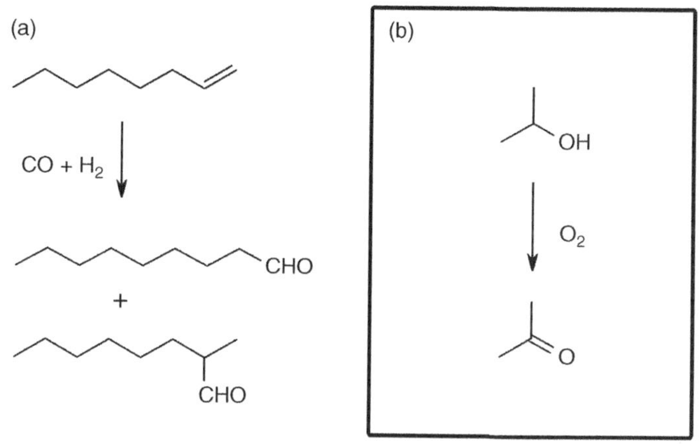

Figure 2.4 *Catalytic reactions in scCO2 using silica-supported Rh phosphine catalysts: (a) hydroformylation; (b) oxidation.*

importance of this barrier. In addition, low-viscosity solvents will typically have good transport properties. One group of solvents that is particularly suitable is the supercritical fluids, of which supercritical carbon dioxide ($scCO_2$) is the most widely used. The properties of this class of solvents are ideal for flow and transport through pore systems, in additional to their other green benefits, and these solvents have been successfully used in continuous catalytic processes, including hydroformylation [27] and oxidation [28] (Figure 2.4).

It should be pointed out that there can be positive effects from pore constraints (shape selectivity in the zeolites being the most pronounced example) or from the surface chemistry of the support itself, which can potentially aid in the catalytic system and enhance reaction rates. For example, amine-functionalized silicas are particularly effective in the Knoevenagel reaction, as the mildly acidic silanols provide H-bonding assistance in the reaction and complement the basic amine centre, providing an acid–base combination that is particularly effective [29]. It is also possible to have both acid and basic sites present on the same material, and bifunctional catalysts are known with independent, non-interacting acid and base sites [30,31,32]. Basic and metal-centred hydrogenation sites are also known [33].

2.4.1 Case Study: Silica as a Catalyst for Amide Formation

The clean production of amides is one of the most important challenges in the pharmaceutical industry, as evidenced by the results of round-table discussions within the pharmaceutical industry [34], and amide formation is also a widely used reaction in other parts of the chemical industry. Standard routes are general and give high yields for a range of substrates but use large quantities of stoichiometric activating agents such as thionyl chloride, dicyclohexylcarbodiimide or benzotriazole, none of which are acceptable on environmental or safety grounds. Boronic acids have recently been shown to be effective catalysts for the reaction, but their synthesis is complex and long-winded, and recovery and reuse is difficult. It has recently been found that chromatographic silica, when heated to 700 °C, catalyses the formation of amides directly from acids and amines with a good level of generality [35]. This simple, cheap, abundant material is very thermally stable and, although substantial quantities are needed for rapid conversion, it is easily recovered, deactivates slowly, and needs only occasional reactivation at elevated temperature to burn off impurities and thus restore its original activity. The E-factor associated with its use is substantially lower than those associated with any of the other routes to amides. Silica is abundant, has very low toxicity and is available in many forms, including chromatographic silica, which is designed to have a particle size and shape that make it ideal for flow applications.

This example illustrates well that a heterogeneous catalyst can be simple, readily available and stable in reaction conditions. Similarly, clays have long been used with good results in synthetic chemistry, usually as mild acid catalysts [36]. The clays are mined, dried, sieved and then acid treated to produce the catalytic material. They tend to work very well in nonhydroxylic solvents for reactions such as acetalization, esterification and some alkylations, but form muds in hydroxylic systems. Their activity is partly dependent on their source, and the nature of their acid treatment, which reduces their long-term stability. Thermal stability is limited and temperatures above ~300 °C will usually collapse the clay structure irreversibly. An exception to this is clayzic, formed by the adsorption of zinc chloride on to a clay (Montmorillonite K10), where activation at 280 °C leads to a delaminated and highly active Friedel-Crafts alkylation catalyst [37].

2.4.1.1 Summary: List of Green Principles Used in Amide Formation

- (1) Prevention.
- (2) Atom Economy.
- (3) Less Hazardous Chemical Synthesis.
- (4) Design for Safer Chemicals.
- (5) Safer Solvents and Auxiliaries.
- (6) Design for Energy Efficiency.
- (9) Catalysis.

Renewable biomass has also been utilized as a source of heterogeneous catalysts or support materials. Chitosan is derived from the shells of crustaceans and is a waste material from the fishing industry. It has been used as a support for a range of catalytic centres, including metal complexes (chitosan has a high affinity for certain metals, and contains a primary amine group, which is ideal for attachment of complexes via amide, imine or amine formation) and enzymes [38–40]. One of its most useful properties is that it can be formed into a range of shapes and forms very readily, so that crosslinked beads, films and fibres are all readily accessible. Bacheva *et al.* have recently demonstrated that functional chitosan films can be prepared on the glass walls of a flask and impregnated with enzymes. In this way, continuous reaction systems can be readily prepared – in this case, peptide hydrolysis and peptide formation were both demonstrated under different conditions, and catalyst separation and reuse was readily accomplished [41] (Figure 2.5).

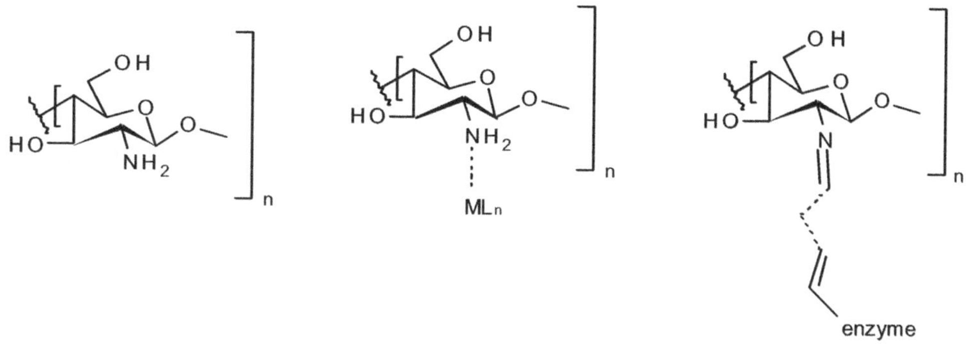

Figure 2.5 *Chitosan and functionalized catalytic chitosan-based materials.*

2.4.2 Case Study: Mesoporous Carbonaceous Material as a Catalyst Support

Starch has also been shown to form an interesting and versatile carbonaceous mesoporous support material, named Starbon. 'Starbon' is a contraction of the words 'starch' and 'carbon', reflecting its formation from the carbonization of expanded starch.

By expansion of starch, a mesoporous aquagel is formed, which can then be retrograded (partly reorganized) by cooling. Subsequent solvent exchange, drying and thermal decomposition leads to a partly carbonized material, which retains its mesoporosity and displays properties of both a polysaccharide and a carbon, the balance shifting towards a carbon as the carbonization temperature increases. This, coupled with its mesoporosity, makes it a unique material (activated carbons, charcoals and other forms of carbon are microporous, which means that only very small molecules (dimensions of a few Angstroms) can enter the pore system). As the starch used for the preparation of the material can be waste starch from food processing, as well as virgin starch, the material has good green credentials. It can be sulfonated to produce a strongly acidic material with the unusual catalytic property of being able to catalyse esterifications in water-rich environments [42]. This is very useful, as fermentation is a potentially very valuable method by which to produce a range of platform molecules, many of which are water-soluble acids and alcohols, whose separation from the complex broth is very challenging. Processing of these molecules within a fermentation broth to form water-insoluble products is therefore an interesting strategy by which to isolate useful compounds.

Starbon synthesis is greener than the standard method of formation of mesoporous carbons, as shown in Figure 2.6, as it utilizes the intrinsic porosity of natural polysaccharides, in contrast to the traditional templating method.

2.4.2.1 Summary: List of Green Principles Used in the Starbon Case Study

- (1) Prevention.
- (2) Atom Economy.
- (3) Less Hazardous Chemical Synthesis.
- (5) Safer Solvents and Auxiliaries.
- (6) Design for Energy Efficiency.

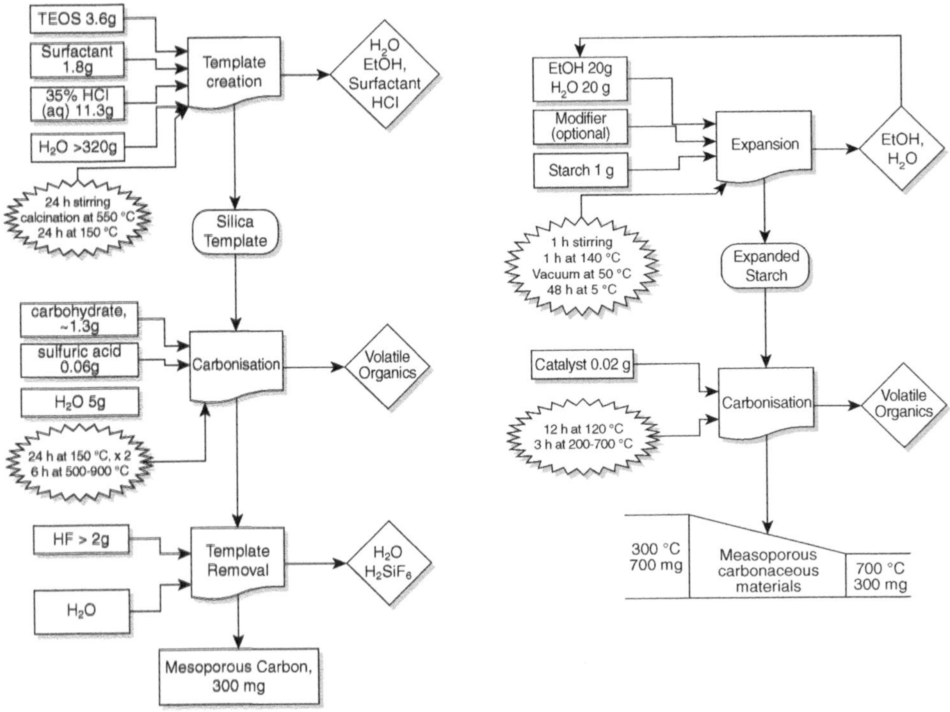

Figure 2.6 *Comparison of Conventional (left) and Starbon®(right) Routes to Mesoporous Carbon.*

- (7) Use of Renewable Feedstocks.
- (9) Catalysis.
- (10) Design for Degradation.

2.5 Renewable Feedstocks and Biocatalysis

Since the 1990s, decreasing fossil reserves, rising oil prices and concerns over security of supply, environmental impacts and sustainability have led to a global policy shift back towards the use of biomass as a local, renewable and low-carbon feedstock. The use of biomass has socioeconomic benefits for rural and coastal communities, giving them valuable and increased markets for their products, byproducts and waste. Biomass offers long-term sustainability through the short cycling of carbon dioxide. Plants carry out biological carbon fixation by converting carbon dioxide from the atmosphere to carbohydrate (biomass) through photosynthesis. This biomass can then be used directly for energy or can be converted by a variety of methods into liquid fuels, energy products or chemicals. At the end of their lifetime, the carbon is returned to the atmosphere as carbon dioxide via burning or microbial degradation, analogous to fossil fuels but on a much shorter timescale.

The source of renewable feedstocks is critical for sustainable development. While there are vast quantities of rice in China, cassava in Africa and corn in the USA, their conversion to chemicals or fuel would be inappropriate as it would drive up food prices and ultimately lead to famine and hunger within these countries. Therefore, biomass wastes such as corn stover [43], wood [44], bagasse [45], wheat straw [46], other food industry wastes and biomasses that are independent of food production, such as seaweed [47], are more appropriate for biorefining. The quantities of food supply-chain waste alone are staggering (over 1 billion MT per year) and represent a carbon equivalent comparable to the total amount of carbon used in organic chemicals, currently largely provided from non-renewable petroleum. There is a wide variety of biomass available to us for use as a renewable feedstock for fuels and chemicals – this is both an issue and an opportunity. The dispersed distribution, seasonality and large number of different biomass types available mean conversion technologies have to be flexible and universally applicable for the conversion of these materials. On the other hand, the wide range of biomasses, the wide growing range and the seasonality of different biomasses also mean that energy and chemical security is less of a problem than with fossil fuels. Each and every area of the globe should be able to grow biomass for local use and be able to survive in the event of crop failure.

The biorefinery concept that has emerged (see Figure 2.7) is analogous to today's petroleum refineries: integrated facilities for the conversion of biomass into multiple value-added products, including energy, chemicals and materials [48,49].

Improvements are needed to create flexible, zero-waste networks, applicable to a variety of low-value local feedstocks. Biorefineries will then be able to challenge the current *status quo* and develop novel interconnecting webs of products capable of

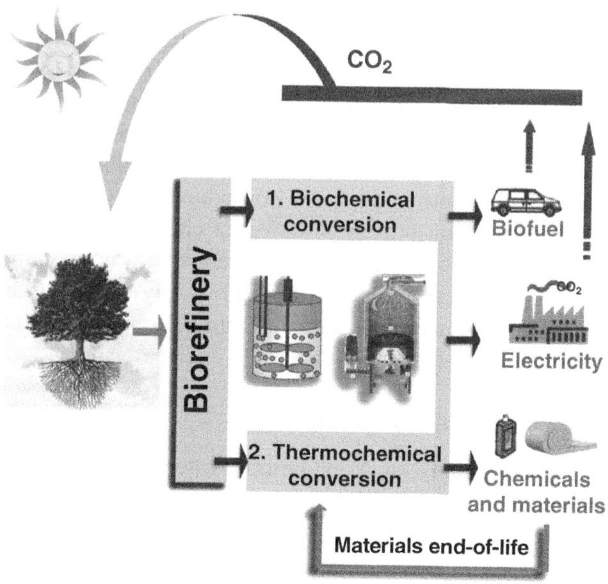

Figure 2.7 *Biorefinery concept.*

Table 2.2 US DOE-identified key platform molecules [52].

Glycerol	Levulinic acid
3-hydroxypropionic acid	Succinic acid
Xylitol/arabinitol	Glutamic acid
3-hydroxybutrolactone	Itaconic acid
Sorbitol	Fumaric acid
Glucaric acid	Malic acid
2,5-furan-dicarboxylic acid	Aspartic acid

meeting the demands of existing and new industries. Initial conversion of renewable feedstocks into chemicals tends to operate by one of two main routes – thermochemical and enzymatic (biocatalysis). Both routes deconstruct the biopolymer structure of the biomass into smaller molecules. Fermentation or biochemical routes tend to rely upon enzymatic (or pre-treatment) routes to sugars, followed by biocatalysis into useful small molecules. Thermochemical routes such as gasification work by near-total destruction of the biomass into very small fragments such as carbon monoxide, hydrogen and methane, followed by reformation via Fischer–Tropsch [50] and other upgrading techniques into useful molecules and fuels. Thermochemical treatment such as pyrolysis, which can be achieved by direct heating or microwave heating, makes a bio-oil consisting of a wide range of decomposition products such as sugars and phenols and rearrangement products such as furans and syngas. Typically, the challenge is to upgrade this bio-oil into chemicals and/or fuels for use in a wide range of techniques and combinations of techniques. This upgrading process has some challenges and issues associated with it [51].

Platform molecules are molecules obtained from the deconstruction of biomass that can be used as the building blocks for the manufacture of more complex chemicals. Acknowledging the wide variety of bio-derived molecules available compared to the limited number of chemicals derived from fossil fuel sources, the US Department of Energy (DOE) screened over 300 compounds for their suitability as platform molecules [52]. Table 2.2 shows a list of suitable molecules identified by this work, which took into account the availability and ease of generation of these compounds, as well as their suitability, flexibility and ease of conversion into more complex chemical products.

A major focus of biorefinery and green chemical research is now on finding methods to generate large quantities of these molecules in high purity with low energy usage and developing synthetic methodologies by which they can be converted into useful molecules and polymers for the chemical-using industries.

2.5.1 Case Study: Wheat Straw Biorefinery

Recently, work at the University of York demonstrated the use of wheat straw as a biorefinery feedstock [46]. Using wheat straw, an abundant, relatively low-value agricultural residue found in Europe, a wide range of products were obtained. As shown in Figure 2.8, first the high-value waxes which coat the outside of the straw were removed using $scCO_2$. This technique was found to be more selective than conventional solvent extractions and capable of splitting the waxes into various fractions with a variety of

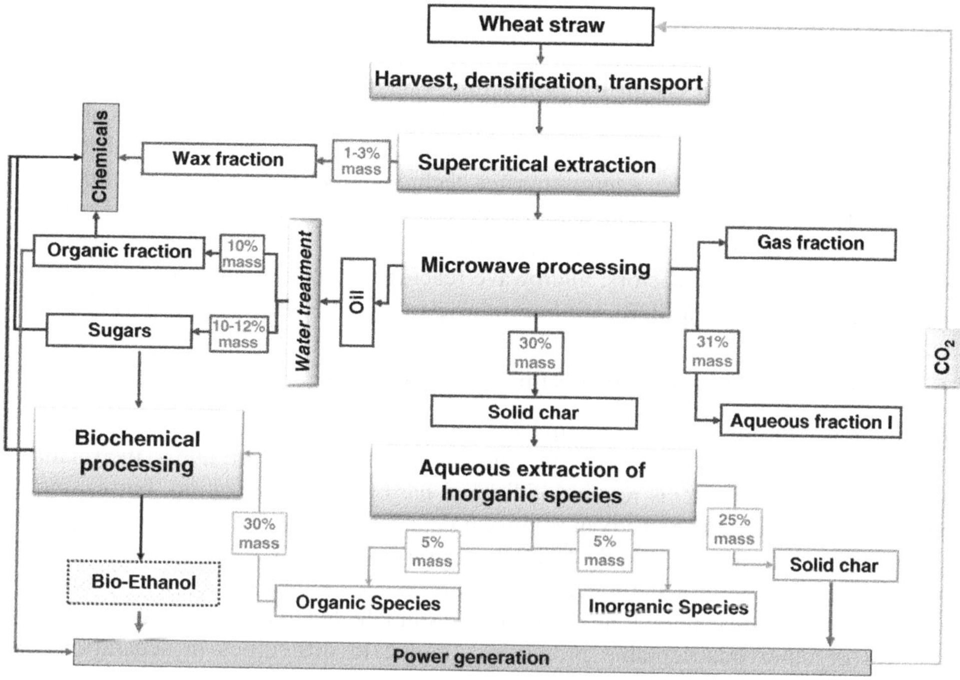

Figure 2.8 *The wheat straw biorefinery.*

properties; their use is anticipated to be as a cosmetic base. The waste material from the extraction was then pyrolysed using a low-temperature microwave system, which generated a gas, a liquid bio oil and a char. The gas can be burned in a gas engine to power the biorefinery. The oil can be split into a sugar-rich fraction for enzymatic fermentation into chemicals and bioethanol, and therefore energy, and an organic fraction for direct extraction of chemicals and energy. The char can be burned for energy generation, co-fired with coal, the ash byproduct of which can be used as a source of biosilicates, which can form a binder for wheat straw as a replacement wood board. Ultimately, the product of the energy generation will be carbon dioxide, which is returned to the plant for growth in a closed-loop fashion.

2.5.1.1 Summary: List of Green Principles Used in Wheat Straw Biorefineries

- (2) Atom Economy.
- (3) Less Hazardous Chemical Synthesis.
- (5) Safer Solvents and Auxiliaries.
- (6) Design for Energy Efficiency.
- (7) Use of Renewable Feedstocks.
- (8) Reduction of Derivatives.
- (10) Design for Degradation.
- (12) Inherently Safer Chemistry for Accident Prevention.

2.6 An Overview of Green Chemical Processing Technologies

The application of green chemistry to reaction systems tends to fall into one of two categories from a hardware perspective. First, there is equipment associated with utilizing alternative reaction solvents, such as supercritical or high-pressure solvents, which allow high mass-transfer and high energy-transfer techniques such as microwave reactors, ultrasonication reactors or microreactors and separation equipment. Second, there is a large family of equipment associated with high energy transfer to the reactants. New technology such as microreactors, microwave reactors, radio frequency heating, electric pulses, ultrasonication and SDRs are becoming part of the technologist's arsenal.

2.6.1 Alternative Reaction Solvents for Green Processing

Ionic liquids, fluorous biphasic systems and supercritical fluids have all been studied as alternatives to conventional organic and aqueous solvents. However, due to their nature, some of these novel systems require additional hardware for utilization. For example, some suppliers have designed advanced mixing paddles and blades to enable polyphasic systems to be intimately mixed at the laboratory scale [53]. There has also been considerable rethinking of the green credentials of some of these alternative solvents, and many ionic liquids are no longer considered suitable due to their complex syntheses, toxicities or other unacceptable properties, or due to difficulties in separation and purification. Fluorous solvents are not considered to be environmentally compatible (as they persist in the environment).

Supercritical solvents are more difficult to manipulate, due to the high pressures and temperatures sometimes employed. In the case of supercritical water, equipment had to be designed which could contain the highly corrosive liquid; this was achieved through the use of alloyed metals. Supercritical water has the unique property of being able to oxidize most organic compounds, and for this reason it is being used in waste water treatment [54]. Vessels for creating supercritical solvents such as $scCO_2$ are now available; these are capable of making fine adjustments to temperature and pressure in order to affect the solvents' properties [55]. Excessive pressure and temperature are thus not required to produce $scCO_2$, as shown in Figure 2.9. $scCO_2$ is an increasingly popular reaction medium, as its properties can be controlled by varying the temperature and pressure or by using a cosolvent. The main environmental benefit of $scCO_2$ lies in the work up, as the product mixture is obtained free from solvent simply by returning to atmospheric conditions. Additionally, carbon dioxide is nontoxic, nonflammable, recyclable and a byproduct of other processes. However, there are energy and safety concerns associated with the elevated temperatures and pressures employed, which must be balanced against the benefits of its use.

Supercritical solvents are finding uses in a number of fields [57]. For example, they are being used as replacements for organic solvents in several conventional reactions [58]. This was considered at commercial scale by Thomas Swan for the multiton synthesis of 3,3,5-trimethylcyclohexanone [59], which is used as a solvent in the manufacture of resins and varnishes, and it has been proven for other compounds [60].

Solvent intensive processes such as decaffeination [61] and extraction of natural products [57] are being carried out with the more gentle technique of supercritical fluid extraction. This enables fragile compounds to be removed selectively without

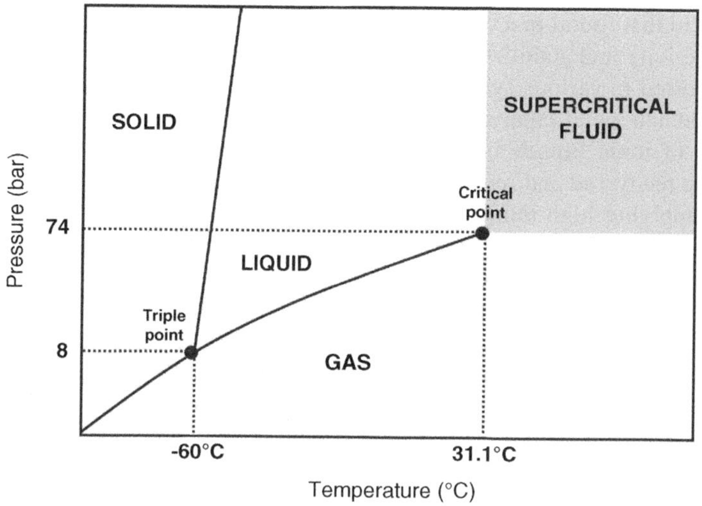

Figure 2.9 *Phase diagram for carbon dioxide. Derived from [56].*

damage, which would occur if high-temperature organic solvent extraction were employed. Botanix [62] and others are carrying out work of this nature on a commercial scale.

Biphasic systems [63] are a novel and effective method by which catalysts, substrates and products can be easily separated into different liquid phases, thereby simplifying and 'greening' reaction work up. One example is the use of fluorous biphasic solvents: conventionally, the homogenous catalyst is soluble within the fluorous phase and reactants are soluble within an immiscible conventional solvent. Upon heating, the two solvents become miscible and the reaction occurs. On completion of the reaction, when cooled, the phases return to being immiscible, with the product partitioning into the conventional solvent phase for isolation. These systems have recently been criticized for lacking green credentials as they persist in the environment and can be hazardous to operators. Phase-transfer catalysts (PTCs) can also be used in biphasic systems for the transfer of species into a phase in which they would not normally be soluble. Although not catalysts in the conventional sense, they aid the reaction by improving availability of the substrates [64]. For example, a reactant soluble in a nonpolar phase can be transferred and reacted in an aqueous phase by use of a PTC. PTCs are usually soluble in both phases of the biphasic system and are based on quaternary alkyl ammonium or phosphonium salts or crown ethers. However, a publication by Paetzold has reported a method for the use of a heterogeneous PTC, formed by immobilizing an amphiphile on to silica [65].

Ionic liquids are molten salts and are liquid at relatively low temperatures. Their characteristic properties include (1) low or no vapour pressure, (2) nonflammability, (3) high thermal stability (although this is queried by some reports), (4) relatively low viscosity, (5) liquid state over extended temperature ranges and (6) high ionic conductivity. When using ionic liquids, as opposed to conventional solvents, the solute is solvated by ions only, which means the reaction can occur in a very different environment to that found in a conventional solvent system. This feature can be used to enhance selectivity and stabilize ionic products/intermediates. The use of ionic liquids has been reported in various synthetic transformations: Friedel–Crafts reaction, Diels–Alder reaction and metal-catalysed asymmetric synthesis [66–72]. The most commonly cited benefit of ionic liquids is their low vapour pressure; using this property, ionic liquids can be recovered and reused after the reaction product is extracted using organic solvents by applying high temperatures or a vacuum and distilling out contaminants.

All these alternative solvents share two common means by which to enhance reaction: first, enhancing diffusion and thereby the availability of reactant, and second, improving product isolation through simplified work-up procedures. It is anticipated that the use of biosolvents will increase in the future. For example, ethanol, succinate esters, lactate esters and limonene – all derived from nonfood biomass using green chemical extraction and production technologies – will become more attractive as direct replacements for conventional solvents as petrochemical prices rise, regulations tighten and public demand for bioderived material increases [73].

2.6.2 Alternative Energy Reactors for Green Chemistry

2.6.2.1 Microchannel Reactors

Among the raft of ideas that came about following the Bhophal tragedy, Trevor Kletz stated that '*What you don't have can't leak*' [74]. By having very small volumes of reactants coming together at any given time, the risk of a thermal runaway is minimized; that is the concept behind a microreactor. This is a different way of thinking for many chemists and its roots lie in process intensification (PI) [7], the many benefits of which include uniform treatment of all molecules and transport rates that match reaction rates (see Chapter 1). These methods lead to improved selectivity, yield and processing time.

By using submillimetre channels etched into quartz or plastic units, highly controlled heat input or removal can be achieved [75]. These units have been successfully demonstrated for liquid–liquid reactions. Work by Jackson [76] has investigated an alternative application of this technology: fixing palladium cross-coupling catalysts to fine glass pipes. This creates an *in situ* heterogeneous catalyst across which the reagents are flowed.

Microchannel reactors cannot conventionally be scaled up to industrial capacities; they can, however, be scaled out, and hundreds of units can be run in parallel. This results in a more flexible production capacity, with the opportunity to rapidly switch product output as market demands change.

2.6.2.2 Microwave Reactors

Microwave reactors are pieces of laboratory equipment that can be used to 'heat' reactions with microwave radiation in exactly the same way that microwave ovens work in the home. Microwave radiation is a high-frequency electric field with a wavelength in the centimetre range, which places it between radio waves and infrared in the electromagnetic spectrum (see Figure 2.10). Microwave energy is very low: around 1 J/mol, which is lower than the bond energy of even a typical hydrogen bond. This means that microwaves cannot directly activate atomic- or molecular-bonding electrons.

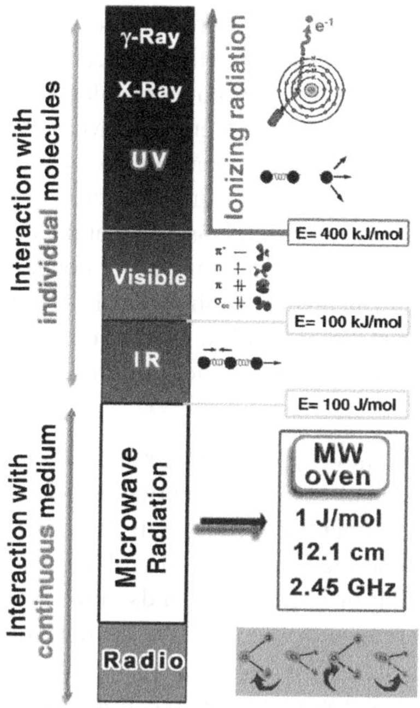

Figure 2.10 *The electromagnetic spectrum.*

Microwave heating operates in a different manner to conventional oil baths. Microwaves interact with dipoles or ions and create 'molecular heating' by causing dipole rotations or ionic conduction. Both of these ways of receiving energy are caused by molecules/ions attempting to align with the rapidly oscillating microwave field. In a simplistic view, microwave reactors are capable of enhancing reaction rates because they allow more molecules to have sufficient energy to overcome the activation barrier of the reaction. These high-energy molecules are created by preventing them from relaxing from the excited state: kinetic relaxation occurs in 10^{-5} seconds, whereas microwaves apply energy in 10^{-9} seconds, which creates a non-equilibrium state [77].

The acceptance of microwave heating as a chemistry tool is demonstrated by the exponential growth in microwave-related publications since the mid 1980s, from initial articles involving organic synthesis in 1986 [78]. Original studies involved the use of domestic ovens: these produced hot and cold spots due to an uneven microwave field, and led to unpredictable results and numerous explosions [79]. Subsequently, a number of companies (CEM Corporation [80], Biotage [81] and Anton Paar Gmbh [82] are among the market leaders) began producing systems designed for chemistry, using monomode microwave generators and laboratory-scale apparatus (5–100 mL). This provided further encouragement to researchers to trial reactions under microwave irradiation. Reported microwave-assisted chemistry benefits include decreased reaction times, reduced overall energy consumption and improved yield and selectivity. This is summarized by Leadbeater, who stated that microwave technology is '*enabling a wide range of reactions to be performed easily and quickly*' [83].

Microwaves are not only useful as alternative methods of heating; there is also a good body of evidence to suggest they can cause specific molecular activations [84]. Microwave processing has now been shown to be effective at pilot scale and at large continuous processing scale, for example in waste treatment [85]. Microwave radiation technology is established in the food industry. It has also been suggested as a method of gasifying food wastes [86]. Many research articles describe the use of microwave radiation to assist chemical reactions: calculations show that microwave-assisted organic chemical reactions can be considerably more energy efficient than reactions using conventional heating [87]. This is another example of PI and should ideally be combined with flow systems due to the limitations of microwave penetration and in order to allow optimum continuous operation [88].

There are a number of methods by which microwaves are thought to enhance reactions [79]:

1. Using a polar solvent to achieve rapid heating rates and elevated temperatures (greater than the boiling point). This leads to conventional-style reactions with rate enhancement through increased temperature. Heating is directly of the solvent, and therefore there are no losses or time lag due to heating of the vessel.
2. Specific activation of polar or ionic compounds in the presence of nonpolar medium. These highly energized reactants will then react with statistical probability when they collide.
3. Acceleration of proton- or ion-catalysed reactions, due to the microwave electromagnetic field interacting with the dipole moment of dissociatable molecules through enhancement of the disoctiation equilibrium.
4. Activation of the active centres of heterogeneous catalysts (so-called 'hot spots'). It must be noted that this is a little-understood reaction and that no detailed explanation or theory has been proposed regarding it.

Furthermore, due to the instantaneous nature of the heating, microwaves offer a major advantage in controllability over conventional heating.

Microwave reactions have been successfully demonstrated for many different organic reactions [89–91], including metal mediated catalysis [92], cycloadditions [93], heterocyclic chemistry [94], rearrangements [95], electrophilic and nucleophilic substitution [86] and reduction [96].

Case Study: Microwave-assisted Organic Synthesis. The vast majority of synthetic transformations that include organic materials and thermochemical conversions have been attempted with microwave heating. In many cases, improved yields or reaction times have been observed.

2.6.2.3 Summary: List of Green Principles Used in Microwave-assisted Organic Synthesis

- (3) Less Hazardous Chemical Synthesis.
- (5) Safer Solvents and Auxiliaries.
- (6) Design for Energy Efficiency.
- (9) Catalysis.
- (11) Real-time Analysis for Pollution Prevention.
- (12) Inherently Safer Chemistry for Accident Prevention.

2.6.2.4 Spinning Disc Reactors

SDRs are another PI development. They are capable of rapid processing times and excellent heat and mass transfer. By flowing reagents over a rapidly spinning disc (100–6000 rpm), a thin film of reactants is created via the centrifugal force. Using grooves, meshes, cones and different disc diameters, different applications can be catered for. As the kinetics can be controlled with high accuracy, SDRs have been successfully employed in polymerization reactions to give polymers with defined characteristics [97]. Furthermore, the disc surfaces have been used to support catalysts such as aluminum chloride for polymerizations [98] and zinc triflate for the rearrangement of α-pinene oxide [99].

This review of some of the more advanced or more promising alternative engineering methods by which cleaner and greener chemistry can be applied has not been exhaustive, and other novel techniques are being developed and scaled-up continually.

2.7 Conclusion

Given the finite nature of fossil fuels and the increasing demand for chemicals and fuel caused by an increased world population, increasingly consumer-based societies and an increased desire for independent travel, the need for sustainable development and green chemistry is clear. Solutions will not come from conventional groupings, and enhanced interdisciplinary collaboration between chemists, chemical engineers, physicists, biochemists and many other disciplines is the only way in which the future energy and chemical needs of the world will be met.

References

(1) P. T. Anastas and J. C. Warner, Green Chemistry: Theory and Practice, Oxford University Press: New York (1998).
(2) http://www.epa.gov/greenchemistry/, last accessed 14 September 2012.
(3) World Commission on the Environment and Development (WCED) , Our Common Future, Oxford University Press, p. 43 (1987).
(4) Ali M. El-Agraa, The European Union: Economics and Policies, 7th edition, Prentice Hall Financial Times (2004).
(5) President Hu Jintao in his keynote speech at the 17th National Congress of the Communist Party of China (CPC), October 2007.
(6) President B. Obama, Development policy and the global climate change initiative, United States of America Policy, 22 September 2010.
(7) T. Kletz, Process Plants: A Handbook for Inherently Safer Design, Taylor & Francis: London (1998).
(8) M. Bennett and P. James, Sustainable Measures, Greenleaf Publishing: Sheffield (1999).
(9) Report – Committee on Industrial Environmental Performance Metrics, Industrial Environmental Performance Metrics, Challenges and Opportunities, National Academy of Engineering, National Research Council (1999).
(10) Corporate Environmental Performance 2000 , volume 1 Strategic Analysis, Haymarket Business Publications (1999).
(11) A.D. Curzons, D. J. C. Constable, D. N. Mortimer and V. L. Cunningham, Green Chem., 3, 1 (2001).

(12) D. J. C. Constable, A. D. Curzons, L. M. Freitas dos Santos, G. R. Geen, R. E. Hannah, J. D. Hayler, J. Kitteringham, M. A. McGuire, J. E. Richardson, P. Smith, R. L. Webb and M. Yu, Green Chem., 3, 7 (2001).
(13) T. Hudlicky, D. A. Frey, L. Koroniak, C. D. Claeboe and L. E. Brammer, Jr, Green Chem., 1, 57 (1999).
(14) B. M. Trost, Science, 254, 1471 (1991).
(15) R. A. Sheldon, Chem. Ind., 903 (1992).
(16) R. A. Sheldon, Chem. Ind., 12 (1997).
(17) P. T. Anastas and R. L. Lankey, Green Chem., 6, 289 (2000).
(18) D. J. C. Constable, A. D. Curzons and V. L. Cunningham, Green Chem., 4, 521 (2002).
(19) P. Riente, C. Mendoza, M. A. Pericas and J. Mater. Chem., 20, 7350–7355 (2011).
(20) Y. H. Zhu, L. P. Stubbs, F. Ho, R. Z. Liu, C. P. Ship, J. A. Maguire and N. S. Hosmane, Chemcatchem., 2, 365–374 (2010).
(21) J. D. Webb, S. MacQuarrie, K. McEleney and C. M. Crudden, J. Catal, 252, 97–109 (2007).
(22) M. Balakrishnan, V. S. Batra, J. S. J. Hargreaves and I. D. Pulford, Green Chem., 13, 16–24 (2011).
(23) J. H. Clark, D. J. Macquarrie and M.de Bruyn, Supported catalysts, in Kirk-Othmer Encyclopaedia of Chemical Technology.
(24) G. L. Athens, R. M. Shayib and B. F. Chmelka, Curr. Opin, Coll. Interf. Sci., 14, 281–292 (2009).
(25) N. Baccile, F. Babonneau, B. Thomas and T. Coradin, J. Mater. Chem, 2009, 19, 8537–8559.
(26) B. Pugin and H.-U. Blaser, Top. Catal., 53, 953–962 (2010).
(27) N. J. Meehan, A. J. Sandee, J. N. H. Reek, P. J. C. Kamer, P. W. M. N.van Leeuwen and M. Poliakoff, Chem. Commun., 1497–1498 (2000).
(28) Z. Hou, N. Theyssen, A. Brinkmann, K. V. Klementiev, W. Grünert, M. Bühl, W. Schmidt, B. Spliethoff, B. Tesche, C. Weidenthaler and W. Leitner, J. Catal., 258, 315–323 (2008).
(29) D. J. Macquarie and D. B. Jackson, Chem. Commun., 1781–1782 (1997)
(30) J. Alauzin, A. Mehdi, C. Reyé and R. J. P. Corriu, J Amer. Chem. Soc., 128, 8718–8719 (2006).
(31) R. K. Zeidan, S. J. Hwang and M. E. Davis, Angew. Chem. Int. Ed., 45, 6332–6335 (2006).
(32) R. K. Zeidan and M. E. Davis, J. Catal, 247, 379–382 (2007).
(33) F. Goettmann, D. Grosso, F. Mercier, F. Mathey and C. Sanchez, Chem. Commun., 1240–1241 (2004).
(34) D. J. Constable, P. J. Dunn, J. D. Hayler, G. R. Humphrey, J. L. Leazer, R. J. Linderman, K. Lorenz, J. Manley, B. A. Pearlman, A. Wells, A. Zaks and T. Y. Zhang, Green Chem., 9, 411–420 (2007).
(35) J. W. Comerford, J. H. Clark, D. J. Macquarrie and S. W. Breeden, Chem. Commun., 2562–2564 (2009).
(36) S. Dasgupta and B. Torok, Org. Preo. Proc. Int., 40, 1–65 (2008).
(37) J. H. Clark, A. P. Kybett, D. J. Macquarrie, S. J. Barlow and P. Landon, J. Chem. Soc., Chem. Comm., 1353 (1989).
(38) J. J. E. Hardy and D. J. Macquarrie, Ind. Eng. Chem. Res., 44, 8499–8520 (2005).
(39) E. Guibal, Prog. Polym. Sci., 30, 71–109 (2005).
(40) B. Krajewska, Enzyme, Microbial. Biotech., 35, 126–139 (2004).
(41) A. Bacheva and D. J. Macquarrie, Green Chem., 10, 692–695 (2008).
(42) V. Budarin, R. Luque, D. J. Macquarrie and J. H. Clark, Chem., Eur. J., 13, 6914 (2007).
(43) http://www.biomassenergycentre.org.uk/, last accessed 14 September 2012.
(44) M. Miura, H. Kaga, T. Yoshida and K. Ando, J. Wood Sci., 47, 502–506 (2001).
(45) M. Garcìa-Pèrez, A. Chaala and C. Roy, J. Anal. Appl. Pyrol., 65, 111–136 (2002).
(46) V. L. Budarin, P. S. Shuttleworth, J. R. Dodson, A. J. Hunt, B. Lanigan, R. Marriott, K. J. Milkowski, A. J. Wilson, S. W. Breeden, J. Fan, E. H. K. Sin and J. H. Clark, Energy Environ. Sci., 4, 471–479 (2011).

(47) V. L. Budarin, Y. Zhao, M. J. Gronnow, P. S. Shuttleworth, S. W. Breeden, D. J. Macquarrie and J H. Clark, Green Chem., 13, 2330–2333 (2011).
(48) National Renewable Energy Laboratory (NREL) : http://www.nrel.gov/biomass/biorefinery.html, last accessed 14 September 2012.
(49) R. Luque, L. Herrero-Davila, J. M. Campelo, J. H. Clark, J. M. Hidalgo, D. Luna, J. M. Marinas and A. A. Romero, Energy Environ. Sci., 1, 542–564 (2008).
(50) http://www.fischer-tropsch.org/primary_documents/patents/US/us1746464.pdf, last accessed 14 September 2012.
(51) D. J. Hayes, Catal. Today, 145, 138–151 (2009).
(52) T. Werpy and G. Pedersen, Top Value Added Chemicals from Biomass, US DOE Report (2005).
(53) http://www.radleys.com/, last accessed 14 September 2012.
(54) H. Schmieder and J. Abeln, Chem. Eng. Technol., 22, 903 (1999).
(55) http://www.supercriticalfluids.com/, last accessed 14 September 2012.
(56) http://web.archive.org/web/20041215124517/http://www.uyseg.org/greener_industry/pages/superCO2/1SuperCO2_intro.htm, last accessed 14 September 2012.
(57) A. J. Hunt, E. H. K. Sin, R. Marriott and J. H. Clark, ChemSusChem, 3(3), 306 (2010).
(58) R. S. Oakes, A. A. Clifford and C. M. Rayner, J. Chem. Soc., Perkin trans., 1, 917 (2001).
(59) http://www.thomas-swan.co.uk/, last accessed 14 September 2012.
(60) P. Licence, J. Kie, M. Sokolova, S. K. Ross and M. Poliakoff, Green Chem., 2, 99 (2003).
(61) K. Zosel, Angew. Chem. Int. Ed. Engl., 17, 702 (1978).
(62) http://web.archive.org/web/20060927073722/http://www.botanix.co.uk/extraction.html, last accessed 14 September 2012.
(63) D. J. Adams, P. J. Dyson and S. J. Tavener, Chemistry in Alternative Reaction Media, John Wiley & Sons: Chichester (2004).
(64) C. M. Starks, C. L. Liotta and M. Halpern, Phase-transfer Catalysis, Chapman & Hall: London (1994).
(65) E. Paetzold, I. Jovel and G. Oehme, J. Mol. Catal. A: Chem, 214, 241 (2004).
(66) T. Welton, Chem. Rev., 99, 2071 (1999).
(67) J. D. Holbrey and K. R. Seddon, Clean Prod. Proc., 1, 223 (1999).
(68) P. Wasserscheid and W. Keim, Angew. Chem. Int. Ed., 39, 3772 (2000).
(69) T. Kitazume, Fain Kemikaru, 30(17), 5 (2001).
(70) T. Kitazume, Fain Kemikaru, 30(18), 15 (2001).
(71) R. Sheldon, Chem. Commun., 2399 (2001).
(72) D. Zhao, M. Wu, Y. Kou and E. Min, Catal. Today, 74, 157 (2002).
(73) R. K. Henderson, C. Jiménez-González, D. J. C. Constable, S. R. Alston, G. G. A. Inglis, G. Fisher, J. Sherwood, S. P. Binks and A. D. Curzons, Green Chem., 13, 854 (2011).
(74) D. Weir, The Bhopal Syndrome: Pesticides, Environment and Health, Earthscan Publications: London (1987).
(75) G. M. Greenway, S. J. Haswell, D. O. Morgan, V. Skelton and P. Styring, Sens. Actuators B: Chemical, 63, 153 (2000).
(76) T. Jackson, J. H. Clark, D. J. Macquarrie and J. H. Brophy, Green Chem., 4, 193 (2004).
(77) R. Gedye, F. Smith, K. Westaway, H. Ali, L. Baldisera, L. Laberge and J. Rousell, Tetrahedron Lett., 27, 279 (1986).
(78) For review see: M. Nuchter, B. Ondruschka, W. Bonrath and A. Gum, Green Chem., 6, 128 (2004).
(79) B. L. Hayes, Microwave Synthesis – Chemistry at the Speed of Light, CEM Publishing: Matthews, NC (2002).
(80) www.cem.com, last accessed 14 September 2012.
(81) www.biotage.com, last accessed 14 September 2012.
(82) www.anton-paar.com, last accessed 14 September 2012.
(83) N. Leadbeater, Chemistry World, 1, 38 (2004).

(84) A.de la Hoz, A. Diaz-Ortiz and A. Moreno, Chem. Soc. Rev., 34(2), 164–178 (2005).
(85) D. E. Clark and W. H. Sutton, Annu, Rev. Mater. Sci., 26, 299–331 (1996).
(86) J. A. Menendez, A. Dominguez, Y. Fernandez and J. J. Pis, Energ. Fuel., 21(1), 373–378 (2007).
(87) M. J. Gronnow, R. J. White, J. H. Clark and D.J. Macquarrie, Org. Proc. Res. Dev., 9, 516 (2005).
(88) Lecture – R. Wood, New wave chemistry, CEM Microwaves in Chemistry Conference, London, September 2004.
(89) J. P. Tierney and P. Lidström, Microwave Assisted Organic Synthesis, Blackwell Publishing: Oxford (2005).
(90) P. Lidström, J. P. Tierney, B. Wathey and J. Westman, Tetrahedron, 57, 9225 (2001).
(91) C. O. Kappe, Angew. Chem. Int. Ed., 43, 6250–6284 (2004).
(92) G. W. Kabalka, L. Wang, R. M. Pagni, C. M. Hair and V. Namboodiri, Synthesis, 217 (2003).
(93) U. Bora, A. Saikia and R. C. Boruah, Org. Lett., 5, 435 (2003).
(94) D. Bogdal and M. Warzala, Tetrahedron, 56, 8769 (2000).
(95) S. Caddick, Tetrahedron, 51, 10403 (1995).
(96) B. K. Banik, K. J. Barakat, D. R. Wagle, M. S. Manhas and A. K. Bose, J. Org. Chem., 64, 5746 (1999).
(97) K. V. K. Boodhoo and R. J. Jachuck, Green Chem., 5, 235 (2000).
(98) K. V. K. Boodhoo, W. A. E. Dunk, R. J. Jachuck, V. Sage, D. J. Macquarrie and J. H. Clark, J. Applied Polymer Sci., 101, 8–19 (2006).
(99) K. Wilson, A. Renson and J. H. Clark, Catal. Lett., 61, 51 (1999).

3

Spinning Disc Reactor for Green Processing and Synthesis

Kamelia Boodhoo

School of Chemical Engineering & Advanced Materials, Newcastle University, UK

3.1 Introduction

Interest in spinning disc technology gradually evolved from the realization during the early 1900s that the flow of thin liquid films on plane surfaces due to gravity was of great practical importance in chemical engineering operations. Such films have been associated with surface wave formation, giving rise to intense mixing action within the film and increased heat- and mass-transfer rates [1–3]. Over the last 2–3 decades, research into thin films generated by rotating surfaces has dominated the scene. The centrifugal force driving the flow is much greater than the force acting on films flowing under gravity, and films produced in this way have been shown to have greatly improved and more controllable characteristics. Much of the work, focused primarily on hydrodynamics and heat and mass transfer, has been driven by intended applications in a microgravity environment as part of the NASA space exploration programme [4,5].

The spinning disc reactor (SDR) has emerged as one of those technologies which exploit the benefits of exerting high centrifugal accelerations to process liquids. The speed of disc rotation may be as high as 4000 revolutions per minute (rpm), creating centrifugal field accelerations of the order of 1000 g at the edge of a typical disc of 0.1 m diameter.

Process Intensification for Green Chemistry: Engineering Solutions for Sustainable Chemical Processing, First Edition. Edited by Kamelia Boodhoo and Adam Harvey.

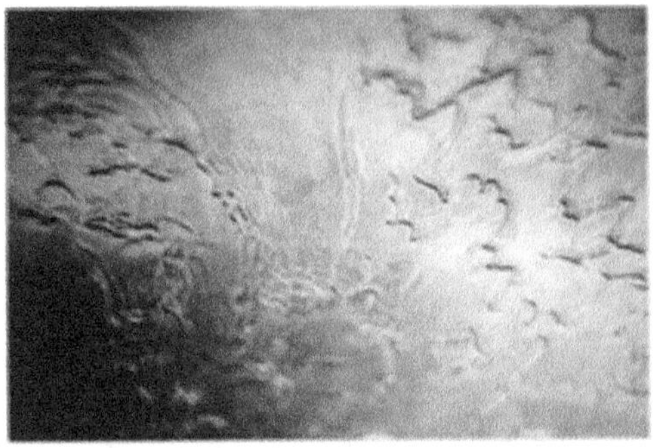

Figure 3.1 Thin, wavy film flow on a rotating disc.

Extremely thin, wavy films, as shown in Figure 3.1, are generated when the liquid is driven by the high acceleration across the rotating disc surface.

In this chapter, the application of an SDR to achieve greener processing is reviewed. The design, operating features and hydrodynamics of thin-film flow are first presented and discussed, then the characteristic features of thin-film flow in the SDR which enable intensification of mixing, heat and mass transfer in chemical processes are appraised. A number of case studies relating to green chemical processing using the SDR are then presented and discussed, and the issues surrounding the implementation of the SDR in industry are analyzed. The chapter concludes with some remarks about the future prospects for this type of reactor.

3.2 Design and Operating Features of SDRs

A schematic of the cross-section of a typical SDR is shown in Figure 3.2. The operating principle of an SDR is based on generating high-acceleration environments by rotating a horizontal disc surface mounted on a shaft. The shaft is driven by an electric motor through a pulley arrangement. Liquid feed streams are introduced to the disc surface through stationary liquid distributors, which are positioned a few millimetres from the surface. Under the action of high centrifugal fields, the liquid is flung towards the disc edges in the form of very thin, highly sheared film. At the disc periphery, the liquid is thrown off and hits the stationary walls of the reactor housing, which provides cooling to quench the product as soon as it comes off the disc. The lid of the reactor has the facility for one or more liquid feed streams and for an optional gas inlet/outlet. Provisions for purging the system with an inert gas may also be incorporated in the reactor setup.

In terms of physical characteristics, SDRs vary mainly in disc size and surface configuration, while operational flexibility depends on variable parameters such as disc speed, feed flow rate, disc temperature and reactor pressure (Table 3.1).

An example of a grooved disc surface is shown in Figure 3.3. The disc material is dependent on the application, but stainless steel and brass are commonly used, with brass being preferred for enhanced heat-transfer applications due to its higher

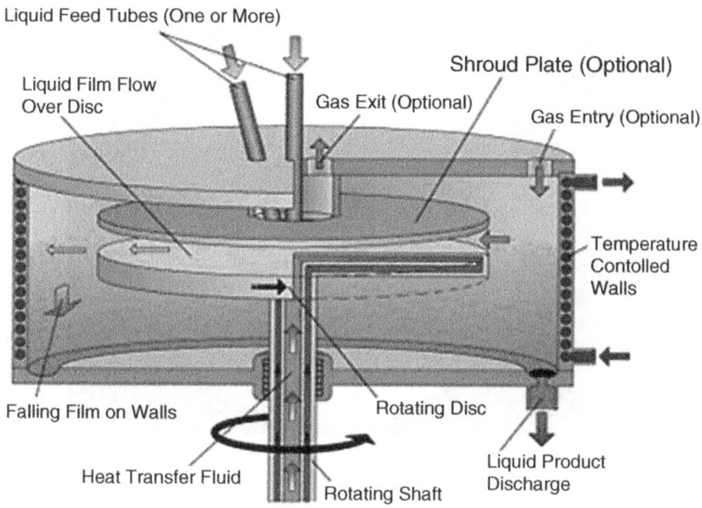

Figure 3.2 *Schematic of an SDR with internal heat transfer. Reprinted from [ref 6] © 2005 Institution of Chemical Engineers.*

thermal conductivity. Teflon-coated discs have also been used in highly fouling applications [7].

Typically, a disc of 500 mm diameter will be capable of processing about 150 kg/hour of material with water-like viscosity. Good temperature control of the disc surface is typically achieved by a heat-transfer fluid circulating underneath the reaction surface, as shown in Figure 3.4, and through a temperature-controlled bath, ideally with both heating and cooling facilities.

Table 3.1 *Typical range of SDR operating parameters.*

Parameter	Typical range
Disc diameter	0.1–1.0 m
Disc speed	200–6000 rpm[a]
Disc surface	Smooth, normal grooves, re-entry grooves, metal sprayed, meshed, non-stick surface
Disc material	Stainless steel, brass, glass
Liquid flow rates	0.5–200 cm^3/s[b]
Liquid viscosity	0.001–10 Pa.s
Operating temperature	−20–300 °C
Operating pressure	≤10 bar

[a]The upper limit is dictated by disc size; small discs of up to 0.15 m diameter can rotate at up to 6000 rpm but larger discs of 1.0 m diameter can typically rotate at up to about 1000 rpm

[b]Upper and lower flow rate limits for a given disc are dictated by optimum film thickness and film breakdown, respectively.

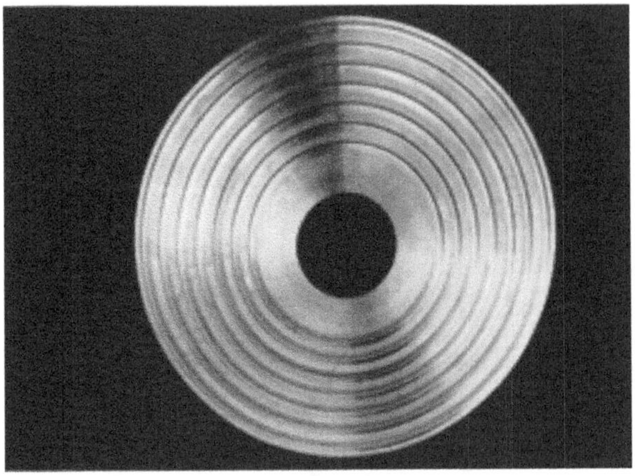

Figure 3.3 A brass disc with concentric grooves machined in its surface.

The SDR is generally suitable for use with gas–liquid, liquid–liquid, liquid–solid and gas–solid reactions. Variations in design exist for each type of application. For example, an SDR for gas–liquid or gas–solid applications will have a stationary shroud immediately above the rotating disc surface to direct the gas flow close to the liquid film/solid substrate on the rotating disc (as shown in Figure 3.2). Mixing of two or more liquid streams on the disc surface requires liquid distributors to be positioned close to the disc surface. In the simplest case, two or more distributors may feed liquids at the centre of the disc, or if a staged mixing process is required at different radial locations, single- or multi-point distributors may be positioned at the desired locations.

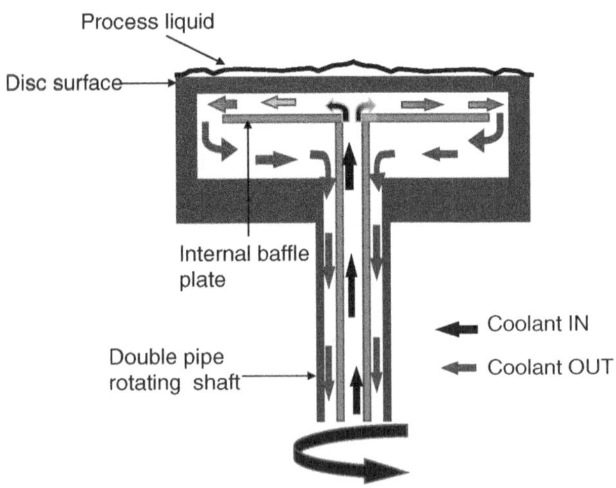

Figure 3.4 Schematic of the internal heat transfer system of SDR.

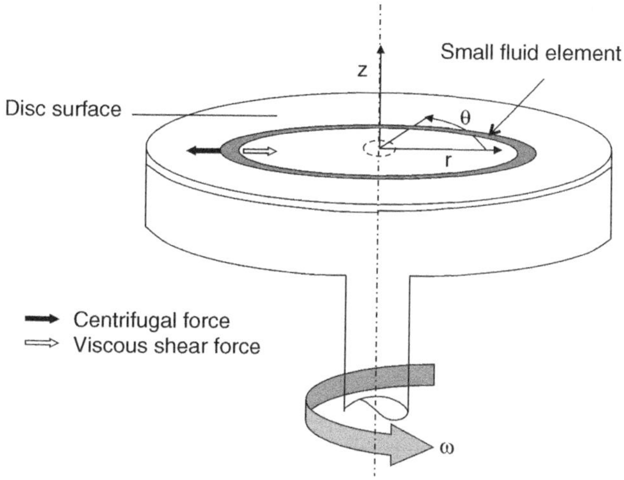

Figure 3.5 *Schematic of film flow on a rotating disc with forces exerted on the fluid in the centrifugal flow model.*

3.2.1 Hydrodynamics

The flow of thin liquid films on a smooth horizontal rotating disc is fully described by the modified Navier–Stokes equations for the conservation of momentum and the equation of continuity for the conservation of mass [8]. The centrifugal model represents a much simplified version of the Navier–Stokes equation, derived on the basis of a number of assumptions applicable to steady-state, symmetrical flow of thin liquid films on the rotating disc [8]. The model, represented by Equation 3.1, is based on a force balance between the body force, which is governed by the centrifugal force directed radially away from the centre of the disc, and the opposing viscous shear force, as shown schematically in Figure 3.5.

$$-\omega^2 \mathrm{r} = \nu \frac{\partial^2 \mathrm{v_r}}{\partial \mathrm{z}^2} \tag{3.1}$$

Expressions for various parameters of interest for SDR processing, such as radial velocity distribution, $\mathrm{v_r}$, mean radial velocity, $\mathrm{u_m}$, film thickness, δ, shear rate, $\dot{\gamma}$, and mean residence time, $\mathrm{t_{res}}$, for flow on the spinning disc can be derived from Equation 3.1 and are as follows [8,9]:

$$v_r(r,z) = \frac{\omega^2 r}{\nu}\left(\delta z - \frac{z^2}{2}\right) \tag{3.2}$$

$$u_m(r) = \left(\frac{Q^2\omega^2}{12\pi^2 r\nu}\right)^{1/3} \tag{3.3}$$

$$\delta(\mathrm{r}) = \left(\frac{3}{2\pi}\frac{\nu Q}{\omega^2 \mathrm{r}^2}\right)^{1/3} \tag{3.4}$$

$$\dot{\gamma}(r,z) = \frac{d\mathrm{v}_r(r,z)}{dz} = \left(\frac{3Q\omega^4 r}{2\pi\nu^2}\right)^{1/3}\left(1-\frac{z}{\delta}\right) \tag{3.5}$$

$$\mathrm{t_{res}} = \left(\frac{81\,\pi^2\,\nu}{16\,\omega^2\,Q^2}\right)^{1/3}\left(r_\mathrm{o}^{4/3} - r_\mathrm{i}^{4/3}\right) \tag{3.6}$$

These parameters are applicable to smooth, fully developed laminar flow of Newtonian liquids where the effects of surface instabilities and inertial forces are neglected. More complex hydrodynamic models have been formulated to describe flow on the rotating disc, taking into consideration, for instance, the Coriolis acceleration in the angular direction [10–13] and the non-Newtonian nature of certain fluids such as polymer melts [14,15]. In contrast to the centrifugal model, the Coriolis model assumes the radial velocity component, $\mathrm{v_r}$, to be fairly significant in magnitude, so that the Coriolis acceleration, $\mathrm{a_{Cor}}$, acting in the angular direction opposite to the direction of rotation of the disc becomes an influential force on the liquid flow:

$$\mathrm{a_{Cor}} = 2\mathrm{v_r}\omega \tag{3.7}$$

The Coriolis effect can be ignored if the Coriolis acceleration is negligible compared to the centrifugal acceleration [10]. That is:

$$2\mathrm{v_r}\omega \ll \omega^2 r \tag{3.8}$$

Equation 3.8 implies that the Coriolis effect can be ignored for very thin films and/or viscous liquids. Films become much thinner the further away the liquid is from its central discharge point on the disc, so the condition in Equation 3.8 can generally be satisfied on relatively large discs.

3.2.2 SDR Scale-up Strategies

This section provides descriptions of available methods by which to scale up the SDR from the lab or pilot plant to the industrial scale [16].

Scaling up is usually employed to estimate the size required for a desired throughput or the throughput possible with a given disc size of a production-scale SDR based on the performance of a lab/pilot-scale SDR with known flow rate, Q, and disc speed, ω. In scaling up an SDR process, it is assumed that liquid properties, such as density and viscosity, are constant. Variable parameters are liquid flow rate and rotational speed.

First, it is necessary to define a scaling factor, S, for the system, as follows:

$$S = \frac{\mathrm{r}_2}{\mathrm{r}_1} \tag{3.9}$$

where r_2 is the disc radius in an industrial plant and r_1 is the disc radius in a pilot or lab scale unit.

There are several scaling methods that can be applied depending on the process requirements. If, for instance, a very fast reaction progresses to completion in a lab-scale SDR under given conditions of mean disc residence time (related to the kinetics of the process) and film thickness at the disc edge (related to the mixing/mass transfer/heat transfer of the process), then for the industrial-scale reactor it is important to keep the same liquid residence time and film thickness at the higher throughput.

If, on the other hand, we assume that for a given reaction of interest with relatively fast kinetics the residence time is limiting, so that a change in conversion of 10% is achieved in one disc pass in a lab SDR, the residence time required for complete conversion in a large SDR will be at least 10 times the residence time of the lab-scale SDR.

Thus, using the expressions for film thickness and residence time given by the centrifugal model (Equations 3.4 and 3.6), and assuming that liquid properties such as the density and viscosity remain constant in the larger reactor, the following equations can be developed for the general case of *X* times the original residence time being required at constant film thickness in the two reactors:

$$Q_2 = \frac{1}{\sqrt{X^3}} S^2 \frac{\omega_1 Q_1}{\omega_2} \tag{3.10}$$

It is to be noted that Equation 3.10 is based on the assumption that the feed is introduced at the centre of the disc; that is, $r_i = 0$.

Also, from the constant film thickness condition:

$$Q_2 = S^2 Q_1 \left(\frac{\omega_2}{\omega_1}\right)^2 \tag{3.11}$$

Combining Equations 3.10 and 3.11, the following can be derived:

$$\omega_2 = \frac{1}{\sqrt{X}} \omega_1 \tag{3.12}$$

$$Q_2 = \frac{1}{X} S^2 Q_1 \tag{3.13}$$

Example: An industrial reactor of 1 m diameter is to be used for a reaction requiring 10 times the lab-scale residence time and constant film thickness. Lab tests in a laboratory-scale reactor of 200 mm disc diameter have revealed that the optimum flow rate Q_1 and disc speed N_1 are 20 cm^3/s and 400 rpm, respectively (where $\omega = 2\pi N/60$). The throughput and disc speed in the industrial-scale reactor, calculated according to Equations 3.12 and 3.13, are given in Table 3.2.

Other scaling methods may be based on a combination of increased residence time and constant/increased shear rate (for mixing-dependent processes) or of increased residence time and constant Reynolds number. The important point to remember is the requirement that two independent parameters (such as disc residence time and film thickness) be used in the scale-up procedure in order to account for two changing variables (typically Q and ω).

Table 3.2 *Scale-up parameters in lab- and industrial-scale SDRs.*

	Disc diameter (m)	Disc speed (rpm)	Throughput (cm^3/s)	Mean disc residence time (s)[a]	Film thickness at disc edge (μm)[a]	Surface area per unit volume (m^2/m^3)
Lab-scale SDR	0.2	400	20	0.2	82	8200
Industrial-scale SDR	1	127	50	2.0	82	8200

[a]Assumes liquid properties similar to those of water.

3.3 Characteristics of SDRs

The characteristic features of fluid flow in the SDR are described in this section.

3.3.1 Thin-film Flow and Surface Waves

Film thicknesses of the order of 50–500 μm are typical for the films flowing on the surface of a horizontal disc rotating at speeds of between 200 and 2000 rpm. As expressed in Equation 3.4, film thickness is dependent on a range of operational parameters, such as the disc speed and diameter, and the liquid flow rate and physical properties (density and viscosity). For a given liquid at a fixed radial position on a disc, the disc speed (represented by the angular velocity, ω) is a more influential parameter than the flow rate, Q, in determining the film thickness, as $\delta \propto \omega^{-2/3}Q^{1/3}$. The effects of increasing the disc speed and liquid flow rate are illustrated in Figure 3.6. It is also clear from Equation 3.4 and Figure 3.6 that, for set values of Q and ω, the film becomes thinner the further away the liquid is from the centre.

The conditions giving rise to the wave-associated flow regimes on rotating discs have been extensively studied. One of the most recent studies was by Woods [18], who carried out a comprehensive theoretical and experimental examination of the formation and

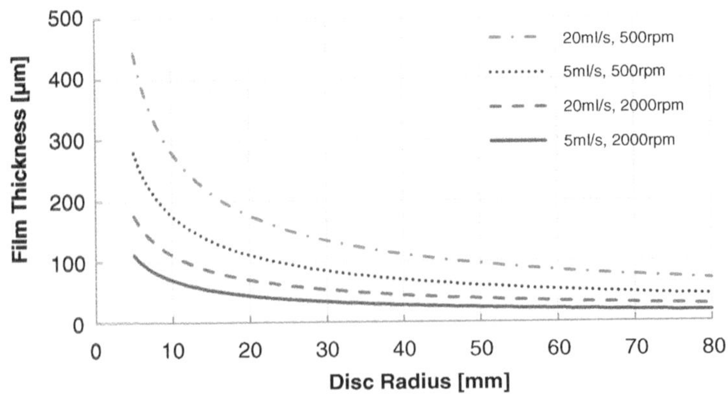

Figure 3.6 *Film thickness variation for water at room temperature on a smooth disc [17].*

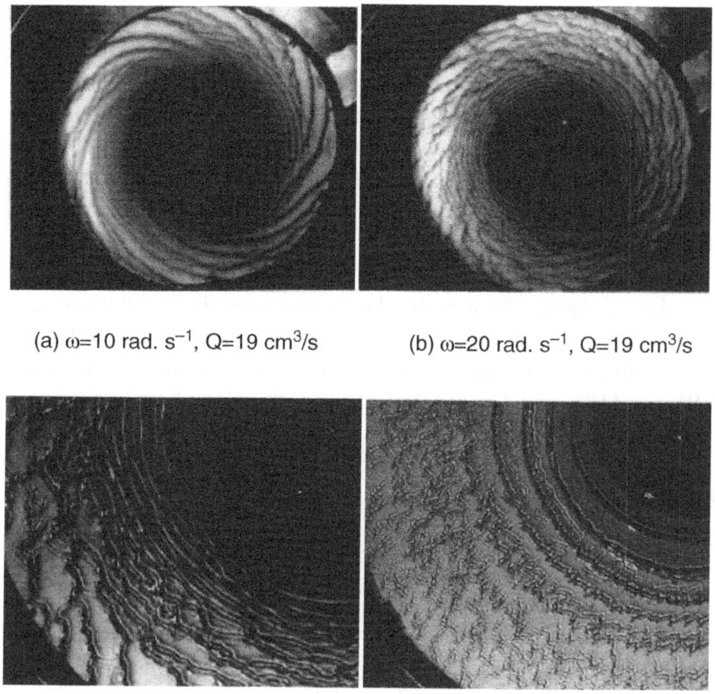

(a) ω=10 rad. s^{-1}, Q=19 cm^3/s (b) ω=20 rad. s^{-1}, Q=19 cm^3/s

(c) ω=20 rad. s^{-1}, Q=19 cm^3/s (d) ω=60 rad. s^{-1}, Q=19 cm^3/s

Figure 3.7 *Wave formation and propagation on a rotating disc under varying operating conditions of flow rate and disc rotational speed. Fig. (a) (b), (c) and (d), Reprinted from [ref 19] © 1999, with permission from Elsevier.*

propagation of waves at the free surfaces of films of water mixed with a dye flowing on a rotating disc. For a range of flow rates between 7 and 19 cm^3/s and disc rotational speeds between 100 and 600 rpm, three distinct flow regions were observed at different radial locations on the disc, as depicted in Figure 3.7:

(1) An inner region, starting from the distributor, with apparent smooth flow.
(2) A middle section, covered with what appeared initially to be a series of concentric waves but was subsequently found to be a single tightly wound spiral, which had a noticeably longer wavelength as the rate of rotation was stepped up.
(3) An outer region right up to the periphery of the disc, where a multitude of spirals with irregularities along the wave fronts eventually disintegrated into random 3D wavelets on propagation across the disc. These wavelets first appeared at a radius that became smaller with increases in the speed of rotation of the disc. The effect of increased flow rate was to cause the wavelets to form more rapidly and hence occupy a greater proportion on the disc surface.

The effect of built-in surface roughness on surface wave formation was investigated by Jachuck and Ramshaw [20]. Three different surface configurations were tested and compared to the flow on a smooth disc: metal sprayed, reentry grooved and normal grooved discs. It was found that for a given flow rate and rotational speed, the tailored surfaces, especially the grooved types, produced a larger number of waves of small amplitude and short wavelength at the film surface than did the smooth disc. The normal grooved disc was found to be more efficient at high speeds of rotation, since it generated more waves and hence more instabilities at the free surface of the film than did the reentry grooved disc. The effect of these textured surfaces on the heat-transfer process was also studied and will be discussed in Section 3.3.2.

Matar and coworkers [21–24] have presented extensive modelling studies for thin film flow over a rotating surface, with a particular focus on predicting the surface instabilities under various operating parameters and their effects on mass transfer. The models are in good agreement with experimentally validated data for wave formation and propagation, obtained by Woods [18], and for gas–liquid mass transfer characteristics, measured by Aoune and Ramshaw [19].

3.3.2 Heat and Mass Transfer

Very high heat and mass transfer rates are associated with the thin films on a rotating disc surface. As shown in Figures 3.8 and 3.9, local film heat-transfer coefficients exceeding 25 kW/m^2K have been achieved for water flowing at the edge of a 0.5 m diameter smooth disc spinning at a relatively modest disc speed of 60 rad/s, while local mass-transfer coefficients in the liquid phase have been measured in the thinnest part of the film at the disc periphery to be as high as 1×10^{-3} m/s for the liquid phase [19]. Increases in disc speed and liquid flow rate have generally been shown to lead to higher transfer coefficients [19,25].

The high heat- and mass-transfer coefficients are attributed to the thin films resulting in reduced path lengths, across which conduction and diffusion occur more rapidly.

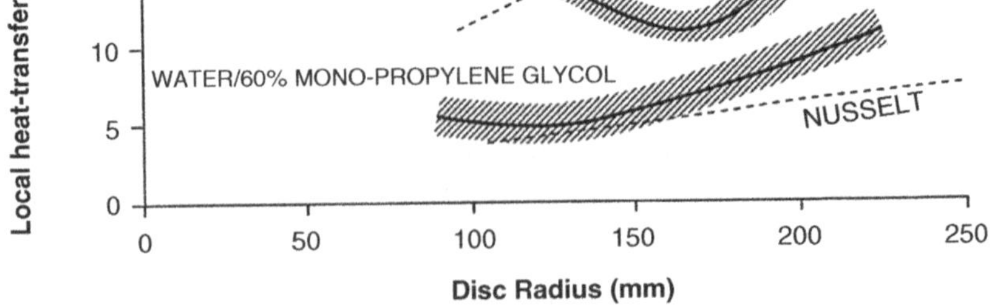

Figure 3.8 *Comparison of measured and predicted local heat-transfer coefficient on a smooth rotating disc for low- and high-viscosity liquid feeds. Reprinted from [ref 19] © 1999, with permission from Elsevier.*

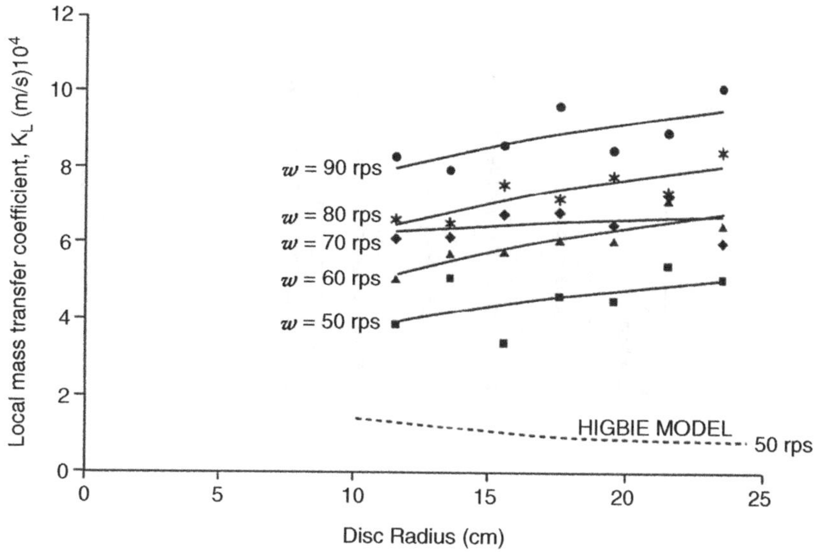

Figure 3.9 *Local liquid phase mass-transfer coefficient under various disc operating conditions. Reprinted from [ref 19] © 1999, with permission from Elsevier.*

Moreover, the wave-induced mixing and increased surface area per unit volume associated with the rippling effects also contribute to improving these transport processes, as highlighted by the enhancements observed in Figures 3.8 and 3.9 when comparing the Nusselt and Higbie model predictions for a smooth film to those for a wavy surface of an actual liquid film travelling across the disc. The enhancement properties of the surface waves on mass transfer have also been demonstrated, in numerical modelling studies undertaken by Matar and coworkers [21]. A number of investigations have shown textured surfaces in the form of grooves and other types of surface roughness to be conducive to further improvements in heat-transfer rates in heating [20] (as depicted in Figure 3.10) as well as in boiling and condensing applications [26,27].

Overall heat-transfer coefficients of up to 5 kW/m^2K have been measured in those SDRs designed to include an internal heat-transfer system, which consists of a heat-transfer fluid recirculation between the underside of the disc surface and a plane surface (as shown in Figure 3.4). The patent literature discloses an improvement in the design of the heat-transfer surface whereby a double spiral disc design made of copper serves as the flow channel for the heat-transfer fluid. It is claimed that this design can double the overall heat-transfer coefficient to 10 kW/m^2K compared to the original plane surface design [28].

The heat-transfer capability of the thin films in the SDR has recently been visually analyzed by Ghiasy *et al.* [17] using a thermal-imaging camera. The thermograms presented in Figures 3.11 and 3.12 illustrate that increasing the disc rotational speed at a constant flow rate results in narrowing of the cold central region and thus in improved disc/film heat-transfer performance. These visual observations confirm that reduced film thickness and thus a reduced conduction path with increasing angular velocity give better heat-transfer rates. It is also apparent that the heat penetrates the more viscous liquid film

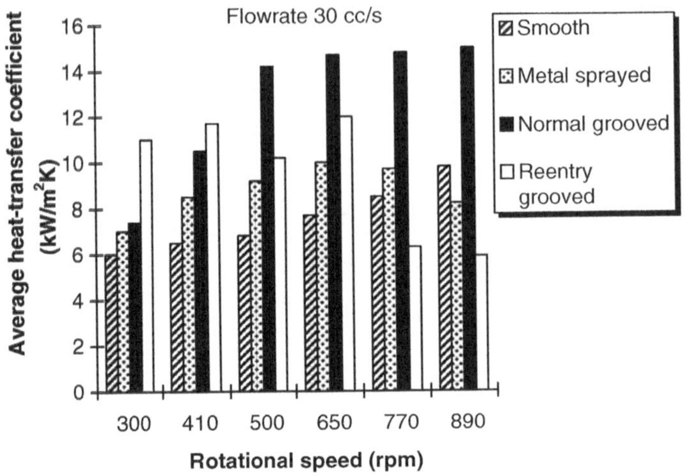

Figure 3.10 Heat-transfer characteristics on various rotating surface configurations applicable to process heating operation. Reprinted from [ref 20] © 1994, with permission from Elsevier.

less efficiently, so that the top film surface remains colder across a larger proportion of the disc area than it does with the water film.

The thermograms also provide a useful insight into the wavy nature of the water film surface and the trajectory of the film path on the disc. In Figure 3.11a, the relatively large-amplitude waves formed on the water film result in overlapping boundaries between annular sections of the film close to the inner region, which is indicative of radial mixing. As the film moves towards the disc edges, these large-amplitude waves appear to either decay completely or give way to waves of such greatly reduced amplitudes that they become less involved in promoting radial mixing. A faint spiral profile is also apparent in

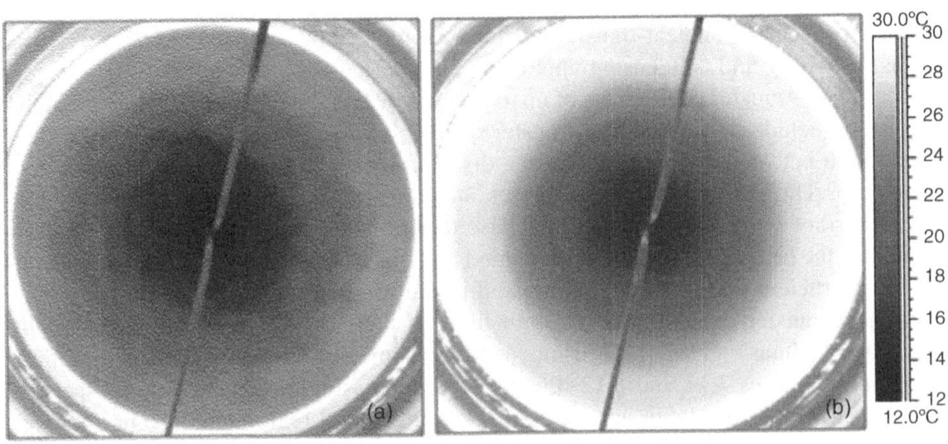

Figure 3.11 Thermograms for Q = 10 ml/s, working fluid: water; (a) N = 500 rpm; (b) N = 2000 rpm. Reprinted from [ref 17] © 2012, with permission from Elsevier.

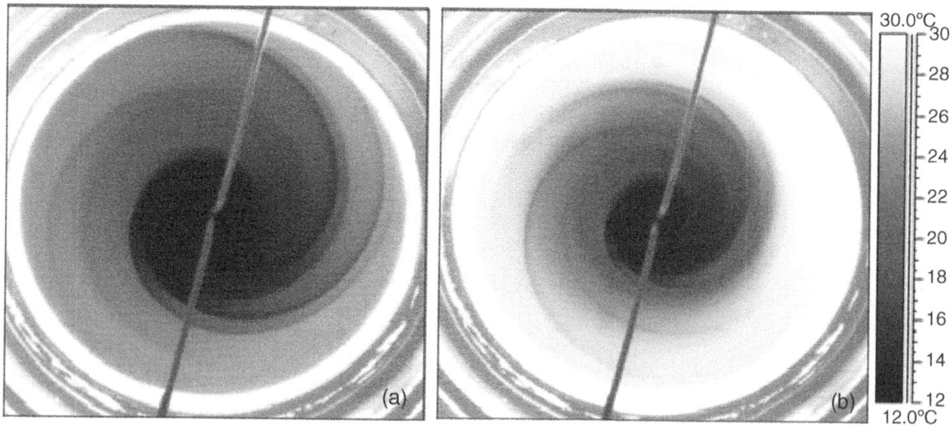

Figure 3.12 Thermograms for $Q = 10\,ml/s$, working fluid: 50% water, 50% glycerol v/v; (a) $N = 500\,rpm$; (b) $N = 2000\,rpm$. Reprinted from [ref 17] © 2012, with permission from Elsevier.

the inner regions of the disc in Figure 3.11a in contrast to Figure 3.11b, which is indicative of the retarding Coriolis forces on the film exerting an increasingly significant influence on the film flow close to the centre of the disc at lower disc speeds. Interestingly, at the much higher rotational speed of 2000 rpm, with water as the liquid (Figure 3.11b), radial mixing induced by the waves appear to be almost non-existent, even close to the disc centre. This may be explained by the fact that not only are the waves of very small amplitude, as described above, but also, more importantly, under the action of the very high centrifugal forces generated at 2000 rpm they are propelled towards the disc edges before they can interact radially within the flow field. At higher liquid viscosity (Figure 3.12), the film travels in a more pronounced spiral path, which unwinds in the direction of rotation; the surface waves observed with water no longer seem to be present even at low disc speeds. This effect is ascribed to the viscous dampening of surface waves, which leads to a smoother film being generated.

3.3.3 Mixing Characteristics

The low film thickness combined with the relatively high radial velocity of the film at its surface gives rise to very high shear stresses/shear rates within the liquid film. The intense mixing capability within the thin films is generally attributed to the combined effects of these high shear rates and the waves generated on the surfaces of the films. The shear rate, expressed in Equation 3.5, varies between its maximum value at the disc surface (at $z = 0$) where the film is stationary relative to the disc and a zero value at the free surface of the film (at $z = \delta$) (assuming no interaction with the gas phase). The highest shear rates would be attained at the disc periphery of any given disc diameter under conditions of high flow rates of low-viscosity liquid and high disc rotational speeds. One application area in which the effects of high shear rates would be very advantageous is the processing of shear thinning fluids such as polymers. The high shear effects in the thin film would cause the viscosity to drop, allowing easier flow and handling of the material, as well as improved mixing/heat

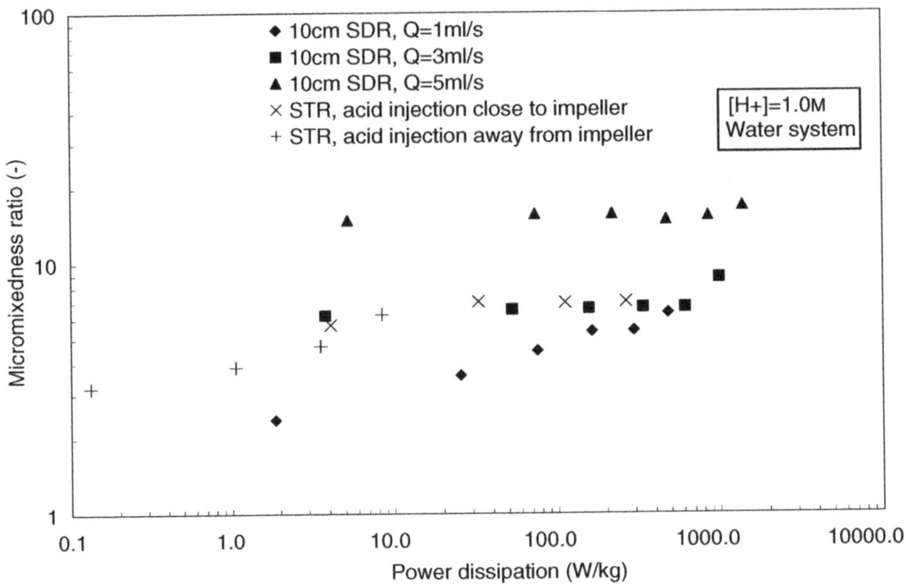

Figure 3.13 *Comparison of micromixing efficiency in an SDR and STR. Reproduced from [ref 29] by permission of John Wiley & Sons. © 2012.*

and mass-transfer/reaction rates, with the possibility of significantly reduced overall energy consumption.

A very recently published study into the micromixing characteristics of thin-film flow in an SDR [29] confirms the predicted conditions for high shear rates and intense mixing. It shows that that the best micromixing conditions, measured in terms of segregation index or micromixedness ratio for the well-established iodide–iodate reaction scheme [30], are generally achieved in a *water* film at *high disc rotational speeds* and *high feed flow rates*. Furthermore, the 10 cm diameter smooth rotating disc used in this work has been demonstrated to be more energy efficient as a micromixer when operating at the highest flow rate of 5 ml/s than is a stirred tank reactor (STR). This is highlighted in Figure 3.13, where the highest micromixedness ratio of 15 in the SDR is more than double that achieved in the STR under similar power dissipation of about 5 W/kg, even with the best possible configuration for acid injection close to the impeller in the STR.

The SDR also outperforms other intensified reactor systems such as the rotating packed bed (RPB) [31,32] and the rotor-stator reactor (RSR) [33], even at a comparatively low flow rate of 5 ml/s, as shown in Figure 3.14.

Further improvements in the micromixing within the thin films on a rotating disc have been noted with a grooved disc and with multi-point liquid distributors as compared to single-point injection systems close to the centre of the disc [34].

3.3.4 Residence Time and Residence Time Distribution

The SDR offers the opportunity of processing under very short residence times of the order of seconds, rather than the minutes or even hours required in conventional reactor

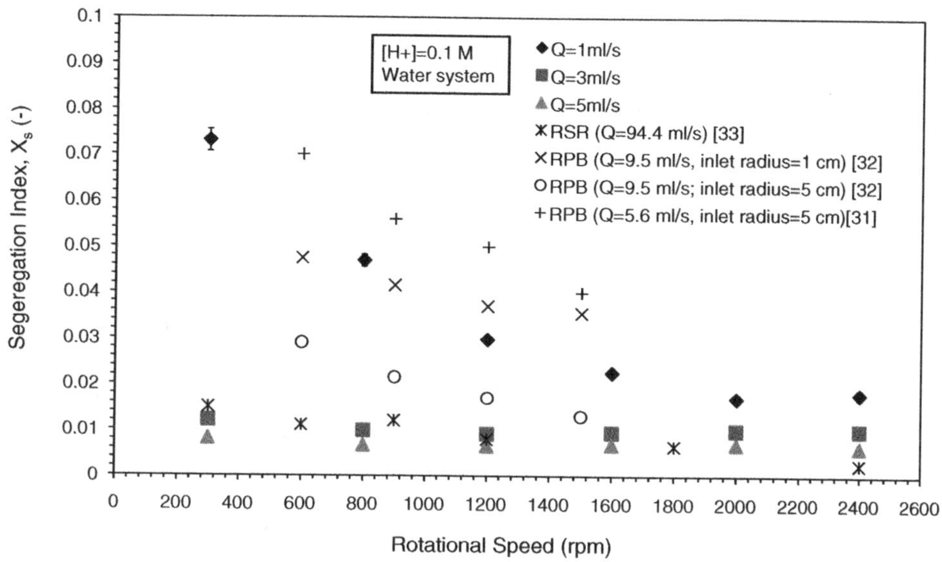

Figure 3.14 *Micromixing efficiency comparison between a 10 cm diameter SDR and other intensified continuous reactor configurations. Reproduced from [ref 29] by permission of John Wiley & Sons. © 2012.*

configurations such as large STRs. This feature is particularly advantageous for processes involving:

(1) Heat-sensitive materials such as food products; for example, the pasteurization of ice cream [35] and fruit juice [36] processing.
(2) Polymers where processing at high temperatures frequently leads to hot spots and polymer degradation.
(3) Competitive reaction schemes in which exceedingly high residence time can significantly reduce selectivity of the desired product if the product molecule can further react to give unwanted byproducts [37].

Often, in continuous reactor configurations such as the continuously stirred tank reactor (CSTR) and the microreactor, the flow rate is the main operating variable by which to manipulate the residence time. In contrast, the SDR has an additional degree of freedom in the disc rotational speed parameter, which can be controlled independently of the process liquid flow rate to achieve the desired residence time without affecting the throughput.

In a recent comprehensive investigation [38], the residence time distribution (RTD) of the thin liquid film was characterized as being close to the ideal plug flow regime, giving each molecule practically the same processing experience on the rotating disc. This is an important consideration in achieving, for example, narrow crystal or particle distributions [7,39] and tight molecular weight distributions (MWDs) in polymerization processes [40,41], as well as high selectivity in competitive organic chemical reactions [37]. It has been demonstrated that near plug-flow conditions are generally attained at higher disc speeds, higher liquid flow rates and lower liquid viscosities [38], as illustrated in

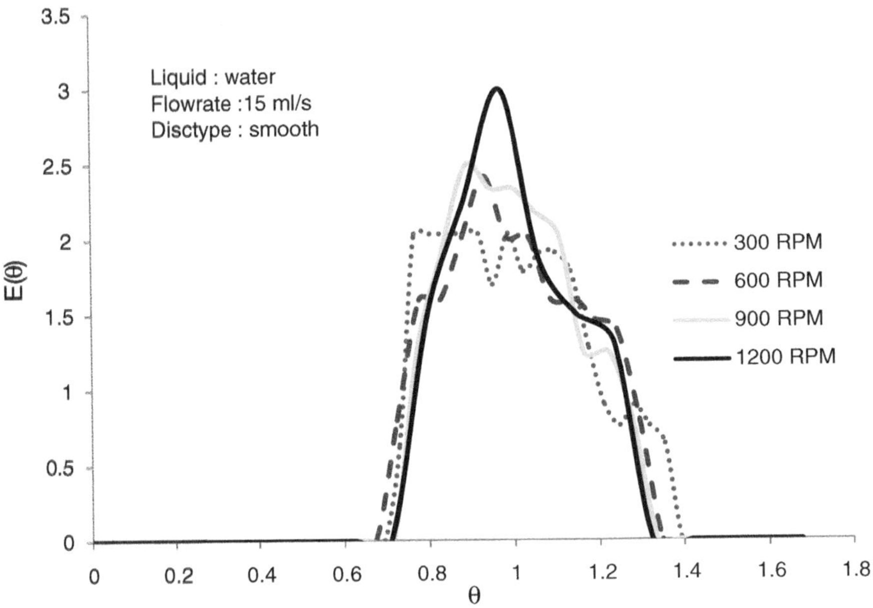

Figure 3.15 Effect of disc speed on RTD. Reproduced from [ref 38] Poster presented at ISCRE22, 2-5 Sept. Maastricht, Netherlands (2012).

Figure 3.15. These RTD results are in agreement with the visual depictions from the thermographic images of the film that were presented and discussed in Section 3.2. The texture of the disc surface also plays an important role, with a grooved disc being better suited to obtaining a narrower RTD than a smooth disc (Figure 3.16).

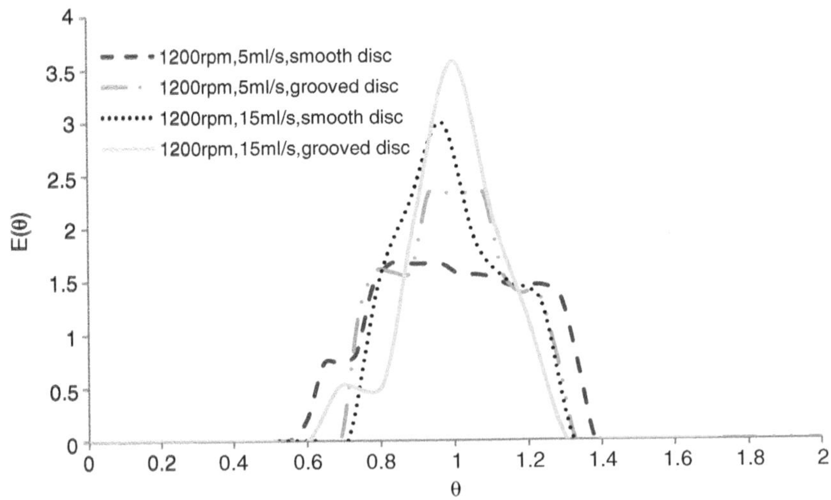

Figure 3.16 Effect of disc surface type on RTD. Reproduced from [ref 38] Poster presented at ISCRE22, 2-5 Sept. Maastricht, Netherlands (2012).

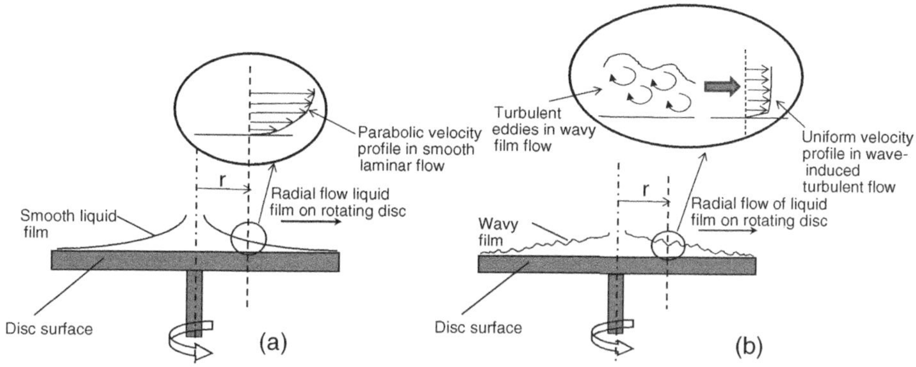

Figure 3.17 *Illustration of differences in velocity profiles in (a) smooth, laminar film and (b) wavy film with induced turbulence. Reproduced from [ref 38] Poster presented at ISCRE22, 2-5 Sept. Maastricht, Netherlands (2012).*

Achieving plug flow behaviour is dependent on a uniform velocity profile in a direction perpendicular to the flow direction and on negligible dispersion in the direction of flow. In the context of film flow on the rotating disc where bulk flow is in the radial direction, the film-surface instabilities or waves are likely to induce turbulence within the layer underneath via the formation of turbulent 'eddies' (as illustrated in Figure 3.17), giving a more uniform velocity profile in the transverse direction, as opposed to the parabolic velocity profile that characterizes waveless laminar flow. Therefore, the more waves are formed, the greater the transverse mixing and the more uniform the radial velocity across a given cross-section in the film. Furthermore, the higher the centrifugal force exerted on the film, the lower the tendency for dispersion in the radial direction away from the direction of flow, which contributes to suppressing deviations from plug flow behaviour.

3.3.5 SDR Applications

The niche applications for SDR technology are kinetically fast, heat- and mass-transfer limited processes, which require good mixing. The thin, wavy films allow high levels of heat transfer across the liquid–solid interface and mass transfer across the liquid film or across the gas–liquid interface. The thin films, which represent a very small inventory of material in the reactor at any instant, also provide a high surface area to volume ratio, typically in the range 5000–15 000 m^2/m^3, to promote high rates of heat transfer from the disc to the liquid or vice versa. High shear rates enable intense local mixing in the film so that inherently fast reactions can proceed at rapid rates dictated only by the chemical kinetics of the process, rather than by mixing intensities. When mixing limitations are thus removed, short residence times can be applied to exploit the very rapid reaction rates. Considerable research has been conducted into the application of SDR technology to numerous reactions conforming to these criteria, as shown in Table 3.3. A selection of these processes pertaining to green chemistry applications will be reviewed in Section 3.4.

Table 3.3 *Reactions investigated in SDR.*

Reaction	SDR characteristics of relevance to the reaction	References
Photo-polymerization of n-butyl acrylate	Thin film for efficient UV penetration High shear for good mixing, even at high viscosity Enhanced heat transfer	[40]
Free radical and cationic polymerization of styrene	High shear rates Enhanced heat and mass transfer High shear and wavy flow mixing	[9,41]
Equilibrium-controlled condensation polymerization	High shear rates High heat and mass transfer for removal of volatile material, e.g. water vapour	[8]
Phase transfer-catalysed Darzen's reaction	Enhanced mass transfer through immiscible liquid phases	[7]
Crystallization	Intense mixing	[7,42]
Preparation of nanoparticles	Intense local mixing Plug flow characteristics	[39,43,44]
Competitive organic reaction (e.g. isomerization of alpha-pinene oxide)	Short residence time and plug flow for high selectivity Intense mixing characteristics	[37]
Food processing	Short residence time for minimal degradation of heat-sensitive food products High micromixing for uniform emulsion droplets	[35,36]

3.4 Case Studies: SDR Application for Green Chemical Processing and Synthesis

3.4.1 Cationic Polymerization using Heterogeneous Lewis Acid Catalysts

Classical cationic polymerization is a well-established industrial procedure for making polymers. The cationic polymerization mechanism involves the formation of an extremely reactive cationic species (normally a proton from a homogeneous acid catalyst), which attacks the double bond in the monomer to initiate the reaction. The highly reactive nature of the cations results in propagation rates in the range of 10^4–10^6 l/mol.s which are about 2 orders of magnitude higher than those in free radical polymerizations [45]. Because of the potentially large exotherms, large volumes of chlorinated organic solvents are typically used in large STRs to dilute the reaction system in order to control the process. Neither these solvents nor the hazardous acids such as aluminum chloride employed as catalysts are environmentally friendly. From an engineering perspective, a reactor technology such as the SDR that allows good control of the cationic polymerization process through enhanced mixing and heat transfer could significantly reduce or even eliminate the use of these solvents, thus allowing greener processing. Furthermore, designing a heterogeneous acid catalyst can offer many advantages compared to its homogeneous counterparts: ease of separation from the reaction mixture by filtration, mild reaction conditions, less harm to the

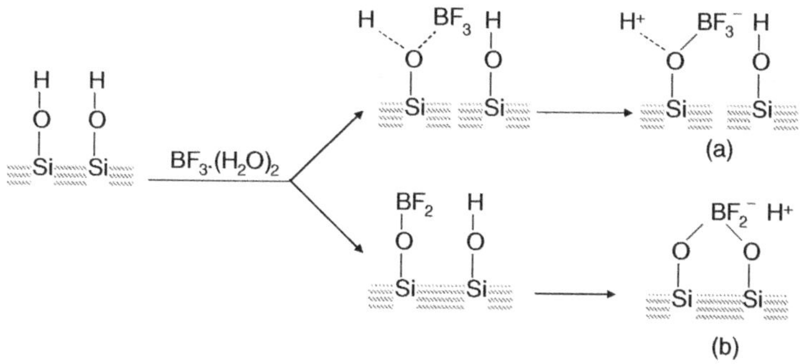

Scheme 3.1 *Grafting of BF3 on silica, yielding two possible structures (a) and (b) for Lewis acid active centres. Reprinted from [ref 9] © 2006 Wiley Periodicals, Inc.*

environment and possible reuse of the catalyst [46]. Through the use of such supported catalysts, the elimination of the water-quenching step required in a homogeneous system at the end of the reaction to neutralize the Lewis acid also reduces the production of hazardous acidic waste.

In order to address these issues, a study involving a performance comparison between the SDR and a conventional STR for the heterogeneous cationic polymerization of styrene was undertaken by Boodhoo *et al.* [9]. A newly developed silica-supported BF_3 Lewis acid catalyst [47] was used to initiate the polymerization, as illustrated in reaction Schemes 3.1 and 3.2. A summary of the comparison is shown in Table 3.4. It was found that the polymerization in the batch at about 50% w/w [St] was difficult to control at the operating temperature of 40 °C (Run B8). The reaction spiralled out of control, giving a maximum temperature of 97 °C within 1 minute into the reaction, with an accompanying detrimental effect on the molecular weight and polydispersity of the polymer formed. These findings highlight the importance of cooling the batch to very low temperatures despite low rates of polymerizations, as seen in Run B4. For safe operation in the STR at 40 °C, it was recommended that the maximum monomer concentration be no greater than 25% w/w. On the other hand, controlled operation at the elevated temperature of 40 °C using 49% w/w [St] was possible in the SDR (Run S14), giving a higher average rate of polymerization and higher molecular weights than the cooled batch at 13 °C (Run B4). More interestingly, the SDR was demonstrated to be capable of handling styrene concentrations up to 75% w/w in a safe and controllable manner, with good control of MWDs (Runs S21, S26, S31), at 20 and 40 °C disc temperature.

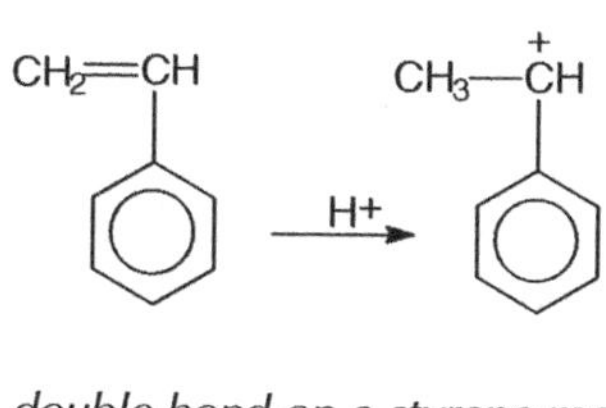

Scheme 3.2 *Proton attack of a double bond on a styrene molecule during the initiation step. Reprinted from [ref 9] © 2006 Wiley Periodicals Inc.*

Table 3.4 Cationic polymerization in batch and SDR using 1.3% w/w BF3 suspended in 1,2-dichloroethane. Reproduced from [ref 9] by permission of Wiley Periodicals, Inc. © 2006.

Reactor type	Run no.	[St] % w/w	Reaction temp. (°C)	Residence time (s)	Conversion (%)	Ave. rate of conversion (%/s)	M_n	PDI
Batch	B4[a]	49	13	1800	10	0.0056	12 032	2.0
Batch	B8[b]	49	97	60	94	1.57	4100	3.12
SDR	S14[c]	49	40	1.0	7.6	7.6	24 679	1.68
SDR	S21[c]	75	40	1.1	7.8	7.1	25 050	1.46
SDR	S26[d]	75	40	0.9	9.5	10.6	14 593	1.60
SDR	S31[d]	75	20	1.0	4.1	4.1	17 045	1.73

[a]Temperature controlled at 7 °C; initial temperature of monomer = 21.5 °C
[b]Temperature controlled at 40 °C; initial temperature of monomer = 36 °C
[c]Disc diameter = 100 mm; disc speed = 200 rpm; flow rate = 0.7 ml/s
[d]Disc diameter = 200 mm; disc speed = 400 rpm; flow rate = 1.5 ml/s

Clearly, the enhanced heat transfer rate in the SDR is critical to achieving control in the reaction, even at high monomer concentrations. The reduction in solvent usage in the SDR is also significant and has important environmental benefits. Overall, the advantages of applying an intensified, continuous reactor technology to a heterogeneously catalysed system for improved control of a rapid, highly exothermic reaction were demonstrated experimentally, emphasizing the combined benefits of process intensification (PI) and green chemistry.

3.4.2 Solvent-free Photopolymerization Processing

Light-induced polymerization, where a brief and intense burst of ultraviolet (UV) radiation upon a monomer/photoinitiator mixture generates a large amount of free radicals, is one of the most efficient methods by which to achieve quasi-instantaneous polymerization, transforming a liquid resin into a solid polymer in a short time. The free radical photopolymerization presents the advantages of a fast reaction (fast initiation rate and reaction rate) and a clean process, as it generally does not involve any solvent [48]. High-quality monomers can also be obtained, with good molecular weight and MWD. However, despite the fact that is a fast and clean process, it is limited by the thickness of the film, due to the very short penetration depth of the UV radiation (typically a few millimetres).

Methods involving a thin layer of monomer on a moving belt passing under UV lights have been disclosed in patents [49,50]. Since there is no agitation, mixing is minimal and the polymerization proceeds in what is essentially a static film. In addition, the removal of the heat of polymerization is limited to the dimensions of the belt. Because of these limitations, such techniques have been difficult to implement in industry.

The ability of the SDR to continuously generate thin, sustainable films under the action of the centrifugal field represents an important advance towards the design of a viable continuous reactor system applicable to bulk photopolymerization reactions. Not only would the thin film (50–300 μm) allow efficient penetration of UV light to initiate the

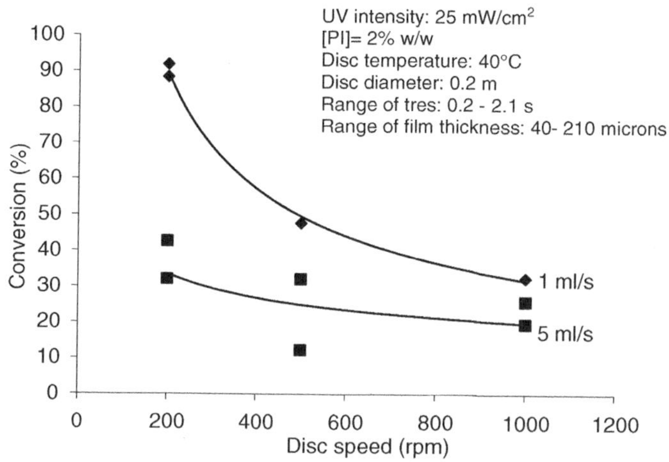

Figure 3.18 *Effect of disc rotational speed on the conversion of n-butyl acrylate in the SDR at different feed flow rates. Reproduced from [ref 40] © 2003 American Chemical Society.*

polymerization but the shear forces developed within the film would promote excellent mixing conditions.

N-butyl acrylate was chosen as the test monomer by which to evaluate the performance of the SDR for photopolymerization [40]. With a propagation rate constant of ~16 000 l/mol.s at 30 °C and a temperature of polymerization of 77.4 kJ/mol, the bulk polymerization of n-butyl acrylate presents significant challenges as a fast, highly exothermic reaction system.

It was observed that the rate of photopolymerization in the SDR was extremely fast. Conversions in excess of 90% were achieved in one SDR pass at 200 rpm disc rotational speed, 1 ml/s feed flow rate (giving a residence time of approximately 2 s) and a moderate UV intensity of 25 mW/cm^2 (Figure 3.18). It has been suggested that the high polymerization rates may be accounted for by the uncoiling of polymer chains subjected to the high shear rates and radially directed centrifugal force on the rotating disc (Figure 3.19). This effect would cause the chains to remain active for longer, thereby encouraging more monomer molecules to be consumed [51]. Narrow MWDs of the polymer, inferred from polydispersity indices (PDIs) in the range of 1.8–2.1, reflect the good temperature control achieved in the SDR.

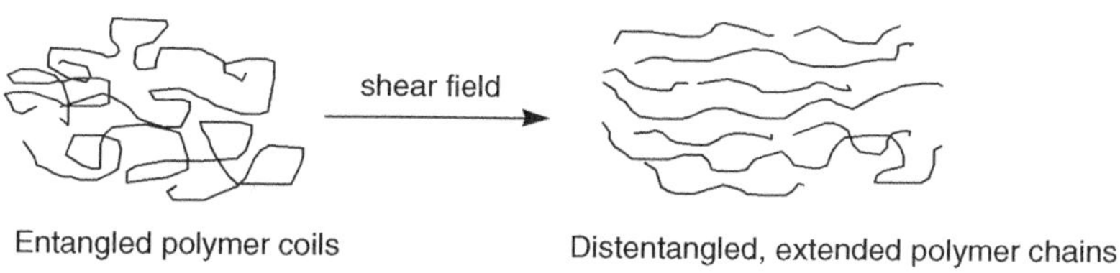

Figure 3.19 *Effect of shear on polymer coils Reproduced from [ref 52] by permission of Oxford University Press, Inc © 1999.*

Table 3.5 *Comparison of static film and SDR film for n-butyl acrylate photopolymerization. Reproduced from [ref 40] © 2003 American Chemical Society.*

Film type	UV intensity (mW/cm^2)	Exposure time (s)	Conversion (%)	M_w	M_n	PDI
Static	75	40	92	35 000	18 000	2.0
Static	25	10	30	52 000	28 000	1.8
SDR (dynamic)	25	2.1	90	70 000	33 000	2.1

These results have been benchmarked against photopolymerization in thin static films (Table 3.5) [53]. It is apparent that thin, well-mixed film in the SDR generally outperforms thin static film, in both conversion and molecular weight properties, when they are exposed to identical UV intensities of 25 mW/cm^2. To obtain polymers of similarly high conversions above 90%, as in the SDR, it is necessary to illuminate the static films at a higher UV intensity of 75 mW/cm^2 and to extend the exposure time to 40 s. Clearly, the static film photopolymerization process proceeds at a much slower rate, even with increased energy input, due to the lack of mixing. It is also possible that temporal variations in the molecular weight properties will exist across the static film thickness, leading to non-uniformity in the polymer product.

The study demonstrates the exciting and industrially viable opportunity to perform continuous photopolymerization of monomers in bulk in the SDR. With its enhanced mixing, high heat-removal capabilities, sustainable thin-film flow for efficient UV penetration and short, controllable residence time, the SDR offers the prospect of rapid polymerization rates and improved polymer product quality. It is noteworthy that a related investigation focusing on *thermally-initiated*, rather than UV-initiated, free-radical polymerization of styrene in bulk in the SDR has achieved similar benefits of rapid polymerization and good control of molecular weight properties in single and multiple passes on the rotating disc [54].

3.4.3 Heterogeneous Catalytic Organic Reaction in the SDR: An Example of Application to the Pharmaceutical/Fine Chemicals Industry

Manufacturing in the pharmaceutical and fine/specialty chemicals sector has traditionally been associated with the accumulation of excessive quantities of hazardous waste, resulting not only from the large amounts of contaminated organic solvents [55–57] but also from widespread use of mineral acids and Lewis acids in stoichiometric proportions [58]. Low selectivity to the desired product, which translates into low atom efficiency, is another factor that can produce significant amounts of unwanted byproducts. In the past 2 decades serious efforts have been made in the search for heterogeneous catalysts that will not only enhance reaction rates and product selectivity but eliminate the problem of separating product from catalyst [46,59].

An important industrial example in the fine chemical industry typified by low-efficiency processing and considerable environmental impact is the rearrangement reaction of α-pinene oxide. The process can lead to the formation of more than 100 different products, depending on the catalyst employed and the reaction conditions [60]. The main products obtained from this reaction are shown in Scheme 3.3. One of these, campholenic aldehyde,

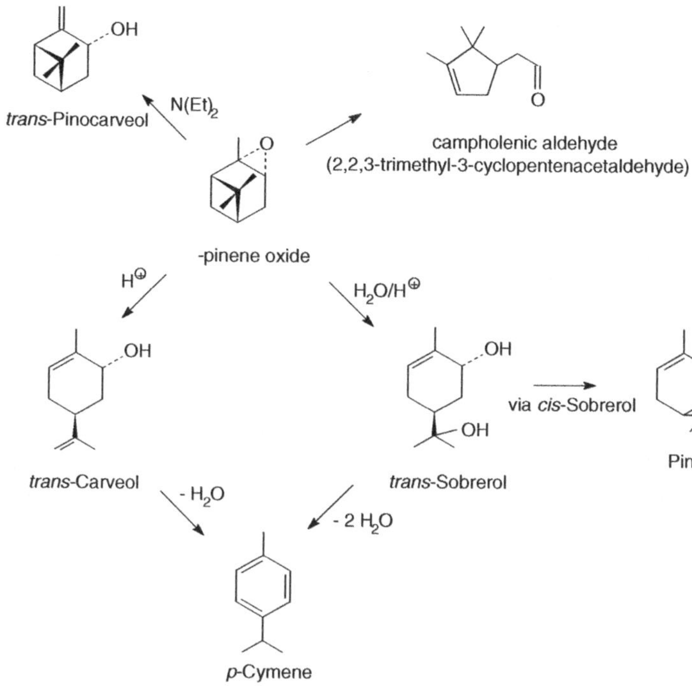

Scheme 3.3 *Main products of the rearrangement of α-pinene oxide [37].*

is of considerable importance as it is a key intermediate in the synthesis of santalol, the main constituent of natural sandalwood oil. Traditionally, campholenic aldehyde is produced on a commercial scale by the Lewis acid ($ZnCl_2$ or $ZnBr_2$) catalyzed isomerization of α-pinene oxide in a homogeneously catalyzed process in batch form [61]. Selectivities as high as 85% have been reported [60], but catalyst deactivation over time is a major issue. The environmental impact of recovering the homogenous catalyst after the process is another drawback of the current processing technique. One solution by which to address these problems is the development of heterogeneous Lewis acid catalysts such as silica-supported zinc triflates, as reported by Wilson *et al.* [62]. It has been observed that the high activity of the supported catalysts results in further reaction of the desired product, causing the selectivity to drop during the course of the reaction [62]. It is therefore very important not only to 'direct' the reaction towards the formation of campholenic aldehyde but to stop the reaction altogether at this point. This in itself is quite difficult to achieve in a batch reactor without adding any reagent to quench the reaction. An ideal processing technique would involve an intensified continuous reactor, especially one in which the residence time can be precisely controlled and direct contact of the reagent with the catalyst can be closely engineered through immobilization of the catalyst.

Vicevic *et al.* [37] investigated the scope for enhancing selectivity and conversion in the rearrangement of α-pinene oxide by employing a continuous flow SDR exhibiting the desirable characteristic of short residence times combined with zinc triflate grafted on silica supports of varied pore sizes and structures. The catalyst was immobilized on the

Table 3.6 *Optimal SDR conditions for each catalyst for maximum productivity of campholenic aldehyde. Reprinted from [ref 37] © 2007, with permission from Elsevier.*

Catalyst	Disc temperature (°C)	Disc speed (rpm)	Feed flow rate (cm^3/s)	Conversion (%)	Selectivity (%)
1[a]	85	1500	6	77	62
2[b]	85	850	4	59	83
3[c]	45	1500	6	85	75

[a]0.05 mmol/g $Zn(OTf)_2$ supported on silica K60
[b]0.01 mmol/g $Zn(OTf)_2$ supported on silica K100
[c]0.05 mmol/g $Zn(OTf)_2$ supported on HMS_{24}

surface of the disc; the optimized conversions and selectivities for each of the catalysts tested are shown in Table 3.6. It is interesting to note that the optimum selectivities were generally obtained at the highest disc speed (1500 rpm) and the highest flow rate (6 cm^3/s) tested. Such operating conditions gave rise not only to the shortest residence times but also to flow behaviour that most closely approaches plug flow. Catalyst 3 gave the best performance overall, with relatively high conversion and selectivity, which was attributed to the regularity and small pores of the support matrix of HMS_{24}.

The observations made on comparing processing in the SDR and a batch STR are summarized below:

- *Reaction Rate and Selectivity Enhancement*: Minimum residence times in the range of 0.2–1.0 s are required for complete consumption of α-pinene oxide (Figure 3.20a), as opposed to at least 5 minutes in the STR (Figure 3.21), giving a reaction rate that is doubled in the SDR after the higher catalyst concentration in the thin continuous film is accounted for. These comparative results are indicative of the more intense mixing in the SDR film, which helps to greatly overcome any mass-transfer limitations within the immobilized supporting matrix. Higher selectivities are generally attained at residence times below 0.3 s in the SDR (as shown in Figure 3.20b for Catalyst 1). These short and controllable disc residence times in the SDR promote the desired reaction while

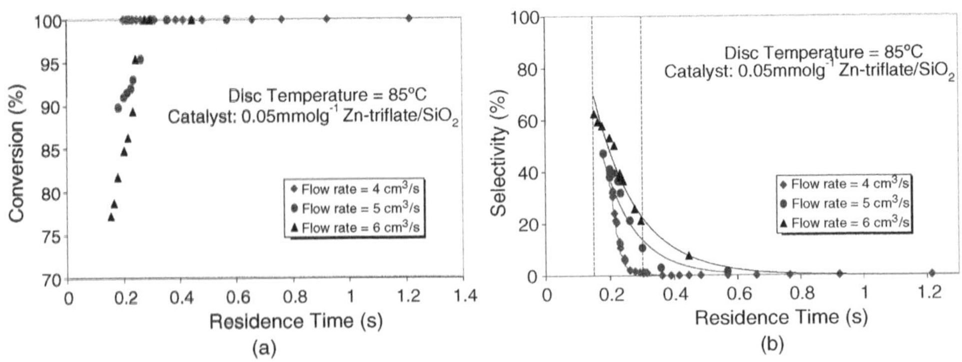

Figure 3.20 *Effect of disc residence time on (a) conversion and (b) selectivity for Catalyst 1. Reprinted from [ref 37] © 2007, with permission from Elsevier.*

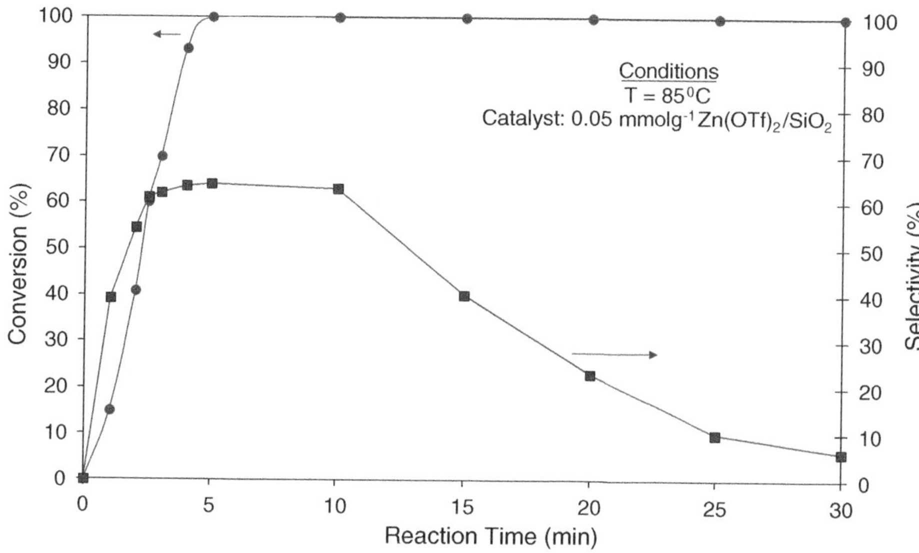

Figure 3.21 *Conversion and selectivity profiles for the catalytic rearrangement of α-pinene oxide in a batch reactor using Catalyst 1. Reprinted from [ref 37] © 2007, with permission from Elsevier.*

minimizing unwanted consecutive reactions of the formed campholenic aldehyde, enabling enhanced process selectivity.

- *Catalyst Activity*: Catalyst deactivation by product accumulation and blocking of active sites was encountered in batch processes, as reported by Wilson *et al.* [62]. Catalyst activity in the SDR, on the other hand, can be prolonged as a result of the high shear rates being able to shift product molecules away from active sites. This has been demonstrated by reuse of the catalyst over a number of runs, with no observable effect on conversion attained [37].
- *Catalyst Immobilization*: With the catalyst fixed on to the surface of the rotating disc, no downstream catalyst separation from the product mixture is required. Leaching of catalyst from the disc surface was not an issue, as confirmed by the observation that no further reaction took place in the collected samples during storage. Although physical immobilization may be possible in the batch reactor, it is deemed not to be advantageous to the process, as severe diffusion limitations may occur under the comparatively poor mixing conditions prevailing in the batch reaction environment.

Overall, the research highlights the exciting and realistic potential of combining green chemistry and PI technologies through the use of supported catalysts in an intensified reactor system to achieve waste minimization at source and minimal downstream processing for environmentally favourable processes.

3.4.4 Green Synthesis of Nanoparticles

Nanoparticles, defined as having diameters between 1 and 100 nm, have received much attention recently due to their superior physical and chemical properties, such as greatly improved mechanical strength of materials (e.g. in polymer nanocomposites, bone cement), thermal stability, catalytic activity and so on. To achieve nanosized crystals from precipitation, high supersaturation levels that are uniform throughout the precipitation system are key processing requirements for high nucleation rates and small particles. The SDR, by virtue of its rapid micromixing in the thin films, has the capability to meet these requirements. By providing the opportunity for a homogeneous reaction environment in terms of concentration, temperature and mass transfer, the SDR should not only enable small particle sizes to be formed but allow the size distribution to be tightly controlled. Several studies of precipitation processes in the SDR have indeed validated these expectations. One involved the liquid–liquid precipitation of barium sulfate crystals of 0.7 μm in size [42]; more recently, Tai and coworkers have demonstrated the synthesis of nanoparticles of magnesium hydroxide [63] and of silver [44]. In the latter work, silver nanoparticles less than 10 nm in diameter were produced by using a benign raw material such as glucose as the reducing agent and were stabilized on formation by polyvinyl pyrrolidone (PVP), commonly employed as a protecting agent to prevent particle agglomeration.

A greener method of silver nanoparticle synthesis involves the use of a more environmentally friendly protecting agent such as starch combined with glucose as the reducing agent [64]. Silver particles with a mean size between 13 and 16 nm were formed in the SDR after 10 minutes of processing, which was much less than the 20 hours needed in an STR for the same result [65].

Another 'green' approach adopted in the synthesis of superparamagnetic magnetite nanoparticles has been reported by Chin *et al.* [39]. Alginic acid, a natural biopolymer originating from algae, was employed as a surfactant to stabilize the Fe_3O_4 particles during processing in the SDR. The ability to inject the alginic acid into the magnetite particle suspension and achieve uniform mixing of the two streams on the rotating disc constituted a considerable processing advantage in producing highly stable particles of about 10 nm and of very narrow distributions.

3.5 Hurdles to Industry Implementation

In spite of the bountiful processing benefits offered by the SDR, as discussed in this chapter in the context of green chemical processing, industrial take up of this technology has been rather slow. This is quite surprising given the great deal of interest expressed in SDR by certain industrial sectors, with some of the multinational companies representing these sectors having even participated in evaluation projects that have produced extremely encouraging performance data, including GlaxoSmithKline, Rhodia and so on. To date, there are no applications of the SDR in the chemical and processing industry, to the best of the author's knowledge. There are clearly important technological barriers that need to be addressed in order to facilitate industrial implementation. These are reviewed briefly in this section, highlighting progress made to date towards tackling some of them.

3.5.1 Control, Monitoring and Modelling of SDR Processes

One of the most important development issues concerns the monitoring and control of processes in the SDR. In fact, the general lack of control-related studies, in particular, is a limitation that affects many intensification technologies currently under development, not just the SDR. While it is recognized that control of intensified processes, especially those in which fast equipment dynamics come into play, may present some specific challenges [66,67], precious little technical information addressing these aspects in the open literature is available with regards to SDRs.

In the context of SDR control, investigations into the selection and development of appropriate control algorithms suitable for controlling its fast dynamics are currently underway at Newcastle University. Simulation studies have so far shown that a PI controller applied to an example intensified system yields superior set-point tracking performance in terms of integral of absolute error (IAE) compared to that achieved in a conventional system [68]. However, it has been found that the controller performance in the intensified system suffers from high overshoots, indicating that the controller parameters require extensive tuning. Work is also ongoing to improve pH control in the SDR by incorporating the relatively novel control concept of 'disturbance observers'. A separate experimental and simulation study of the control of a precipitation reaction using the disc rotational speed as the control variable is also underway.

Online monitoring of the SDR process is highly desirable and will invariably require the development of miniaturized probes or sensors for measuring, for example, the temperature of the thin film and process parameters such as pH [67]. In order to track the progress of chemical reactions on the surface of the disc, the development of non-invasive monitoring systems would be advantageous. One such method is Raman spectroscopy, which has been applied recently to monitor in real time the polymerization of styrene, methyl methacrylate and n-butyl acrylate in the SDR. Promising spectral data relating to monomer concentration were captured at various radial positions on the disc during styrene polymerization, as shown in Figure 3.22. This study was undertaken in conjunction with Clairet Scientific Ltd, a spectroscopy instrument provider. More such collaborations between academia and instrument providers/manufacturers are strongly recommended

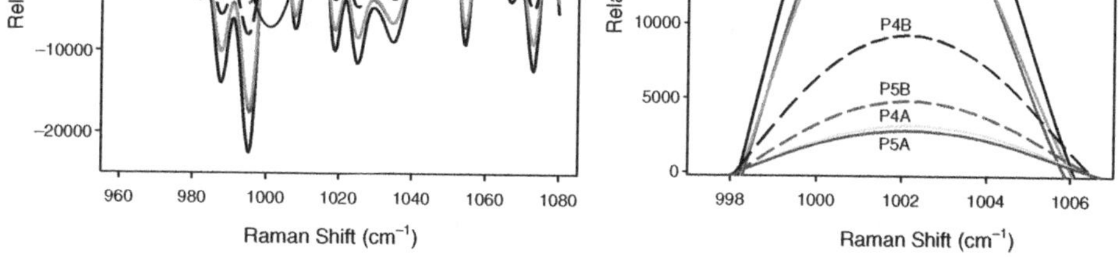

Figure 3.22 *Raman spectra for online monitoring of styrene polymerization in an SDR: (a) full processed spectra at various radial positions; (b) processed spectra focusing on a selected region between 997 and 1007 cm^{-1} (where 1, 2, 3, 4, 5 represent radial positions moving away from the centre of the disc, 1 being the closest to the centre).*

if significant progress towards developing and implementing online monitoring tools in challenging reactor environments such as miniaturized, low-volume processing systems is to be achieved.

Mechanistic modelling of SDR processes is quite challenging due to the inherently unstable nature of the films. Although some effort has been directed at developing models for mass transfer [21], polymerization [69] and photocatalytic reations [70], there is clearly great scope for further work in this area. In particular, more generally applied models should be developed for heat- and mass-transfer and chemical-reaction systems.

3.5.2 Limited Process Throughputs

Limitations on flow rate in an SDR are imposed in order to achieve the benefits of thin-film flow on a given disc size. Throughputs of 300 kg/hour on a 1 m diameter disc may be more than adequate for pharmaceutical and fine chemicals industries but they are not sufficient for the bulk chemicals/petrochemicals industry, where annual throughputs of the order of 100 000 tonnes are the norm. To address this limitation, multi-disc designs operating in parallel could allow high production capacities. Such systems have yet to be designed, built and tested.

3.5.3 Cost and Availability of Equipment

SDRs are still relatively expensive compared to conventional, established technologies such as stirred tank vessels. A small-scale reactor (10–15 cm diameter) complete with a temperature-controlled bath and feed pumps has been known to cost more than £150 000. More equipment manufacturers should be involved in developing this technology and making off-the-shelf units more readily available to industry users. Costs can also be expected to be significantly reduced if more reactors are supplied on the market.

3.5.4 Lack of Awareness of SDR Technology

There is a need for industrial chemists and development engineers to familiarize themselves with SDR and PI technologies in general and to be aware of the potential benefits of such technologies for their specific applications. This should help them match the capability of the technology in question with their process requirements and better select reactors for a given process. Short courses on SDR principles and potentials should be offered to people in industry across a wide range of sectors, in order to generate more widespread awareness of the opportunities presented by this technology.

3.6 Conclusion

The SDR represents a step change in reactor technology development. It offers a combination of highly desirable characteristics for a niche of applications involving fast, highly exothermic reactions. Some of its characteristics, such as extremely short and controllable residence times, are hard to replicate in any other technologies without incurring severe penalties, rendering the SDR unique in some aspects. The SDR holds much promise for highly efficient processing in many applications particularly relevant to the pharmaceutical and fine chemicals industry. Combining highly efficient continuous processing technologies, such as the SDR, and tools for achieving green chemistry, such as heterogeneous catalysts and environmentally benign reagents, can open up new avenues to attaining a truly sustainable process, as has been highlighted in this chapter. Moreover, the improved energy efficiency offered by intensified equipment such as the SDR provides a distinct advantage in the current bid to cut down on energy use, as elaborated on in Chapter 15.

There are nevertheless a number of important technical issues and challenges to be addressed before this emerging technology can find widespread commercial application. Resolving these issues will require a good deal of collaborative effort between many stakeholders in academia and industry, and could pave the way for greater uptake of SDRs in the chemical and processing industries in the next 10–15 years.

Nomenclature

a_{cor}	Coriolis acceleration (m/s^2)
Q	Volumetric flow rate (m^3/s)
r	Radial position from centre of disc (m)
S	Scaling factor (–)
t_{res}	Mean disc residence time (s)
u_m	Mean radial velocity (m/s)
v_r	Radial velocity (m/s)
z	Vertical distance from disc surface (m)

Greek Letters

δ	Film thickness (m)
$\dot{\gamma}$	Shear rate (s^{-1})
ν	Kinematic viscosity (m^2/s)
θ	Angular direction (–)
ρ	Liquid density (kg/m^3)
ω	Angular velocity (rad/s)

Subscripts

i	inner
o	outer

References

(1) C. G. Kirkbride, Heat transfer by condensing vapour on vertical tubes. Ind. Eng. Chem., 26(4), 425–428 (1934).

(2) R. E. Emmert and R. L. Pigford, A study of gas absorption in falling liquid films. Chem. Eng. Prog., 50(2), 87–93 (1964).

(3) G. D. Fulford, The flow of liquids in thin films, in Advances in Chemical Engineering, vol. 5, Academic Press: New York, pp. 151–236 (1964).

(4) M. M. Rahman, A. Faghri, W. L. Hankey and T. D. Swanson, Prediction of heat transfer to a thin liquid film in plane and radially spreading flows. J. Heat Transfer, 112(3), 822–825 (1990).
(5) M. M. Rahman and A. Faghri, Gas absorption and solid dissolution in a thin liquid film on a rotating disk. International Journal of Heat and Mass Transfer, 36(1), 189–199 (1993).
(6) C. Ramshaw and S. Cook, Spinning around. The Chemical Engineer, 774(5), 42–44 (2005).
(7) P. Oxley, C. Brechtelsbauer, F. Ricard, N. Lewis and C. Ramshaw, Evaluation of spinning disc reactor technology for the manufacture of pharmaceuticals. Ind. Eng. Chem. Res., 39, 2175–2182 (2000).
(8) K. V. K. Boodhoo and R. J. Jachuck, Process intensification: spinning disc reactor for condensation polymerisation, Green Chem., 2, 235–244 (2000).
(9) K. V. K. Boodhoo, W. A. E. Dunk, M. Vicevic, R. J. Jachuck, V. Sage, D. J. Macquarrie and J. H. Clark, Classical cationic polymerisation of styrene in a spinning disc reactor using silica-supported BF_3 catalyst. J. App. Pol. Sci., 101(1), 8–19 (2006).
(10) A.G. Emslie *et al.* Flow of a viscous liquid on a rotating disk. J. Appl. Phys., 29(5), 858–862 (1958).
(11) S. T. Lim, Hydrodynamics and mass transfer processes associated with the absorption of oxygen in liquid films flowing across a rotating disc. PhD Thesis, Newcastle University (1980).
(12) E. Momoniat and D. P. Mason, Investigation of the effect of the Coriolis force on a thin fluid film on a rotating disk. Int. J. Non-linear Mechanics, 33(6), 1069–1088 (1998).
(13) T. G. Myers and M. Lombe, The importance of the Coriolis force on axisymmetric horizontal rotating thin film flows. Chem. Eng. Process., 45, 90–98 (2006).
(14) A. Acrivos, M. J. Shah and E. E. Petersenet, On the flow of a non-Newtonian liquid on a rotating disk. J. Appl. Phys., 31(6), 963–968 (1960).
(15) N. K. Zinnatullin *et al.*, The two-dimensional flow of a non-Newtonian fluid over the open surface of a rapidly rotating flat disc. J. Eng. Physics, 15, 701–704 (1968).
(16) R. J. J. Jachuck and J. R. Burns, Spinning disk reactor, in Encyclopedia of Chemical Processing, S. Lee (Ed.), vol. 5, Taylor and Francis: New York, pp. 2847–2858 (2006).
(17) D. Ghiasy, K. V. K. Boodhoo and M. T. Tham, Thermographic analysis of thin liquid films on a rotating disc: approach and challenges. Appl. Therm. Eng., 44, 39–49 (2012).
(18) W. Woods, The hydrodynamics of thin liquid films flowing over a rotating disc. PhD Thesis, Newcastle University (1995).
(19) A. Aoune and C. Ramshaw, Process intensification: heat and mass transfer characteristics of liquid films on rotating discs. Int. J. Heat Mass Tran., 42, 2543–2556 (1999).
(20) R. J. J. Jachuck and C. Ramshaw, Process intensification – heat transfer characteristics of tailored rotating surfaces. Heat Recovery Systems & CHP, 14, 475–491 (1994).
(21) O. K. Matar, G. M. Sisoev and C. J. Lawrence, Evolution scales for wave regimes in liquid film flow over a spinning disc. Phys. Fluids, 16, 1532–1545 (2004).
(22) O. K. Matar, C. J. Lawrence and G. M. Sisoev, The flow of a thin film over a spinning disc: hydrodynamics and mass transfer. Phys. Fluids, 17, 52–102 (2005).
(23) G. M. Sisoev, O. K. Matar and C. J. Lawrence, Modelling of film flow over a spinning disk. J. Chem. Technol. Biot., 78, 151–155 (2003).
(24) G. M. Sisoev, O. K. Matar and C. J. Lawrence, Axisymmetric wave regimes in viscous liquid film flow over a spinning disc. J. Fluid Mech., 495, 385–411 (2003).
(25) J. R. Burns and R. J. J. Jachuck, Monitoring of $CaCO_3$ production on a spinning disc reactor using conductivity measurements. AIChE J., 51, 1497–1507 (2005).
(26) S. Yanniotis and D. Kolokotsa, Experimental study of water vapour condensation on a rotating disc. Int. Commun. Heat Mass, 23(5), 721–729 (1996).
(27) S. Yanniotis and D. Kolokotsa, Boiling on the surface of a rotating disc. Journal of Food Engineering, 30(3/4) 313–325 (1996).

(28) J. R. Burns and I. Henderson, Spinning disc with spiral passage way for heat transfer fluid. GB Patent 2416316 (2006).

(29) K. V. K. Boodhoo and S. Al-Hengari, Micromixing characteristics in a small scale spinning disk reactor. Chem. Eng. Technol., 35(7), 1229–1237 (2012).

(30) P. Guichardon and L. Falk, Characterisation of micromixing efficiency by the iodide-iodate reaction system. Part I: experimental procedure. Chem. Eng. Sci., 55(19), 4233–4243 (2000).

(31) Y.-S. Chen, H.S. Liu, C.C. Lin and W.T. Liu, Micromixing in a rotating packed bed, J. Chem. Eng. Jpn, 37(9), 1122–1128 (2004).

(32) Y. S. Chen, C. Y. D. Tai, M. H. Chang and H. S. Liu, Characteristics of micromixing in a rotating packed bed. J. Chin. Inst. Chem. Engrs., 37(1), 63–69 (2006).

(33) G.-W. Chu, Y. H. Song, H. J. Yang, J. M. Chen, H. Chen and J.F. Chen, Micromixing efficiency of a novel rotor–stator reactor. Chem. Eng. J., 128(2–3), 191–196 (2007).

(34) S. Al-Hengari, Micromixing and residence distribution characteristics in a spinning disc reactor. PhD Thesis, Newcastle University (2012).

(35) M. Akhtar, I. Blakemore, G. Clayton and S. Knapper, The use of spinning disc reactor for processing ice cream base – effect of ageing in making model ice cream. Int. J. Food Sci. Tech., 44, 1139–1145 (2009).

(36) M. Akhtar, P. Chan, N. Safriani, B. Murray and G. Clayton, Concentration of apple juice using spinning disc reactor technology. J Food Process. Technol., 2(2), 1000108 (2011).

(37) M. Vicevic, K. V. K. Boodhoo and K. Scott, Catalytic isomerisation of alpha-pinene oxide to campholenic aldehyde using silica supported zinc triflate catalysts: II. performance of immobilised catalysts in a continuous spinning disc reactor. Chem. Eng. J., 133(1–3), 43–57 (2007).

(38) S. Mohammadi and K. V. K. Boodhoo, Online conductivity measurement of residence time distribution of thin film flow in the spinning disc reactor. Chem. Eng. J., 207–208, 885–894 (2012).

(39) S. F. Chin, K. S. Iyer, C. L. Raston and M. Saunders, Size selective synthesis of superparamagnetic nanoparticles in thin fluids under continuous flow conditions. Adv. Funct. Mater., 18, 922–927 (2008).

(40) K. V. K. Boodhoo, W. A. E. Dunk and R. J. Jachuck, Continuous photo-polymerisation in a novel thin film spinning disc reactor, in Photoinitiated Polymerisation, K. Belfield and J. Crivello (Eds.), ACS Symp. Ser. 847, American Chemical Society: Washington DC (2003).

(41) K. V. K. Boodhoo and R. J. Jachuck, Process intensification: spinning disc reactor for styrene polymerization. Appl. Therm. Eng., 20, 1127–1146 (2000).

(42) L. M. Cafiero, G. Baffi, A. Chianese and R. J. Jachuck, Process intensification: precipitation of barium sulphate using the spinning disc reactor. Ind. Eng. Chem. Res., 41(21), 5240–5246 (2002).

(43) N. Anantachoke, M. Makha, C. L. Raston, V. Reutrakul, N. C. Smith and M. Saunders, Fine tuning the production of nanosized β-carotene particles using spinning disc processing. J. Am. Chem. Soc., 128(42), 13847 (2006).

(44) C. Y. Tai, Y. H. Wang, Y. W. Kuo, M. H. Chang and H. S. Liu, Synthesis of silver particles below 10 nm using spinning disk reactor. Chem. Eng. Sci., 64, 3112–3119 (2009).

(45) K. Matyjaszewski, in Cationic Polymerization: Fundamentals and Applications, R. Faust and T.D. Shaffer (Eds.), ACS Symp. Ser. No. 665, American Chemical Society: Washington, DC (1997).

(46) J. H. Clark, Solid acid catalysts for green chemistry. Acc. Chem. Res., 35, 791–797 (2002).

(47) K. Wilson and J. H. Clark, Synthesis of a novel supported solid acid BF_3 catalyst. Chem. Commun., 2135–2136 (1998).

(48) A. Ledwith, Photoinitiation of polymerization. Pure & Appl. Chem., 49, 431–441 (1977).

(49) J. Boutin and J. Neel, German Patent 2,716,606 (1977).

(50) P.J. Arndt *et al.*, German Patent 3,208,369 (1983).

(51) K. V. K. Boodhoo, W. A. E. Dunk and R. J. Jachuck, Influence of centrifugal field on polymerisation kinetics. J. Appl. Polym. Sci., 85(11), 2283–2286 (2002).
(52) M. P. Stevens, Polymer Chemistry: An Introduction, Oxford University Press: Oxford (1999).
(53) K. V. K. Boodhoo, W. A. E. Dunk, M. S. Jassim and R. J. Jachuck, Thin film solvent free photopolymerisation of n-butyl acrylate. I. Static film studies. J. Appl. Polym. Sci., 91, 2079–2095 (2004).
(54) M. R. Moghbeli, S. Mohammadi and S. M. Alavi, Bulk free-radical polymerization of styrene on a spinning disc reactor. J. Appl. Polym. Sci., 113, 709–715 (2009).
(55) R. A. Sheldon, Organic synthesis – past, present and future. Chem. Ind., 23, 903–906 (1992).
(56) R. A. Sheldon, Consider the environmental quotient. Chem. Tech., 24, 38–47 (1994).
(57) D. J. C. Constable, C. Jimenez-Gonzalez and R. K. Henderson, Perspective on solvent use in the pharmaceutical industry. Org, Proc. Res. Dev., 11, 133–137 (2007).
(58) R. A. Sheldon, The E Factor: fifteen years on. Green Chem., 9, 1273–1283 (2007).
(59) R. A. Sheldon and H.van Bekkum, Fine Chemicals through Heterogeneous Catalysis, Wiley-VCH: Weinheim (2001).
(60) W. F. Holderich, J. Roseler, G. Heitmann and A. T. Liebens, The use of zeolites in the synthesis of fine and intermediate chemicals. Catal. Today, 37(4), 353 (1997).
(61) J. B. Lewis and G. W. Hedrick, Reaction of alpha-pinene oxide with zinc bromide and rearrangement of 2,2,3-trimethyl-3-cyclopentene products derived therefrom. J. Org. Chem., 30, 4271 (1965).
(62) K. Wilson, A. Renson and J. H. Clark, Novel heterogeneous zinc triflate catalysts for the rearrangement of alpha-pinene oxide. Catal. Lett., 61, 51–55 (1999).
(63) C. Y. Tai, C. T. Tai, M. H. Chang and H. S. Liu, Synthesis of magnesium hydroxide and oxide nanoparticles using a spinning disk reactor. Ind. Eng. Chem. Res., 46, 5536–5541 (2007).
(64) C. Y. Tai, Y. H. Wang and H. S. Liu, A green process for preparing silver nanoparticles using spinning disc reactor. AIChE J., 54, 445–452 (2008).
(65) P. Raveendran, J. Fu and S. L. Wallen, Completely 'green' synthesis and stabilization of metal nanoparticles. J. Am. Chem. Soc., 125, 13 940–13 941 (2003).
(66) N. M. Nikacevic, A. E. M. Huesman, P. M. J.Van den Hof and A. I. Stankiewicz, Opportunities and challenges for process control in process intensification. Chem. Eng. Process., 52, 1–15 (2012).
(67) R. Barzin, S. R. Abd Shukor and A. L. Ahmad, New spectrophotometric measurement method for process control of miniaturized intensified systems. Sensors and Actuators B- Chemical, 146, 403–409 (2010).
(68) D. Ghiasy, K. V. K. Boodhoo and M. T. Tham, Control of intensified equipment: a simulation study for pH control in a spinning disc reactor. Chem. Eng. Process., 55, 1–7 (2012).
(69) M. Vicevic, K. Novakovic, K. V. K. Boodhoo and A. J. Morris, Kinetics of styrene free radical polymerisation in the spinning disc reactor. Chem. Eng. J., 135(1–2), 78–82 (2008).
(70) I. Boiarkinaa, S. Pedron and D. A. Patterson, An experimental and modelling investigation of the effect of the flow regime on the photocatalytic degradation of methylene blue on a thin film coated ultraviolet irradiated spinning disc reactor. Appl. Catal. B-Environ., 110, 14–24 (2011).

4

Micro Process Technology and Novel Process Windows – Three Intensification Fields

Svetlana Borukhova and Volker Hessel

Department of Chemical Engineering and Chemistry, Micro Flow Chemistry & Process Technology, Eindhoven University of Technology, Eindhoven, The Netherlands

4.1 Introduction

Micro process technology offers a great potential platform for flow operation that embraces principles of green chemistry and process intensification (PI) [1]. Pioneering devices such as micropumps, micromixers, microreactors, micro heat exchangers, microseparators and microanalysers are discussed in Gravesen *et al.* [2], Laser and Santiago [3] and Wiles and Watts [4]. Microchannels of an internal diameter in the range 10–500 μm minimize transport limitations due to the increased exchange area when dimensions are minimized. Minimization of transport limitations results in high diffusion and convection rates, thus enhancing mass transfer through mixing and heat transfer through the high heat-exchange capacity [5]. Small scale results in a lower consumption of chemicals and energy, higher safety and lower waste generation, obeying one of the green chemistry principles [6,7]. High surface-to-volume ratio due to the small dimensions offers a great control over the processes taking place within the system. Accurate temperature control via provision of the exact amount of energy needed maintains isothermal conditions within the reactor. Thus, side reactions are minimized, resulting in a higher selectivity in case of competing reaction

Process Intensification for Green Chemistry: Engineering Solutions for Sustainable Chemical Processing, First Edition. Edited by Kamelia Boodhoo and Adam Harvey.

pathways [8]. Improved product selectivity requires less consumption of raw materials and recycling of unconverted materials, reducing the operating costs. Enhanced heat transfer increases the control over highly exothermic reactions and prevents thermal runaways [9, 10]. Apart from providing an excellent heat transfer, high surface-to volume-ratio stabilizes generation rate of unstable species within microreactors. Microchannel walls terminate radicals upon their contact, thus stabilizing their propagation rate [11]. In addition, inherent safety and modularity of microsystems provide an opportunity for synthesizing chemicals on demand at the point of use [12], eliminating the need for storage or transport of dangerous or toxic chemical species such as cyanides, peroxides and azides [13–16]. Process simplification is achieved through fast and simultaneous processes combined with integration of unit operations, and results in a higher yield and better energy utilization. Scale-up of micro processes is realized by increasing the number of parallel operating units in x- and y-directions [17]. Stacking items of equipment on top of each other rather than enlarging the capacity through dimension enlargement guarantees the same optimal operation achieved during the development of the process at the lab scale. This eliminates the risk of drastic changes upon scaling up due to the fact that the physics and channel-flow hydrodynamics stay the same [9]. Thus, transition period from lab to industry is shortened and does not entail any additional safety permissions, as long as the one for lab scale is provided.

Microreactors are particularly useful for PI for the following reasons [18]:

- Pressure-resistant microstructures allow the application of reaction temperatures above the normal boiling points of solvents, enabling them to be used at harsher than reflux conditions.
- By connecting several microstructured flow reactors in bilateral fashion, a continuous multistep process can be realized. Reactants can then be added from several inlets at one or many positions, depending on the speed of previous reaction.
- Photochemical reactions can be realized much more efficiently (alternative catalytic activation).
- Toxic, hazardous or highly labile species can be utilized or generate *in situ.*

Microflow chemistry offers continuous operation in micro- or miliscale tubing, with a relatively high throughput comparable to industrial productivity [19]. Rethinking of synthetic approaches is necessary in order to adapt batch chemistry to flow. Combinatorial synthesis and screening of small molecules and systems of nucleic acids are frequently employed in pharmaceutical drug discoveries [20,21].

Microflow systems allow parallel synthesis of almost 100 different molecules, shortening the time required for an innovation and minimizing the amounts used in the syntheses [22]. A new perspective on continuous microflow reactors was recently presented, following the examples of the organic synthesis by Wiles *et al.* [23].

Hessel stated that novel process windows aim at the intensification of intrinsic reaction kinetics performed at the microscale under harsh process conditions and continuous flow operation, as well as at the exploitation of synergetic effect in the integrated systems [24]. Synergy of microdimensions and harsh operating conditions results in a three-step intensification within a reaction system: these are transport, chemical and process design intensification. The latter two constitute novel process windows. Efficient mixing and accurate control over residence time and heat transfer, accompanied by 'right' handling of chemicals within microchannels, have already been demonstrated by numerous successful research groups [25–28]. Process design intensification is new in the micro world, but is quite well known at larger scales. Examples are the integration of membranes, extraction in a reactor and much-simplified separation processes due to an improved reaction [29–32].

In this chapter, we will first discuss fundamental principles of transport intensification, covering the classification of micromixer designs and available heat-exchange units. We then look at chemical intensification, with an investigation of the opportunities offered by novel process windows in terms of new chemical transformations and the routes taken at high temperatures and pressures, at increased concentrations or under neat conditions, at explosive or runaway regimes and simplified through integration of synthetic transformations. The novel concept of process design intensification will be discussed, examining its prospect for microflow operations, partly mimicking integration from large-scale equipment, partly developing our own proprietary solutions. Taken together, we present the fundamentals along with the opportunities provided by synchronized applications of novel process windows and microprocess technology. This leads to a novel classification concerning PI with microreactors.

While the chemical intensification presented here may be one foundation of green chemistry, process design intensification is a pillar of green engineering. It promotes the interdisciplinary integration of chemistry, micro process technology, process engineering and process control. The decreased number of apparatuses needed and thus its decreased size results in a smaller footprint, which when combined with the decreased number of process development steps constitutes a concept of process design intensification. In the following sections, the three intensification fields are introduced, along with their roles and strengths in PI.

4.2 Transport Intensification

4.2.1 Fundamentals

A molecule entering any reaction medium has to be activated through energy gain and meet with a catalyst or another molecule of sufficient energy to react. As Górak and Stankiewicz [33] stated:

'Process intensification follows four main goals, namely

i. to maximize the effectiveness of intra- and intermolecular events,
ii. to give each molecule the same processing experience,
iii. to optimize the driving forces/maximize specific interfacial areas, and
iv. to maximize the synergistic effects of partial processes.'

(ii) and (iii) refer to transport intensification, as defined here, (i) and (iv) to chemical and process design intensification, respectively, in the context of this review.

Consequently, the values of reaction rate, conversion and selectivity are highly dependent on the reaction environment provided. Micro process technology provides several advantages when compared to equipment of a larger scale [24]:

- *Flow Intensification:* At microscale, the flow is laminar, resulting in a parabolic profile flow or quasi-plug flow in the case of large radial diffusion, eliminating dead zones, hot-spot formation and axial dispersion (narrow residence time distribution, RTD). Multiscale flows such as gas/liquid or liquid/liquid are highly periodic, with characteristic patterns and defined interfaces.

- *Mass-transfer Intensification:* Shortened diffusion path and higher exchange area accelerate the mass transfer in flows and enhance mixing, providing more homogeneous reactant distribution and thereby minimizing byproduct formation in consecutive reactions.
- *Heat-transfer Intensification:* Decreased dimensions increase heat-transfer coefficient and temperature gradient, in addition to surface area, thus increasing total heat transfer.

In recent years, significant interest has been given to the development of optimal micromixers. Extensive elaboration on the performances of various types of micromixer is available in recently published reviews [34]. Micromixers can provide complete mixing in milliseconds, or sometimes even in as little as 1 microsecond [35]. When integrated into the microchannel, they result in a flow with narrow RTD [36–39].

4.2.2 Mixing Principles

Transport phenomena can be described on two levels: molecular and continuum. The discrimination comes from the characteristic size of the system: for molecular it ranges within the nanometer scale, while for continuum it can range across micro- and centimetres [40]. A fluid within a microchannel is considered to be a continuum [41]. At the continuum level, transport phenomena of a given fluid can be described by a set of conservation equations for mass, momentum and energy, while viscosity, density and conductivity are considered as material properties [40]. Fluid properties are considered to be identical throughout the microspace.

In general, the hydrodynamics of fluid flow within any confined space can be described by either laminar or turbulent flows. Laminar flow is a constant unidirectional flow of fluid elements moving with a constant velocity at any location with respect to the stream axis. Viscous forces dominate over inertial forces, resulting in a very well-defined parabolic flow – unless there is no counteraction taken, such as having mixing elements all along the microchannel, which promotes axial diffusion. Thus, the convective mass transfer takes place along the direction of the fluid only, while mixing of the species on the molecular level can only be achieved through diffusion [42]. Turbulent flow is governed by inertial forces, which decrease the fluid elements through vorticity and eddy formation, where engulfment and deformation result in smaller fluid elements, and thus an increased interfacial area and shortened mixing path. The deformation involves shear and elongation, which dissipates energy irreversibly to heat and enhances the diffusion [43]. Thus, turbulent flow relies on both convective and diffusive mixing [44,45].

The ratio of inertial to viscous forces is defined as the Reynolds number, with values below 2000 associated with laminar flow [46]. The Reynolds number can be expressed through:

$$Re = \frac{uD_h}{\nu} \tag{4.1}$$

Due to small dimensions of microchannels, Re values are particularly low, implying laminar flow. Hence, molar diffusive flux dominates the mass transport within the microchannels. Molar diffusive flux is based on Fick's law and is a function of the diffusion coefficient of a certain molecule, interfacial area and concentration gradient:

$$D = D_{diff} A \nabla C \tag{4.2}$$

While the diffusion mass coefficient can be maximized through temperature, both the interfacial area and the concentration gradient can be enhanced by narrowing the size of the fluid elements. Segregation is then decreased, while molecular diffusion rate is increased. Thus, there are two transport modes contributing to mixing in a laminar flow regime, relying on two principles [47]:

1. *Convection:* Transport of fluid elements within the microflow due to deformation with elongation of the fluid elements, which increases their length and decreases their width, yielding higher interfacial area.
2. *Diffusion:* Transport of species on a molecular level among fluid elements via splitting of fluid streams into narrower segments and redistribution within the mixing unit.

Mixing time ranges with respect to flow rate, channel geometry and mixing length. Diffusion time decreases quadratically with decreasing size of fluid aggregates. A typical dimensional size for the feed channels is 10–50 μm, and 500–5000 μm for mixing chamber channels in multilamination mixers, most often used for diffusion mixing [48,49]. Micromixers can be compared and characterized by three dimensionless fluid parameters: Reynolds number, Re, Peclet number, Pe, and Strouhal number, St.

The Peclet number is defined as:

$$Pe = \frac{uL}{D} \tag{4.3}$$

and indicates the prevailing mass transport means, to be performed by either convection or diffusion.

The Strouhal number is defined as:

$$St = \frac{fD_h}{u} \tag{4.4}$$

and represents the ratio between the residence time and the time period of disturbance. It is usually used to describe mixing in active micromixers, where f is the frequency of disturbance action that facilitates the mixing [50–52].

Convective transport occurs in the intermediate Re number regime, starting at about Re > 10 and being most often fully developed at about Re > 100. Convection stands for the transport of molecules with flow energy. It can be created within laminar flow when a certain disturbance is applied. Secondary flow – such as that formed in helical microchannels – can be described by the Dean number:

$$De = \mathrm{Re}\sqrt{\frac{a}{R}} \tag{4.5}$$

There exists a critical Dean number $De_{cr} = 150$ where the secondary flow pattern changes [53]. For $De < 150$, there is only a pair of counter-rotating vortices. At $De > 150$, the centrifugal force is dominant, leading to the formation of two additional vortices at the outer channel wall.

The most recent micromixer designs and corresponding experimental results for single-phase fluid flow applications have been presented and compared by Kumar *et al.* [34]. The sequential presentation of micromixers here is the same as that followed by Kumar *et al.* and Capretto *et al.* [34,54]. The fundamentals and logic behind most of the micromixer designs were successfully brought down to engineering and physics basics by Nguyen [40]. Furthermore, the mathematical description is accompanied by an explanation of the importance and nature of the molecular diffusion, Taylor dispersion, chaotic advection and viscoelastic, electrokinetic, magnetic and electromagnetic effects. The variety of available designs and their fabrications, characterizations and applications have been reviewed by Hessel *et al.* [51]. Particular focus is given here to scale-up and practical applications.

4.2.3 Micromixers

Micromixers can be divided into two categories, depending on the means for the facilitation of mixing: passive and active. Passive micromixers facilitate mixing by utilizing the flow energy to promote molecular diffusion and chaotic advection via either special channel geometry or the introduction of a secondary flow [55]. This brings about a requirement for efficient mixing, involving short diffusion paths and flow fragmentation (stretching, folding, etc.). Passive mixers do not need an additional energy source other than that provided from the pump and are thus easier to fabricate than active mixers. Active micromixers, on the other hand, demand an external disturbance in addition to the flow energy, such as electrical, pumping or ultrasound energy. The biggest advantage of active micromixers is that they can provide complete mixing at extremely low flow rates and low Re. This is more typical for lab-chip systems, which also often have stop-flow motion modes, than for microstructured reactors. However, the need for the integration of an external power source into the actively mixed microdevice, as well as for sometimes complex and costly fabrication processes, is the biggest disadvantage of their employment.

4.2.3.1 Lamination-based Micromixers

Lamination-based micromixers rely purely on diffusion in promoting mixing. Thus, the aim is to maximize the diffusive flux by maximizing the interfacial area and concentration gradient.

T- and Y-shaped Micromixers. The simplest of all the mixers with respect to fabrication and design is the T- or Y-shaped mixer [56–59]. Two fluids flow through two inlets to merge in the common mixing channel, increasing the mixing region at the interface by 20–30% [40].

Mixing after merging of the two streams depends solely on the diffusion of the species at the interface of the two liquids. Even then, the interface provided is usually not sufficient for rapid mixing, so long mixing channels are required. However, obstacles and roughening of the walls can be implemented within the channels in order to induce vortices, enhancing convection and thus mixing [46,57,60]. Another even simpler method is to largely increase

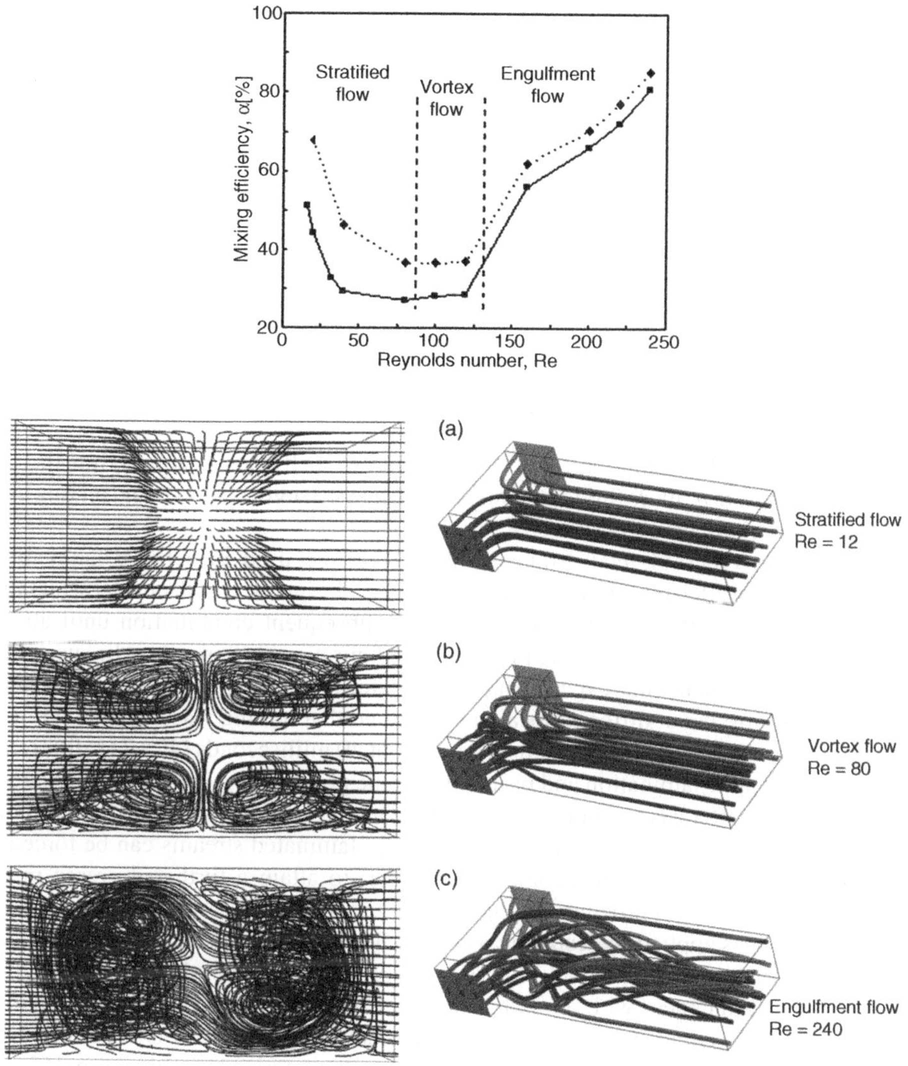

Figure 4.1 *(a) Mixing efficiency ranging with Re calculated for liquid mixing (square legends $D = 2 \times 10^{-7}\,m^2/s$; diamond legends $D = 1 \times 10^{-6}\,m^2/s$). (b) Streamlines at the entrance of the mixing channel (a–c) corresponding to stratified (Re = 12), vortex (Re = 80) and engulfment flow (Re = 240), respectively (courtesy of Elsevier B.V.) [59].*

the flow rates, which induces vortex formation. Figure 4.1a demonstrates the numerical results of mixing efficiency with increasing Re. Two streams get entangled in the so-called intertwining flow regime at high Re (see Figure 4.1b) [59].

Parallel Lamination. Further decrease in the diffusion length and increase in interfacial area can be achieved when incoming streams are split into two for *n* number of times and then

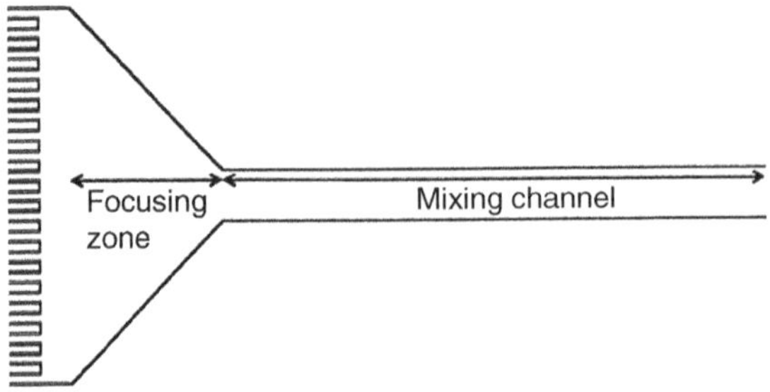

Figure 4.2 *Interdigital micromixer with triangular focusing zone. Reproduced with kind permission of Springer-Verlag from [62].*

reunited to form a multilamellae stream. The efficiency of such a design comes from the fact that splitting results in n^2 more fluid elements; thus diffusion length is shorter and mixing time decreases by the same factor [47].

There are two types of feed arrangement that can be used to achieve lamination: bifurcation and parallel interdigital. Bifurcation-type design is based on the sequential splitting of the fluid streams introduced and their subsequent combination until all the segments are collected in one channel [47]. This configuration was proven to be successful when fluorescence quenching experiments were performed, and quantitative information on the course of mixing at different locations indicated that mixing was completed in less than 100 milliseconds, while 95% mixing needs about 40 milliseconds [47]. The difference between parallel-flow lamination and bifurcation is the pressure loss-triggered distribution of the feed; in the latter case, symmetry is exploited and pressure loss results only from the liquid flowing in the microchannel. After contacting, laminated streams can be forced to pass through a narrow channel (geometric focusing) to attain a further decrease in the diffusion path and facilitate better mixing (see Figure 4.2).

This concept was applied in designing a special liquid-focusing interdigital mixer with multiple parallel feed flows, called SuperFocus (Figure 4.3). This device resulted in a lamellae of 4 μm thickness with a mixing time of 4 milliseconds at 95% efficiency [63].

A widely used feeding concept in parallel lamination-type mixers is a parallel-flow interdigitated feed. This comprises a feeding structure characterized by either a co- [63–67] or a counterflow [68–73] interdigitated array of microchannels. Lamination (combined with convection) can also be carried out in circular fashion, in coined cyclone micromixers [74]. Mixing takes place when the rotating fluid stream is injected into the circular mixing chamber in a quasi-tangential manner, inducing vortex formation. Mixing is efficient at both high (Re = 150) [72] and low (Re = 4) Reynolds numbers [71,72]. Further examples of similar mixer designs can be found elsewhere [75,76].

Hydrofocusing. As the name suggests, the main principle used here is to focus one stream between two others within a three-inlet channel flow configuration [77,78]. The decrease in stream width can be achieved to a widely variable extent via manipulation of the flow rates of incoming streams. A five-inlet port mixer, shown in Figure 4.4, was designed by Park *et al.* [79], in which two diagonal flows create a barrier between the streams flowing from the centre

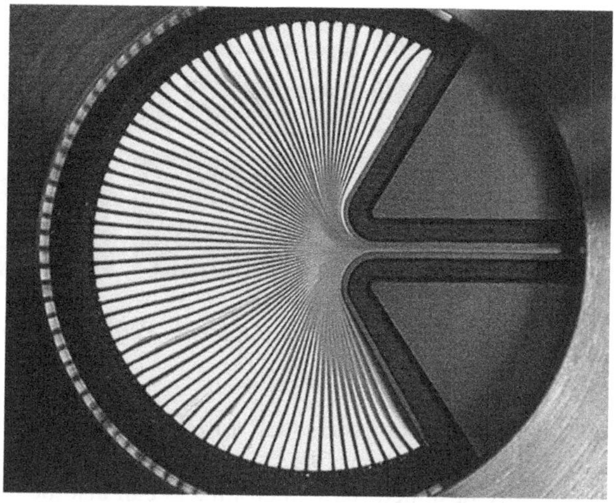

Figure 4.3 SuperFocus interdigital mixer, containing 138 microchannels made of stainless steel (flow: 350 l/hour at 3.5 bar). Reproduced with kind permission of Wiley-VCH from [63].

and the two side channels. Three-dimensional focusing is achieved when one incoming stream is vertically focused by two others [80–85].

4.2.3.2 Split-and-Recombine Micromixers

In the creeping laminar flow regime ($Re \leq 10$), split-and-recombine (SAR) micromixers rely on the same principle as do multilamination mixers, yet they differ in the fact that they incorporate flow splitting, recombination and rearrangement (Figure 4.5). Using

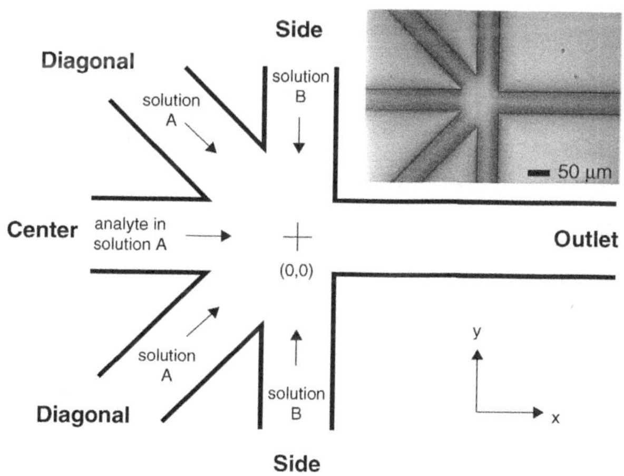

Figure 4.4 Five-inlet port micromixer. Reproduced with kind permission of ACS 2006 from [79].

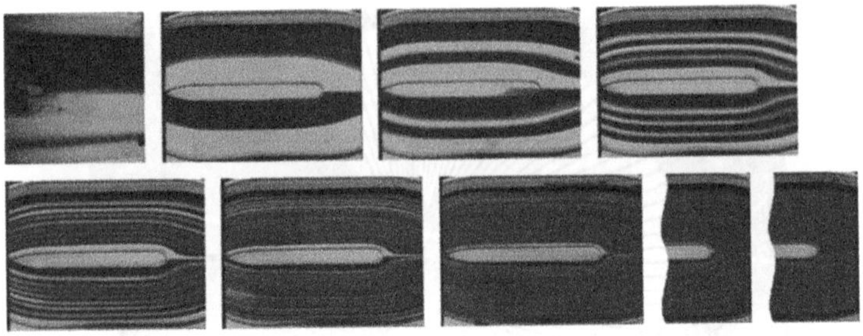

Figure 4.5 Cross-section of laminated flow in an SAR mixer. Permission granted by RSC, [87].

microchannel sizes typical for micromixers, such a flow regime is only used for higher-viscosity liquids such as glycerol solutions (Re $\leq$ 10) and not for water or organic solvents. Above this regime, specially designed SAR micromixers can also be operated at a wide range of Re by inducing chaotic advection within the channels. The main disadvantage of the SAR mixer is the often-complex fabrication required to generate the necessary 3D internal structure. The design introduced by Lim *et al.* consists of a 3D micromixer, named the crossing manifold micromixer, in which almost complete mixing is achieved in 250 μm [86].

The flow rearrangement pattern can be induced via incorporation of intersecting microchannels within the main stream. The design of such a mixer (Figure 4.6) was undertaken by He *et al.*, and complete mixing was achieved in 0.4 seconds [55]. In a design introduced by Sudarsan *et al.*, centrifugal force can ensure splitting and recombining of the flow [88]. This first allows vortices to be generated within a stream along the vertical plane, then splits them into multiple substreams and later recombines them, yielding a multi-compartment arrangement.

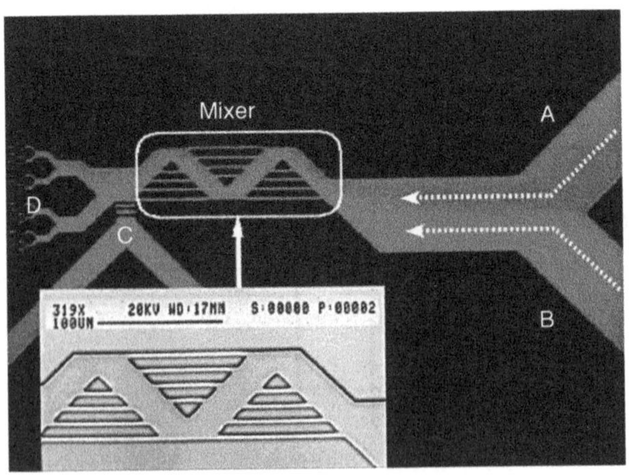

Figure 4.6 Planar SAR micromixer design based on constantly intersecting changing flow patterns. Courtesy of ACS, [55].

4.2.3.3 Chaotic Advection-based Micromixers

Manipulation of streamlines within a flow in a microchannel, as already discussed, can enhance mixing. In such cases, diffusion dominates along the regularly structured flow. If geometric obstacles or anything else disturbs the ordered flow, streamlines may be made to cross each other periodically with respect to time or space, creating chaos.

Channel shaping can generate a swirling transverse flow that stretches, folds, breaks and splits laminar flow over cross-section of the channel. Creativity over the exact shapes of the channels and their possible substructures has led to a large variety of micromixers over the years [34,54]. This collection comprises 2D and 3D curved channels, channels with obstacles inserted into their walls and many more; each with its own optimal performance in its own flow regime, leading to a classification of low-, medium- and high-Re micromixers. Chaotic mixing was first achieved by alternately arranged slanted grooves – so-called herringbone structures – in a microchannel [89,90]. In a staggered herringbone micromixer, only 1–1.5 cm mixing length is required for complete mixing, in contrast to 1–10 m for a T-mixer at Pe numbers of 104–105.

Unlike their geometrically modified, passive counterparts, active chaotic advection-based micromixers are based on external temporal and spatial disturbance, which affects intrinsic properties of the fluid and thus streamlines its behaviour. External sources of disturbance include electrokinetic, dielectrophoretic, electrowetting, magneto-hydro-dynamic and ultrasound forces, thermal (buoyancy-induced) motion, centrifugal forces in a rotating device and actively induced pressure fluctuations [51].

4.2.3.4 Chaotic Advection-based Segmented Multiphase Flow in Microchannels

Mixing within droplet-like segments flowing along the multiphase flows, composed of immiscible liquids or gas–liquid systems, provides a great advantage over single-phase micromixers because the axial dispersion is almost completely suppressed (Figure 4.7) [94]. This can lead to narrow RTD, which can cause even faster mixing compared to the single-phase versions [91]. In addition, particulate flow can be handled and mixed without absorption of species on to the channel walls (clogging) when a thin film of one immiscible liquid covers the other solid-containing liquid. The flow segments slide along the microchannel wall upon which the recirculation and thus enhanced mixing within the segment's volume takes place [92]. There are several ways of creating segmented flow, the most common of which is to bring two immiscible fluids into contact within a T-flow arrangement [93].

4.2.4 Micro Heat Exchangers

4.2.4.1 Fundamentals

Micro heat exchangers usually also serve as reactors in chemical transformations. Thermal energy is a significant parameter in every chemical reaction. Tight control of the energy (heat) possessed by any system of interest results in more efficient energy use, enhancing yield and selectivity and preventing meltdowns, runaways and explosions. It is also important in areas where rapid cooling or heating of process fluids is required.

Rapid and efficient cooling is possible within microreactors, allowing minimization of large exotherms. Reactions such as nitrations, for instance, can be more safely performed within microstructured devices than at the large scale. Both selectivity and yield can be

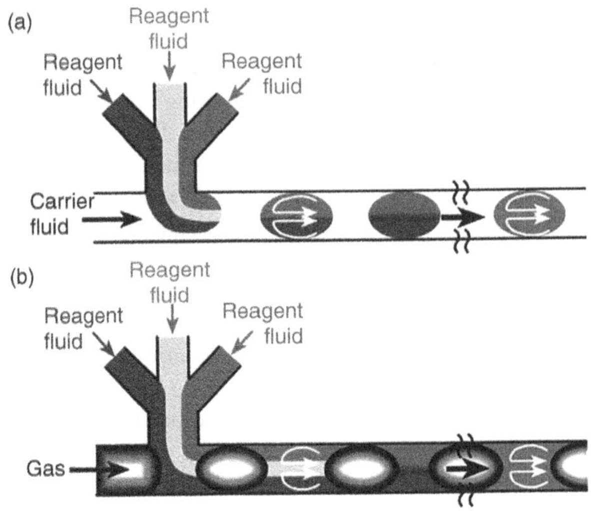

Figure 4.7 Multiphase flows composed of (a) immiscible liquids or (b) gas–liquid systems. Droplets consist of finely mixed reagents and prevent axial dispersion. Reproduced with kind permission of Wiley-VCH, Weinheim from [94].

tightly controlled. Roberge *et al.* showed that during nitration of phenolic compounds, a temperature increase of 55 °C is observed in batch, in contrast to only 5 °C in a microreactor [95]. The latter leads to an increase of the yield, rising from 55 to 75%. There are many more successful examples of exothermic and autocatalytic reactions safely performed within microstructured reactors [96–104].

To illustrate the effect of the microchannel characteristic dimension on heat-transfer intensification, a single-phase fully developed laminar flow is assumed, for which the Nusselt number (ratio of convective to conductive heat transfer) is constant. The relation of the heat-transfer coefficient to the dimension of the heat-exchange medium can be expressed by:

$$h = \frac{Nu \cdot k}{d} \tag{4.6}$$

Decreasing the channel diameter thus increases heat-transfer coefficient. Apart from being dependent on the heat-transfer coefficient, total heat flux is a function of the surface area of the medium and the temperature gradient:

$$Q = hA\Delta T \tag{4.7}$$

Therefore, intensified heat exchange leads to a compact design with small characteristic dimensions. This resembles initial definition of PI, which aimed at decreasing the size of the plant. Overall heat-transfer coefficients achievable with a micro heat exchanger reach values of more than 20 kW/m^2/K, which considerably exceed those of macroscopic devices [105].

There are several micro heat exchanger designs available, depending on the heat-transfer method. Cross- and counterflow micro heat exchangers rely on the principles of macro heat exchangers. There are also electrically powered micro heaters and evaporators, and microwave and inductive heaters. It also seems relevant to mention another common alternative energy input: light.

4.2.4.2 Crossflow and Counterflow Micro Heat Exchangers

A generic design is provided by a crossflow micro heat exchanger, which consists of plates stacked periodically in a crosswise fashion to yield a cube of connected plates through diffusion bonding. Figure 4.8a demonstrates such a design, realized by Schubert *et al.* In one typical example, 4000 microchannels with a $100 \times 80\,\mu m$ cross-section provide a total surface area of $300\,cm^2$ and a total heat-exchange area of $150\,cm^2$. Such a microdevice achieved an overall heat-transfer coefficient of $20\,kW/m^2/K$. The total volume of the cubic unit was $1\,cm^3$, yielding a 6 bar pressure drop when a water flow of 700 kg/hour was applied. In a later optimization, overall heat-transfer coefficients of up to $54\,kW/m^2/K$ were achieved for a water-based system [105].

Counterflow heat exchangers are the most efficient, since they create the largest temperature gradients. They are especially useful during heat integration, when an exothermic process needs to be cooled by an endothermic one, where they provide room for intensification via integration [105,106]. A typical design consists of a stack of parallel plates encompassed into an outer frame to form a plate-type heat exchanger [107]. The plates can be arranged in such a way that inlet and outlets for the fluid are in diagonal vicinity, so that

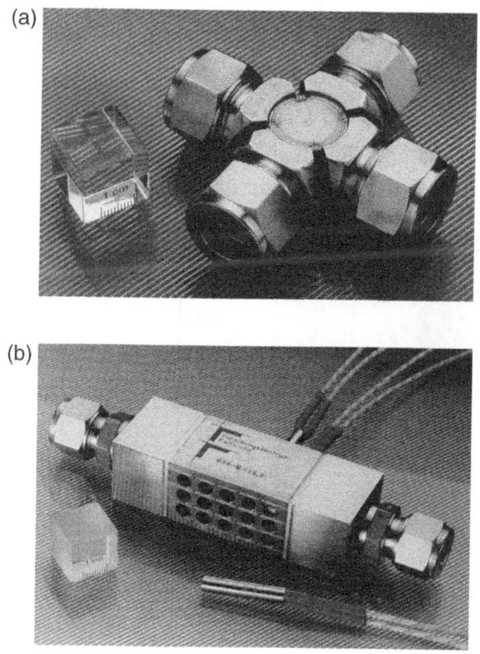

Figure 4.8 *(a) Crossflow heat exchanger (permission granted by Z-Tec Publishing). (b) Electric microheating device with thermal power of 15 kW. Permission granted by Z-Tec Publishing, [106].*

the diverging paths in the feed are mostly compensated for by flow mirroring. While the heating fluid enters one plate, the cooling liquid flows on the plate underneath. The total amount of heat transferred is determined by the number of plates stacked and limited by the pressure loss generated at the inlets and outlets, in addition to the loss along the channel.

4.2.4.3 *Fin-type Micro Heat Exchangers*

Further increase in heat-transfer area can be achieved by installation of fins into the heat exchanger unit. Such devices are common in cooling electronics, air conditioners and refrigerating units [108–110]. A new boundary layer develops along the heat exchanger on every subsequent pin fin where it makes contact with fluid, increasing heat transfer and absorbing more heat, and thus acting as a heat sink. Fins can take the shape of pins, cylinders, ellipsoids or squares placed on plates. Numerical simulation results suggest that Nu number can increase by a factor of 10 when microfin arrays are used (see Figure 4.9) [111].

A power transfer of 12.4 kW to large mass flows of 300 kg/hour is possible in fin-type heat exchangers. This doubles the capacity of a straight-channel micro heat exchanger [110]. A numerical study performed by Wang *et al.* demonstrated that thermal convection can be quadrupled when pin-type fins made of conductive SiC are installed in a micro heat exchanger. [112]. Moreover, this study showed an elegant microfabrication approach for a special chromatographic application requiring a thermally high and electrically low conducting material (SiC). The fin-type heat exchanger was split into two parts: one with a typical microfabrication material and carrying the fins, and one with the specialty material (SiC) but just given as a simple plate.

4.2.4.4 *Microchannel Electrical Heaters and Evaporators*

Microchannel electrical heaters and evaporators are also available [113]. A principal application is where fast and precise heating of sensitive liquids or gases is needed. Heat

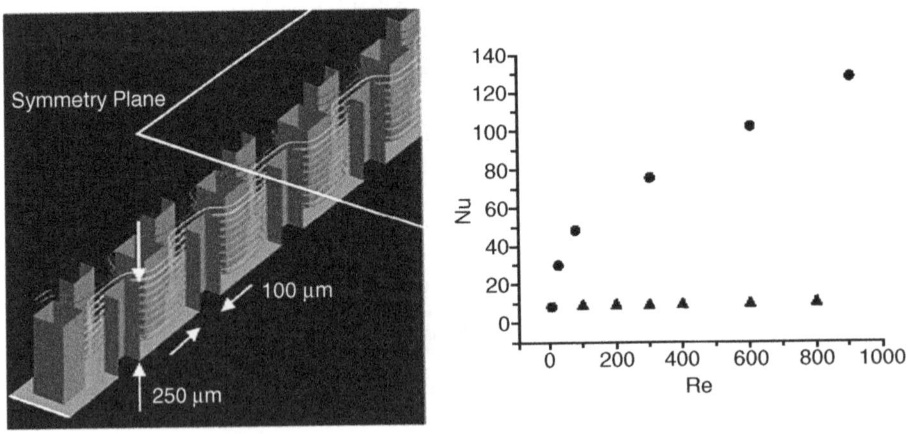

Figure 4.9 *(a) Streamlines along the channel of a fin-type heat exchanger. (b) Nusselt number comparison for parallel-plate and fin-type heat exchangers, where the same length and hydraulic diameter are used for calculations. Permission granted by Taylor & Francis, [111].*

exchangers driven by heating fluids have an upper temperature limit that depends on the fluid used, with 350 °C being the maximal operation point.

Cartridge resistor heaters, shown in Figure 4.8b, can be used as heating elements in electrically heated microstructure devices. These can be operated up to 900 °C, as needed for example in diesel reforming [105]. The design of such a heater has microstructured stainless-steel foils and spacer blocks arranged in alternating order between two stainless-steel plate end caps. This stack is then welded to one block. Heater cartridges are incorporated into holes in the spacer blocks. Depending on the number of cartridges and their power, the overall power of the heater device can range from several hundred watts up to several kilowatts. A heater equipped with 15 cartridges reaches a maximum electrical power of 15 kW. The calculated heat transfer coefficient is about 17.5 kW/m^2/K, much higher than the heat-transfer coefficient of conventional electrical heaters, which is around 2.5 kW/m^2/K.

Electrical heaters can also take the form of a heating plate, incorporated in a heating block on to which glass micro chips are placed. These can be heated up to 200 °C [114].

4.2.4.5 Electromagnetic Radiation Heating

Electromagnetic radiation is an alternative way of introducing energy into reacting systems. Molecules with permanent dipoles align themselves with respect to the electric field applied. Thus, oscillations of the electric field cause molecules to oscillate. Such molecular oscillations create an intense heat that can reach a 10 °C increase per second. Microwave heating of sealed vessels under superheated conditions can reduce reaction times from hours to minutes [115,116].

Incorporation of microwave heating into continuous synthesis using microflow has been done several times in past years [117]. Various setups for microwave radiation are discussed in detail by Wegner *et al.* [18]. This review discriminates microwave-assisted flow synthesis that uses flow coils versus glass vials versus superheated metal particles that promote inductive heating. Recently, an esterification reaction of acetic acid and acetic acid and ethanol to yield an ethyl ester was studied as a model reaction within the continuously operated microwave flow setup by Patil *et al.* [118]. The setup consisted of two parts: a microwave cavity and a reactor part, as shown in Figure 4.10.

Such a splitting was proposed in order to study temperature profiles in greater detail. A pseudo 2D model describing a microwave cavity was designed in COMSOL to estimate the dimensions of the reactor and the range of operating conditions, whereas a 1D model was used for the reactor part. The model and experimental results are consistent and in good agreement, having only 2.5% deviation in predicted conversion and ±3 K deviation in temperature estimations. Average efficiency of electromagnetic energy conversion into heat was determined to be 96%, which shows the high potential in continuous flow applications of microwave technology.

4.2.4.6 Inductive Heaters

Inductive heating has been known in industrial applications for surface hardening, melting, brazing, soldering and welding [119]. It is used in medicinal applications for thermal destruction of tumours, by means of the application of a magnetic field to a magnetic or conductive particle, which heats up to relatively high temperatures [120]. Application of inductive heating within organic synthesis is a relatively recent development, but it has already proven to give enhanced yields in intensified ways [121]. The basic principle is

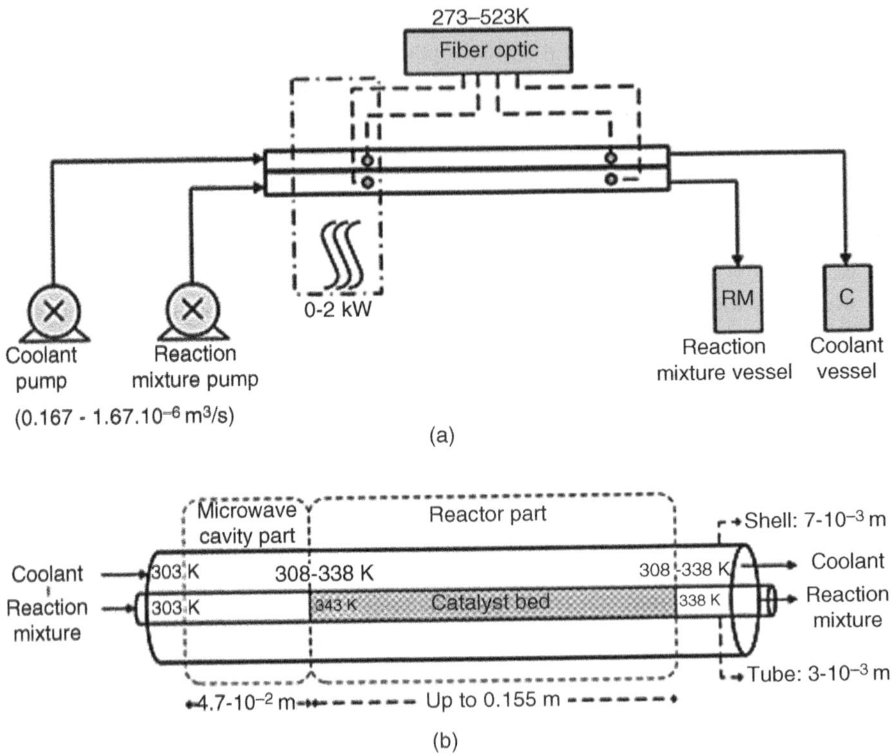

Figure 4.10 *(a) Process flow diagram. (b) Schematic illustration of the integrated reactor–heat exchanger system, consisting of a microwave cavity part and a reactor part. Permission to reprint is granted by AIChE, [118].*

similar to that used in medical treatments involving superparamagnetic particles coated with silica gel or conductive particles (see Figure 4.11a) and heated within an applied electromagnetic field (see Figure 4.11b). Inductive heating was used in the Claisen rearrangement by Kirschning *et al.* and high yields were reached. MagSilica particles were coated with chromium and nickel oxides to facilitate oxidations in flow. This was most effective when reaction time, temperature and solvent were kept constant [122].

4.2.5 Exothermic Reactions as Major Application Examples

Fast and exothermic reactions constitute a class of reactions that quickly provide sufficient amounts of required substances at the expense of generating great heats; often more than can be managed by conventional reactors, so that the reaction is slowed down in one way or another. The issue of safety comes up whenever an operating vessel's ability to release energy is lower than a reacting system's capacity for generation. Thermal runaways and explosions are the consequences of such shortcomings. This frequently stops promising fine chemistry and pharmaceutical drug-making syntheses from being implemented, when calorimetric analysis gives a negative decision. Microreactors, with their superb heating removal capabilities, open processing windows that were formerly closed.

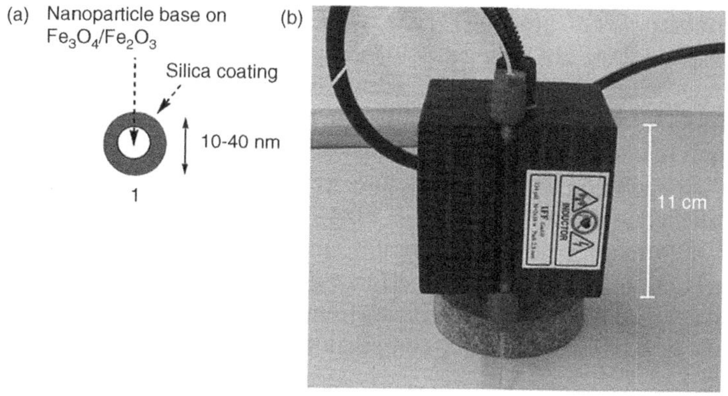

Figure 4.11 *(a) Structure of functionalized magnetic particle. (b) Flow tube reactor filled with catalyst-coated magnetic particles, shown within an inductor. Permission granted by 2008 Wiley-VCH, Weinheim, [121].*

The same thing is found in the field of safety and explosions as a whole. Thermal explosions take place when a system generates positive feedback in response to the reaction mechanism. There are two possible types of explosion: kinetic and thermal. Kinetic explosions take place in gas phase-containing systems, whenever propagation leads to a generation of more highly reactive radicals. Thermal explosions occur when heat evolved in exothermic reactions facilitates such reactions further through the temperature-dependent rate coefficients [123].

In order to quench radical formation and extensive heat generation, quenching materials, inerts and special solvents can be utilized on a macroscale. However, these measures are not in line with green chemistry principles, and further are not completely reliable, due to fluctuations in local temperature and concentrations within reactors. Microreactors, as discussed earlier, facilitate the homogeneous distribution of both chemical species and heat within the stream.

Fischer *et al.* demonstrated a method of determining the maximum safe capillary diameter for microreactors in the oxidation process of ethane by pure oxygen [11]. A stoichiometric amount of oxygen was safely used to afford 88% selectivity in the oxidation of cyclohexane under 16 bar and 200 °C reaction conditions within a capillary microreactor [124]. Struempel *et al.* produced benzoic acid methyl ester in a 75% yield by reacting benzoic acid with diazomethane in the temperature range 0–50 °C [125]. Maleic anhydride [126] and ethylene oxide [127,128] are further successful examples of safety-relevant direct oxidations in microchannels. In both reactions, full oxidation to carbon dioxide can set in, which releases much more heat than the already exothermic selective oxidation – almost an order of magnitude more. Highly pyrophoric trimethylaluminium was used to produce a yield of 49% of the antiobesity drug rimonabant, in 2 minutes' reaction time [129]. Roberge *et al.* demonstrated an improved selectivity and yield in the nitration of aromatic compounds, rising to 79 and 77%, respectively, under almost solvent-free conditions for particular compounds [95]. While formerly prohibited due to safety hazards, the utilization of ethyl diazoacetate was enabled through microreactor technology by Zhang *et al.* in a reaction with N-Boc-4-piperidine (yield: 49%, resolution time: 1.8

minutes). Pennemann *et al.* produced H_2O_2 synthetic conditions that resulted in an overall yield of 76.5% and 2.0 g H_2O_2/(g cat hour) as space yield [130].

Another green chemistry principle that is closely related to safety is the minimization of the use of hazardous and toxic substances. Due to their small scale, microchannels process limited amounts of chemicals when a particular point in space or time is considered. Thus, exposure during operation and system failure are minimized. Reactions including phosgene [131], diethylaminosulfur trifluoride (DAST) (1-chloromethyl-4-fluoro-1,4-diazoniabicyclo[2.2.2]octane), bis(tetrafluoroorate), andtrimethyl-silyl trifluoromethane [132] and hydride [133] have been safely performed within microreactors. A much more extensive and detailed review of safety issues from a process-control engineering perspective was given by Klais *et al.* in four compilations [134–137].

4.3 Chemical Intensification

4.3.1 Fundamentals

Novel process windows are based on two pillars: one with direct emphasis on the process chemistry and one with indirect emphasis through the application of new reaction schemes. The first is called 'chemical intensification', which exploits typically intensified, harsh chemistries that set the main parameter of a chemical protocol to extreme values, such as temperature (high-T), pressure (high-p) and concentration (high-c; solvent-free); the other is called 'process design intensification', which starting from new chemistries proposes and develops new flow-based process designs with entirely new cost, sustainability and energy structures (see Figure 4.12) [8,24]. Both, but particularly the latter, are related to complexity and system integration; that is, a holistic viewpoint.

4.3.2 New Chemical Transformations

Microreactors facilitate the development of direct, one-step syntheses, along with those syntheses which so far have been hindered in batch technology by rapid heat removal or other safety concerns [138]. Insertion of functional groups can be carried out starting from simple molecules or even directly from elements, in order to produce complex molecules.

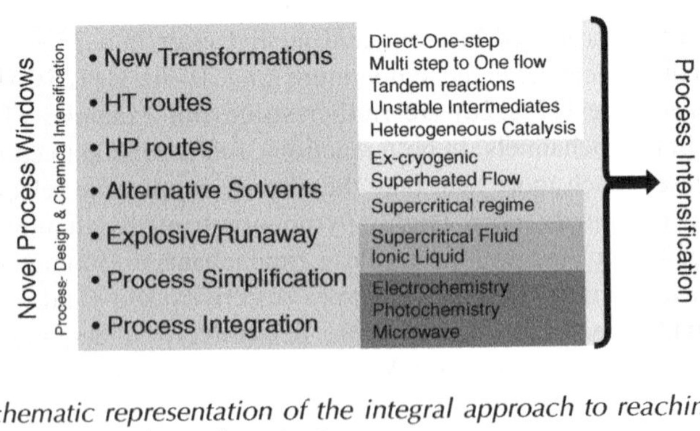

Figure 4.12 Schematic representation of the integral approach to reaching PI provided by novel process windows. Redrawn from [24].

Such synthesis is atomically more efficient than multiple circumventing steps yielding the same end product [103]. Through better selectivity and reduction of byproducts, waste generation is decreased for one-step approaches. Finally, solvent use and corresponding environmental impacts (solvent-mass index) are much reduced. Naturally, this has to be counterbalanced by the development of new chemistries and all its attendant risks during later process development. Usually, new chemistries are more costly in reagents, so single-step costs may be high, but in total their cost will be lower than that for alternative routes, which entail all the accumulated costs of multiple steps.

Beyond this new synthetic-driven approach, there is a second method based on bio-inspired synthesis. Fast multistep reaction schemes can be classified as tandem, cascade or other, depending on the choice and sequence of reagents and catalysts introduced. The spatial and temporal axes of flow operations ideally match such readily coupled reactions and, unlike one-pot batch synthesis, can afford optimized conditions for each of these single steps. This creates new opportunities to connect chemistries and their intermediates that would not be possible in batch, such as the flash chemistry-derived generation of unstable intermediate species *in situ*, which can be converted directly through tightly connected microreactor units with typical residence times of a few milliseconds [139,140]. In this way, reagents made by one initial reaction can be introduced in the next reaction, which normally would not have formed under equilibrium conditions. Such fast chemical connections are best carried out in a single chip with suitable short characteristic dimensions. However, for more moderate fast reactions, coupling is often carried out by a series of connected microdevices, which are intersected, where needed, by flow-separation devices. The modularity of microflow equipment facilitates such combinations of reaction steps [141]. In order to complement a synthetic process network, it is often advisable to substitute homogeneous catalysis with heterogeneous catalysis [142–144].

4.3.2.1 One- versus Multi-step Synthesis

Fluorinations. The high exothermicity of fluorination reactions with elemental fluorine is the major obstacle encountered in large-scale processes and has prevented the widespread use of such a direct approach. Microreactors have advanced heat-transfer properties and a demonstrated ability to remove the reaction temperatures found in fluorination. Pilot-scale operation of direct fluorination has been investigated and proven for several reacting systems, on both aliphatic (Scheme 4.1a) [145] and aromatic side (Scheme 4.1b,c) [146].

An example of the former is the fluorination of β-dicarbonyl compounds, of which a great diversity can be reacted, usually at high yields [145,146]. Toluene was the preferred model compound for aromatic fluorination in several investigations [103]. A relatively high selectivity of 80% towards the monofluorinated toluene species was achieved, unlike in a few batch processes reported [147]. Here, additions (under ring opening), multiple substitutions (in the core and side chain) and even unselective polymerizations occur. The substitution pattern of the *ortho*-, *meta*- and *para*-isomers was determined to be 4 : 1 : 2, which proves that a selective reaction path was undergone [147]. This is clearly indicative of an electrophilic reaction, while the multiple-product mixture of the batch processing clearly points to a radical path. The consequence of the latter is a high heat release, which in turn pushes fluorine to dissociate into fluorine radicals. This gives an autostimulated increase in reactivity, which can finally result in an explosion.

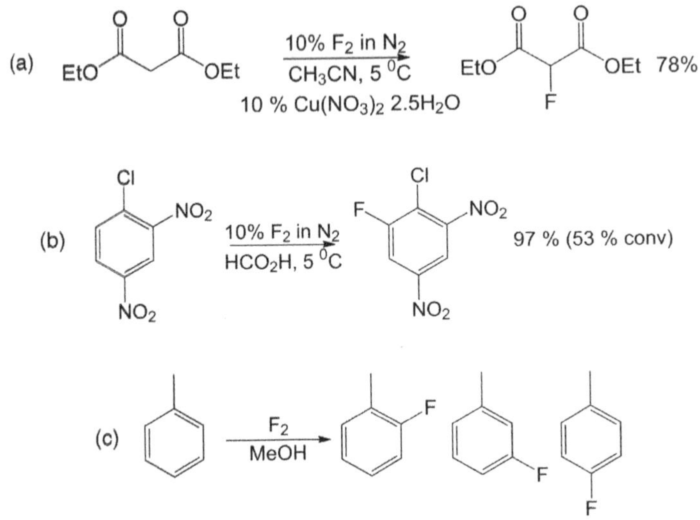

Scheme 4.1 *Fluorinations performed with elemental fluorine.*

Oxidations. One of the key hazardous yet useful reactions in the synthesis of pharmaceuticals is ozonolysis [148]. This is used quite extensively to yield important intermediates such as vitamin D. O'Brien *et al.* constructed a semipermeable Teflon tube within a tube-in-tube gas–liquid flow reactor for the ozonolysis of ketones at yields as high as 95% (see Figure 4.13) [149].

Glasnov *et al.* developed a procedure for the ozonolysis of styrenes and thioanisole in an O-cube, compact-benchtop setup, suitable for the preparation of up to 10 g of oxidized material per day in a quantitative yield (Figure 4.14 and Scheme 4.2) [150]. Roberge *et al.* intensified the process and increased the scale to half a ton per day in the ozonolysis of chrysanthemic ester [148].

Molecular oxygen was recently utilized as a reactant by Peterson *et al.* [151] in the Glaser–Hay oxidative acetylene coupling reaction to form 1,3-butadiynes. A tube-in-tube flow reactor was used for the ozonolysis. It was shown that increasing the ratio of the oxygenated stream mixture to the substrate (alkyne)/catalyst(CuOTf) stream increased the conversion of the reaction. Positive effects of both temperature and pressure were observed, yielding optimum conditions at 8 bar O_2 and 100 °C, resulting in quantitative yields.

Carboxylations. Carboxylic acids (and their derivatives) are significant as final and intermediate compounds in the production of fine chemicals [152–154]. Polyzos *et al.* made highly sustainable use of carbon dioxide as a reagent in the carboxylation of Grignard reagents to yield versatile carboxylic acids [155]. A tube-in-tube flow reactor was used to give complete conversion at 4 bar of CO_2 at a residence time of 42 seconds. Moreover, a scale-up was demonstrated without the need for significant reactor modification, yielding 3 g 2,4,6-trimethyl benzoic acid (see Scheme 4.3).

It can thus be concluded that synthetic pathways developed in flow and microreactors demonstrate the possibility of allowing direct, one-step functional integration of relevant groups such as F_2, CO, H_2, ammonia, ethylene, SO_2 and NO. In most of these cases, this is

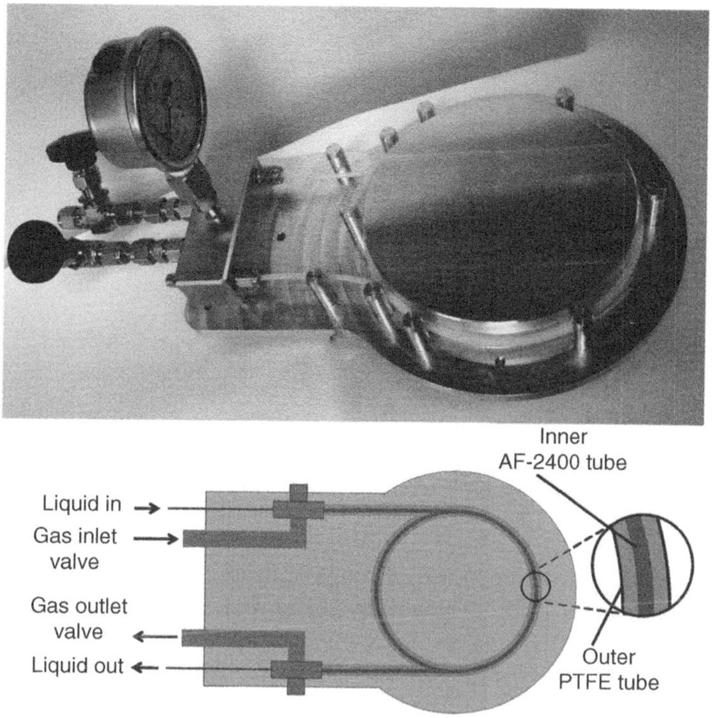

Figure 4.13 *Image and schematic representation of the tube-in-tube reactor/injector employed in ozonolysis. Permission granted by 2012 John Wiley & Sons, [151].*

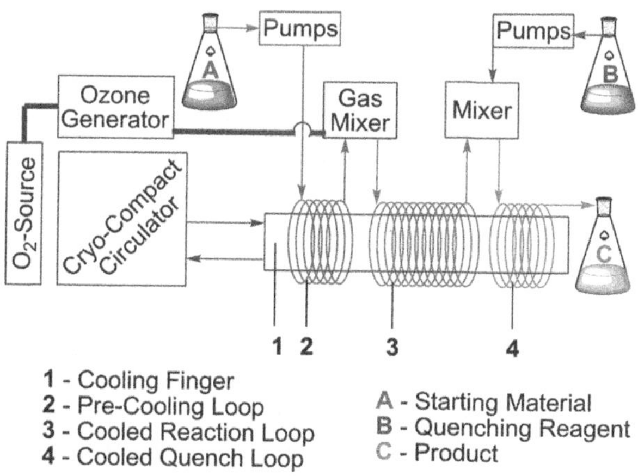

Figure 4.14 *Schematic illustration of an O-cube reactor used for ozonolysis. Courtesy of ACS [150].*

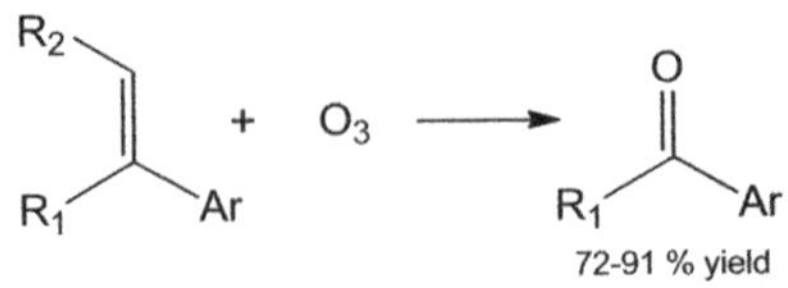

Scheme 4.2 *Ozonolysis of styrenes.*

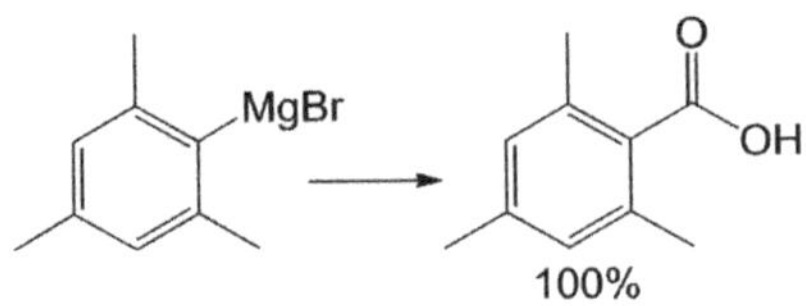

Scheme 4.3 *Synthetic route to 2,4,6-trimethyl benzoic acid.*

achieved via gas–liquid reaction, mostly through utilization of hazardous gases. There are probably also options for direct synthesis by liquid routes.

4.3.2.2 One-flow versus One-pot Synthesis: Multistep Synthesis

Microreactors contribute to chemical intensification by allowing continuous operation. Connecting microstructured modules in a bilateral fashion can result in a synthetic network that incorporates reactors, separation units and analytical equipment. The possibility of the spatial and temporal separation of single reactions (unlike in one-pot synthesis) creates chances for optimization at every single step and for maximization of synergy. Different pressures, temperatures and solvents can be realized along the flow axis, by integrating heaters, back-pressure regulators and intermediate separation units such as scavengers, columns and membranes, for example. These are extensively applied by Ley *et al.* [164].

Multistep synthesis in 'one flow' can be carried out in various ways, for example through:

- One microchannel, as a kind of cascaded reaction.
- One device (micro chip), by adding new reactant streams to the initial microchannel at a point where the first reaction is considered to be finished; that is, through a multi-microchannel network.
- Several micro/flow devices connected to each other, possibly with inserted micro/flow separation units.

In the last 4 years, there has been an ever-rising growth in the number of papers on multistep synthesis carried out in flow, mostly in the field of natural products and active pharmaceutical ingredients (APIs). Usually the third route, based on the device connection, is used, but there are also examples of the second route. Table 4.1 summarizes the documented performances, and the interested reader can use the references provided for detailed information. An update on commercially available equipment used for such purposes can be found elsewhere [155].

Table 4.1 *Overview of recently performed multistep syntheses in continuous microflow.*

Product name	Structure	Importance	No. of steps	Yield/purity	Remarks
Ciprofloxacin [161]	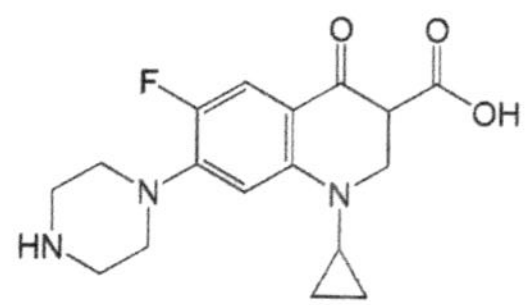	Fluroro-quinoline antibiotic (a potent gyraze inhibitor) for the treatment of infections of the respiratory tract (pneumonia, bronchitis), some types of gonorrhoea, diarrhoea caused by bacteria, typhoid fever and bone, joint, skin, prostate, sinus and urinary tract infections	5	57/99	A one-pot batch procedure was split and adapted to a continuous mode
Pristane [162]	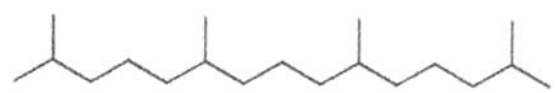	Adjuvant for monoclonal antibody Production in mouse ascites	4	50–55/99	Multikilogram scale, no byproducts
Efaproxiral [129]	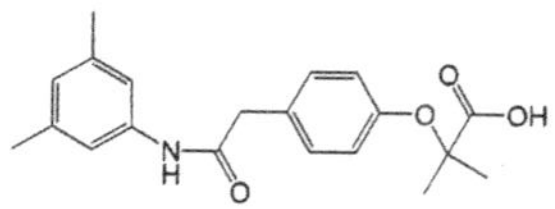	Enhancement of radiation therapy	3	51/-	
Rimonabant [129]	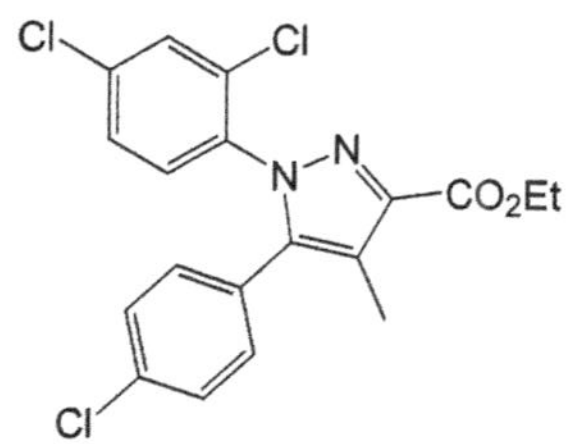	An antiobesity drug that acts as a central cannabinoid receptor antagonist	3	49/-	Offline work-up and purification of intermediates
Ibuprofen [163]	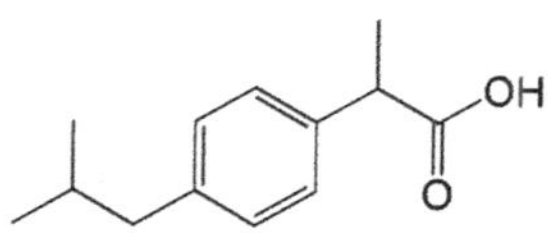	Nonsteroidal anti-inflammatory drug	3	51/99	Safer saponification

(*Continued*)

Table 4.1 *(Continued)*

Product name	Structure	Importance	No. of steps	Yield/purity	Remarks
Imatinib [164]	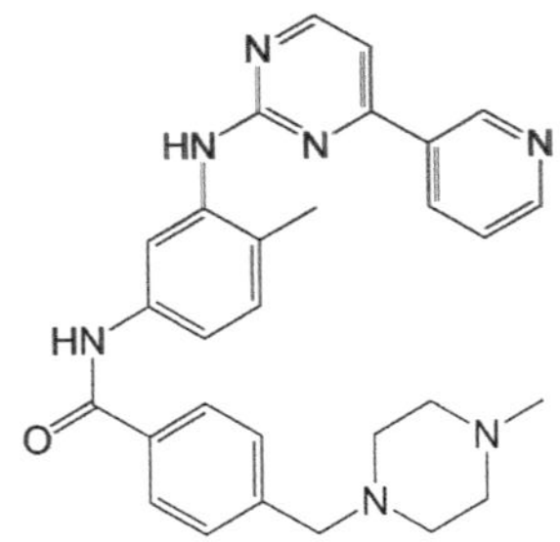	Tyrosine kinase inhibitor, developed by Novartis AG, used for the treatment of chronic myeloid leukaemia and gastrointestinal stromal tumours	3	32/95	In-line solvent-switching procedure
6-hydroxy-buspirone [165]	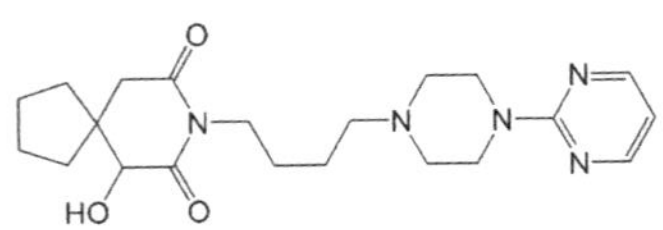	A major metabolite of buspirone, a potent and selective 5-HT1A receptor partial agonist that has been prescribed for the treatment of generalized anxiety disorders	3	85–92/97	Scale-up of enolization, numbering-up of oxidation reaction steps, 300 g/day
5HT1B antagonist [166]	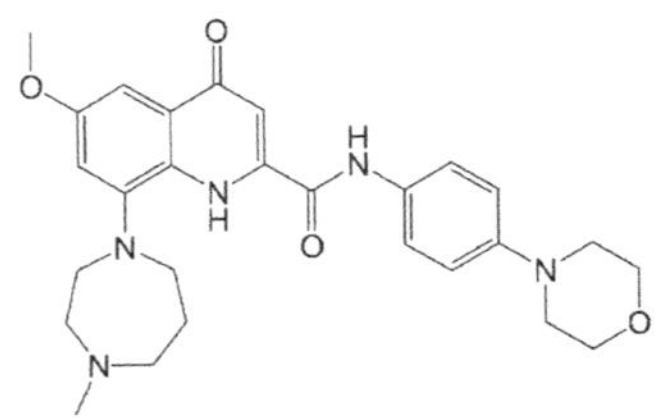	Fast antidepressant	5	18/98	'Catch-and-release' principle, high T (250 °C), used for cycloconden-sation reaction

(±)-oxomaritidine [168]	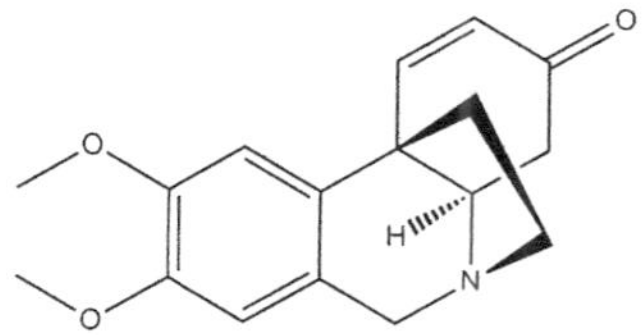	Natural alkaloid	7	40/90	Solid-supported reagents, catalysts, scavengers and catch-and-release agents Phenolic oxidation a problematic step, 50% conversion
Grossamide [169]	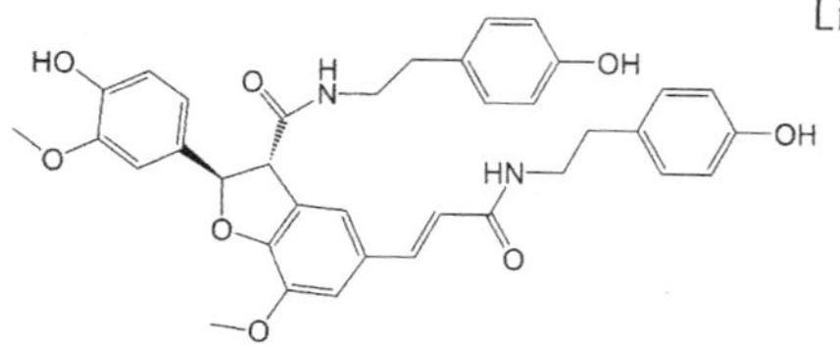	Lignanamide, antimicrobial compound regulating biological functions in both plants and microorganisms	3	90/92	Enantioselective continuous synthesis involving immobilized reagents
1,2,3-triazole [167]	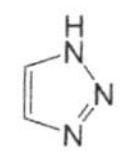	Aromatic heterocycle building block of various pharmaceuticals	3	55–82/99	Bestmann–Ohira reagent for preparation of terminal acetylenes, UV-triggered liquid handling automation affords 24/7 operation
Bis-substituted oxadiazoles [170]	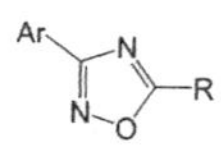	1,2,4-oxadiazole scaffold found in several drugs and drug leads, including the potent S1P1 agonist 120 and the metabotropic glutamate subtype 5 (mGlu5) receptor antagonist 2.21,22	3	45/99	Complete dehydrative cyclization achieved in 10 minutes under 20 °C and 7.5–9.0 bar in DMF
Chemokine receptor CCR8 [171]	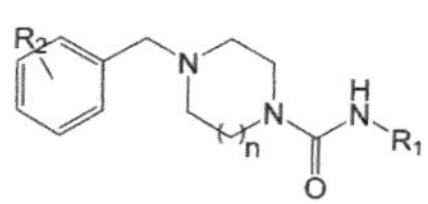	Anti-inflammatory and -allergic	3	10–96/99	—

4.3.2.3 Two-step Synthesis involving an Unstable Intermediate

Two-step syntheses are often based on the generation of a labile intermediate, which is then further converted to a stable product. At best, a fast ('hand-in-hand') generation, accumulation and reaction of the intermediate compound is achieved. A too-slow generation results in a slow accumulation and will suffer from decomposition if the latter process is relatively faster, leading to a low yield. Under cryogenic conditions, the generation is slowed by the low temperature, although this is still not low enough to stabilize the intermediate sufficiently. By virtue of improved mixing, the cryo conditions are not needed when microflow processing is used – extremely fast processing at ambient temperature can be utilized and the intermediates are consumed almost instantly (eventually after a few milliseconds (ms)) after being created. For such ms-processing at ambient temperature of normally cryogenic reactions, operating within microchannels and highly efficient mixing with a help of micromixers are needed to result in high-yields.

The Swern–Moffat oxidation generates highly unstable intermediates that can undergo Pummerer rearrangement at temperatures above −30 °C. Using a three-step cascade of micromixers (minimizing the residence time to 0.01 seconds), Kawaguchi *et al.* showed that the carbonyl products could be achieved in quantitative yields even at 20 °C [156].

Br/Li exchange reactions through Muharashi coupling (Scheme 4.4), conventionally performed at −78 °C, were successfully undertaken at much higher temperatures of 0 and −28 °C. By using an integrated flow microreactor system, Nagaki *et al.* sequentially introduced two electrophiles into dibromopyridines [157]. Recent advances in cross-coupling performed in continuous flow are reviewed in Noël and Buchwald [158].

De Mello *et al.* demonstrated safe diazonium chemistry within monolithic microreactors using electroosmotic flow for flow movement [159]. The diazonium salt intermediate generated is highly reactive and thus unstable. The results point to microreaction technology having the potential for large-scale production of such highly unstable reaction intermediates. An efficient method for the preparation of unstable tetramethylzirconium was realized by Uehata *et al.* [160].

4.3.2.4 Heterogeneous Catalytic Routes

Catalysts often have an enabling function in the chemistries mentioned in this section and can in addition allow the use of less toxic materials. Microreactors typically need more active catalysts than are required for batch operation, simply because of their shorter reaction times. The requirement for readily separable and recyclable catalysts often leads to the application of heterogeneous catalysts. This has led to tailored catalyst coating techniques on the walls of microchannels [172,173]. Recently the immobilization of enzymes on microchannels or on porous supports coated on microchannels has opened the path to biochemical reactions and often to the generation of chiral products [174–178].

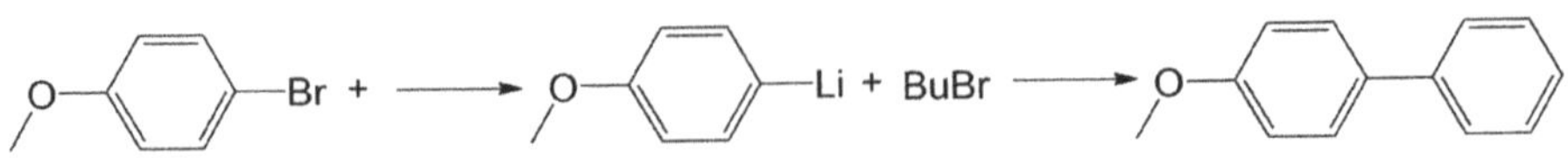

Scheme 4.4 *Lithiation and Muharashi coupling performed in flow.*

Some reviews of catalytic reactions in microchannels have recently been published, including overviews of heterogeneously catalysed hydrogenation and oxidation reactions, as well as technical papers with more individual achievements such as the combination of both types in the alkylation of alcohols by amines [179]. Irfan *et al.* provide a recent update based on green chemistry principles, presenting the advantages of using catalytic flow devices [180]. Various catalysts are applied to various types of reaction, such as the hydrogenation of double, triple and carbon–carbon bonds and carbonyl, imine and nitrile groups, as well as enantioselective hydrogenations. Chinnusamy *et al.* presents an overview of the application of metal-based reagents and catalysts in microreactors [181].

Novel ways of intensifying catalyst use have recently been achieved. Zotova *et al.* have shown that by increasing pressure (up to 50 bar) and temperature (up to 180 °C) the conversion of aniline with benzylic alcohol over Au/TiO_2 can result in 99% conversion at 97% selectivity [179]. This is an apparent intensification compared to the batch operation under reflux conditions in inert atmosphere, which shows 35% conversion and 68% selectivity.

4.3.2.5 Photochemistry

Photochemistry enables reactions such as [2 + 2] cycloadditions [182], cyanations, cyclizations, dimerizations, halogenations, oxidations [183], reductions and alkylations [184]. Photochemically catalysed transformations are green because no additives are needed, an easy workup is usually given and they present an opportunity to circumvent multiple steps. Photochemistry in flow benefits from the thin liquid layers and the short characteristic times of the microreactors. The thin paths minimize losses in photoefficiency, which normally come from solvent absorption and reflection. The resulting high quantum efficiency provides a highly uniform exposure with photoenergy. In addition, thin layers enable a very efficient cooling, preventing both the thermal overheating common in batch photochemistry and the danger of explosion, which is most often caused by an accumulation of highly reactive photointermediates such as peroxides or singlet oxygen.

Photochemistry is not used much in industry because of the complexity and inefficiency of the special reactors required, which are equipped with expensive light sources such as medium- and high-pressure mercury lamps, xenon lamps and halogen lamps [185]. By virtue of newly available small, lightweight light sources such as LEDs, a compact integration can be achieved, preferably in a simple stacking mode.

As was mentioned earlier, one of the advantages of the microflow reactors is the opportunity they provide to generate labile intermediates *in situ* that can be consumed upon generation. Recently, successfully performed reactions have been reviewed [155].

The enhancing effect of visible light on the catalytic activities of $Ru(bpy)_3^{2+}$ in continuous microflow was demonstrated by Seeberger *et al.* [186]. Catalytic ring opening of epoxides and bromination of alcohol were complete in less than 30 minutes, at low catalytic loading and in quantitative yield. A revolutionary achievement has been realized in the synthesis of the antimalaria drug artemisinin [187]. Using singlet oxygen generated *in situ* in a photochemical reactor, dihydroartemisinic acid was converted into tertiary allylic hydroperoxide (see Scheme 4.5), which is one of the key steps in making artemisinin.

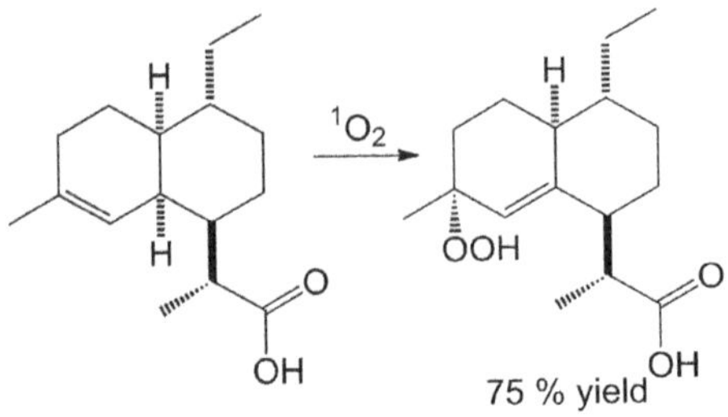

Scheme 4.5 Ene reaction in flow performed via photochemically generated singlet oxygen.

4.3.2.6 *Electrochemistry*

Electrochemistry is considered to be a 'green' technique in synthetic applications because of its ability to generate reactive species, anions and cations *in situ* and to transfer electrons via application of a current across the reaction medium [188]. Electrodes can be incorporated as plates or else imprinted on (microstructured) plates being assembled within a microstructured device. The Friedel–Crafts alkylation was performed by Yoshida *et al.* in flow using the 'cation pool' methodology [189], which is part of the governing concept of 'flash chemistry' [139,140]. A yield of >70% was achieved when reacting N-acyl iminium ions with allylsilanes. A detailed overview of related transformations is given by Wiles and Watts [155].

4.3.3 High Temperature

4.3.3.1 *Fundamentals*

The temperature of a reaction affects its rate according to the well-known Arrhenius equation. An increase in temperature increases the energy of the system and facilitates the shift from reactants to products. Microreactors provide an opportunity for rapid heat transfer and thus safe operation under nonclassical conditions, such as elevated temperatures and pressures [190,191]. The term 'high temperature' (high-T) as used here relates to the ambient temperature flow operation of reactions formerly processed in batch under cryogenic conditions. In the majority of cases, however, it relates to reactions carried out under superheated conditions – that is, at temperatures much higher than normal – which most often involves reflux processing, and here the boiling point of the solvent is decisive. Superheated processing – at temperatures beyond the boiling point in a single liquid phase – also demands an increase in pressure (high-p). This opens novel process windows and leads to chemical intensification.

Normally, simple capillaries are sufficient for such operation, because high-T reactions performed at room temperature are so slow that premixing can be done and no mixing element is required. If for any reason, however, this is not allowed or is unwanted, one must use high-T proprietary chip microreactor designs that can withstand high ps and Ts [192]. Usually pressurization is ensured by either high-performance liquid chromatography (HPLC) or high-p syringe pumps and a back-pressure regulator after the microreactor [193,194].

After demonstrating the initial feasibility, a major push towards the use of high-T superheated flow technology was given by research taking a similar approach in batch microwave technology, using sealed vials. Extensive investigations demonstrated order-of-magnitude decreases in reaction time [115]. Seeing the limitations of scaling up microwave equipments, especially under such harsh conditions, a paradigm shift occurred – mainly initiated by Kappe – in which high-p,T microwave reactions were transformed into high-p,T flow reactions, almost all being done in capillaries. There are numerous examples demonstrating such transformations and the associated massive speed up in reactivity under superheated conditions [116,117].

4.3.3.2 Application Examples

Jensen *et al.* constructed a silicon/Pyrex microreactor (shown in Figure 4.15) that can withstand temperatures up to 450 °C and pressures up to 250 bar [191]. Silicon etching techniques, combined with bonding to Pyrex, allowed tight thermal control, good chemical compatibility and high visibility. Since connections capable of withstanding high T,p are rare, a circumventing solution was provided by incorporating a 'mild'-conditions platform separate from the high-T zone platform. Reactor was operated under supercritical conditions. Formation of one supercritical fluid (SCF) from two immiscible liquid–liquid phases was visualized for a water/hexane mixture; synthesis of iron oxide nanoparticles in liquid toluene was accomplished at 100 bar and 300 °C; and supercritical water oxidation of methanol was performed by hydrogen peroxide at 250 bar and 400 °C. At a fixed residence time of 2 seconds, conversion increased from 36.5 at 20 MPa and 300 °C to 87 at 25 MPa and 380 °C [192].

The same group also used another high-p,T chip microreactor made pressure-tight by the aid of solder techniques known to give strong, irreversible joints between microstructured plates [129]. This microreactor was applied in the Heck aminocarbonylation reaction at pressures up to 100 bar. The reaction had dual reaction paths: inserting one molecule of

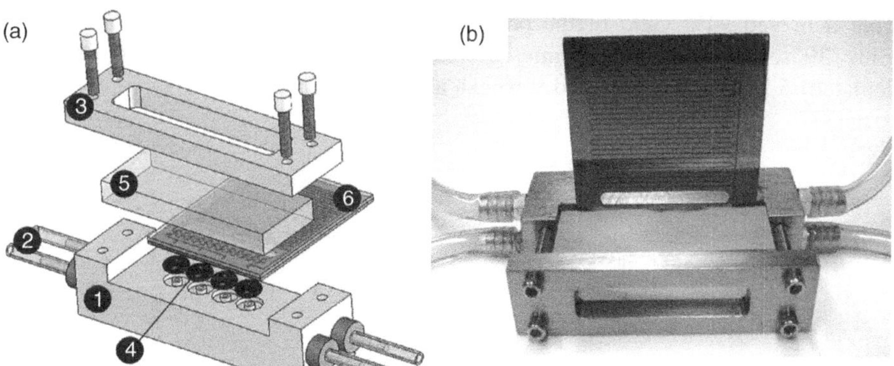

Figure 4.15 *(a) Schematic illustration of a high-T,p reactor assembled from six parts: (1) compression platform; (2) cooling fluid inlet/outlet; (3) compression part; (4) O-rings and grooves; (5) Pyrex plate; and (6) microreactor. (b) Photograph of the assembled reactor. Permission granted by American Chemical Society, [192].*

carbon monoxide gave the amide, while two carbon monoxide molecules yielded the α-ketoamide. Tuning operating conditions meant tuning the majorly formed product: with increased temperature, the amide was obtained in higher yield, while an increase in CO pressure resulted in a higher content of α-ketoamide.

Carboxylated phenols can be synthesized by following the Kolbe–Schmitt reaction protocol, where strong base is used to form phenolate with electron pair stabilized at ortho position that attacks CO_2, usually under 100 atm pressure and at temperatures around 125 °C [196]. The activation barrier of the reaction is lower where more than one electron-donating substituent is present, allowing milder operating conditions, lower temperatures and a weaker base. When resorcinol is used, the Kolbe–Schmitt reaction proceeds to completion in 2 hours at 100 °C under reflux and batch conditions in the presence of bicarbonate as base. In 2005, investigating the first high-p,T example of a microreactor, Hessel *et al.* achieved a significant reduction in residence time to 4 seconds at a yield of 40% under mildly basic superheated conditions of 200 °C and 40 bar [112,190]. Raising temperature to 200 °C further shortened the reaction time, resulting in a space–time yield of ~18.20 t/(h m^3), which is about a thousand times more than can be gained from a 1 l flask. A further 1.5 times increase in the space yield offered by micro processing was observed when the temperature was increased to 270 °C. Combination with another PI technology – reactive ionic liquids (ILs) for solvent-free operation – results in further improvements of space–time yield up to 64.23 t/(h m^3). As a second PI combination, flow chemistry and microwave heating (for the improvement of temperature profiles) lead to an increase in yields of up to 59% (at 180 °C, 130 seconds), which is at least 10% above the best literature values for batch processing.

Claisen rearrangement of allyl phenyl ether (see Scheme 4.6a) is known to take hours under reflux conditions in batch. Kappe *et al.* demonstrated that this reaction was completed in 1–2 hours at 13 bar of autogenic pressure inside a sealed vial heated to 250 °C under microwave conditions. The same reaction was completed at residence time of 4 minutes in flow at 100 bar and 240 °C [197]. Only 5% byproduct formation was observed.

Flow operation of the reaction between acetic acid and o-phenylendiamine at 270 °C and 130 bar achieved 94% yield of the product 2-methylbenzimidazole (see Scheme 4.6b) in only 30 seconds' residence time. This constitutes a significant acceleration over (calculated) 5 hours at 100 °C and 9 weeks at room temperature.

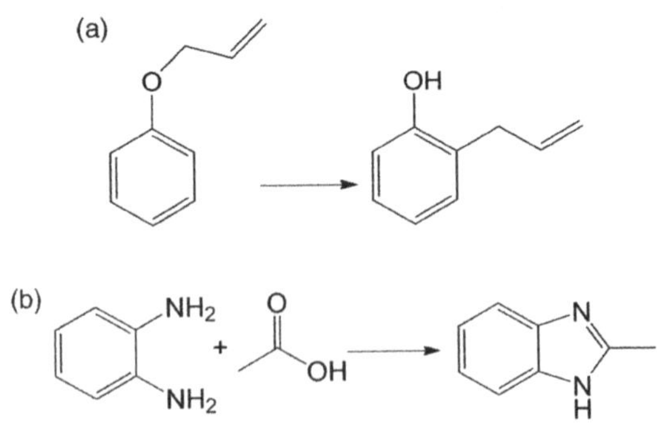

Scheme 4.6 (a) Claisen rearrangement of allyl phenyl ether. (b) Synthesis of 2-methylbenzimidazole.

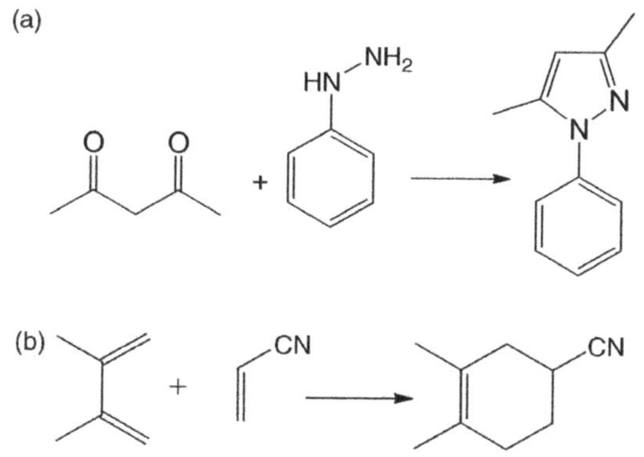

Scheme 4.7 *(a) Synthesis of 3,5-dimethyl-1-phenylpyrazole. (b) Diels–Alder cycloaddition.*

Condensation of acetylacetone with phenyl hydrazine (see Scheme 4.7a) in ethanol under acidic conditions is completed in 8 hours when performed under ambient conditions. Acceleration up to 1 seconds to yield 90% of isolated product was achieved when the reaction was superheated to 180 °C at 17 bar. The Diels–Alder cycloaddition of 2,3-dimethyl butadiene and acrylonitrile was accelerated from 5 days at 100 °C to 2 minutes at 280 °C and 130 bar pressure, yielding 82% product (see Scheme 4.7b) [117].

Benzannulation of aromatic carbonyls and alkynes was intensified via microwave heating over a gold film as catalyst. At 190 °C, 90% conversion was achieved, in comparison to 68% when the temperature was increased to 240 °C. Repeating the experiment in an oil bath at 190 °C resulted in 14% conversion. Large difference in the microwave experiment suggests the possibility of local superheating under latter conditions [198].

A Dimroth rearrangement substituting 2-amino-6H-1,3-thiazines into 3,4-dihydropyrimidine-2-thiones was demonstrated to take place in 66 minutes' residence time when performed under microwave flow conditions at 200 °C, with a yield of 88% [199].

Increased pressure allows operation at higher temperatures than the boiling point under ambient conditions. This opportunity avoids the need to use high-boiling solvents, which are usually difficult to separate after chemical transformation. For the Diels–Alder cycloaddition of 2,3-dimethylbutadiene and acrylonitrile mentioned in Section 4.3.4.1, high-p operation allowed switching from a high-boiling solvent, toluene, to low-boiling solvents such as MeCN, THF and DME. Inverse electron-demand hetero-Diels–Alder reactions were performed in flow. A pyrimidine alkyne was rearranged to a corresponding annulated pyridine with 96% conversion at 250 °C and 50 minutes' residence time [200].

In an even more far-reaching and generic high-T approach, Kappe *et al.* transformed flash vacuum pyrolysis (FVP) to flash flow pyrolysis. The demonstration example was the pyrolysis of Meldrum's acids, which generates reactive intermediates such as ketenes. Reaction rates were shortened to the range of several seconds to minutes at quantitative yields [201].

On the route to the synthesis of 4-aryl-2-butanone, a precursor of the anti-inflammatory drug nabumetone, three paths were investigated. A Wittig reaction variant, the olefination of aldehydes, at 200 °C in acetonitrile or dimethyl formamide needed 10 minutes to achieve

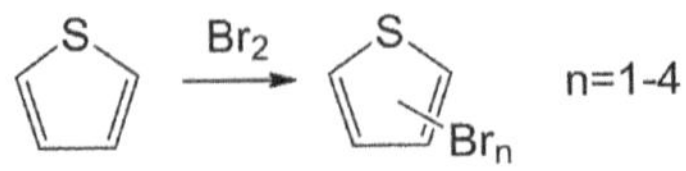

Scheme 4.8 *Bromination of thiophene.*

100% conversion [202]. Many more such high-T examples are available in Obermayer *et al.* and Razzaq *et al.* [203,204].

Flow-based operation at much higher temperatures than are conventionally used in batch operations was also established for organometallic or similar reactions, usually carried out under cryogenic conditions [8,24]. The use of cryogenic conditions has three motivations. Most notably, they 'freeze' the reaction in order to make it slower than the mixing; otherwise, the very fast reactions tend to be severely mixing-masked, which results in much lower yields. Second, unstable intermediates must be handled. Third, some of the fast reactions are highly exothermic, so again 'freezing' is the right concept under batch conditions with less improved heat transfer. Intensified mixing and heat and mass transfer allow the limits mentioned to be overcome, so that the former cryoreactions can be processed at ambient temperature without any compromises in yield – often with even better yields – and always with much improved space–time productivity.

The bromination of thiophene (see Scheme 4.8) is conventionally performed at 0 °C due to the high reactivity. Through variation of the molar ratio of the reactants under short reaction times in a microreactor, full control over the distribution of the products was achieved, ranging from mono- to tetra-substituted. No change in product distribution was observed in the range 0–60 °C, keeping the molar ratio of bromine to thiophene 2.0 and favouring the formation of 2,5-dibromothiophene. The selectivity increased to 80% [205].

Beyond many organometallic reactions investigated in microreactors, ambient–cryo shift was also carried out for an autocatalytic and explosive nitration reaction of phenol. The nitration was safely performed under tight temperature control at room temperature and with high concentrations in flow. The highest combined flow-derived yield of p- and o-nitrophenols was 77% greater than those obtained in batch [95].

4.3.4 High Pressure

4.3.4.1 Reaction Volume Effects

Apart from facilitating superheated conditions, high p affects reaction through its influence on reaction equilibrium and rate constant and affects reaction medium through its effect on its physical properties. In cases where the volume occupied by the product following the reaction is lower than that given for the two separate reactants, the overall volume change is negative and application of high p will have an acceleration effect. Thus, cycloadditions are good candidates for intensifcation by elevated pressures.

Kappe *et al.* [204] performed a Diels–Alder cycloaddition of 2,3-dimethylbutadiene and acrylonitrile under high-T (300 °C) and high-p (200 bar) conditions in a stainless-steel tubular (1000 μm, id[1]) reactor. Upon raising the temperature from 250 to 280 °C, the reaction was accelerated from 5 to 2 minutes' residence time at a yield of 82%, total throughput of 80.4 g/hour and space yield of 1.4 kg/m^3/s.

[1]internal diameter

Pressure effects are also expected for reactions that undergo cyclic transition states on their course to the product, such as Claisen and Cope rearrangements. Sato *et al.* [206] achieved a conversion of 98% upon Claisen rearrangement of allyloxybenzene to 2-allylphenol in aqueous solvent at 265 °C and 50 bar. The lower concentration compared to batch (0.27 versus 0.77 mol/kg) resulted in a longer residence time (149 versus 81 seconds) but in a higher yield (98 versus 73%).

Apart from the effect upon the reaction volume, pressure also affects the concentration of charge and ionization, accelerating reactions of the following types [207]:

- Those that proceed via dipolar transition states (Menshutkin reactions and electrophilic aromatic substitution reactions).
- Those with steric hindrance.

4.3.4.2 Density and Viscosity Effects

Increased pressure affects both density and viscosity. The viscosity of liquids increases approximately by a factor of two every 100 MPa; thus, diffusion control of the reaction is important [208]. Increased viscosity at high p (kbar range) has been found to accelerate chemical processes such as Diels–Alder reactions, 1,3-dipolar cycloadditions and Claisen rearrangements [209]. It has been demonstrated that apart from the acceleration due to the negative reaction volume, a certain portion of the reaction-rate increase is associated with the increase in viscosity; this is given in cases where the activation volume is more negative than the overall reaction volume.

Transition state theory deviates for highly viscous media, where diffusion processes play an important role. Since pressure increases viscosity coefficient exponentially, such effects apparently must be taken into account. To illustrate this, in Z–E isomerization of 4-(dimethylamino)-4′-nitroazobenzene, the rate constant initially increases with pressure, then decreases at higher pressures when the medium is a viscous silicone oil [210].

Packing and the corresponding interaction of the solvent molecules around the reactants and around the transition state complex have relevance for the reaction outcome when high pressure is applied. Electrostriction and steric effects have impact. Even the regio- or stereoselectivity can be enhanced by using bulky groups as blocking units at one site while enhancing the addition and substitution at the less bulky site.

4.3.4.3 Dielectric Constant Effects

Dielectric constant, ε, of any liquid is a weak function of temperature but strongly dependent on pressure.

$$\frac{\epsilon_{p0}}{\epsilon_p} = 1 - A \ln\left[\frac{(B+p)}{(B+1)}\right] \tag{4.8}$$

A and B are characteristic parameters for a liquid and are not dependent on pressure, while B decreases with increasing temperature. Dielectric constant determines the magnitude of electrostriction induced by the solvent in the vicinity of charged species, thus playing a significant role in activation volume calculations. To illustrate the effect, the dielectric constant of supercritical water varies from 1 (i.e. nonpolar) to 30 at high p, in contrast to being 80 at ambient conditions.

4.3.4.4 Solubility Effects

Finally, the solubility of gases and liquids increases with pressure. Thus, gas–liquid reactions can ultimately be performed as a quasi single-phase medium with high concentrations of dissolved gases. Miscibility tends to change as well, due to the change in density and polarity of liquids.

Probably by virtue of such solubility effects, CO_2 was incorporated in n-benzylmethylamine to yield carbamic acid in 8 seconds at 300 bar using a specially arranged microchip [211]. The low mechanical strength of the polymethylsiloxane microreactor required it to be embedded into a high-pressure Parr reactor (45 bar), thereby showing a way towards the use of high-p polymer microreactors. A 10-fold increase in the reaction rate of cyclohexene hydrogenation over Pd catalyst was achieved upon going from ambient conditions to 71 °C and 51 bar [212]. No byproduct was formed at 20% conversion of acetone to isopropanol on Ru/C catalyst [213].

Until now, high T has been used largely as an accelerating tool in continuous flows at the microscale. However, with increasing technological abilities, equipment can be developed for a higher-pressure performance that will assist in the investigation of the complexity of the processes under high-p,T conditions.

Tiggelaar *et al.* tested the performance of several designs based on the in-plane connection of the fibre interfaces to the glass chip [213]. Connections at the inlet/outlets were shown to withstand pressures as high as 690 bar when connected in a tubular fashion. The chip was tested in carbamation of N-benzylmethylamine up to pressures of 300 bar; at higher pressures no product formation was observed, due to the lower residence time available.

4.3.5 Alternative Reaction Media

Solvents mainly derived from petroleum [214] are used as bystanders in chemical reactions to facilitate the encounter of reactants. They must either be reprocessed after each use, which usually requires high energy consumption, or else sent to waste, which is less environmentally friendly. Thus, eliminating the use of these solvents, or using alternatives that require less energy consumption and are easier to recover seems to be a sound green idea. Solvents such as water [215], ILs [216] and SCFs [217] are the most desirable options. As has already been mentioned, microfluidics offers a tight control over operating conditions, such as temperature, pressure and residence time, and create a platform for the investigation and hopefully large-scale implementation of alternative solvents, such as SCFs and ILs.

4.3.5.1 Supercritical Fluids

At conditions above their critical pressure and temperature, substances become SCFs. The possibility of using microfluidics as an operating platform for supercritical processing has recently been reviewed by S. Marre *et al.* [218]. Figure 4.16 illustrates the operating conditions of recently developed microfluidic devices that can be operated

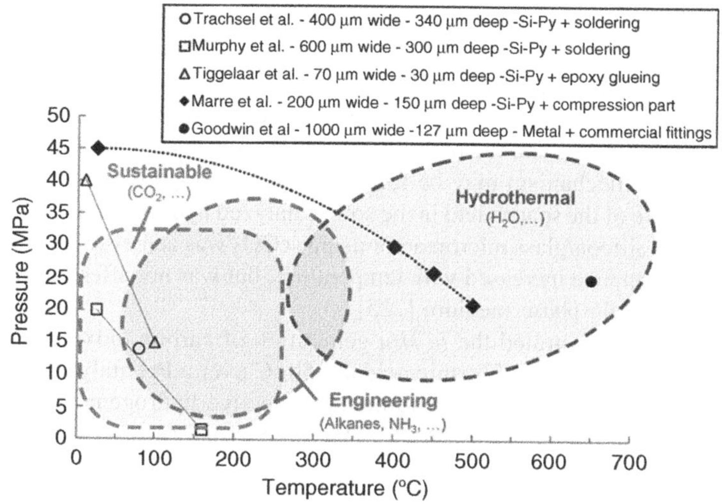

Figure 4.16 *Overview of several high-p,T SCF-compatible microsystems from Trachsel* et al. *[212], Murphy* et al. *[195], Tiggelaar* et al. *[213], Marre* et al. *[192] and Goodwin et al. [219] and of the operating regions corresponding to sustainable, engineering and hydrothermal fluids [218].* © *2011, with permission from Elsevier.*

supercritically. The typical range of pressure and temperature conditions for the use of three kinds of fluids (sustainable, engineering and hydrothermal) is also given in the figure.

A positive effect of this is the transition from laminar to turbulent flow, achieved simply through changes in fluid density and viscosity, with a significant corresponding increase in Reynolds number. This is much in contrast to the liquid microfluidic case [220]. Thus, a tight control over temperature and pressure can be used to tune the hydrodynamics within the channels of microdevices. Furthermore, since the diffusion coefficient is much higher for gas-like fluids, rapid mixing can be achieved without the need for specifically designed mixers. Low viscosities and high diffusivities reduce the dispersion within the microchannels, facilitating narrow RTD. The two main solvents studied under supercritical conditions are water and carbon dioxide, due to their high abundance, low cost and easy recovery. Besides these, alcohols often find use.

Supercritical $\boldsymbol{CO_2}$. Supercritical CO_2 ($scCO_2$) is inexpensive, recyclable, easily obtainable as a byproduct of fermentation and combustion, nontoxic, nonflammable, relatively inert, non-oxidizable and provides simple separation via evaporation. Despite being quite inert, even to free radicals, $scCO_2$ can react with nucleophiles. The list of reactions includes aldol reactions, carbonylations, cyclizations, epoxidations, esterifications, carbon–carbon cross-coupling reactions, hydrogenations, hydroformylations and polymerizations. Continuous flow processes on the microscale provide unique opportunities for working with green $scCO_2$ due to the small scale and improved safety at the required operating pressures.

The esterification reaction of phthalic anhydride with methanol was performed in a glass microreactor by Verboom *et al.* [207,213,221–224]. ScCO_2 was used as a cosolvent (0.45 mol fraction) and resulted in a 5400-fold increase at 110 bar and 60 °C. An interesting fact is that both activation energy and preexponential factors determined, varied with pressure values and the presence of scCO_2 as a cosolvent. This indicates that a different reaction mechanism may be followed under changing operating conditions.

A 10 fold increase of the space yield in the solid catalyzed hydrogenation of cyclohexene over Pd/Al_2O_3 in a silicon/glass microreactor using scCO_2 was achieved by von Rohr et al. The reaction performance increased with temperature, but was not affected by changes in pressure within a single phase medium [225].

Poliakoff *et al.* demonstrated the *in situ* generation of carbon dioxide and hydrogen upon catalytic decomposition of formic acid at 450 °C over a Pt catalyst [226,227]. The reaction mixture was then used in a subsequent selective hydrogenation of cinnamaldehyde to hydrocinnamaldehyde at 200 °C. The same transformation was performed by Kawanami *et al.* in a high-pressure membrane reactor with 100% selectivity at 50 °C and 10 bar hydrogen pressure [228].

An easy separation through the use of scCO_2 was demonstrated by Patcas *et al.* in the cobalt-catalysed hydroformylation of n-octene. Higher selectivity was achieved in this way than with supercritical toluene. Finally, when cooled down the catalyst became insoluble at low temperatures and could be filtered and recycled several times [229].

Super- and subcritical H_2O. Water reaches its criticality at 374.2 °C and 218.3 atm – much higher than scCO_2. Certain properties of the water change drastically, including the relative static dielectric constant, the specific heat capacity and the pKw value.

Poliakoff *et al.* performed several industrially relevant conversions in a flow device that utilized supercritical water [230]. One example is the one-pot hydrolysis followed by cyclization of 6-aminocapronitrile to ε-caprolactam.

The disadvantages of using supercritical H_2O (scH_2O), however, are the high energy input required and the high corrosive power. There are also insufficient data available on multicomponent system behaviour in scH_2O; near- or subcritical water can be used as an alternative. Ikushima *et al.* [231] developed a high-speed, highly selective and hydrolysis-free acylation of various alcohols through acetic anhydride in the absence of catalyst under subcritical water conditions. High yields (>93%) with 100% selectivity at temperatures of 200 and 250 °C were afforded.

Ikushima *et al.* [232,233] reported that for the noncatalytic benzaldehyde disproportionation and noncatalytic Beckmann and pinacol rearrangements, selectivity was low unless the reaction flow was rapidly heated above the critical temperature. Lower temperature values facilitated side reactions.

Supercritical Alcohols. Supercritical solvents such as methanol and ethanol were also utilized in flow syntheses. Kappe *et al.* performed esterification of benzoic acid in absence of catalyst under flow conditions in supercritical ethanol at 300 °C and 120 bar, with a residence time of 12 minutes, to afford ethyl benzoate in 87% yield [204,234]. High-T and supercritical conditions are required, because below 200 °C no esterification to ethyl ester occurs. Transesterification to methyl-3-phenylpropanoate was performed at 85% yield in 350 °C supercritical methanol in 8 minutes.

Several Newman–Kwart rearrangements were performed by Kappe *et al.* in supercritical 1,2-dimethoxyethane (300 °C, 80 bar) to yield products with high purity (>90%) and in short reaction times (<8 minutes) [235].

4.3.5.2 Ionic Liquids

ILs are widely considered to be green due to their low toxicity, nonflammability and nonvolatility. They can stand for the solvent, catalyst or catalyst support within the reaction medium [236], as demonstrated for the polycondensation of 4-(3-hydroxynaphthalene)-1,2,4-triazolidine-3,5-dione with aliphatic diacid chlorides, eliminating the need for low-volatile solvent and catalyst [237]. In the Michael addition of N-heterocycles to α,β-unsaturated compounds, the IL 1-methyl-3-butylimidazolium hydroxide could be reused for several cycles with consistent activity [238]. However, if recycle of the solvent is not employed, the greenness of ionic liquids is more questionable, as shown by a more holistic evaluation through life cycle assessment (LCA) [239].

A multiscaled flow system was developed and used to produce ethylmethylimidazole ethylsulfate ([EMIM][$EtSO_4$]) under solvent-free conditions via highly exothermic alkylation of methylimidazole with diethylsulfate [240]. Precise temperature and conversion control through multistaged flow reactor assembly was required in order to achieve high-quality products and avoid temperature runaway of the reactor [241]. Recovery and reuse was achieved for the IL ([bmim]I and [bmim]PF_6) in the selective alkylation of amino groups within amine derivatives with a variety of alkyl halides [242].

Polycondensation of 4-(3-hydroxynaphthalene)-1,2,4-triazolidine-3,5-dione with aliphatic diacid chlorides, circumventing the need for an organic solvent and catalyst via the use of ILs and tetrabutylammonium bromide, was realized [237]. A significant reduction in reaction time from hours to seconds was achieved when 1-butyl-3-methylimidazolium methylsulfate and $InCl_3{\cdot}3H_2O$ were used in the synthesis of optically active furan diol from d-glucal in quantitative yields (78%) [243].

4.3.5.3 Solvent-free or High-concentration Processing

Apart from using master solvents, the reduction of the load of 'normal' solvents is still an industrially and enviromentally relevant issue (high-concentration (high-c) processing). Solvents constitute 80 and 90% of mass utilization in a typical pharmaceutical/fine chemicals batch chemical operation [244]. Applicable harsh conditions in microreactors in terms of temperature and pressure suggest a similar boost is achievable under solventless and solvent-free high-c conditions.

The impact of high-c operation was demontrated by Wu *et al.* in the peroxidation of methyl ethyl ketone. The process cannot be speeded up by the use of high temperature, due to the favourable decomposition of the peroxides. Higher H_2O_2 and H^+ concentration increased the active oxygen content and led to a higher yield [245].

Loeb *et al.* studied the bromination of toluene, m-nitro toluene and thiophene under solvent-free conditions. Aromatic compounds are pretty unreactive under normal conditions, especially in absence of catalyst and photoinitiation. The operation conditions were intensified by the use of undiluted elemental bromine and toluene mixed in an interdigital triangular micromixer. In this way, fast reaction in flow could be achieved for each of the compounds, either at room temperature (thiophene) or under superheated conditions (aromatics) [104].

Solvent-free Claisen rearrangement was performed to yield 2-allyl-4-methoxyphenol in 20 minutes in the microreactor. The yield in batch processing conversion was only 32% in 20 minutes and rose to 80% after 3 hours' reaction time. A higher purity was suggested, due to the transparency of the crude product [246].

Solvent-free operation of 2,4-dihydroxybenzoic acid from resorcinol in flow was developed using carboxylate-containing reactive ILs to increase the yield to 59% at 180 °C in 130 seconds. The overall space–time yield was increased to 64.2 t/(h m^3) [112].

Michael addition reactions of several α,β-unsaturated carbonyl compounds were performed under solvent-free conditions by Loewe *et al.* [247]. Due to the exothermicity of these reactions, the addition of the olefin proceeds had to be performed slowly into a diluted amine solution in order to avoid detrimental accidents. The total reaction time in these cases ranged from 17 to 25 hours, but could be brought down to seconds when the batch procedure was transferred to continuous and performed in a microstructured reactor. The space–time yields were increased by up to 650 times.

4.4 Process Design Intensification

4.4.1 Fundamentals

Third intensification field is a holistic approach that targets chemical processes as a whole (on a flow-sheet dimension), called 'process design intensification'. It is related to large scientific streams of complexity and systems engineering [248,249] and follows a request by the European Union and leading expert advisory groups to promote strong interdisciplinary integration among chemistry, microreactors, process engineering, process control and so on [250].

In a way, it makes use of transport and chemical intensification on a full-process level, but not simply by transferring the benefits to a higher level. The direct, transfer-related effect of transport and chemical intensification on process design intensification is:

- A decreased number of apparatuses and a decreased size of apparatus.
- Decreased expenditure for the plant environment with decreased footprint.
- A decreased number of development steps [251].

All of which, in principle, can also lead to a significantly decreased overall development time. On the other hand, additional 'rethinking' and switching to flow processes such as the exploration of new reaction pathways and new ways of implementing harsh process conditions will require time [4].

Microflow process design is currently not covered or supported by existing education/-training or plant and process computing platforms, such as the Aspen software. It is different from industrial batch and flow processing (large-tubing based) for the following reasons [252]:

- No standard modules exist. The intensified (by orders of magnitude) operating conditions of the diverse components often do not match each other (there is no completely settled set) or existing process control equipment.

- As a result, no additive (plug-and-play) design from modules into a chained plant is possible. Rather, design is interactive: each flow module development has the potential to change the whole plant design, forcing development on all other components.
- Flow process design is reaction-dominant, whereas today's plants are utility- and purification-dominated – with totally different cost, process control and performance structures.
- Integrated design (process systems) has to be considered a new interactive platform step between the classical component and plant levels.

The most evident effect on a full process level is that higher selectivity and lower solvent loads of the processes due to the intensified operation can lead to a reduction in separation requirements and equipment sizes. In a more advanced but not so distant step, due to the compactness of the basic elements within the units and high speed of processing, integrated reactors will partly replace consecutive arrangement of otherwise separated units. Today, heating and reaction or exothermic and endothermic reactions are combined in microreactors [253–256]. This is largely given for the field of fuel processing, utilizing mainly integrated microreactors that comprise energy and mass flow integration; it is less prominent in fine chemistry [257]. The next logical step is to combine separation and reaction, and actually that has started [251]. Another eminent integration issue is to incorporate functionalities such as process control and analysis [258]. Usually, the plant size is defined by pre- or post-reaction units. Corresponding equipment number and size reduction has a direct impact on the plant footprint. Pioneering vision of process intensification was given by Ramshaw [259–263]. Beyond the equipment size, the decrease is also aimed for the CAPEX costs, environmental footprint, and even the process development time [263]. Thus, there is extra large room and impact for flow separation developments, in a period of ten years or more. Even processes such as crystallization, not per se suited to be handled in flow, are nowadays investigated for their potential in continuous small-scaled processing.

The possibilities of step change may be realized via exploration of new reaction designs (pathways), which involves circumvention of additional steps such as the introduction of protecting groups, safer and longer routes and so on. Such single-step chemistries have the potential to reduce the cascades of separation units otherwise needed for multistage process chemistries [264]. Consequently, they can have a major impact on process design. Time spent on the development of separation units or completely new chemistries has to be the criterion on which to choose the 'correct' route in each particular case.

Apart from such unit or system improvements, not directly related anymore to process-design intensification, is the transformation from batch to continuous operational mode, i.e. the vision is to have a modular plant environment for all the development stages, lab, pilot and finally production [265–269].

Process design intensification will require the design of new equipment and its synchronization. The reaction part and the coupling of chemistry in integrated reactors will particularly benefit from new reaction design, as already discussed. A single-step synthesis can simplify the process via elimination of the separation equipment needed for otherwise consecutive reactions. If they cannot be eliminated, there is a chance of replacing separation units with simpler and less expensive ones. Recycle looping can be introduced to increase efficiency and the extent of upstream reaction. Other options for process simplification are provided by the exchange of heating/cooling units through room-temperature processing and by the reduction of expenditure on process control units due to self-control/-sustainment.

The combination of process steps can have a time dimension, where fast, simultaneous operation of several steps occurs within one process (such as multiple reactant dosing) or where several reactions are combined in one process through the flow chemistry adaptation of a multi-step one-pot synthesis. It may also have a spatial dimension through the integration of function and system; that is, combined reaction and separation or combined reaction and heat exchange. The elimination of process steps can be achieved by the elimination and exchange of heating/cooling units through room-temperature processing and of purification units through a decrease in the level of byproducts to trace amounts.

4.4.2 Large-scale Manufacture of Adipic Acid – A Full Process Design Vision in Flow

Process design intensification was introduced by Hessel *et al.* and is still pretty new in the literature [263]. We shall thus focus here on one reaction example, which was the first we investigated and for which publication is currently underway [270].

This first-hour investigation concerns the manufacture of adipic acid, which is an important chemical for the nylon industry. About 80% of worldwide adipic acid consumption is used for the manufacture of nylon 6,6 fibres and resins. It is also largely used in the manufacture of polyesters, for use in plasticizers, lubricants and a variety of polyurethane resins [271].

Without exception, current commercial production processes for adipic acid are carried out in two stages. The first involves the production of so-called KA oil (a mixture of cyclohexanone, the ketone or K component, and cyclohexanol, the alcohol or A component) and the second is its subsequent oxidation to adipic acid using an excess of strong nitric acid in the presence of copper and vanadium catalysts. In the second step, adipic acid is obtained in 92–96% yield [272]. By far the most common route to the KA oil needed for adipic acid production is the air oxidation of cyclohexane (see Scheme 4.9); this is either cobalt-catalysed, boric acid promoted or performed as a high-peroxide process [273].

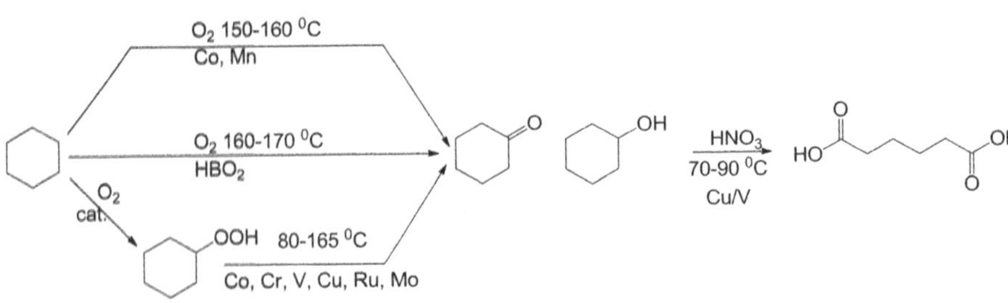

Scheme 4.9 *Air oxidation of cyclohexane.*

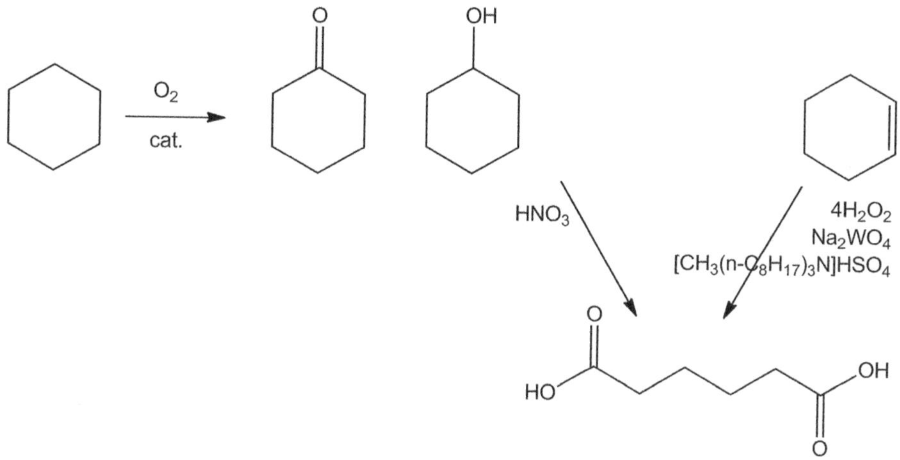

Scheme 4.10 *Two-step (left) and direct (right) routes to adipic acid.*

In order to promote selectivity (85–90%) in the first reaction step, low conversion levels per pass are maintained (4–8%). Unconverted cyclohexane must be recovered by distillation and recycled. In the second step, high c (50–60 wt%) and large consumption of nitric acid cause serious problems. Nitric acid is highly corrosive and produces significant quantities of nitrous oxide, N_2O and NOx. The N_2O emission from the adipic acid process causes global warming and ozone, and corresponds to 5–8% of the total amount released worldwide by humans [274].

Single-step all-air oxidation is economically attractive and has been heavily researched [285]. Along these lines, direct oxidation (see Scheme 4.10) of cyclohexene to adipic acid with 30% hydrogen peroxide (H_2O_2) over Na_2WO_2 and $[CH3(n\text{-}C_8H_{17})3N]HSO_4$ as a phase-transfer catalyst has been reported [275,276]. This aqueous, organic biphasic reaction enables high yield (93%) under conditions that are entirely free of organic solvents and halides. H_2O_2 is a clean oxidant, with water its sole byproduct.

This new route thus has theoretical potential (but currently no more than that) for massive process simplification – provided that the new reaction design works on a large scale. Issues to be solved, even when using microreactor technology, are the limits in interfacial transfer, the problem of safely handling H_2O_2, exploration of new, harsher process chemistries and how to achieve better selectivity at much reduced reaction times (first transport and second chemical intensification field).

In order to compare the commercial route with the direct synthesis route, process simulations have been. The commercial route was modelled with data from the industrial production process (see Figure 4.17a) and the direct route was modelled as a continuous process, essentially requiring microreaction technology for the reasons already mentioned (see Figure 4.17b). As can be seen from the process flow diagrams, the direct route leads to a more compact plant design. Some energy-intensive separation steps are omitted, which otherwise might impact human toxicity potential. In the nitric acid post-treatment, NOx gases need to be collected, reabsorbed and converted back to nitric acid. Moreover, low cyclohexane conversion requires large separation expenditure and a high

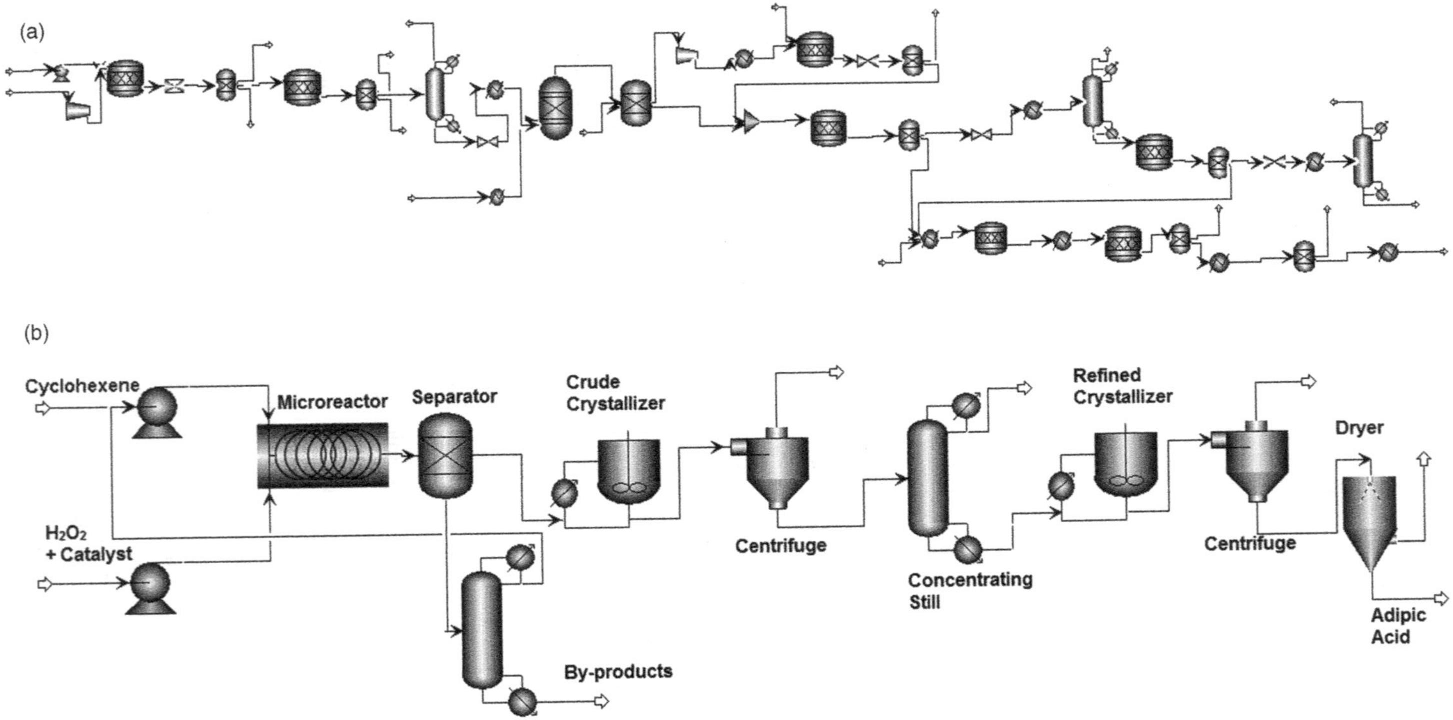

Figure 4.17 Commercial (two-step) (a) that involves oxidation of cyclohexanol and subsequent cleavage of double bond (not shown here) under acidic conditionsvs direct adipic acid synthesis route (b) that is based on cyclohexene oxidation with hydrogen peroxide.

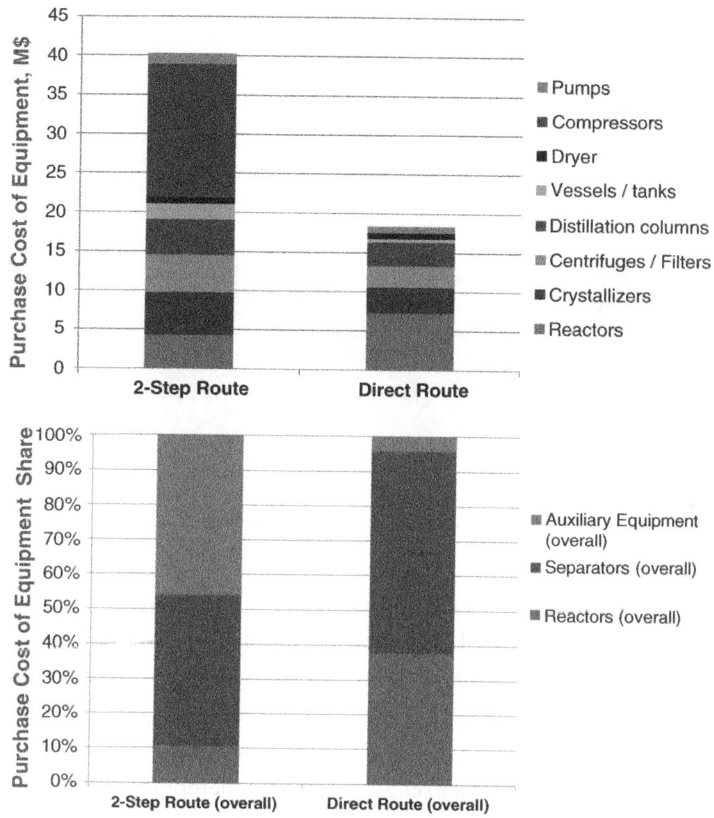

Figure 4.18 *Cost splitting for commercial, two-step versus direct adipic acid production.*

recycle ratio. On the other hand, in the direct flow route conversions per pass are much higher and separation/recycling is reduced. Water is derived as the only byproduct, facilitating waste treatment.

In sum, this means that engineering, procurement and construction phases can be performed significantly faster with much less equipment required. The total capital investment is significantly reduced as well, while the costs of the more advanced (flow) reactor are somewhat increased. The proportion of reaction and separation costs shifts from separation dominance to reaction dominance (see Figure 4.18).

4.4.3 Process Integration – From Single Operation towards Full Process Design

Beyond such full process design analysis, intensification beyond the transport and chemical level can be applied to parts of the flow process sheet by process integration of various scales.

4.4.3.1 Integration of Reaction and Cooling/Heating, Separation or Other

Integration of (unit) operations has been demonstrated quite widely for microchips, with their diverse single microfluidic processes. The small-scale operation can lead to complexities in the

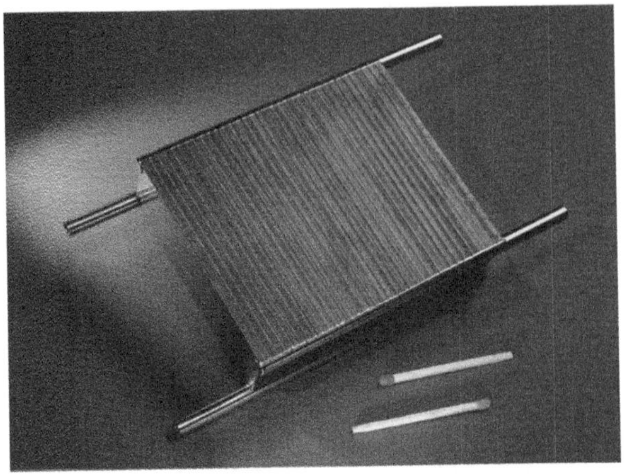

Figure 4.19 *Integrated microstructured unit for steam reforming and catalytic combustion. Courtesy of Institut für Mikrotechnik Mainz GmbH. Courtesy of 2007 Institut für Mikrotechnik Mainz GmbH.*

underlying physics and chemistry. The differences in the time constants of the participating units also need to be approached [277]. Beyond the microchip level, the integration of operations (mostly just two) has also been carried out for the microstructured counterparts.

Micro Fuel Processing. Micro fuel processing systems are the forerunners in flow integration of reactor–reactor or reactor–process unit integration into one system; process units are afterburners, evaporators, micro heat exchangers or even membranes. With the exception of the latter case, thermal coupling to improve efficiency is the goal, which leads to a compact overall system design. Micro fuel processors consist of several such thermally coupled systems, typically four to six.

Integration of a microstructured steam reformer and a catalytic combustor was carried out for methane-based hydrogen generation (see Figure 4.19). Under static load conditions methane conversion rates were as high as about 70%. Dynamic tests show that less than 20 seconds is required to reach static performance after the introduction of methane to the reformer [278].

An integrated microburner/thermoelectric device allows self-start at room temperature by catalytic self-ignition of methanol over a platinum/alumina catalyst. The thermoelectrical power is 0.65 W with a thermal efficiency of 1.1%. At high flow rates, steady-state operation is rapidly obtained, lowering fuel utilization. All these measures lead to a compact integration of the microburner/thermoelectric system [279].

Fine Chemical Processing. Beyond such fuel processing applications, reactor–process unit integration has been introduced into fine chemical applications.

A microreaction system, consisting of an enzyme-immobilized microreactor and a microextractor, was used for the production of an optically pure amino acid from initially racemic mixture. An acetyl-D,L-amino acid mixture was hydrolysed enantioselectively at the acylase-CEM to give the L-amino acid and the unhydrolysed acetyl-D-amino acid. A high enantioselectivity (99.2–99.9%[2] ee) to the L-isomer was achieved.

[2]enantiomeric excess

Through addition of HCl, the acetyl-D-amino acid was neutralized, increasing its ethyl acetate solubility, while the L-amino acid was separated by water. Separation was carried out through a membrane on the microchannel surface, which enabled the selective extraction of the product [280].

Knoevenagel reactions between benzaldehyde and ethyl cyanoacetate, ethyl acetoacetate and diethyl malonate were carried out in a membrane microreactor. Water generated as a result of condensation was removed from the reaction mixture by pervaporation through a zeolite micromembrane. The best reaction yield was obtained by locating the CsNaX catalyst adjacent to the membrane. For microreactor operation with powder catalyst, a selectivity of 58% is found; the presence of a thin CsNaX film led the selectivity increase to 78% [281,282].

4.4.3.2 Integration of Process Control and Sensing

The integration of microreactors with sensors, actuators and automated fluid handling is useful for fast, continuous automation of reaction screening and optimization using online monitoring [283].

A microreactor system for self-contained and standalone portable energy generation contained integrated microfabricated sensors and actuators. The latter were used for effective control strategy, especially concerning system start-up and controllability for rapid load changes. On-chip sensing was done with a MEMS-based thin-film flow sensor to measure the rate of hydrogen production. In this way, 100% conversion, rapid turn-on and responsiveness to varying hydrogen loads were achieved [284].

An integrated gas-phase microreactor system comprised a 1 μm thin platinum film catalyst coated underneath a silicon nitride membrane. This system was assembled as a computer chassis with modular boards. Seven platinum heaters and temperature sensors were placed on top of the silicon nitride membrane. Various pins were used as electrical contacts to allow onboard controlling and sensing [285].

The same group, in cooperation with the DuPont Company, further constructed a MEMS-based system from first principles that had the same functionalities as a more conventional laboratory-scale system used by DuPont. The components were designed and fabricated as prototype electromechanical boards [286].

4.4.3.3 Thermal Integration on the Process Level

Complete thermal integration of a methanol micro fuel processor and fuel cell system is essential to reducing full system size and enabling portability. With such a goal in mind, heat losses were experimentally determined for various pathways from the planar microreactor structure for catalytic methanol steam reforming. This allowed fundamental heat management issues to be addressed, such as the transfer of heat between reactor components, control of temperature, insulation and heat losses. In this way, a design/packaging and scale-up proposal with reduced convective and radiative losses was made [287,288].

4.4.3.4 Integration of Units on Racks, Backbones, Frames, Interfaces or Similar

A commercial plastic socket, normally used for integrated circuit testing, was adapted for installation of the reactor chip to allow alignment with the electrical contacts.

A heated fluidics interface connected the nonmetallic feed and product gas ports on the microreactor chip to some metal tubing. The microreactor was heated to about 600 °C. Step-response tests showed that the temperature could be increased by 100 °C in less than 10 milliseconds [285].

In the next step on the plant hierarchy, modular circuit-like boards that fit the slots of a computer chassis were used to carry a silicon reactor chip. In this way, control functions for the gas flow rate and the reaction temperature were provided. System automation was performed using a process logic controller. A closed-loop proportional–integral–derivative control was operated at 500 Hz with 24 microreactor heaters [289].

4.4.4 Process Simplification

4.4.4.1 Fast Processing

Both exothermicity and the high speed of the reaction of dichloro or dibromo alkyl-aromatics with concentrated sulfuric acid and the respective aldehydes require slow dropwise addition of the sulfuric acid. The high viscosity of reactants further complicates this process due to more problematic mixing and the decreased heat transfer, which results in a higher probability of hot spot formation. The continuous operation in a microreactor allowed the introduction of benzal chloride in the form of finely dispersed droplets into the continuous phase of concentrated sulphuric acid [130]. A yield of 69%, comparable to the batch yield, was achieved in 8 seconds, whereas the batch reaction time is 3 hours.

4.4.4.2 Simplification on the Separation Part

Separation simplification has already been discussed for adipic acid (Section 4.4.2). In a work by both Clariant GmbH and IMM, such simplification on the separation side was achieved for the phenyl boronic acid process using a microreactor. Phenyl boronic acid is made from phenyl magnesium bromide and trimethyl boronic acid ester.

The processing is done in batch at cryogenic temperature in order to avoid a negative impact of mixing (masking) on the selectivity of the organometallic reaction. The industrial batch synthesis is operated at −40 to −50 °C with 65% yield. Even slight temperature increases suffice to reduce the yield by some 10%. Flow processing with a micromixer-tube reactor resulted in 20% higher yields at much increased temperatures, with a maximum yield of 89% at 20 °C. Moreover, a much higher purity (99%) was achieved for the raw product as compared to the conventional process (82%). Phenyl-boronic acid synthesis can thus be conducted at ambient temperature instead of under cryo conditions, which saves energy. In addition, the higher purity of the raw product eliminates the need for capital- and energy-intensive distillation, which further saves energy. The rather simpler extraction or crystallization of the product suffices to generate the final pure product [290].

4.4.4.3 Catalyst Omission

Process simplification can also be achieved by turning a formerly catalytic reaction into a noncatalytic one. The uncatalysed selective oxidation of liquid cyclohexane with air was performed in a microcapillary reactor. A temperature increase from 453 to 533 K sped up the reaction by a factor of 30, allowing a reduction of the inner reactor hold-up from 500 to 8 m^3 and a decrease in the power of the explosion by factor of 6.3 [291].

Process simplification was achieved for the core bromination of toluene and for photoirradiation in the side-chain bromination of toluene and m-nitro toluene through avoidance of the need for Lewis acid catalysts in core substitution and in photointiation for side-chain substitution, respectively. Catalyst-free Toluene bromination was completed by Loeb *et al.* within 2.0–3.9 minutes, and at high Ts practically all brominated species were core-substituted. The photoinitiation-free bromination of m-nitrotoluene in the side-chain position was performed at elevated temperatures (170–230 °C). The space–time yield was multiplied by a factor of 18 and the turnover of nitrotoluene increased from 40 to 95% [292].

The click chemistry synthesis of 5-substituted tetrazoles from cyanides and NaN_3 in a microreactor was performed safely and efficiently and with a broad substrate scope [293]. Process simplification was achieved, since no metal promoter was needed and a near-equimolar amount of NaN_3 was sufficient, with no need for an excess. The synthesis was also safer because shock-sensitive metal azides such as $Zn(N_3)_2$ are avoided (as HN_3 is formed *in situ*) and unreacted NaN_3 is quenched in line with $NaNO_2$. High-T (190 °C) processing leads to high reaction rates and short reaction times. Moreover, compared to batch processing, the flow synthesis is scaleable.

High p is currently being investigated as a candidate for Cu-catalyst replacement in 1,3-dipolar cycloaddition with preservation of regioselectivity under continuous flow conditions. Hessel and coworkers developed a home-built high-p,T setup (see Figure 4.20) with limits of 300 °C and 400 bar [193,194]. Multistep synthesis of the anticonvulsant drug Rufinamide (see Scheme 4.11) is to be speeded up and brought to a commercial scale.

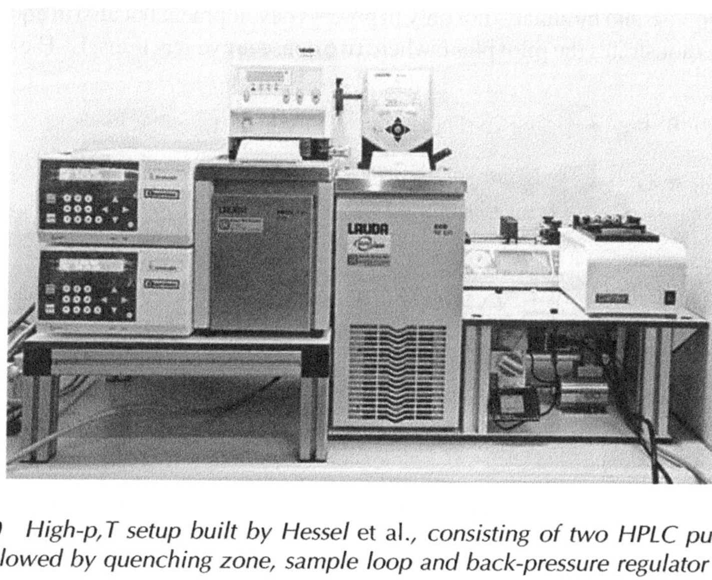

Figure 4.20 *High-p,T setup built by Hessel* et al., *consisting of two HPLC pumps, high-T reaction followed by quenching zone, sample loop and back-pressure regulator [193,194].*

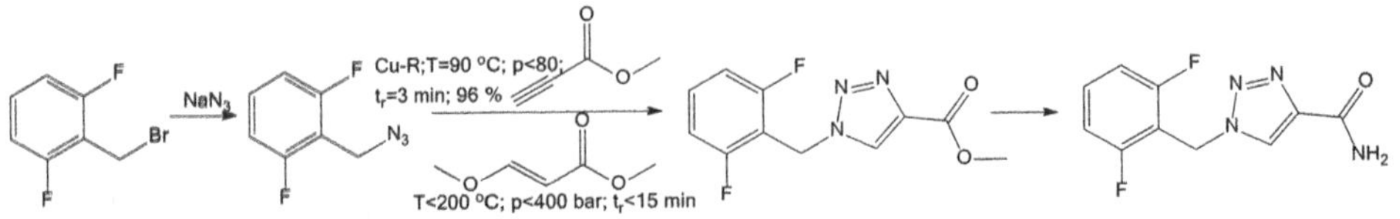

Scheme 4.11 *Multistep synthesis of Rufinamide, consisting of nucleophilic substitution, [3+2] cycloaddition and amidation.*

4.5 Industrial Microreactor Process Development

In the literature, one can find a number of statements by industrialists demonstrating the use of microreactors and intensified flow processes in industrial production [294]. Compilations are available which summarize this in the scientific literature [295–297]. Five recent widely published industrial demonstration activities concerning flow manufacture give some insight into exploitation, completing this chapter – two in specialty/pharma chemistry, two in fine chemistry and one in bulk chemistry.

4.5.1 Industrial Demonstration of Specialty/Pharma Chemistry Flow Processing

Lonza, Visp/Switzerland, has probably given the most insight into a chemical company's approach and success in the process development of continuous flow production with micro- and other flow reactors; to sum up all reports from Lonza, at least 10 reactions were transferred to production, but there are probably more [298–301]. Lonza describes four development phases, with increasing time and complexity [302]. The first phase is for proof of concept, the second is an optimization study, the third is a long-run study (kg-production) and the fourth finishes with pilot production and commercial manufacture. This involves kg-production for 1–20 kg campaigns and pilot production of from 20 kg to 2.5 tons. Lonza's concept for scale-up is stepwise and modular – not only in process development but also in equipment use. Figure 4.21 demonstrates the pilot plant, where two consecutive reactions, Li-H exchange and

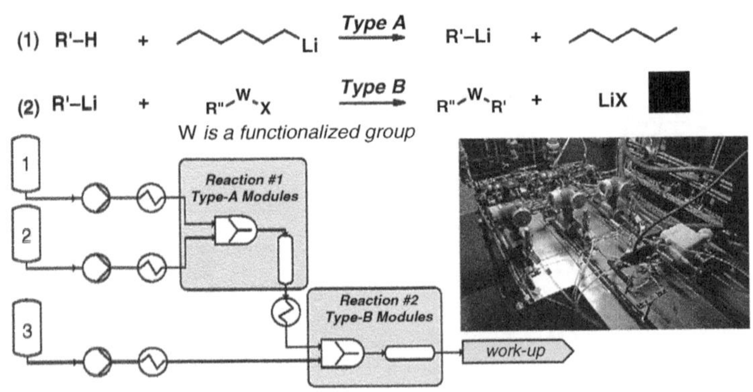

Figure 4.21 *Reaction scheme and process diagram of a modular plant based on a microreactor system, combining three process steps – lithiation, coupling and purification. Courtesy of 2010 Elsevier, [302]. Reprinted from [ref 212] © 2011, with permission from Elsevier.*

coupling, are carried out in a continuous manner and followed up by work-up. Three classes of microstructured reactors are used, each assigned to one development stage [303].

DSM Company in Austria and NicOx Company in France developed a pilot-scale process based on numbering-up multiple parallel microreactors to make an arthritis drug that is the first in a new class of cyclooxygenase-inhibiting nitric oxide donators. The drug, naproxcinod, is the nitroxybutyl-substituted form of naproxen, the well-known non-steroidal anti-inflammatory drug [304]. This compound-containing nitro group is difficult to make, since the corresponding nitration reaction is highly exothermic. With Corning, Inc., a microreactor system was designed that combines three process steps: the nitration reaction, neutralization and work-up. The pilot-scale process by DSM, with a throughput of several hundred kilograms, was achieved after about six months' work. NicOx reported positive phase III trial results for the treatment of osteoarthritis with naproxcinod. The time from feasibility studies to large-scale production was only about 18 months.

4.5.2 Industrial Demonstration of Fine Chemistry Flow Processing

The Ritter reaction to a polymer intermediate was performed at 3600 kg/hour throughput in a microstructured reactor. This was achieved by cooperation between the Karlsruhe Institute of Technology and DSM in Linz/Austria [305]. Heat removal of this exothermic issue after fast reaction initiation by micromixers was the key to the development. Thus, an integrated microstructured reactor heat exchanger was developed and applied to production. To the best of our knowledge, this microreactor was inserted in a retrofit manner in the existing production plant, which facilitated development time and industrial acceptance of the new technology.

The same retrofit strategy was applied to an industrial process whose reaction could not be named for reasons of confidentiality; still, a large number of process details were shared with the public by the process developer Microinnova [306]. Here, the microreactor key component was an even simpler device, namely a microstructured mixer (based on multi-injection dosing). Very fast reactant contacting replaced the former much slowed-down dosing of one reactant into another. The reason for the latter was to avoid evaporation into the gas atmosphere above the liquid due to local overheating. Faster mixing favours even faster heat release, but evaporation can be suppressed by a pressurized encased environment and the heat transfer accomplished directly behind the mixing zone by standard tubing. In this way, a production of 3000 l/hour was achieved.

In both cases, fast development times through to production of no more than 3 years were achieved, which shows that the predictions made in this chapter are corroborated by achievements in practice.

4.5.3 Industrial Demonstration of Bulk Chemistry Flow Processing

Velocys (Oxford Catalysts Group PLC) and Petróleo Brasileiro S.A. (Petrobras), alongside Toyo Engineering Corporation and MODEC Construction, explored the use of micro process technology in the field of mobile gas–liquid conversion [307–309]. An integrated gas–liquid demonstration plant (CompactGTL) couples steam methane reforming (SMR) and Fischer–Tropsch (FT) synthesis in an integrated two-stage system with a series of minichannels (2.5–5 cm). The module weight is <25 tonnes, with production of 200 bbl/day. The FT microchannel reactors (see Figure 4.22a) have footprints of

(a)

(b)

Figure 4.22 *(a) FT microchannel reactor. (b) Concept of the Velocys FPSO GTL concept to be applied on FPSO vessels. Courtesy of Oxford Catalysts Group, [307–309]. Image courtesy of Oxford Catalysts Group.*

0.61×0.61 m and the combined SMR/FT system will amount to 2.4×8 m at a production of 300 barrels/day. The technology is to be applied on floating production, storage and offloading (FPSO) vessels (see Figure 4.22b) used by the offshore industry for the processing of hydrocarbons produced at nearby platforms or subsea templates.

4.6 Conclusion

Micro process technology has become a widely accepted reaction engineering tool for transport intensification – that is, for the improvement of mixing (including multiphase mass transfer) and heat exchange, each combined with a defined setting of residence times. These speed-ups of transport are especially useful in challenging ultrafast metallorganic reactions and highly exothermic reactions, respectively. Recently, chemists approached this technology as a versatile tool in their quest for process automation and optimization and coined such scientific endeavours 'flow chemistry'. It was the chemists, together with the authors of this chapter, who – instead of 'only' releasing the intrinsic reaction potential of a chemical reaction – aimed to use the microreactor as a tool by which to actively push the chemistry to its limits.

In this way, the idea of novel process windows was developed. These represent a gateway to entirely new operating conditions, intensifying the reaction kinetics at the small scale and changing the overall process scheme at the large. A favourite tool with which to accomplish novel process windows is microstructured equipment, with its capabilities and modularity, which jointly lead to a higher selectivity, lower energy input per product, lower amount of byproduct (and thus lower waste generation) and higher yields. It is proposed that chemical intensification be explored via harsh conditions attained by the application of high T, high p and high c (favourably under the use of designer solvents). Successful examples promoting the concept are demonstrated in this chapter.

Beyond making reactions much faster, microreactors also allow new transformations such as direct or single-flow multistep syntheses. This, together with the reduction of separation expenditure and the integration of processing functions, yields a completely different process scheme for a flow process, called 'process design intensification'.

To conclude, three intensification fields – transport, chemical and process design – for microreactors have been presented. The latter two constitute novel process windows and enrich the role of micro process technology for future chemical manufacture. Via boosted productivity and other measures, the cost structure of flow processes is affected, as is the sustainability footprint. Novel process windows provide even greater degrees of freedom to meet the demands which green chemistry and green engineering issues place on the chemical industry.

Acknowledgement

Funding by the Advanced European Research Council Grant 'Novel Process Windows – Boosted Micro Process Technology' under grant agreement number 267443 is gratefully acknowledged.

References

(1) J.-C. Charpentier, Process intensification by miniaturization. Chem. Eng. Technol., 28, 255–258 (2005).

(2) P. Gravesen, J. Branebjerg and O. Sondergard Jensen, Microfluidics – a review. J. Micromech. Microeng., 3, 168–182 (1993).

(3) D. J. Laser and J. G. Santiago, A review of micropumps. J. Micromech. Microeng., 14, R35–64 (2004).

(4) C. Wiles and P. Watts, Continuous flow reactors: a perspective. Green Chem., 14, 38–54 (2012).

(5) W. Ehrfeld, V. Hessel and H. Lowe, Microreactos: New Technology for Modern Chemistry, Wiley-VCH (2000).
(6) D. C. Hendershot, Process minimization: making plants safer. Chem. Eng. Prog. 23 (2000).
(7) R. Jachuck, Process intensification for responsive processing. Chem. Eng. Res. Dev., 80, 233–238 (2002).
(8) V. Hessel, B. Cortese and M. H. J. M.de Croon, Novel process windows – concept, proposition and evaluation methodology, and intensified superheated processing. Chem. Eng. Sci., 66, 1426–1448 (2011).
(9) K. Geyer, J. D. C. Codée and P. H. Seeberger, Microreactors as tools for synthetic chemists – the chemists' round-bottomed flask of the 21st century? Chem. Eur. J., 12, 8434–8442 (2006).
(10) H. Pennemann, V. Hessel and H. Löwe, Chemical microprocess technology – from laboratory-scale to production. Chem. Eng. Sci., 59, 4789–4794 (2004).
(11) J. Fischer, C. Liebner, H. Hieronymus and E. Klemm, Maximum safe diameters of microcapillaries for a stoichiometric ethene/oxygen mixture. Chem. Eng. Sci., 64, 2951–2956 (2009).
(12) D. M. Roberge, L. Ducry, N. Bieler, P. Cretton and B. Zimmermann, Microreactor technology: a revolution for the fine chemical and pharmaceutical industries? Chem. Eng. Technol., 28, 318–323 (2005).
(13) K. F. Jensen, Microreaction engineering – is small better? Chem. Eng. Sci., 56, 293–303 (2001).
(14) K. F. Jensen, Microchemical systems: status, challenges, and opportunities. AIChE J., 45, 2051–2054 (1999).
(15) J. C. Charpentier, In the frame of globalization and sustainability, process intensification, a path to the future of chemical and process engineering (molecules into money). Chem. Eng. J., 134, 84–92 (2007).
(16) J. C. Charpentier, Four main objectives for the future of chemical and process engineering mainly concerned by the science and technologies of new materials production, Chem. Eng. J., 107, 3–17 (2005).
(17) A. Tonkovich, D. Kuhlmann, A. Rogers, J. Mcdaniel, S. Fitzgerald, R. Arora and T. Yuschak, Microchannel technology scale-up to commercial capacity. Chem. Eng. Res. Des., 83, 634–639 (2005).
(18) J. Wegner, S. Ceylan and A. Kirschninga, Flow chemistry – a key enabling technology for (multistep) organic synthesis. Adv. Synth. Catal., 354, 17–57 (2012).
(19) S. Kee and A. Gavriilidis, Design and performance of a microstructured PEEK reactor for continuous poly-l-leucine-catalysed chalcone epoxidation. Org. Process Res. Dev., 13, 941–951 (2009).
(20) P. S. Dittrich and A. Manz, Lab-on-a-chip: microfluidics in drug discovery. Nat. Rev. Drug Discovery, 5, 210–218 (2006).
(21) J. Khandurina and A. Guttman, Microchip-based high-throughput screening analysis ofcombinatorial libraries. Curr. Opin. Chem. Biol., 6, 359–366 (2002).
(22) R. A. Potyrailo, R. J. Wroczynski, J. P. Lemmon, W. P. Flanagan and O. P. Siclovan, Fluorescence spectroscopy and multivariate spectral descriptor analysis for high-throughput multiparameter optimization of polymerization conditions of combinatorial 96-microreactor arrays. J. Comb. Chem., 5, 8–17 (2003).
(23) C. Wiles and P. Watts, Recent advances in micro reaction technology. Chem. Commun, 47, 6512–6535 (2011).
(24) V. Hessel, Novel process windows – gate to maximizing process intensification via flow chemistry. Chem. Eng. Technol., 32, 1655–1681 (2009).
(25) P. Watts and C. Wiles, Micro reactors: a new tool for the synthetic chemist. Org. Biomol. Chem, 5, 727–732 (2007).

(26) B. Ahmed-Omer, J. C. Brandt and T. Wirth, Advanced organic synthesis using microreactor technology. Org. Biomol. Chem., 5, 733–740 (2007).
(27) P. P. Lange, L. J. Gooßen, P. Podmore, T. Underwood and N. Sciammetta, Decarboxylative biaryl synthesis in a continuous flow reactor, Chem. Commun., 47, 3628–3630 (2011).
(28) F. Venturoni, N. Nikbin, S. V. Ley and I. R. Baxendale, The application of flow microreactors to the preparation of a family of casein kinase I inhibitors. Org. Biomol. Chem., 8, 1798–1806 (2010).
(29) J. de Jong, R. G. H. Lammertink and M. Wessling, Membranes and microfluidics: a review. Lab Chip, 6, 1125–1139 (2006).
(30) Y. Seike, E. Kamio, T. Ono and H. Yoshizawa, Extraction of ethyl ester of polyunsaturated fatty acids by utilizing slug flow prepared by microreactor. J. Chem. Eng. Jpn, 40, 1076–1084 (2007).
(31) G. M. Rios, M. P. Belleville, D. Paolucci and J. Sanchez, Progress in enzymatic membrane reactors – a review. J. Membr. Sci., 242, 189–196 (2004).
(32) S. M. Lai, C. P. Ng, R. Martin-Aranda and K. L. Yeung, Knoevenagel condensation reaction in zeolite membrane microreactor. Microporous and Mesoporous Mater., 66, 239–252 (2003).
(33) A. Górak and A. Stankiewicz, Intensified reaction and separation systems. Annu. Rev. Chem. Biomol. Eng., 2, 431–451 (2011).
(34) V. Kumar, M. Paraschivoiu and K. D. P. Nigam, Single-phase fluid flow and mixing in microchannels. Chem. Eng. Sci., 66, 1329–1373 (2011).
(35) S. Yao and O. Bakajin, Improvements in mixing time and mixing uniformity in devices designed for studies of protein folding kinetics. Anal. Chem., 79, 5753–5759 (2007).
(36) A. Cantu-Perez, Shuang Bi, S. Barrass, M. Wood and A. Gavriilidis, Residence time distribution studies in microstructured plate reactors. Appl. Therm. Eng., 31, 634–639 (2011).
(37) M. Worner, Approximate residence time distribution of fully develop laminar flow in a straight rectangular channel. Chem. Eng. Sci., 65, 3499–3507 (2010).
(38) J. Aubin, M. Ferrando and V. Jiricny, Current methods for characterising mixing and flow in microchannels. Chem. Eng. Sci., 65, 2065–2093 (2010).
(39) J. Aubin, L. Prat, C. Xuereb and C. Gourdon, Effect of microchannel aspect ratio on residence time distributions and the axial dispersion coefficient. Chem. Eng. Process., 48, 554–559 (2009).
(40) N. Nguen, Micromixers: Fundamentals, Design and Fabrication, 2nd edition, Elsevier: Waltham, MA, 10–22 (2012).
(41) I. Papautsky, J. Brazzle, T. Ameel and A. B. Frazier, Laminar fluid behavior in microchannels using micropolar fluid theory. Sens. Actuators, A, 73, 101–108 (1999).
(42) D. J. Beebe, G. A. Mensing and G. M. Walker, Physics and application of microfluidic in biology. Annu. Rev. Biomed. Eng., 4, 261–286 (2002).
(43) J. Baldyga and J. R. Bourne, Simplification of micromixing calculations. I. Derivation and application of new model. Chem. Eng. J., 42, 83–92 (1989).
(44) B. Weigl, R. Bardell and C. Cabrera, Lab-on-a-chip for drug development. Adv. Drug Deliv. Rev., 55, 349–377 (2003).
(45) S. Taghavi-Moghadam, A. Kleemann and K. Georg Golbig, Microreaction technology as a novel approach to drug design, process development and reliability. Org. Process Res. Dev., 5, 652–658 (2001).
(46) T. J. Johnson, D. Ross and L. E. Locascio, Rapid microfluidic mixing. Anal. Chem., 74, 45–51 (2002).
(47) F. G. Bessoth, A. J. deMello and A. Manz, Microstructure for efficient continuous flow mixing. Anal. Commun., 36, 213–215 (1999).
(48) L. Falk and J. M. Commenge, Characterization of mixing and segregation in homogeneous flow systems, in Micro Process Engineering, Vol. 1: Fundamentals, Operations and Catalysts, V. Hessel, A. Renken and J. C. Schouten (Eds.), Wiley-VCH, pp. 147–170 (2009).

(49) A. Bertsch, S. Heimgartner, P. Cousseaub and P. Renauda, Static micromixers based on large-scale industrial mixer geometry. Lab Chip, 1, 56–60 (2001).
(50) N. Nguyen and Z. Wu, Micromixers – a review. J Micromech Microeng, 15, R1 (2005).
(51) V. Hessel, H. Lowe and F. Schonfeld, Micromixers – a review on passive and active mixing principles. Chem Eng Sci, 60, 2479–2501 (2005).
(52) G. Jeong, S. Chung, C. Kim and S. Lee, Applications of micromixing technology. Analyst, 135, 460–473 (2010).
(53) S. Dreher, N. Kockmann and P. Woias, Characterization of laminar transient flow regimes and mixing in T-shaped micromixers. Heat Transfer Eng., 30, 91–100 (2011).
(54) L. Capretto, W. Cheng, M. Hill and X. Zhang, Micromixing within microfluidic devices. Top. Curr. Chem., 304, 27–68 (2011).
(55) B. He, B. J. Burke, X. Zhang, R. Zhang and F. E. Regnier, A picoliter-volume mixer for microfluidic analytical systems. Anal. Chem., 73, 1942–1947 (2001).
(56) A. Kamholz, B. Weigl, B. Finlayson and P. Yager, Quantitative analysis of molecular interaction in a microfluidic channel: the T-sensor. Anal. Chem., 71, 5340–5347 (1999).
(57) D. Gobby, P. Angeli and A. Gavriilidis, Mixing characteristics of T-type microfluidic mixers. J. Micromech. Microeng., 11, 126 (2001).
(58) R. Ismagilov, A. Stroock, P. Kenis, G. Whitesides and H. Stone, Experimental and theoretical scaling laws for transverse diffusive broadening in two-phase laminar flows in microchannels. Appl. Phys. Lett., 76, 2376–2379 (2000).
(59) A. Soleymani, E. Kolehmainen and I. Turunen, Numerical and experimental investigations of liquid mixing in T-type micromixers. Chem. Eng. J., 135, S219–S228 (2008).
(60) S. Wong, P. Bryant, M. Ward and C. Wharton, Investigation of mixing in a cross-shaped micromixer with static mixing elements for reaction kinetics studies. Sens. Actuators. B Chem, 95, 414–424 (2003).
(61) S. Wong, M. Ward and C. Wharton, Micro T-mixer as a rapid mixing micromixer. Sens. Actuators. B Chem, 100, 359–379 (2004).
(62) S. Hardt, K. S. Drese, V. Hessel and F. Schonfeld, Passive micromixers for applications in the microreactor and μTAS field. Microfluid. Nanofluid., 1, 108–118 (2005).
(63) K. S. Drese, Optimization of interdigital micromixers via analytical modeling – exemplified with the SuperFocus mixer. Chem. Eng. J., 101, 403–407 (2004).
(64) P. Lob, K. Drese, V. Hessel, S. Hardt, C. Hofmann, H. Lowe, R. Schenk, F. Schonfeld and B. Werner, Steering of liquid mixing speed in interdigital micro mixers-from very fast to deliberately slow mixing. Chem. Eng. Technol., 27, 340–345 (2004).
(65) S. Ehlers, K. Elgeti, T. Menzel and G. Wießmeier, Mixing in the offstream of a microchannel system. Chem. Eng. Process., 39, 291–298 (2000).
(66) S. Hardt and F. Schonfeld, Laminar mixing in different interdigital micromixers: II. Numerical simulations. AIChE J., 49, 578–584 (2003).
(67) V. Hessel, S. Hardt, H. Lowe and F. Schonfeld, Laminar mixing in different interdigital micromixers: I. Experimental characterization. AIChE J., 49, 566–577 (2003).
(68) W. Ehrfeld, V. Hessel and V. Haverkamp, Microreactors, in Ullmann's Encyclopedia of Industrial Chemistry, Wiley-VCH: Weinheim (1999).
(69) T. Veenstra, T. Lammerink, M. Elwenspoek and A.van den Berg, A characterization method for a new diffusion mixer applicable in micro flow injection analysis systems. J. Micromech. Microeng., 9, 199 (1999).
(70) J. Cha, J. Kim, S. Ryu, J. Park, Y. Jeong, S. Park, H. Kim and K. Chun, A highly efficient 3D micromixer using soft PDMS bonding. J. Micromech. Microeng., 16, 1778 (2006).
(71) C. Lin, C. Tsai and L. Fu, A rapid three-dimensional vortex micromixer utilizing self-rotation effects under low Reynolds number conditions. J. Micromech. Microeng., 15, 935 (2005).

(72) Y. Chung, Y. Hsu, C. Jen, M. Lu and Y. Lin, Design of passive mixers utilizing microfluidic self-circulation in the mixing chamber. Lab. Chip., 4, 70–77 (2004).

(73) K. Jahnisch, V. Hessel, H. Lowe and M. Baerns, Chemistry in microstructured reactors. Angew. Chem. Int. Ed., 43, 406–446 (2004).

(74) S. Hardt, T. Dietrich, A. Freitag, V. Hessel, H. Loewe, C. Hofmann, A. Oroskar and F. Schonfeld, Radial and tangential injection of liquid/liquid and gas/liquid streams and focusing thereof in a special cyclone mixer, in Sixth International Conference on Microreaction Technology, IMRET 6, New Orleans, USA, I. Rinard and B. Hoch (Eds.), AIChE Publications, 164, 329–344 (2002).

(75) J. Lee and S. Kwon, Mixing efficiency of a multilamination micromixer with consecutive recirculation zones. Chem. Eng. Sci., 64, 1223–1231 (2009).

(76) S. Chang and Y. H. Cho, Static micromixers using alternating whirls and lamination. J. Micromech. Microeng., 15, 1397–1405 (2005).

(77) J. Knight, A. Vishwanath, J. Brody and R. Austin, Hydrodynamic focusing on a silicon chip: mixing nanoliters in microseconds. Phys. Rev. Lett., 80, 3863–3866 (1998).

(78) T. Vestad, D. Marr and J. Oakey, Flow control for capillary-pumped microfluidic systems. J. Micromech. Microeng., 14, 1503 (2004).

(79) H. Park, X. Qiu, E. Rhoades, J. Korlach, L. Kwok, W. Zipfel, W. Webb and L. Pollack, Achieving uniform mixing in a microfluidic device: hydrodynamic focusing prior to mixing. Anal. Chem., 78, 4465–4473 (2006).

(80) C. Chang, Z. Huang and R. Yang, Three-dimensional hydrodynamic focusing in two-layer polydimethylsiloxane (PDMS) microchannels. J. Micromech. Microeng., 17, 1479 (2007).

(81) C. Simonnet and A. Groisman, Two-dimensional hydrodynamic focusing in a simple microfluidic device. Appl. Phys. Lett., 87, 114104 (2005).

(82) N. Sundararajan, M. Pio, L. Lee and A. Berlin, Three-dimensional hydrodynamic focusing in polydimethylsiloxane (PDMS) microchannels. J. Microelectromech. Syst., 13, 559–567 (2004).

(83) R. Yang, D. Feeback and W. Wang, Microfabrication and test of a three-dimensional polymer hydro-focusing unit for flow cytometry applications. Sens. Actuators. A Phys., 118, 259–267 (2005).

(84) X. Mao, S. Lin, C. Dong and T. Huang, Single-layer planar on-chip flow cytometer using microfluidic drifting based three-dimensional (3D) hydrodynamic focusing. Lab Chip, 9, 1583–1589 (2009).

(85) X. Mao, J. Waldeisen and T. Huang, 'Microfluidic drifting' – implementing three-dimensional hydrodynamic focusing with a single-layer planar microfluidic device. Lab Chip, 7, 1260–1262 (2007).

(86) T. Lim, Y. Son, Y. Jeong, D. Yang, H. Kong, K. Lee and D. Kim, Three-dimensionally crossing manifold micro-mixer for fast mixing in a short channel length. Lab Chip, 11, 100–103 (2010).

(87) F. Schönfeld, V. Hessel and C. Hofmann, An optimised split-and-recombine micro-mixer with uniform 'chaotic' mixing. Lab Chip, 4, 65–69 (2004).

(88) A. Sudarsan and V. Ugaz, Multivortex micromixing. Proc. Natl Acad. Sci. USA, 103, 7228–7233 (2006).

(89) A. D. Stroock, S. K. Dertinger, G. M. Whitesides and A. Ajdari, Pattering flow using grooved surfaces. Anal. Chem., 74, 5306–5312 (2002).

(90) A. D. Stroock, S. K. W. Dertinger, A. Ajdari, I. Mezic, H. A. Stone and G. M. Whitesides, Chaotic mixer for microchannels. Science, 295, 647–651 (2002).

(91) A. Gunther and K. F. Jensen, Multiphase microfluidics: from flow characteristics to chemical and materials synthesis. Lab Chip, 6, 1487–1503 (2006).

(92) K. Handique and M. Burns, Mathematical modeling of drop mixing in a slit-type microchannel. J. Micromech. Microeng., 11, 548 (2001).

(93) L. Yobas, S. Martens, W. Ong and N. Ranganathan, High-performance flow-focusing geometry for spontaneous generation of monodispersed droplets. Lab Chip, 6, 1073–1079 (2006).
(94) H. Song, D. L. Chen and R. F. Ismagilov, Reactions in droplets in microfluidic channels. Angew. Chem. Int. Ed., 45, 7336–7356 (2006).
(95) L. Ducry and D. M. Roberge, Controlled autocatalytic nitration of phenol in a microreactor. Angew. Chem. Int. Ed., 44, 7972–7975 (2005).
(96) T. Iwasaki, J. Yoshida, V. Hessel, C. Sera and H. Loewe, Free radical polymerization in microreactors: significant improvement in molecular weight distribution control. Macromolecules, 38, 1159 (2005).
(97) J. L. Steinbacher and D. T. McQuade, Polymer chemistry in flow: new polymers, beads, capsules, and fibers. J. Polym. Sci., 44, 6505–6533 (2006).
(98) R. D. Chambers and R. C. H. Spink, Microreactors for elemental fluorine. Chem. Commun. 883–884 (1999).
(99) R. D. Chambers, D. Holling, R. C. H. Spink and G. Sandford, Elemental fluorine. Part 13. Gas–liquid thin film microreactors for selective direct fluorination. Lab Chip, 1, 132–137 (2001).
(100) R. D. Chambers, M. A. Fox, D. Holling, T. Nakano, T. Okazoe and G. Sandford, Elemental fluorine. Part 16. Versatile thin-film gas–liquid multi-channel microreactors for effective scale-out. Lab Chip, 5, 191–198 (2005).
(101) R. D. Chambers, M. A. Fox, D. Holling, T. Nakano, T. Okazoe and G. Sandford, Versatile gas/liquid microreactors for industry. Chem. Eng. Technol, 28, 344–352 (2005).
(102) K. Kawai, T. Ebata and T. Kitazume, The synthesis of fluorinated materials in microreactors. J. Fluor. Chem., 126, 956–961 (2005).
(103) K. Jahnisch, M. Baerns, V. Hessel, W. Ehrfeld, V. Haverkamp, H. Lowe, C. Wille and A. Guber, Direct fluorination of toluene using elemental fluorine in gas/liquid microreactors. J. Fluor. Chem., 105, 117–128 (2000).
(104) P. Lob, H. Lowe and V. Hessel, Fluorinations, chlorinations and brominations of organic compounds in micro reactors. J. Fluor. Chem., 125, 1677–1694 (2004).
(105) K. Schubert, J. Brandner, M. Fichtner, G. Linder, U. Schygulla and A. Wenka, Microstructure devices for applications in thermal and chemical process engineering. Microscale Thermophys. Eng., 5, 17–39 (2001).
(106) U. Schygulla, A. Wenka, J. J. Brandner, W. Benzinger, U. Schygulla and K. Schubert, Microstructure devices for efficient heat transfer. Microgravity Sci. Technol., XIX–3/4 (2007).
(107) W. Ehrfeld, C. Grartner, K. Golbig, V. Hessel, R. Konrad, H. Lowe, T. Richter and T. Schulz, Fabrication of components and systems for chemical and biological microreactors, in Microreaction Technology, Proc. of the 1st Int. Conf. on Microreaction Technology, W. Ehrfeld (Ed.), Springer-Verlag: Berlin, 72–90 (1997).
(108) D. B. Tuckerman and R. F. W. Pease, High-performance heat sinking for VLSI. IEEE Electron Device Lett., 2, 126–129 (1981).
(109) J. J. Brandner, E. Anurjew, L. Bohn, E. Hansjosten, T. Henning, U. Schygulla, A. Wenka and K. Schubert, Concepts and realization of microstructure heat exchangers for enhanced heat transfer. Exp. Therm. Fluid Sci., 30, 801–809 (2006).
(110) W. Qu and A. Siu-Ho, Liquid single-phase flow in an array of micro-pin-fins: part II – pressure drop characteristics. J. Heat Transfer, 130, 124501-1–124501-4 (2008).
(111) S. Hardt, W. Ehrfeld, V. Hessel and K. M. Vanden Bussche, Strategies for size reduction of microreactors by heat transfer enhancement effects. Chem. Eng. Commun., 190, 540–559 (2003).
(112) F. Benaskar, V. Hessel, U. Krtschil, P. Loeb and A. Stark, Intensification of the capillary-based Kolbe-Schmitt synthesis from resorcinol by reactive ionic liquids, microwave heating, or a combination thereof. Org. Process Res. Dev., 13, 970–982 (2009).

(113) T. Henning, J. J. Brandner and K. Schubert, Characterisation of electrically powered micro heat exchangers. Chem. Eng. J., 101/1-3, 339–345 (2004).

(114) Labtrix Start: http://www.chemtrix.com/products/1/LABTRIX%C2%AE-START, last accessed 14 September 2012.

(115) C. O. Kappe, Microwave dielectric heating in synthetic organic chemistry. Chem. Soc. Rev., 37, 1127–1139 (2008).

(116) C. O. Kappe, Controlled microwave heating in modern organic synthesis. Angew. Chem. Int. Ed., 43, 6250–6284 (2004).

(117) M. Damm, T. N. Glasnov and C. Oliver Kappe, Translating high-temperature microwave chemistry to scalable continuous flow processes. Org. Process Res. Dev., 14, 215–224 (2010).

(118) N. G. Patil, A. I. G. Hermans, F. Benaskar, E. Rebrov, J. Meuldijk, L. A. Hulshof, V. Hessel and J. C. Schouten, Energy efficient and controlled flow processing under microwave heating by using a milli reactor-heat exchanger. J. Chem. Eng. Res. Dev. (submitted).

(119) D. Liedtke, in Warmebehandlung von Eisenwerkstoffen I, Vol. 8, D. Liedtke (Ed.), Expert Verlag: Renningen (2008).

(120) D. Zhao, H. Zhang, X. Zeng, Q. Xia and J. Tang, Inductive heat property of Fe_3O_4/polymer composite nanoparticles in an ac magnetic field for localized hyperthermia. Biomed. Mater., 1, 198–201 (2006).

(121) S. Ceylan, C. Friese, C. Lammel, K. Mazac and A. Kirschning, Inductive heating for organic synthesis by using functionalized magnetic nanoparticles inside microreactors. Angew. Chem. Int. Ed., 47, 8950–8953 (2008).

(122) J. Wegner, S. Ceylan, C. Friese and A. Kirschning, Inductively heated oxides inside microreactors – facile oxidations under flow conditions. Eur. J. Org. Chem., 2010, 4372–4375 (2010).

(123) G. Veser, Experimental and theoretical investigation of H_2 oxidation in a high-temperature catalytic microreactor. Chem. Eng. Sci., 56, 1265–1273 (2001).

(124) A. Leclerc, M. Alamé, D. Schweich, P. Pouteau, C. Delattre and C.de Bellefon, Gas–liquid selective oxidations with oxygen under explosive conditions in a micro-structured reactor. Lab Chip, 8, 814–817 (2008).

(125) M. Struempel, B. Ondruschka, R. Daute and A. Stark, Making diazomethane accessible for R&D and industry: generation and direct conversion in a continuous micro-reactor set-up. Green Chem., 10, 41–43 (2008).

(126) A. Kursawe and D. Hönicke, Selective oxidation of 1-butene to maleic anhydride- comparison of the performance between microchannel reactors and fixed bed reactors, in Proc. of IMRET 5: 5th Int. Conf. on Microreaction Technology, M. Matlosz, W. Ehrfeld and J. P. Baselt (Eds.), Springer-Verlag: Berlin, 240 (2001).

(127) H. Kestenbaum, A. L.de Oliveira, W. Schmidt and F. Schüth, Silver-catalyzed oxidation of ethylene to ethylene oxide in a microreaction system. Ind. Eng. Chem. Res., 41, 710 (2000).

(128) H. Kestenbaum, A. Lange de Oliveira, W. Schmidt, F. Schüth, W. Ehrfeld, K. Gebauer, H. Löwe and Th. Richter, Synthesis of ethylene oxide in a catalytic microreactor system. Stud. Surf. Sci. Catal., 130, 2741 (2000).

(129) T. Gustafsson, F. Pontén and P. H. Seeberger, Trimethylaluminium mediated amide bond formation in a continuous flow microreactor as key to the synthesis of rimonabant and efaproxiral. Chem. Commun. 1100 (2008).

(130) H. Pennemann, S. Hardt, V. Hessel, P. Löb and F. Weise, Micromixer based liquid/liquid dispersion. Chem. Eng. Technol., 28, 501 (2005).

(131) S. K. Ajmera, M. W. Losey, K. F. Jensen and M. A. Schmidt, Microfabricated packed-bed reactor for phosgene synthesis. AIChE J., 47, 1639 (2001).

(132) M. Baumann, I. R. Baxendale, L. J. Martin and S. V. Ley, Development of fluorination methods using continuous-flow microreactors. Tetrahedron, 65, 6611–6625 (2009).

(133) T. Fukuyama, M. Kobayashi, M. T. Rahman, N. Kamata and I. Ryu, Spurring radical reactions of organic halides with tin hydride and TTMSS using microreactors. Org. Lett., 10, 533 (2008).
(134) O. Klais, F. Westphal, W. Benaissa and D. Carson, Guidance on safety/health for process intensification including MS design; part I: reaction hazards. Chem. Eng. Technol., 32, 1831–1844 (2009).
(135) O. Klais, F. Westphal, W. Benaissa and D. Carson, Guidance on safety/health for process intensification including MS design; part II: explosion hazards. Chem. Eng. Technol., 32, 1966–1973 (2009).
(136) O. Klais, F. Westphal, W. Benaissa, D. Carson and J. Albrecht, Guidance on safety/health for process intensification including MS design; part III: risk analysis. Chem. Eng. Technol., 33, 444–454 (2010).
(137) O. Klais, J. Albrecht, D. Carson, M. Kraut, P. Lob, C. Minnich, F. Olschewski, C. Reimers, A. Simoncelli and M. Uerdingen, Guidance on safety/health for process intensification including MS design; part IV: case studies. Chem. Eng. Technol., 33, 1159–1168 (2010).
(138) H. Usutani, Y. Tomida, A. Nagaki, H. Okamoto, T. Nokami and J. Yoshida, Generation and reactions of o-Bromophenyllithium without benzyne formation using a microreactor. J. Am. Chem. Soc., 129, 3046–3047 (2007).
(139) J. Yoshida, Flash Chemistry: Fast Organic Synthesis in Microsystems, Wiley-Blackwell, (2008).
(140) J. Yoshida, A. Nagaki and T. Yamada, Flash chemistry: fast chemical synthesis by using microreactors. Chem. Eur. J., 14, 7450–7459 (2008).
(141) R. L. Hartman and K. F. Jensen, Microchemical systems for continuous-flow synthesis. Lab Chip, 9, 2495–2507 (2009).
(142) B. Xu, R. Zhang, X. Liu, H. Wang, Y. Zhang, H. Jiang, L. Wang, Z. Ma, J. Ku, F. Xiao and H. Sun, On-chip fabrication of silver microflower arrays as a catalyticmicroreactor for allowing in situ SERS monitoring. Chem. Commun., 48, 1680 (2012).
(143) T. McCreedy and N. G. Wilson, Microfabricated reactors for on-chip heterogeneous catalysis. Analyst, 126, 21 (2001).
(144) N. G. Wilson and T. McCreedy, On-chip catalysis using a lithographically fabricated glass microreactor – the dehydration of alcohols using sulfated zirconia. Chem. Commun. 733 (2000).
(145) R. D. Chambers and J. Hutchinson, Elemental fluorine: part 9, catalysis of the direct fluorination of 2-substituted carbonyl compounds. J. Fluor. Chem., 92, 45–52 (1998).
(146) R. D. Chambers, M. A. Fox, G. Sandford, J. Trmcic and A. Goeta, Elemental fluorine: part 20, direct fluorination of deactivated aromatic systems using microreactor techniques. J. Fluor. Chem., 1228, 29–33 (2007).
(147) N.de Mas, A. Günther, M. A. Schmidt and K. F. Jensen, Increasing productivity of microreactors for fast gas-liquid reactions: the case of direct fluorination of toluene. Ind. Eng. Chem. Res., 48, 1428–1434 (2009).
(148) M. Nobis and D. M. Roberge, Mastering ozonolysis: production from laboratory to ton scale in continuous flow. Chem. Today, 29, 56–58 (2011).
(149) M. O'Brien, I. R. Baxendale and S. V. Ley, Flow ozonolysis using a semipermeable teflon AF-2400 membrane to effect gas-liquid contact. Org. Lett., 12, 1596–1598 (2010).
(150) M. Irfan, T. N. Glasnov and C. O. Kappe, Continuous flow ozonolysis in a laboratory scale reactor. Org. Lett., 13, 984–987 (2011).
(151) T. P. Petersen, A. Polyzos, M. O'Brien, T. Ulven, I. R. Baxendale and S. V. Ley, The oxygen-mediated synthesis of 1,3-butadiynes in continuous flow: using teflon AF-2400 to effect gas/liquid contact. ChemSusChem, 5, 274–277 (2012).
(152) L. Gooßen, N. Rodríguez and K. Gooßen, Carboxylic acids as substrates in homogeneous catalysis. Angew. Chem. Int. Ed., 47, 3100–3120 (2008).

(153) T. Sakakura, J. C. Choi and H. Yasuda, Transformation of carbon dioxide. Chem. Rev., 107, 2365–2387 (2007).

(154) A. Metzger, S. Bernhardt, G. Manolikakes and P. Knochel, $MgCl_2$-accelerated addition of functionalized organozinc reagents to aldehydes, ketones, and carbon dioxide. Angew. Chem. Int. Ed., 49, 4665–4668 (2010).

(155) C. Wiles and P. Watts, Micro Reaction Technology in Organic Synthesis, CRC Press (2011).

(156) A. Nagaki, A. Kenmoku, Y. Moriwaki, A. Hayashi and J. Yoshida, Cross-coupling in a flow microreactor: space integration of lithiation and murahashi coupling. Angew. Chem. Int. Ed., 49, 7543–7547 (2010).

(157) A. Nagaki, S. Yamada, M. Doi, Y. Tomida, N. Takabayashi and J. Yoshida, Flow microreactor synthesis of disubstituted pyridines from dibromopyridines via Br/Li exchange without using cryogenic conditions. Green Chemistry, 13, 1110 (2011).

(158) T. Noël and S. L. Buchwald, Cross-coupling in flow. Chem. Soc. Rev., 40, 5010–5029 (2011).

(159) R. C. R. Wootton, R. Fortt and A. J.de Mello, On-chip generation and reaction of unstable intermediates – monolithic nanoreactors for diazonium chemistry: Azo dyes. Lab Chip, 2, 5 (2002).

(160) K. Uehata, M. Nishida and A. Nishida, A novel and efficient method for the preparation of unstable tetramethylzirconium and its application using a microflow system. Chem. Let., 41, 73–75 (2012).

(161) T. Schwalbe, D. Kadzimirsz and G. Jas, Synthesis of a library of ciprofloxacin analogues by means of sequential organic synthesis in microreactors. QSAR Comb. Sci., 24, 758–768 (2005).

(162) K. Tanaka, S. Motomatsu, K. Koyama, S. Tanaka and K. Fukase, Large-scale synthesis of immunoactivating natural product, pristane, by continuous microfluidic dehydration as the key step. Org. Lett., 9, 299–302 (2007).

(163) A. R. Bogdan, S. L. Poe, D. C. Kubis, S. J. Broadwater and D. T. McQuade, The continuous-flow synthesis of ibuprofen. Angew. Chem. Int. Ed., 48, 8547–8550 (2009).

(164) M. D. Hopkin, I. R. Baxendale and S. V. Ley, A flow-based synthesis of Imatinib: the API of Gleevec. Chem. Commun., 46, 2450–2452 (2010).

(165) T. L. LaPorte, M. Hamedi, J. S. DePue, L. Shen, D. Watson and D. Hsieh, Development and scale-up of three consecutive continuous reactions for production of 6-hydroxybuspirone. Org. Process Res. Dev., 12, 956–966 (2008).

(166) Z. Qian, I. R. Baxendale and S. V. Ley, A flow process using microreactors for the preparation of a quinolone derivative as a potent 5HT1B antagonist. Synlett, 4, 505–508 (2010).

(167) I. R. Baxendale, S. V. Ley, A. C. Mansfield and C. D. Smith, Multistep synthesis using modular flow reactors: Bestmann–Ohira reagent for the formation of alkynes and triazoles. Angew. Chem. Int. Ed., 48, 4017–4021 (2009).

(168) I. R. Baxendale, J. Deeley, C. M. Griffiths-Jones, S. V. Ley, S. Saaby and G. K. Tranmer, A flow process for the multi-step synthesis of the alkaloid natural product oxomaritidine: a new paradigm for molecular assembly. Chem. Commun. 2566–2568 (2006).

(169) I. R. Baxendale, C. M. Griffiths-Jones, S. Ley and G. K. Tranmer, Preparation of the neolignan natural product grossamide by a continuous-flow process. Chem. Inform., 37, 2566–2568 (2006).

(170) D. Grant, R. Dahl and N. D. P. Cosford, Rapid multistep synthesis of 1,2,4-oxadiazoles in a single continuous microreactor sequence. J. Org. Chem., 73, 7219–7223 (2008).

(171) T. P. Petersen, A. Ritzen and T. Ulven, A multistep continuous-flow system for rapid on-demand synthesis of receptor ligands. Org. Lett., 11, 5134–5137 (2009).

(172) A. Stefanescu, A. C. van Veen, C. Mirodatos, J. C. Beziat and E. Duval-Brunel, Wall coating optimization for microchannel reactors. Catal. Today, 125, 16–23 (2007).

(173) V. Sebastian, O. de la Iglesia, R. Mallada, L. Casado, G. Kolb, V. Hessel and J. Santamarí, Preparation of zeolite films as catalytic coatings on microreactor channels. J. Microporous and Mesoporous Materials, 115, 147–155 (2008).

(174) E. A. S. Doherty, R. J. Meagher, M. N. Albarghouthi and A. E. Barron, Microchannel wall coatings for protein separations by capillary and chip electrophoresis. Electrophoresis, 24, 34–54 (2003).
(175) M. S. Thomsen and B. Nidetzky, Coated-wall microreactor for continuous biocatalytic transformations using immobilized enzymes. Biotechnol. J., 4, 98–107 (2009).
(176) J. M. Bolivar, J. Wiesbauer and B. Nidetzky, Biotransformations in microstructured reactors: more than flowing with the stream. Trends in Biotechnol., 29, 333 (2011).
(177) D. A. Cowan and R. Fernandez-Lafuente, Enhancing the functional properties of thermophilic enzymes by chemical modification and immobilization. Enzyme Microb. Technol., 49, 326 (2011).
(178) A. Schwarz, M. S. Thomsen and B. Nidetzky, Enzymatic synthesis of β-glucosylglycerol using a continuous-flow microreactor containing thermostable β-glycoside hydrolase CelB immobilized on coated microchannel walls. Biotechnol. Bioeng., 103, 5 (2009).
(179) N. Zotova, F. J. Roberts, G. H. Kelsall, A. S. Jessiman, K. Hellgardt and K. Hii, Catalysis in flow: Au-catalysed alkylation of amines by alcohols. Green Chem., 14, 226 (2012).
(180) M. Irfan, T. N. Glasnov and C. O. Kappe, Heterogeneous catalytic hydrogenation reactions in continuous-flow reactors. ChemSusChem, 4, 300–316 (2011).
(181) T. Chinnusamy, S. Yudha, M. Hager, P. Kreitmeier and O. Reiser, Application of metal-based reagents and catalysts in microstructured flow devices. ChemSusChem, 5, 247–255 (2012).
(182) N. Hoffmann, Photochemical reactions as key steps in organic synthesis. Chem. Rev., 108, 1052–1103 (2008).
(183) P. Esser, B. Pohlmann and H. D. Scharf, The photochemical synthesis of fine chemicals with sunlight. Angew. Chem. Int. Ed., 33, 2009–2023 (1994).
(184) M. Fagnoni, D. Dondi, D. Ravelli and A. Albini, Photocatalysis for the formation of the C–C bond. Chem. Rev., 107, 2725–2756 (2007).
(185) A. M. Braun, M. Maurette and E. Oliveros, Photochemical Technology, John Wiley & Sons: Chichester (1991).
(186) F. R. Bou-Hamdanab and P. H. Seeberger, Visible-light-mediated photochemistry: accelerating $Ru(bpy)_3^{2+}$-catalyzed reactions in continuous flow. Chem. Sci., 3, 1612–1616 (2012).
(187) F. Lévesque and P. H. Seeberger, Continuous-flow synthesis of the anti-malaria drug artemisinin. Angew. Chem. Int. Ed., 51, 1706–1709 (2012).
(188) W. B. Zimmerman, Electrochemical microfluidics. Chem. Eng. Sci., 66, 1412–1425 (2011).
(189) A. Nagaki, M. Togai, S. Suga, N. Aoki, K. Mae and J. Yoshida, Control of extremely fast competitive consecutive reactions using micromixing: selective Friedel–Crafts aminoalkylation. J. Am. Chem. Soc., 127, 11 666–11 675 (2005).
(190) V. Hessel, C. Hofmann, P. Leb, J. Lehndorf, H. Lewe and A. Ziogas, Aqueous Kolbe–Schmitt synthesis using resorcinol in a microreactor laboratory rig under high-p,T conditions. Org. Process Res. Dev., 9, 479 (2005).
(191) K. F. Jensen, Silicon-based microchemical systems: characteristics and applications. Mater. Res. Soc. Bull., 31, 101–107 (2006).
(192) S. Marre, A. Adamo, S. Basak, C. Aymonier and K. F. Jensen, Design and packaging of microreactors for high pressure and high temperature applications. Ind. Eng. Chem. Res., 49, 11 310–11 320 (2010).
(193) S. Borukhova, A. C. Varas, V. Hessel, Q. Wang, P. Watts and C. Wiles, 1,3-dipolar Huisgen cycloaddition to rufinamide and the impact of pressure on reaction environment and kinetic parameters, in Proceedings of the 12th International Conference on Microreaction Technology, IMRET12, February 20–22, 2012, Lyon, France (2012).
(194) http://www.rsc.org/chemistryworld, last accessed 14 September 2012.

(195) E. R. Murphy, J. R. Martinelli, N. Zaborenko, S. L. Buchwald and K. F. Jensen, Accelerating reactions with microreactors at elevated temperatures and pressures: profiling aminocarbonylation reactions. Angew. Chem., 119, 1764–1767 (2007).

(196) R. Schmitt, Beitrag zur Kenntniss der Kolbe'schen Salicylsäure Synthese. J. Prakt. Chem, 31, 397 (1885).

(197) T. Razzaq, T. N. Glasnov and C. O. Kappe, Accessing novel process windows in a high-temperature/pressure capillary flow reactor. Chem. Eng. Technol., 32, 1702–1716 (2009).

(198) G. Shore, M. Tsimerman and M. G. Organ, Gold film-catalysed benzannulation by microwave-assisted, continuous flow organic synthesis (MACOS). Beilstein J. Org. Chem., 5, 35 (2009).

(199) T. N. Glasnov, D. J. Vugts, M. M. Koningstein, B. Desai, W. M. F. Fabian, R. V. A. Orru and C. O. Kappe, Microwave-assisted Dimroth rearrangement of thiazines to dihydropyrimidinethiones: synthetic and mechanistic aspects. QSAR Comb. Sci., 25, 509–518 (2006).

(200) R. E. Martin, F. Morawitz, C. Kuratli, A. M. Alker and A. I. Alanine, Synthesis of annulated pyridines by intramolecular inverse-electron-demand hetero-Diels–Alder reaction under superheated continuous flow conditions. Eur. J. Org. Chem. 47–52 (2012).

(201) D. Cantillo, H. Sheibani and C. O. Kappe, Flash flow pyrolysis: mimicking flash vacuum pyrolysis in a high-temperature/high-pressure liquid-phase microreactor environment. J. Org. Chem., 77, 2463–2473 (2012).

(202) M. Viviano, T. N. Glasnov, B. Reichart, G. Tekautz and C. O. Kappe, A scalable two-step continuous flow synthesis of nabumetone and related 4-Aryl-2-butanones. Org. Process Res. Dev., 15, 858–870 (2011).

(203) D. Obermayer, T. N. Glasnov and C. O. Kappe, Microwave-assisted and continuous flow multistep synthesis of 4-(pyrazol-1-yl)carboxanilides. J. Org. Chem., 76, 6657–6669 (2011).

(204) T. Razzaq, T. N. Glasnov and C. O. Kappe, Continuous-flow microreactor chemistry under high-temperature/pressure conditions. Eur. J. Org. Chem. 1321–1325 (2009).

(205) P. Loeb, V. Hessel, H. Klefenz, H. Loewe and K. Mazanek, Bromination of thiophene in microreactors. Lett. Org. Chem., 2, 767–779 (2005).

(206) M. Sato, H. Kawanami, M. Chatterjee, N. Otabe, T. Tuji, Y. Ikushima, T. Yokoyama and T. M. Suzuki, Highly selective non-catalytic Claisen re-arrangement in a high pressure and high pressure and high temperature water microreaction system. 11th International Conference on Microreaction Technology, Kyoto, Japan, 50–51 (2011).

(207) F. Benito-Lopez, R. J. M. Egberink, D. N. Reinhoudt and W. Verboom, High pressure in organic chemistry on the way to miniaturization. Tetrahedron, 64, 10 023–10 040 (2008).

(208) C. McKee, M. Mortimer, in Chemical Kinetics and Mechanism: The Molecular World, M. Mortimer and P. G. Taylor (Eds.), Royal Society of Chemistry: Cambridge, UK, p. 86 (2002).

(209) K. A. Swiss and R. A. Firestone, Phantom activation volumes. J. Phys. Chem. A, 104, 3057–3063 (2000).

(210) A. Shito, T. Takahashi, Y. Ohga, T. Asano, H. Saito, K. Matsuo and H. D. Z. Luedemann, Pressure effects on the thermal Z/E isomerization of 4-(dimethylamino)-4′-nitroazobenzene in a liquid polymer: a comparison of dynamic solvent effects in polymeric and monomeric solvents. Z. Naturforsch., 55a, 616–622 (2000).

(211) K. P. Pimparkar, D. J. Miller and J. E. Jackson, Hydrogenation of amino acid mixtures to amino alcohols. Ind. Eng. Chem. Res., 47, 7648–7653 (2008).

(212) F. Trachsel, C. Hutter and P. R. Rohr, Transparent silicon/glass microreactor for high-pressure and high-trmperature reactions. Chem. Eng. J., 135, S309–S316 (2007).

(213) R. M. Tiggelaar, F. Benito-Lopez, D. C. Hermesa, H. Rathgen, R. J. M. Egberink, F. G. Mugele, D. N. Reinhoudt, A.van den Berg, W. Verboom and H. J. G. E. Gardeniers,

Fabrication, mechanical testing and application of high-pressure glass microreactor chips. Chem. Eng. J., 131, 163–170 (2007).

(214) D. J. C. Constable, C. Jimenez-Gonzalez and R. K. Henderson, Perspective on solvent use in the pharmaceutical industry. Org. Proc. Res. Dev., 11, 133–137 (2007).

(215) C. Li and T. Chan, Water can act as green and safe reaction media for organic reactions and thus can in some instances replace organic solvents: diverse applications have been explored in both industry and academic laboratories. Comprehensive Organic Reactions in Aqueous Media, 2nd Edition, July 2007.

(216) L. Á.de Cienfuegos, R. Robles, D. Miguel, J. Justicia and M. Juan, Reduction reactions in green solvents: water, supercritical carbon dioxide, and ionic liquids. ChemSusChem, 4, 1035–1048 (2011).

(217) S. Liu and J. Xiao, Toward green catalytic synthesis – transition metal-catalyzed reactions in non-conventional media. J. Mol. Cat. A: Chem., 270, 1–43 (2007).

(218) S. Marre, Y. Roig and C. Aymonier, Supercritical microfluidics: opportunities in flow-through chemistry and materials science. J. Supercrit. Fluids (2011).

(219) A. K. Goodwin and G. L. Rorrer, Conversion of glucose to hydrogen-rich gas by supercritical water in a microchannel reactor. Industrial & Engineering Chemistry Research, 47, 4106–4114 (2008).

(220) S. Marre, J. Baek, J. Park, M. G. Bawendi and K. F. Jensen, High-pressure/high-temperature microreactors for nanostructure synthesis. JALA, 14, 367–373 (2009).

(221) E. A. Oosterbroek, D. C. Hermes, M. Kakuta, F. Benito-Lopez, J. G. E. Gardeniers, W. Verboom, D. N. Reinhoudt and A.van den Berg, Fabrication and mechanical testing of glass chips for high-pressure synthetic and analytical chemistry. Microsyst. Technol., 12, 450 (2006).

(222) F. Benito-Lopez, R. M. Tiggelaar, K. Salbut, J. Huskens, R. J. M. Egberink, D. N. Reinhoudt, H. J. G. E. Gardeniers and W. Verboom, Substantial rate enhancements of the esterification reaction of phthalic anhydride with methanol at high pressure and using supercritical CO_2 as a co-solvent in a glass microreactor. Lab Chip, 7, 1345 (2007).

(223) W. Verboom, Selected examples of high-pressure reactions in glass microreactors. Chem. Eng. Technol., 32, 1695 (2009).

(224) F. Benito-Lopez, A. J. Kettelarij, R. J. M. Egberink, A. H. Velders, R. M. Tiggelaar, H. J. G. E. Gardeniers, D. N. Reinhoudt and W. Verboom, A capillary micro-reactor as a tool to study pressurized reactions: the influence of pressure on the stereoselectivity of the Diels-Alder reaction of 2- and 3-furylmethanol. Chim. Oggi, 28, 16 (2010).

(225) F. Trachsel, B. Tidona, S. Desportes and R.von Rohr, Solid catalyzed hydrogenation in a Si/glass microreactor using supercritical CO_2 as the reaction solvent. J. Supercriti. Fluids, 48, 146–153 (2009).

(226) J. R. Hyde and M. Poliakoff, Supercritical hydrogenation and acid-catalysed reactions 'without gases'. Chem. Commun. 1482–1483 (2004).

(227) J. R. Hyde, B. Walsh, J. Singh and M. Poliakoff, Continuous hydrogenation reactions in supercritical CO2 'without gases'. Green Chem., 7, 357–361 (2005).

(228) N. M. Islam, M. Chatterjee, Y. Ikushima, T. Yokoyama and H. Kawanami, Development of a novel catalytic membrane reactor for heterogeneous catalysis in supercritical CO2. Int. J. Mol. Sci., 11, 164–172 (2010).

(229) F. Patcas, C. Maniut, C. Ionescu, S. Pitter and E. Dinjus, Supercritical carbon dioxide as an alternative reaction medium for hydroformylation with integrated catalyst recycling. Appl. Catal. B, 70, 630–636 (2007).

(230) C. Yan, J. Fraga-Dubeuil, E. Garcia-Verdugo, P. A. Hamley, M. Poliakoff, I. Pearson and A. Suart Coote, The continuous synthesis of ε-caprolactam from 6-aminocapronitrile in high-temperature water. Green Chem., 10, 98 (2008).

(231) M. Sato, K. Matsushima, H. Kawanami and Y. Ikuhsima, A highly selective, high-speed, and hydrolysis-free O-acylation in subcritical water in the absence of a catalyst. Angew. Chem. Int. Ed., 46, 6284–6288 (2007).

(232) Y. Ikushima, K. Hatakeda, O. Sato, T. Yokoyama and M. Arai, Structure and base catalysis of supercritical water in the noncatalytic benzaldehyde disproportionation using water at high temperatures and pressures. Angew. Chem. Int. Ed. Engl., 40, 210 (2001).

(233) Y. Ikushima, K. Hatakeda, O. Sato, T. Yokoyama and M. Arai, Acceleration of synthetic organic reactions using supercritical water: noncatalytic Beckmann and Pinacol rearrangements. J. Am. Chem. Soc., 122, 1908 (2000).

(234) G. Anitescu, A. Deshpande and L. L. Tavlarides, Integrated technology for supercritical biodiesel production and power cogeneration. Energy Fuels, 22, 1391 (2008).

(235) T. Razzaq and C. O. Kappe, Continuous flow organic synthesis under high-temperature/-pressure conditions. Chem. Asian J., 5, 1274–1289 (2010).

(236) N. V. Plechkova and K. R. Seddon, Ionic liquids: 'designer' solvents for green chemistry, in Methods and Reagents for Green Chemistry: An Introduction, P. Tundo, A. Perosa and F. Zecchini (Eds.), John Wiley & Sons: Hoboken, NJ, pp. 103–130 (2007).

(237) S. Mallakpour and Z. Rafiee, Ionic liquids as novel and green media for clean synthesis of soluble aromatic-aliphatic poly(amide-ester)s containing hydroxynaphthalene urazole moiety. Polym. Adv. Technol., 19, 1015–1023 (2008).

(238) J. Xu, C. Qian, B. Liu, Q. Wu and X. Lin, A fast and highly efficient protocol for Michael addition of N-heterocycles to α,β-unsaturated compound using basic ionic liquid [bmIm]OH as catalyst and green solvent. Tetrahedron, 63, 986–990 (2007).

(239) Y. Zhang, B. R. Bakshi and E. S. Demessie, Life cycle assessment of an ionic liquid versus molecular solvents and their applications. Environ. Sci. Technol., 42, 1724–1730 (2008).

(240) D. A. Waterkamp, M. Heiland, M. Schlüter, J. C. Sauvageau, T. Beyersdorff and J. Thöming, Synthesis of ionic liquids in micro-reactors – a process intensification study, Green Chem., 9, 1084–1090 (2007).

(241) A. Renken, V. Hessel, P. Löb, R. Miszczuk, M. Uerdingen, L. Kiwi-Minsker, Ionic liquid synthesis in a microstructured reactor for process intensification. Chem. Eng. Process., 46, 840–845 (2007).

(242) Y. Gök, İ. Özdemir and E. Çetinkaya, Ionic liquids as solvents/catalysts for selective alkylation of amines with alkyl halides. Chin. J. Catal., 28, 489–491 (2007).

(243) M. Teijeira, Y. Fall, F. Santamarta and E. Tojo, Efficient and rapid experimental procedure for the synthesis of furan diol from d-glucal using ionic liquid. Tetrahedron Lett., 48, 7926–7929 (2007).

(244) D. J. C. Constable, C. Jimenez-Gonzalez and R. K. Henderson, Perspective on solvent use in the pharmaceutical industry. Org. Process Res. Dev., 11, 133–137 (2007).

(245) W. Wu, G. Qian, X. Zhou and W. Yuan, Peroxidization of methyl ethyl ketone in a microchannel reactor. Chem. Eng. Sci., 62, 5127–5132 (2007).

(246) L. Kong, Q. Lin, X. Lv, Y. Yang, Y. Jia and Y. Zhou, Efficient Claisen rearrangement of allyl para-substituted phenyl ethers using microreactors. Green Chem., 11, 1108 (2009).

(247) H. Loewe, V. Hessel, P. Lob and S. Hubbard, Addition of secondary amines to r,â-unsaturated carbonyl compounds and nitriles by using microstructured reactors. Org. Process Res. Dev., 10, 1144–1152 (2006).

(248) R. F. Ludlow and S. Otto, Systems chemistry. Chem. Soc. Rev., 37, 101 (2008).

(249) U. Sauer, Systems biology: getting closer to the whole picture. Science, 316, 550 (2007).

(250) High Level Group on the Competitiveness of the European Chemicals Industry, Final Report: European Chemical Industry, Enabler of a Sustainable Future, European Communities (2009).

(251) V. Hessel and I. Vural-Gursel, Potential analysis of smart flow processing and micro process technology for fastening process development – use of chemistry and process design as intensification fields. Chem. Eng. Technol. (2012).
(252) D. Kralisch, E. Santacesaria, B. Cortese, M. H. J. M.de Croon and V. Hessel, Transfer of the epoxidation of soybean oil from batch to flow chemistry guided by cost and environmental issues. ChemSusChem, 5, 300–311 (2012).
(253) S. R. Deshmukh and D. G. Vlachos, CFD Simulations of coupled, countercurrent combustor/reformer microdevices for hydrogen production. Ind. Eng. Chem. Res., 44, 4982 (2005).
(254) G. Kolios, B. Gloeckler, A. Gritsch, A. Morillo and G. Eigenberger, Heat-integrated reactor concepts for hydrogen production by methane steam reforming. Fuel Cells Bull., 5, 52 (2005).
(255) M. S. Mettler, G. D. Stefanidis and D. G. Vlachos, Scale-out of microreactor stacks for portable and distributed processing: coupling of exothermic and endothermic processes for syngas production. Ind. Eng. Chem. Res., 49, 10 942–10 955 (2010).
(256) R. C. Ramaswamy, P. A. Ramachandran and M. P. Duduković, Coupling exothermic and endothermic reactions in adiabatic reactors. 63, 1654–1667 (2008).
(257) C. Bramsiepe, S. Sievers, T. Seifert, G. D. Stefanidis, D. G. Vlachos, H. Schnitzer, B. Muster, C. Brunner, J. P. M. Sanders, M. E. Bruins and G. Schembecker, Low-cost small scale processing technologies for production applications in various environments – mass produced factories. Chem. Eng. Process., 51, 32–52 (2012).
(258) Technische Universitat Darmstadt, Center of Smart Interfaces (CSI), www.csi.tu-darmstadt.de, last accessed 14 September 2012.
(259) C. Ramshaw, Higee distillation – an example of process intensification. Chem. Eng., 389, 13–14 (1983).
(260) W. T. Cross and C. Ramshaw, Process intensification – laminar-flow heat-transfer. Chem. Eng. Res. Des., 64, 293–301 (1986).
(261) A. Aoune and C. Ramshaw, Process intensification: heat and mass transfer characteristics of liquid film on rotating discs. Int. J. Heat Mass Trans., 42, 2543–2556 (1999).
(262) I. Vural-Gursel, Q. Wang, V. Hessel, T. Noel and J. Lang, Microreactors for faster and eco-efficient process development – process design as 3. intensification field. Chem. Eng. Technol. (2012).
(263) N. M. Nikacevic, A. E. M. Huesman, P. M. J.van den Hof and A. I. Stankiewicz, Chemical engineering & technology; opportunities and challenges for process control in process intensification. Chem. Eng. Process., 52, 1–15 (2012).
(264) K. Sato, M. Aoki and R. Noyori, A 'green' route to adipic acid: direct oxidation of cyclohexenes with 30 percent hydrogen peroxide. Science, 281, 1646–1647 (1998).
(265) V. Hessel, J. Lang and D. Kralisch, Shining a light: new routes in process chemistry and plant conception, http://www.chemanager-online.com/en/topics/plant-engineering-components/shining-light, last accessed 14 September 2012.
(266) M. Henig, Invite and F^3 factory, the European chemical factory of the future is taking shape at the Invite Research Center in Germany. Process (2011). http://www.process-worldwide.com.
(267) A. Geipel-Kern, F^3 project what will tomorrow's chemical plant look like? Process (2010). http://www.process-worldwide.com.
(268) F^3 Factory Newsletter, Issue 1 (2010). http://f3factory.com/scripts/pages/en/newsevents/F3_FactoryNewsletter_1.pdf, last accessed 14 September 2012.
(269) F^3 Factory Newsletter, Issue 2 (2011). http://f3factory.com/scripts/pages/en/newsevents/F3_FactoryNewsletter_2.pdf, last accessed 14 September 2012.
(270) I. Vural, V. Hessel, J. Lang, T. Noel and Q. Wang. Green Proc. Synth (submitted).
(271) M. T. Musser and E. I. DuPont de Nemours & Co. Adipic Acid, in Ullmann's Encyclopedia of Industrial Chemistry, Wiley-VCH (2003).

(272) Adipic Acid, in Kirk-Othmer Encyclopedia of Chemical Technology, John Wiley & Sons: Hoboken, NJ, pp. 553–581 (2007).

(273) A. Castellan, J. C. J. Bart and S. Cavallaro, Industrial production and use of adipic acid. Catal Today, 9, 237–254 (1991).

(274) R. A. Reimer, C. S. Slaten, M. Seapan, M. W. Lower and P. E. Tomlinson, Abatement of N_2O emissions produced in the adipic acid industry. Environ Prog, 13, 134–137 (1994).

(275) Y. Q. Deng, Z. F. Ma, K. Wang and J. Chen, Clean synthesis of adipic acid by direct oxidation of cyclohexene with H2O2 over peroxytungstate – organic complex catalysts. Green Chem., 1, 275–276 (1999).

(276) K. Sato, M. Aoki, M. Ogawa, T. Hashimoto and R. Noyori, A practical method for epoxidation of terminal olefins with 30% hydrogen peroxide under halide-free conditions. J. Org. Chem., 61, 8310–8311 (1996).

(277) P. R. Patnaik, Multi-unit integration in microfluidic processes: current status and future horizons. Int. J. Bioautomation, 15, 77–84 (2011).

(278) C. Cremers, A. Pelz, U. Stimming, K. Haas-Santo, O. Goerke, P. Pfeifer and K. Schubert, Micro-structured methane steam reformer with integrated catalytic combustor. Fuel Cells Bull., 7, 91–98 (2007).

(279) A. M. Karim, J. A. Federici and D. G. Vlachos, Portable power production from methanol in an integrated thermoeletric/microreactor system. J. Power Sources, 179, 113–120 (2008).

(280) T. Honda, M. Miyazaki, Y. Yamaguchi, H. Nakamura and H. Maeda, Integrated microreaction system for optical resolution of racemic amino acids. Lab Chip, 7, 366–372 (2007).

(281) W. N. Lau, K. L. Yeung and R. Martin-Aranda, Knoevenagel condensation reaction between benzaldehyde and ethyl acetoacetate in microreactor and membrane microreactor. Microporous and Mesoporous Mater., 115, 156–163 (2008).

(282) W. N. Lau, K. L. Yeung, X. F. Zhang and R. Martin-Aranda, Zeolite membrane microrectors and their performance. Stud. Surf. Sci. Catal., 170B, 1460–1465 (2007).

(283) J. P. McMullen and K. F. Jensen, Integrated microreactors for reaction automation: new approaches to reaction development. Annu. Rev. Anal. Chem., 3, 19–42 (2010).

(284) W. C. Shin, An integrated micro chemical reactor system with effective control strategies. Dissertation Abstracts International, DAI/B 68-03 (2007).

(285) D. J. Quiram, K. F. Jensen, M. A. Schmidt, P. L. Mills, J. F. Ryley, M. D. Wetzel and D. J. Kraus, Integrated microreactor system for gas-phase catalytic reactions. 1. Scale-up microreactor design and fabrication, Ind. Eng. Chem. Res., 46, 8292–8305 (2007).

(286) D. J. Quiram, K. F. Jensen, M. A. Schmidt, P. L. Mills, J. F. Ryley, M. D. Wetzel and D. J. Kraus, Integrated microreactor system for gas phase reactions, in Micro Instrumentation, M. V. Koch, K. VandenBusche and R. W. Chrisman (Eds.), pp. 363–406 (2007).

(287) K. Shah and R. S. Besser, Understanding thermal integration issues and heat loss pathways in a planar microscale fuel processor: demonstration of an integrated silicon microreactor-based methanol steam reformer. Chem. Eng. J., 135, S46–S56 (2008).

(288) K. Shah and R. S. Besser, Key issues in the microchemical systems-based methanol fuel processor: energy density, thermal integration, and heat loss mechanisms. J. Power Sources, 166, 177–193 (2007).

(289) D. J. Quiram, K. F. Jensen, M. A. Schmidt, P. L. Mills, J. F. Ryley, M. D. Wetzel and D. J. Kraus, Integrated microreactor system for gas-phase catalytic reactions. 3. Microreactor system design and system automation. Ind. Eng. Chem. Res., 46, 8319–8335 (2007).

(290) V. Hessel, C. Hofmann, H. Löwe, A. Meudt, S. Scherer, F. Schönfeld and B. Werner, Selectivity gains and energy savings for the industrial phenyl boronic acid process using micromixer/tubular reactors. Org. Proc. Res. Dev., 8, 511–523 (2004).

(291) J. Fischer, T. Lange, R. Boehling, A. Rehfinger and E. Klemm, Uncatalyzed selective oxidation of liquid cyclohexane with air in a microcapillary reactor. Chem. Eng. Sci., 65, 4866 (2010).

(292) P. Loeb, V. Hessel, H. Loewe, C. Hofmann, K. Mazanek and H. Klefenz, Brominations in micro reactors using high and low p, T- processing, in Proceedings of the 8th International Conference on Microreaction Technology (IMRET 8), Atlanta, GA, p. 1341 (2005).

(293) P. B. Palde and T. F. Jamison, Safe and efficient tetrazole synthesis in a continuous-flow microreactor. Angew. Chem. Int. Ed., 50, 3525 (2011).

(294) http://europic.jeppix.eu/. Last visited April 13, 2012.

(295) P. Poechlauer, M. Vorbach, M. Kotthaus, S. Braune, R. Reintjens, F. Mascarello and G. Kwant, Microreactor plant for the large-scale production of a fine chemical intermediate: a technical case study, in Micro Process Engineering: A Comprehensive Handbook, vol. 1, V. Hessel, A. Renken, J. C. Schouten and J.-I. Yoshida (Eds.), Wiley-VCH: Weinheim, pp. 249–254 (2009).

(296) D. Kralisch, U. Krtschil, D. M. Roberge, V. Hessel and D. Schmalz, The economic potential of microreaction technology, in Micro Process Engineering: A Comprehensive Handbook, Vol. 3: System, Process and Plant Engineering 3: System, V. Hessel, A. Renken, J. C. Schouten and J. Yoshida (Eds.), Wiley-VCH: Weinheim, pp. 279–294 (2009).

(297) V. Hessel, P. Lob and H. Lowe, Industrial microreactor process development up to production, in: Microreactors in Organic Synthesis and Catalysis, T. Wirth (Ed.), Wiley-VCH: Weinheim, pp. 211–275 (2008).

(298) M. Nobis and D. M. Roberge, Mastering ozonolysis: production from laboratory to ton scale in continuous flow. Chim Oggi, 29, 56–58 (2011).

(299) N. Kockmann and D. M. Roberge, Harsh reaction conditions in continuous-flow microreactors for pharmaceutical production. Chem. Eng. Technol., 32, 1682–1694 (2009).

(300) P. Barthe, C. Guermeur, O. Lobet, M. Moreno, P. Woehl, D. M. Roberge, N. Bieler and B. Zimmermann, Continuous multi-injection reactor for multipurpose production – part I. Chem. Eng. Technol., 31, 1146–1154 (2008).

(301) D. M. Roberge, N. Bieler, M. Mathier, M. Eyholzer, B. Zimmermann, P. Barthe, C. Guermeur, O. Lobet, M. Moreno and P. Woehl, Development of an industrial multi-injection microreactor for fast and exothermic reactions – part II. Chem. Eng. Technol., 31, 1155–1161 (2008).

(302) D. M. Roberge, N. Bieler and M. Thalmann, Microreactor technology and continuous processes in the fine chemical and pharmaceutical industries. PharmaChem, 28, 14 (2006).

(303) N. Kockmann, M. Gottsponer and D. M. Roberge, Scale-up concept of single-channel microreactors from process development to industrial production. Chem. Eng. J., 167, 718–726 (2011).

(304) W. B. White, T. J. Schnitzer, R. Fleming, B. Duquesroix and M. Beekman, Effects of the cyclooxygenase inhibiting nitric oxide donator naproxcinod versus naproxen on systemic blood pressure in patients with osteoarthritis. Am. J. Cardiol., 104(6), 840–845 (2009).

(305) V. Hessel, A. Renken, J. C. Schouten and J. Yoshida, Micro Process Engineering: A Comprehensive Handbook, 1st edition, Wiley-VCH: Weinheim (2009).

(306) D. Kirschneck and G. Tekautz, Integration of a microreactor in an existing production plant. Chem. Eng. Technol., 30, 305–308 (2007).

(307) Green Car Congress, Velocys signs microchannel GTL agreement with Petrobras and Partners: targeting GTL deployment on FPSO, www.greencarcongress.com, last accessed 14 September 2012.

(308) Velocys Inc., GTL demonstration at Petrobras, Financial Announcement. www.velocys.com, last accessed 14 September 2012.

(309) Toyo Engineering Cooperation, TOYO signs joint development agreement with MODEC and Velocys for offshore GTL plant, in Newsletter, Tec Communications, vol. 13, www.toyo-eng.co.jp, last accessed 14 September 2012.

5

Green Chemistry in Oscillatory Baffled Reactors

Adam Harvey

School of Chemical Engineering & Advanced Materials,
Newcastle University, UK

5.1 Introduction

5.1.1 Continuous versus Batch Operation

Generally, continuous processing is more efficient than batch processing. The reasons for this are explained in detail in Section 1.4.3, but briefly they are that for the same process continuous reactors will tend to be smaller, better controlled and more energy efficient, and will exhibit improved product quality. However, batch processing is still widely used, for reasons of flexibility, economy (at small scale) and batch integrity (in the pharmaceuticals industry), and for use with long-residence-time processes, where continuous processing is perceived to be difficult/impractical.

5.1.2 The Oscillatory Baffled Reactor's 'Niche'

Most of the reasons given in Chapter 1 for using batch reactors can be countered to some extent by the use of oscillatory baffled reactor (OBRs). Various other continuous reactors also somewhat mitigate these points, but only OBRs meet the challenge of long residence times.

Process Intensification for Green Chemistry: Engineering Solutions for Sustainable Chemical Processing, First Edition. Edited by Kamelia Boodhoo and Adam Harvey.

There are two main types of continuous reactor: the continuously stirred tank reactor (CSTR) and the plug flow reactor (PFR):

1. *CSTRs* are essentially stirred batch vessels with flows in and out. They exhibit very poor control of residence time, the residence time distribution (RTD) being essentially an exponential decay, meaning that much of the material will spend too long in the reactor, leading to undesired side reactions. However, some of the material will not spend enough time in the reactor, as it is able to 'shoot through' from inlet to outlet, and will therefore not reach the desired conversion. This is usually addressed by oversizing the reactor.
2. The solution to the problem of poor control of residence time is to use a *PFR*. In its simplest incarnation, the PFR is simply a tube through which the reactants flow at a high enough velocity to achieve turbulence, and with it a flat velocity profile; that is, the material flows as a 'plug'. Each element of fluid entering the reactor can be said to experience the same 'processing history', and the conversion in any one element of fluid is the same as that in any other. These reactors can therefore be designed for the exact volume required. However, an important limitation of PFRs is that a high velocity must be maintained. This is a problem for 'long' reactions, as maintaining a high velocity for a long period of time necessarily results in an extremely long, narrow reactor, engendering a variety of practical problems, particularly control, footprint and pumping. The OBR is the solution to this.

It should be noted that there are other possible solutions to these problems:

- PFRs can be fitted with various inserts to facilitate earlier onset of turbulence, thereby reducing the velocity required and shortening the reactor. However, this does not entirely solve the problem, as there is still a linear dependence of operation on velocity.
- CSTRs can be arranged in series to give an approximation to plug flow, but this tends to be an expensive solution, as each new CSTR incurs the same costs. The increasing cost must be balanced against the increasing quality of the plug flow with each additional tank.

OBRs can perform relatively long reactions continuously, without requiring impractically long, narrow reactor designs. This is because unlike in conventional PFRs the mixing is decoupled from the net flow velocity, as it is achieved by oscillating fluid through orifice plate baffles, each acting as a small CSTR and thereby achieving plug flow, but in a much more integrated and inexpensive manner, which produces extremely high degrees of plug flow. The quality of plug flow can be expressed as the number of perfect CSTRs in series. In practice, a value of 10 is adequate for any reaction, but this is easily exceeded, and values >100 have been achieved.

A typical OBR configuration can be seen in Figure 5.1. As for most PFR designs, changes to the operation, such as staged addition and integration of different reaction stages, can readily be achieved. Furthermore, there are a range of features that cannot necessarily be achieved by other forms of PFR: they can accommodate different forms of mixing (liquid–liquid, liquid–gas etc.) where necessary, and suspension of catalyst particles.

The improved control of mixing opens up the possibility of operation in the absence of a solvent, as the OBR has been demonstrated to operate well as a two-phase mixing device [1].

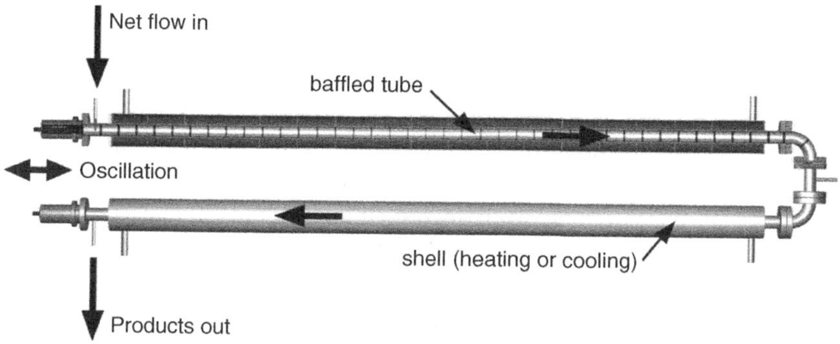

Figure 5.1 *Oscillatory baffled reactor. Reprinted from Stonestreet and Harvey © 2002, with permission from Elsevier.*

5.1.2.1 Oscillatory Mixing

Mixing is provided by oscillation of the entire volume of the fluid, via its interaction with the orifice plate baffles. Mixing is decoupled from the net flow of the fluid (which dictates the mixing in conventional PFRs), so the high length-to-diameter ratio (L/D) of conventional PFRs is no longer necessary. As the fluid is oscillated through the baffles, large vortices are formed. These represent good mixing patterns, as they transport material radially and axially in a uniform manner. Axial and radial velocities are of similar magnitude, so the mixing is less directional (as can be observed in Figure 5.2). The mixing produced is more uniform than that observed in conventional agitated vessels.

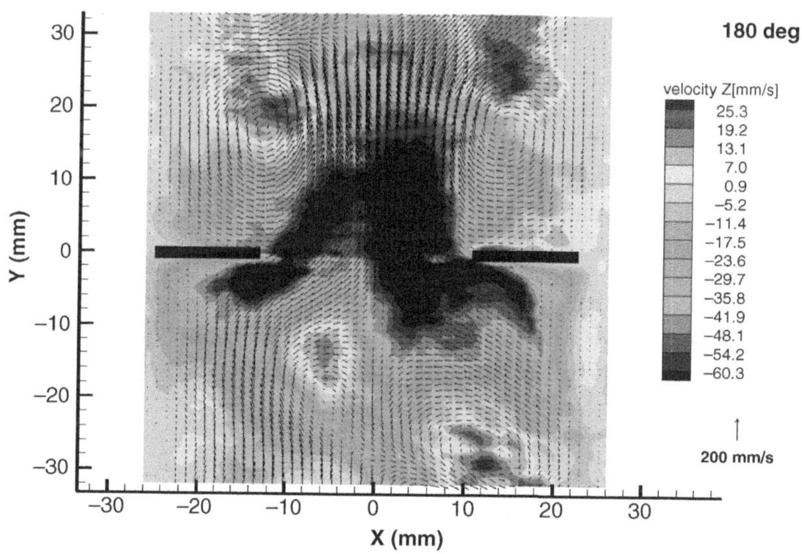

Figure 5.2 *Particle image velocimetry (PIV) around a single baffle in an OBR, showing early development of vortex above it.*

Vortices are formed on either side of the orifice plate baffles as the fluid is pushed through. Under the correct conditions, the vortices will fill the cross-section of the tubes, but will not be propagated from one baffle cavity to another (as this will produce unwanted axial dispersion). These mixing patterns have similar radial and axial components.

In stirred tanks there are areas of very high shear around the blade, and areas of very little agitation (stagnation points). This can present a problem when mixing shear-thinning fluids, as an extremely well mixed, thinned core of material can develop around the agitator while the rest of the material remains stagnant, at its original viscosity.

Furthermore, the mixing in STRs can be very directional. In an OBR the mixing is much more uniform throughout, which is an important advantage for some applications.

5.1.2.2 *Ensuring Plug Flow*

As well as the conventional 'net flow' Reynolds number, here given the symbol Re_n, we define an 'oscillatory Reynolds number', Re_o:

$$Re_o = \frac{\rho.2\pi f x_o.D}{\mu} \quad \text{Oscillatory Reynolds Number} \tag{5.1}$$

$$Re_n = \frac{\rho.v.D}{\mu} \quad \text{'Net Flow'Reynolds Number} \tag{5.2}$$

Re_o is simply Re_n with the velocity, v, replaced by the term $2\pi f x_0$, which is the maximum velocity due to oscillation. f is frequency, in Hz, and x_0 is the centre-to-peak amplitude of the oscillation (i.e. half the distance from the peak to the trough of the oscillation). A 'velocity ratio' is also defined:

$$\varphi = \frac{Re_o}{Re_n} \quad \text{Velocity Ratio} \tag{5.3}$$

This is simply the ratio of the oscillatory to the net flow velocities, and is used as a measure of the degree of plug flow attained. A velocity ratio below 12 is usually acceptable (see Figure 5.3), as this means that the RTD of the reactor is equivalent to more than 10 theoretical tanks-in-series, which approximates plug flow well. However, the lower the value of ψ, the higher the L/D, and the lower the reduction in L/D compared to conventional PFRs the smaller the design advantage. These two constraints loosely define an operating window of design advantage (in terms of reduced L/D) and acceptable plug flow of $2 < \psi < 12$.

5.1.2.3 *Solids Suspension*

One of the key advantages of OBRs over conventional PFRs is their use with solid suspensions. The toroidal vortex flow structures produced in OBRs can be used to uniformly suspend solid particles in liquids. This would be difficult to achieve in a conventional PFR due to the turbulent conditions and high net flow required to achieve

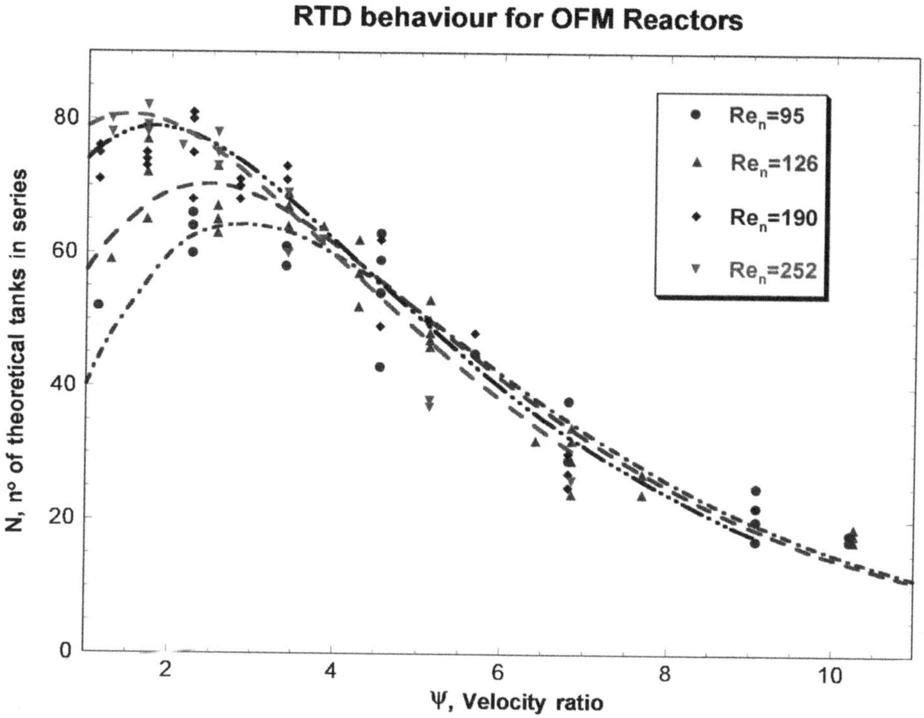

Figure 5.3 *Tanks-in-series versus velocity ratio. Reprinted from Stonestreet and Van Der Veeken © 1999, with permission from Elsevier.*

a flat velocity profile (i.e. plug flow). Mackley *et al.* [3] developed an empirical correlation to describe solid suspension in liquids:

$$\gamma = \left[1 - \exp\left(-R\frac{V_m}{V_s}\right)\right] \tag{5.4}$$

where V_m is the maximum oscillation velocity, V_s is the settling velocity of the particles and R is an empirical constant. γ is the ratio of solids concentration in one baffle cavity to that in the next and is found to be a constant for any given reactor and set of conditions. When $\gamma = 1$, a uniform suspension is achieved in the tube.

It should be noted that the degree of solids suspension can also be controlled, most easily *in situ* by simply varying V_m. That is to say: a particle concentration gradient can be imposed. This allows suspensions to be stratified according to size and density, a technique known as 'wet sieving'.

5.1.2.4 Heat-transfer Enhancement in OBRs

Due to the significant radial component of oscillatory flows, transport of heat can be greatly enhanced. The tube-side Nusselt number, a measure of the convective heat transfer inside an OBR, is increased by a factor of between 10 and 30 (Figure 5.4). This indicates that *heat*

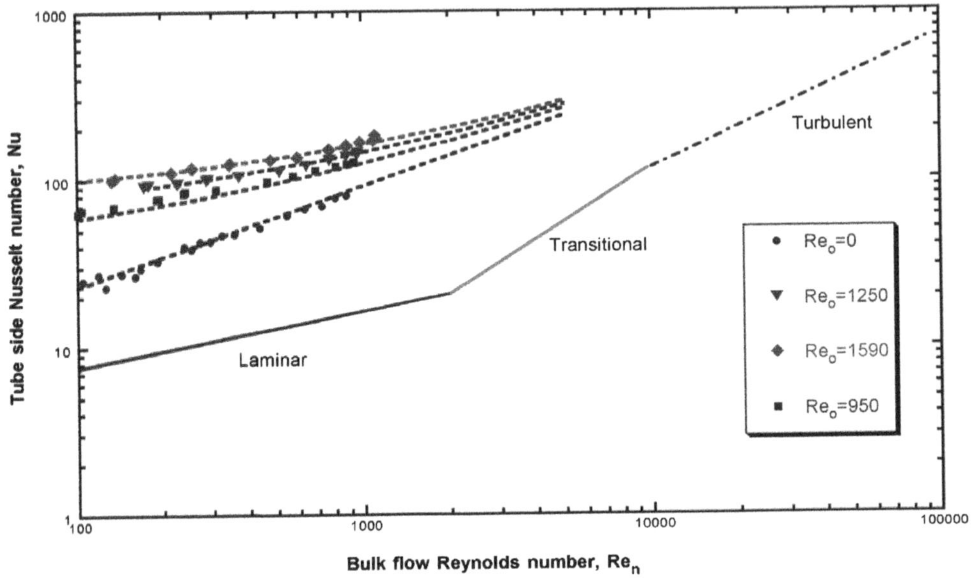

Figure 5.4 *Heat-transfer enhancement of OBRs. Reprinted from Mackley and Stonestreet © 1995, with permission from Elsevier.*

transfer coefficients associated with turbulent flows can be achieved at much lower net flow rates. The lower dashed line, corresponding to $Re_o = 0$, on the graph illustrates the heat transfer enhancement due to the incorporation of baffles in the absence of oscillation, and indeed orifice plate baffles, and many other designs of 'flow inserts' are often used to enhance heat transfer. Comparison of the $Re_o = 0$ line with the other experimental data illustrates the positive effect of the vortex generation in oscillatory flow on radial heat transfer.

5.1.2.5 OBR Design

Broadly, OBRs are designed along similar lines to shell-and-tube heat exchangers. The visualization in Figure 5.5 is a concept design for a small industrial unit, illustrating this principle.

The design procedure for OBRs was developed by Stonestreet and Harvey [5]. It was based on following shell-and-tube design protocols, plus ensuring that the correct mixing conditions are in place. The starting information required is: residence time, physical properties of the reaction mixture – such as viscosity and density – and an estimate or knowledge of the required Re_o (to overcome any density differences in the system). The design flowchart is given in Figure 5.6.

To briefly explain the design steps:

(1) Re_o and ψ define the Re_n. The value of ψ comes from knowledge of the required degree of plug flow and the velocity ratio 'operating window'.
(2) Together with τ, various values of D, the tube diameter, are input to generate outputs of u, the velocity. This is best implemented on a spreadsheet.
(3) Outputs for z (the length), m (the number of baffles), Q (the volumetric flowrate) and εt (the power) are then calculable (see Ref. [5] for more detail on these calculations).

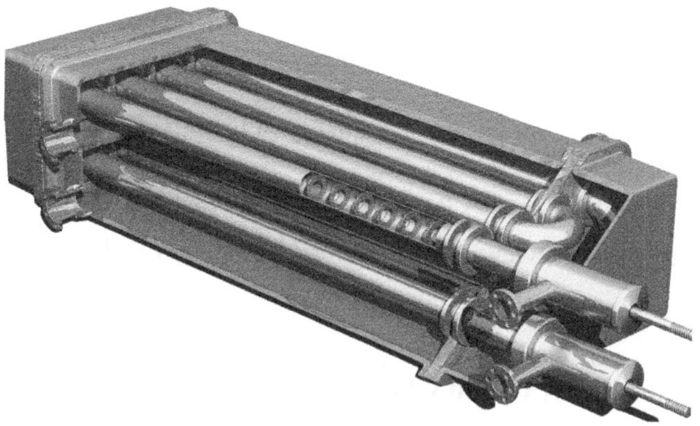

Figure 5.5 OBR visualization (based on shell-and-tube heat exchanger design). Reprinted from Stonestreet and Harvey © 2002, with permission from Elsevier.

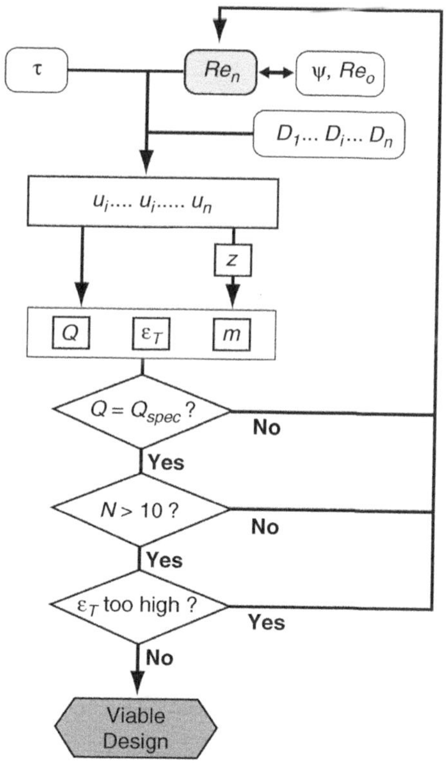

Figure 5.6 OBR design protocol. Reprinted from Stonestreet and Harvey © 2002, with permission from Elsevier.

(4) The remaining actions are a number of checks:
 (i) Is Q close to the desired value? The pipe diameter closest to the desired Q should be chosen.
 (ii) Is N > 10? This is a measure of the quality of the plug flow. N > 10 is satisfactory for any application.
 (iii) Is ε_T too high? This is the power consumption, and is calculated.

For a more comprehensive review of OBR history, technology and applications, the reader is directed to [6].

5.2 Case Studies: OBR Green Chemistry

5.2.1 A Saponification Reaction

A 1.3 l OBR (see Figure 5.7) was evaluated for an industrial batch saponification process in which conversion to continuous processing in conventional tubular reactors was considered unfeasible due to the long residence time required (2 hours). The saponification reaction was the hydrolysis of a complex natural mixture of esters in an ethanol/water solvent.

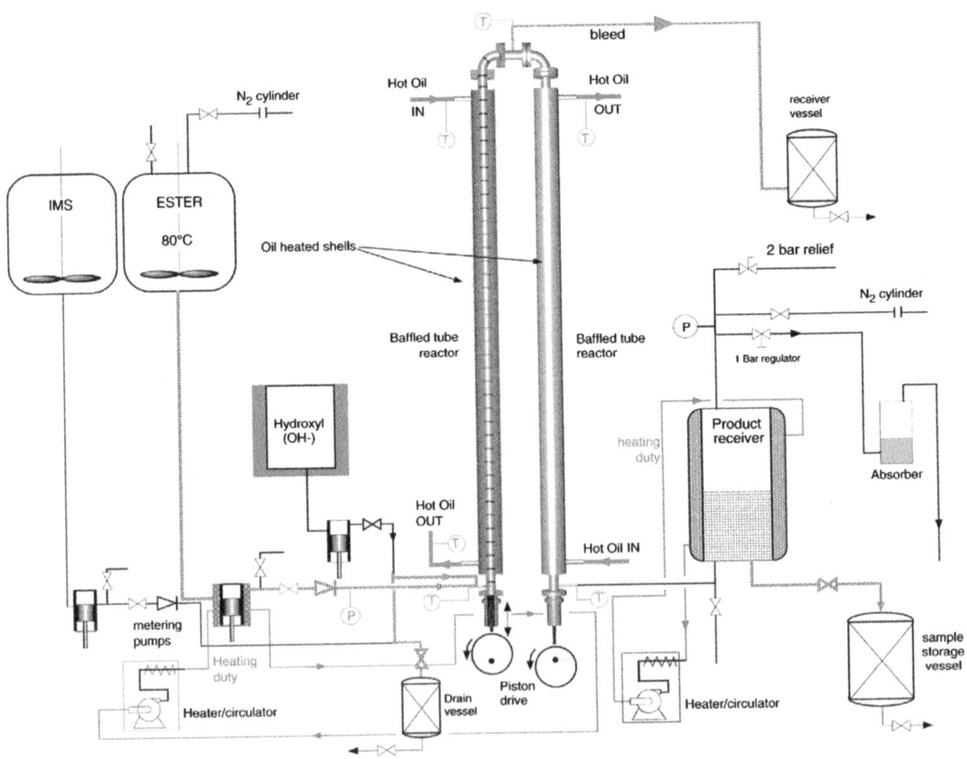

Figure 5.7 Pilot plant OBR. Reprinted with permission from Harvey et al. © 2001 American Chemical Society.

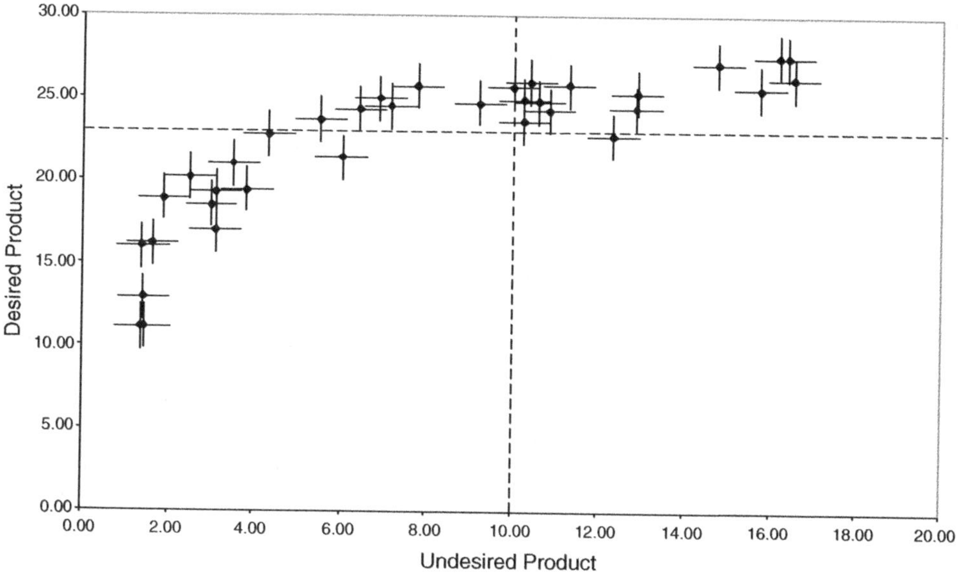

Figure 5.8 *Product quality improvement. Reprinted with permission from Harvey et al. © 2001 American Chemical Society.*

It was demonstrated that the OBR could achieve the batch product specification in a residence time one-tenth that of the batch reactor. This was because the reaction was able to proceed at its inherent kinetic rate due to the improved mixing characteristics of the OBR; that is, the reaction was no longer *mixing-controlled.*

A 'flow-conversion' model, incorporating a tanks-in-series RTD and the reaction kinetics, agreed well with the experimental results, predicting that the reaction could indeed be performed at relatively low residence times when kinetically controlled [1]. The model led to a number of insights into the optimal operation of the OBR. The most significant prediction was that the OBR could give the desired conversion and selectivity at lower temperatures, without significant alteration to other process variables such as residence time and molar ratio. The predictions were verified by experiment. Furthermore, at these lower temperatures an improved product quality specification could be achieved, if required (Figure 5.8).

The greatest incentive for conversion of this saponification reaction to continuous processing was safety, as the conditions at which the current batch reactor was operated involved maintaining a large volume of solvent at a temperature above its boiling point at ambient pressure. The switch to continuous operation would greatly lower the inventories of solvent at high temperature, thereby significantly improving safety (Table 5.1). In

Table 5.1 *Comparison of operating conditions for batch and OBR.*

	Batch	OBR
Temperature	115 °C	85 °C
Pressure	1.0 barg	1.7 barg
Volume	75 m^3 (50 m^3 fill)	0.5 m^3
Residence time	2 hours	12 minutes

this case, the safety was further improved by the fact the new operating temperature of 85 °C was below the ambient-pressure boiling point of the solvent. The lower temperature operation would also result in some energy savings, although the reduction would be primarily due to the more efficient heat transfer and reduced residence time.

The size reduction in Table 5.1 was due to the decrease in required residence time and the increase in occupancy time (100% for a continuous reactor, less than 10% for this particular batch reactor).

A subsequent, similar study showed that similar benefits could be achieved for the 'biodiesel reaction' – that is, transesterification of vegetable oil using methanol [7] – as the reaction time could be reduced to 15 minutes. More recent studies in mesoscale OBRs have demonstrated that even this can be outdone, as it has become clear that the residence time can be made as low as 2 minutes [8], resulting in a further 7.5-fold reduction in reactor size. In so doing, the mesoreactor was used to determine the kinetics of the biodiesel reaction and competing saponification reactions to show exactly why this was possible: under the correct conditions the saponification is relatively slow compared to the transesterification. This was only realizable in a reactor with extremely well-defined mixing.

5.2.1.1 *Green Chemistry Elements*

This case study illustrates four aspects of green chemistry:

(1) *Atom Economy*: Greater selectivity for the desired product was demonstrated, leading to reduced overall process waste and a reduced load on downstream separation steps.
(2) *Design for Energy Efficiency*: The energy costs of constructing and running the reactor were substantially reduced, as it was less than one-hundredth the volume of the reactor that it could replace.
(3) *Real-time Analysis for Pollution Prevention*: As this was a continuous reactor, it made sense to develop real-time analysis. An online infrared monitoring technique was thus implemented, which was able to monitor the concentration of the main reactant and product.
(4) *Inherently Safer Chemistry for Accident Prevention*: The OBR-based process no longer required large inventories of solvent, nor that the solvent be above its boiling point in operation.

5.2.2 A Three-phase Reaction with Photoactivation for Oxidation of Waste Water Contaminants

At the University of Cambridge in the mid to late 1990s, an OBR was developed for an application involving gas, liquid and solid phases, with *in situ*ultraviolet (UV) irradiation (Figure 5.9). The liquid was water containing hydrocarbons, mimicking typical waste water contaminants. The solid was a titanium dioxide catalyst that required UV activation to catalyse oxidation of the hydrocarbons in the water. The oxidizing agent was air, bubbled through the column. This illustrates a number of potential advantages of OBRs, and that an OBR can be designed to exploit many of them simultaneously:

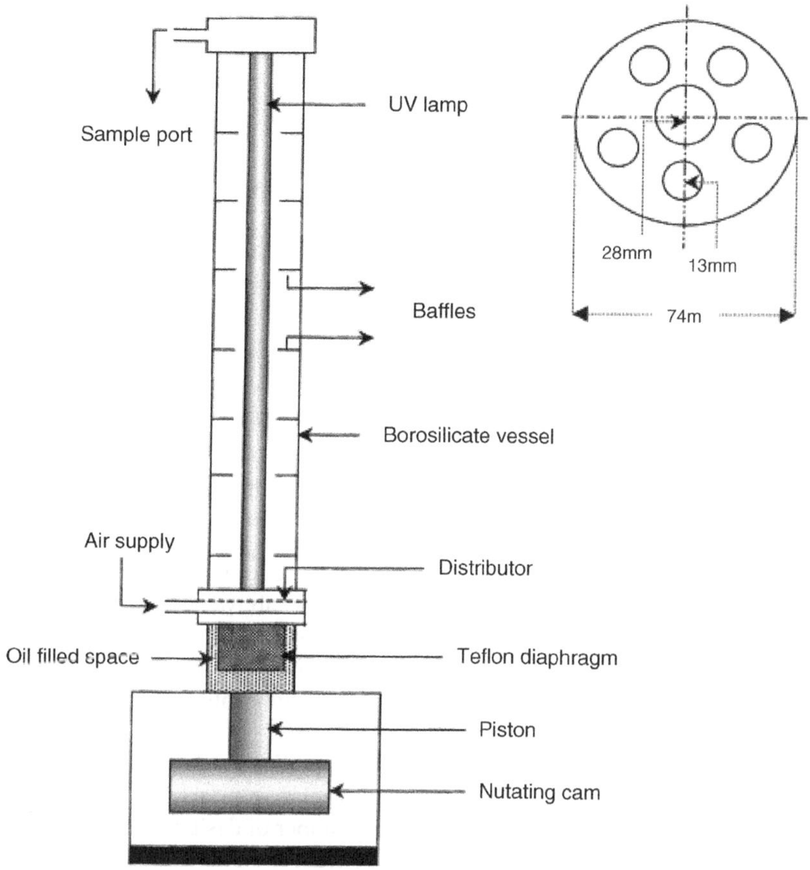

Figure 5.9 *OBR for waste water oxidation. Reprinted from Fabiyi and Skelton © 2000, with permission from Elsevier.*

(1) *Mass Transfer*: The oscillation of the fluid enhanced mass transfer of the oxygen into the water, first by reducing bubble size, which increased the surface area of the interface, and second by increasing bubble residence time.
(2) *Solids Suspension*: The titanium dioxide-coated particles were uniformly suspended in the water by the oscillatory mixing patterns. A further point to note is that the particles were not entrained out of the reactor when a net flow was imposed, by the simple expedient of having a baffle-free section at the top (the net flow was upward). Without the baffles, the particles in simply settled under gravity.
(3) *Uniform Irradiation*: The reactor in this case was composed of one vertical annular glass section with a UV lamp positioned centrally within it. The greatest challenge in designing any kind of photoreactor is that the penetration depth of the radiation is very small, and many designs centre on ensuring that the liquid flows in a thin film under the UV source such that the entire volume is irradiated. This severely limits the reactor design. The approach was different here: the fluid was constantly brought past the radiation source.

In addition to the advantages conferred by continuous operation, this reactor was shown to have higher photon utilization efficiencies than conventional photoreactors and to be less prone to fouling [9].

5.2.2.1 Green Chemistry Elements

This case study illustrates the following green chemistry principles:

(1) *Design for Energy Efficiency*: The OBR will be considerably more efficient in terms of its mixing than comparable conventional technologies for this duty, and this design is also energy efficient in terms of photon utilization.
(2) *Catalysis*: This is an excellent illustration of the use of heterogeneous catalysts in OBRs, and how it can be relatively easily achieved.

5.2.3 'Mesoscale' OBRs

Since 2004, efforts have been ongoing to investigate scale-down of OBRs to 'mesoscale', referring to OBRs of a few millimetres' diameter (typically 4–5 mm) [10]. In some areas, these reactors are referred to as millireactors, but here the prefix 'meso' is used to distinguish between these and micro- (sub-millimetre diameter) and macroreactors (the conventional-scale OBRs of 12 mm diameter and above). The initial idea behind this was to facilitate laboratory-scale process development by rapidly producing data required for scale-up, such as inherent kinetics or mass-transfer information, or else simply to try out process options in an easily reconfigurable reactor or to determine process optima.

At mesoscale it is difficult to fabricate geometrically similar baffles to those used in conventional reactors, as the supports become relatively large, perturbing the flow, and baffle thicknesses become more important. A number of designs have been developed, a few of which are shown in Figure 5.10 to give the reader an idea of the variety and scale.

The use of a continuous reactor allows for certain tests to be performed, such as determination of heterogeneous catalyst lifetime, that would be difficult in batch. The effect of recycles on the process can also be evaluated. Various designs of continuous flow screening reactor do exist, but the 'meso-OBR' has some specific advantages:

- The ability to suspend solids, whether they are reactants or catalyst particles.
- The decoupling of mixing from net velocity allows very long reactions (of the order of hours) to be performed in continuous mode with plug flow in a reactor of only a few metres length.

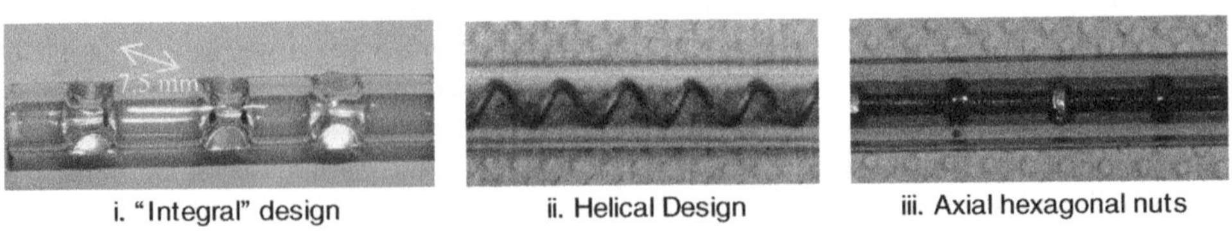

Figure 5.10 *Some typical mesoreactor configurations. Reprinted from Phan and Harvey © 2010, with permission from Elsevier.*

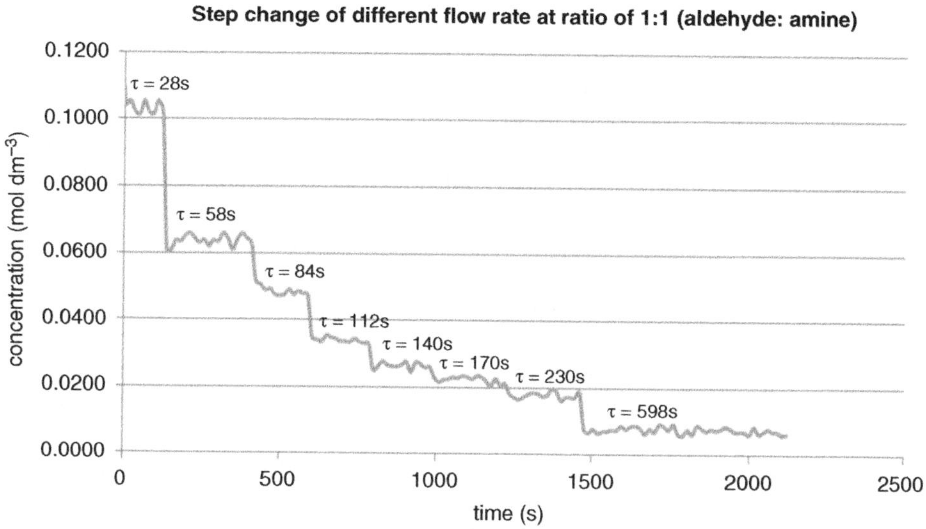

Figure 5.11 *Rapid screening of residence times in an imine synthesis using a meso-OBR. Reprinted from Mohd Rasdi et al © 2012, with permission from Elsevier.*

This new scale of operation has opened up a number of new applications for OBRs as devices for the green chemist. They allow process development with minimal use of reactants and minimal creation of waste. A recent study has demonstrated the following advantages for a screening/kinetic study of an imine formation reaction [12]:

- A 75% reduction in the amount of reagent required.
- A 50% reduction in process development time (versus conventional screening using batch stirred beakers).
- The rate constant data gathered from the mesoscale system is more reproducible, with a ~20% decrease in the standard deviation.

Figure 5.11 illustrates rapid screening of the imine synthesis for residence time, in a step-by-step fashion. This can also be achieved by dynamically altering variables. The concentration is monitored by online infrared (IR).

5.2.3.1 Green Chemistry Elements

This case study illustrates the following green chemistry principles:

(1) *Prevention*: Process development in these reactors should facilitate the development of more efficient, less wasteful processes in the laboratory. It is often the case that process inefficiencies begin in the laboratory when processes are developed in beakers and so on.
(2) *Atom Economy*: As Point 1.
(3) *Less Hazardous Chemical Synthesis*: Again, laboratory-scale flow reactors are valuable tools in investigating process options for reducing chemical hazards within processes.

(4) *Safer Solvents and Auxiliaries*: Processes are already in development, although not currently in press, in which various reactions are conducted solvent-free in mesoreactors. This is possible because of the tight control of mixing achievable in OBRs, and is easiest to investigate at mesoscale.
(5) *Use of Renewable Feedstocks*: Conventional biodiesel production has been investigated at mesoscale, with surprising results [8]. Due to working at this scale, while screening the reaction for residence time and catalyst concentration it was observed that biodiesel could be produced at a residence time of below 5 minutes at acceptable yield, when using relatively high catalyst concentrations. This had always been assumed to be impossible, as high concentrations of the usual catalysts (NaOH and KOH) cause the saponification side reaction to reach problematic levels, reducing yield and rendering downstream water washing very difficult. However, with tight control of the mixing it was observed that this was not a problem as the saponification was significantly slower.
(6) *Catalysis*: One of the mesoreactor's key attributes is the ability to suspend solid particles uniformly. This is an advantage over other laboratory-scale reactor designs, and was one of the main reasons for its development. Heterogeneous catalyst trials are underway for various reactions in which the catalyst is used in powder form; that is, it does not need to be immobilized to fit the reactor. In such trials the mesoreactor has been shown to be useful for *in situ* leaching/deactivation tests, which would be very difficult to achieve with batch reactors.
(7) *Real-time Analysis for Pollution Prevention*: IR monitoring has been developed in the imine synthesis case study which allows rapid screening of conditions, with a view towards reaching 'dynamic design of experiments'.

5.3 Conclusion

The OBR's beneficial characteristics are: plug flow, enhanced heat transfer, enhanced mass transfer, uniformity of mixing, two-phase mixing and predictable scale-up. Its key design advantage lies in allowing 'long' batch processes to be converted to continuous: this is the 'process intensification' that it brings about.

The fits between use of oscillatory flow reactors for processes and process development and the 'Twelve Principles of Green Chemistry' are listed in this section. Many of the advantages belong to all continuous reactors, or all PFRs, but they can only be realized if the reaction can be performed in the reactor, and conventional continuous reactors are difficult to implement for a range of process conditions, including solid suspension and long residence times, under which the OBR may be a viable alternative, allowing the change to continuous processing.

(1) *Prevention*: The greater control and uniformity of process conditions achieved in PFRs such as the OBR leads to reduced waste, as the tighter control of the processing history of reactants leaves less scope for byproduct formation. There are in general fewer hotspots than are found in stirred tank reactors, which reduces by-product formation. Furthermore, the use of such reactors often leads to innovations in processing, such as solvent-free processing (which greatly reduces waste), and

contacting patterns, such as staged addition of reactions (which is commonly used as a strategy for byproduct reduction for certain reaction schemes).

(2) *Atom Economy*: All PFRs exhibit greater control over the conditions to which the reactant mix is exposed. This results in greater selectivity, and therefore less waste, and a greater inclusion of the feed in the product. This should be one of the key advantages of the use of PFRs.

(3) *Less Hazardous Chemical Synthesis*: Generally this depends upon synthesis route development rather than reactor engineering, but the meso-OBR platform can be used to aid rapid development of such processes, particularly when they have solid catalysts and so on.

(4) *Design for Safer Chemicals*: This cannot be influenced by the reactor design *per se*, but the development of new processes can be accelerated by dynamic process screening achieved in platforms, such as the meso-OBR.

(5) *Safer Solvents and Auxiliaries*: Reactors exhibiting better mixing have the potential to allow the removal of solvents from the process altogether. Again, this is due to better control of mixing.

(6) *Design for Energy Efficiency*: Intensified reactors such as the OBR are usually smaller than analogous conventional reactors. This reduces the carbon footprint of their manufacture. Furthermore, their smaller size and increased efficiency reduces the energy running costs. The OBR typically consumes less energy for the same mixing/reaction duties compared to a comparable STR.

(7) *Use of Renewable Feedstocks*: Numerous studies have been performed on chemistries based on renewable feedstocks. Examples biodiesel [7], bioethanol [13], biobutanol [14] and biopolymers (PHA) [15]. OBRs for biofuel production are reviewed in Ref. [16]. The 'fit' between renewable feedstocks is based upon (i) the fact that many of these processes are too long for conventional PFRs and (ii) the ability of OBRs to process solid–liquid mixtures.

(8) *Reduction of Derivatives*: There are no known illustrations of this, as it is a function of the chemistry rather than the reactor.

(9) *Catalysis*:
   (i) Mesoscale OBRs are proving to be extremely useful for the screening and process development of solid-catalysed liquid reactions, as the solids can be simply and uniformly suspended in the reactor by the oscillatory flow and kinetic/mass-transfer studies can be rapidly performed, as can *in situ* leaching and lifetime tests.
   (ii) One interesting study in this area concerned phase-transfer catalysis, in which the degree of control of droplet size was an advantage, as it allowed control of the overall rate of reaction and reduced energy consumption [17].

(10) *Design for Degradation*: It is unlikely that the choice of reactor design will have a bearing on this.

(11) *Real-time Analysis for Pollution Prevention*: Mesoscale OBRs have been developed very much with real-time analysis in mind: real-time analysis is required to allow them to realize their full potential for rapid screening of process conditions via 'dynamic design of experiments'. This is an issue for any continuous reactor: it would be expected that the development of a continuous reactor will involve the development of some form on on- or at-line monitoring and control.

(12) *Inherently Safer Chemistry for Accident Prevention*: Continuous intensified reactors such as the OBR generally reduce hazards by reducing the inventories of material in a hazardous state. This is an example of implicit safety. The case study in Section 5.2.1 provides a clear example of this: the inventory of material in a hazardous state was reduced 100-fold (as the reactor was reduced in size 100-fold). Furthermore, the material was in a less hazardous state, as the study showed that the process could be operated at a lower temperature in the OBR.

References

(1) A. P. Harvey, M. R. Mackley and P. Stonestreet, Operation and optimization of an oscillatory flow continuous reactor. Industrial and Engineering Chemistry Research, 40(23), 5371–5377 (2001).

(2) P. Stonestreet and P. M. J.Van Der Veeken, The effects of oscillatory flow and bulk flow components on residence time distribution in baffled tube reactors. Chemical Engineering Research and Design, 77(8), 671–684 (1999).

(3) M. R. Mackley, K. B. Smith and N. P. Wise, Mixing and separation of particle suspensions using oscillatory flow in baffled tubes. Chemical Engineering Research and Design, 71(A6) 649–656 (1993).

(4) M. R. Mackley and P. Stonestreet, Heat transfer and associated energy dissipation for oscillatory flow in baffled tubes. Chemical Engineering Science, 50(14), 2211–2224 (1995).

(5) P. Stonestreet and A. P. Harvey, A mixing-based design methodology for continuous Oscillatory Flow Reactors. Chemical Engineering Research and Design, 80, 31–44 (2002).

(6) X. Ni, M. R. Mackley, A. P. Harvey, P. Stonestreet, M. H. I. Baird and N. V. Rama Rao, Mixing through oscillations and pulsations – a guide to achieving process enhancements in the chemical and process industries. Chemical Engineering Research and Design, 81(3), 373–383 (2003).

(7) A. P. Harvey, M. R. Mackley and T. Seliger, Process intensification of biodiesel production using a continuous oscillatory flow reactor. Journal of Chemical Technology and Biotechnology, 78(2–3), 338–341 (2003).

(8) A. N. Phan, A. P. Harvey and V. Eze, Rapid production of biodiesel in mesoscale oscillatory baffled reactors. Chemical Engineering & Technology, 35(7), 1214–1220 (2012).

(9) M. E. Fabiyi and R. L. Skelton, Photocatalytic mineralisation of methylene blue using buoyant TiO2-coated polystyrene beads. Journal of Photochemistry and Photobiology A: Chemistry, 132(1–2), 121–128 (2000).

(10) N. Reis, A. P. Harvey, M. R. Mackley, A. A. Vicente and J. A. Teixeira, Fluid mechanics and design aspects of a novel oscillatory flow screening mesoreactor. Chemical Engineering Research and Design, 83(4A), 357–371 (2005).

(11) A. N. Phan and A. Harvey, Development and evaluation of novel designs of continuous mesoscale oscillatory baffled reactors. Chemical Engineering Journal, 159(1–3), 212–219 (2010).

(12) F. R. Mohd Rasdi, A. N Phan and A. P. Harvey, Rapid determination of the reaction kinetics of an N-butylbenzaldimine synthesis using a novel mesoscale oscillatory baffled reactor, 20th International Congress of Chemical and Process Engineering CHISA 2012, 25-29 August 2012, Prague, Czech Republic, Procedia (2012).

(13) J. Ikwebe and A. P. Harvey, Intensification of Bioethanol Production by Simultaneous Saccharification and Fermentation (SSF) in an Oscillatory Baffled Reactor (OBR), World Renewable Energy Conference, Sweden May 2011.

(14) M. S. Takriff, N. Masngut, A. A. H. Kadhum, M. S. Kalil and A. W. Mohammad, Solvent fermentation from palm oil mill effluent using clostridium acetobutylicum in oscillatory flow bioreactor. Sains Malaysiana, 38(2), 191–196 (2009).
(15) C. N. Troeger and A. P. Harvey, The production of polyhydroxyalkanoates using an oscillatory baffled bioreactor. Chemical Product and Process Modeling, 4(5) (2009).
(16) N. Masngut, A. P. Harvey and J. Ikwebe, Potential uses of oscillatory baffled reactors for biofuel production. Biofuels, 1(4), 605–619 (2010).
(17) B. Wilson, X. Ni and D. C. Sherrington, A study of a phase transfer catalytic reaction between N-butyl bromide and sodium phenolate in an oscillatory baffled reactor. Studies in Surface Science and Catalysis, 133, 481–487 (2001).

6

Monolith Reactors for Intensified Processing in Green Chemistry

Joseph Wood

School of Chemical Engineering, University of Birmingham, Birmingham, UK

6.1 Introduction

Monolithic catalysts were initially developed for use in automotive exhaust emission control but during the past 10–15 years increasing interest has focused on their application to a more diverse range of areas in the chemical and allied industries [1]. Structured honeycomb monolith catalyst supports usually consist of parallel, straight capillary channels, upon which the catalyst is coated within a washcoat layer of porous material. In the field of energy, monolith reactors have been applied to hydrogen production from steam reforming [2], water gas shift reactions [3], methanation of carbon dioxide to produce methanol [4] and the manufacture of synthetic fuels using Fischer–Tropsch (FT) synthesis [5]. The purification of air or water by monolithic catalysts has been studied [6,7], as has the catalytic reduction of pollutants such as NO_x from stationary sources [8,9]. Although originally applied to gas-phase systems, new applications of monoliths in three-phase *gas–liquid–solids* have been considered [10,11]. Reaction systems of interest include hydrogenations – for example of 2-butyne-1,4-diol [12], nitrobenzoic acid [13], α-methyl styrene [14], benzaldehyde [15] and pyrolysis gasoline [16] – and oxidations – such as of glycerol [17], glucose [18] and ethanol [19]. However, one of the few known scaled-up industrial applications is in the production of hydrogen peroxide [20].

Process Intensification for Green Chemistry: Engineering Solutions for Sustainable Chemical Processing, First Edition. Edited by Kamelia Boodhoo and Adam Harvey.

Process intensification (PI) seeks to utilize novel chemical engineering concepts and equipment in order to overcome mass-transfer limitations, enhance reaction rates and therefore miniaturize process equipment. This often requires the use of an increased input of mechanical energy in order to mix or spread the reacting fluids into thin layers with a high interfacial area. However, the monolith reactor has the advantage that intensified performance can be achieved with much lower power inputs than some alternative PI devices. Key to this enhancement of performance is the length scales over which reaction and mixing occur. Although a range of flow patterns may take place within the channel, one of the most desirable regimes for enhanced mass transfer is 'Taylor flow' or bubble train flow, in which liquid slugs separated by gas bubbles flow along each channel. This creates a thin film of liquid at the sides of the gas bubbles in contact with the catalyst coating on the channel wall, the thickness of which may range from 5 to 50 μm for fluids of low viscosity [21]. Typically the channel side length in monoliths is of the order of 1–2 mm and optimum washcoat thickness is of the order of 70 μm, such as in the case of Syngas production [22]. These scales mean that the momentum balance and bulk mass transfer occur over a characteristic length of channel diameter of the order of millimetres, while diffusion across the liquid film and into the washcoat occurs over distances of the order of tens of microns [23]. Reaction occurs at catalytic sites in pores of the order of tens of nanometres, within the washcoat. These small length scales can be contrasted with the much larger sizes in traditional reactors such as fixed and trickle beds, which utilize extrudate pellets of several millimetres' length within reactors on the scale of several metres. By minimizing the transport and diffusion distances, utilizing a thin catalyst washcoat, enhancing mixing within the channels and bringing the reactants together within intimate contact, monolith reactors represent an advantageous technology for PI. They can lead to improved reactant conversion, selectivity towards desirable products and improved process safety [24].

In this chapter, PI in monoliths is reviewed for green chemistry applications. Monolith manufacture, operating principles, theory, flow regimes and mass transfer are discussed and explained. Since three-phase monolith reactors present greater design challenges for the contact of the gas–liquid–solid, Sections 6.2 and 6.3 highlight some of the particular considerations, reactor designs and flow regime studies that need to be taken account of by the designer. In Section 6.4 the advantages for PI of the monolith reactor are described, and in Section 6.5 several case studies in green chemistry applications are presented. Finally, the future outlook for this type of reactor is discussed.

6.2 Design of Monolith Reactors

6.2.1 Monolith and Washcoat Design

In designing a structured reactor process, the design of the monolith itself must be considered in terms of the channel and cell geometry, the dimensions, the thickness and characteristics of the catalyst washcoat and the loading and dispersion of catalytic metal within the washcoat pores. Monolith supports are typically manufactured industrially by extrusion processes and are often made from cordierite ($2MgO.2Al_2O_3.5SiO_2$). Some typical designs and properties of monoliths supplied by Corning, New York are shown in

Cell Density [cells/in²]	25	25	50
Shape	Square	Square finned	Square
Void fraction [%]	66	73	68
Surface/volume ratio [m²/m³]	640	1040	920
Hydraulic diameter [mm]	4.14	2.81	2.96
Number of full channels	44	44	88

Figure 6.1 *Structured monoliths and their characteristics. Reprinted from Heibel et al © 2001, with permission from Elsevier.*

Figure 6.1. The central monolith in the figure illustrates the use of fins to increase the available surface area for catalyst coating and improve the interfacial area for flowing films that coat the surface during operation [25].

Having selected a suitable design for the support structure, the channels must be washcoated with a thin layer of porous material, which will act as a high-surface-area support for the catalyst metals. In preparing a washcoat, a powder such as γ-alumina is milled to the appropriate particle size, possibly mixed with a structure improving agent such as dispersable boehmite and then mixed with water to form a slurry [26]. The pH of the slurry is controlled using dilute HNO_3 or NaOH. Washcoating is achieved by dipping the monolith into the slurry at a controlled rate, blowing excess slurry off the structure with compressed air, then drying and calcining. Stutz and Poulikakos [22] studied the optimum thickness of washcoat, for which it was noted that a thin washcoat gives a smaller, limiting amount of catalyst, leading to a low methane reactant conversion in Syngas production. For a thicker washcoat, the limiting effect is the reduced residence time, which occurs when the channel cross-sectional area open to flow of fluid becomes narrower, resulting in a higher flow velocity for the constant throughput of gas.

Mogalicherla and Kunzru [26] studied whether impregnating palladium on γ-alumina and using the resultant $Pd/\gamma Al_2O_3$ powder to prepare the slurry for washcoating was more effective than γ-alumina washcoating followed by impregnation of palladium. Well-adhered washcoats were observed using slurry prepared with γ-alumina, but the washcoat prepared from $Pd/\gamma Al_2O_3$ powder was very poor and led to a lower rate of reaction in the hydrogenation of α-methyl styrene than the former catalyst at low washcoat loadings. The addition of binders could be used to improve the adhesion of the $Pd/\gamma Al_2O_3$ powder washcoat.

Generally it is desirable to produce a uniform washcoat loading rather than an uneven one. However, in contrast with uneven channels and washcoats, some novel designs have been proposed which attempt to increase the surface area. For example, using angled microgrooves leads to enhanced mass-transfer rates relative to the flat microchannel [27].

6.2.2 Reactor and Distributor Design

A range of designs for three-phase monolith reactor systems have been proposed, particularly of methods for distributing the gas and liquid across the channels. Heiszwolf *et al.* [28] proposed the monolith loop reactor, in which liquid is circulated around a loop and is distributed over the monolith channels using a 'showerhead'-type distributor. Gas flows in co-current mode and may be recycled via a low-pressure gas recycle line (Figure 6.2a). The authors claimed that the advantages over a traditionally stirred slurry reactor included: (1) the catalyst and product remaining separated, (2) no attrition of the catalyst, (3) independently scalable heat transfer, (4) high gas–liquid mass transfer, (5) simple reactor construction and (6) energy-efficient operation. An alternative design is the cocurrent downflow contactor reactor (CDCR) [29,30], in which gas and liquid are supplied to the top of the column via an orifice that induces a swirling motion (Figure 6.2b). This gives rise to a stable dispersion of small bubbles with a high interfacial surface area. The advantages of the CDCR include many of those of the monolith loop reactor, plus 100% gas utilization, 97% approach to equilibrium in very short times, good control of the bubble interfacial area (1000–6000 m^2/m^3) and high gas hold-up (0.5–0.6) [31]. Further designs include the monolith stirrer reactor, which is similar to a stirred tank [32], and the monolith froth reactor [33].

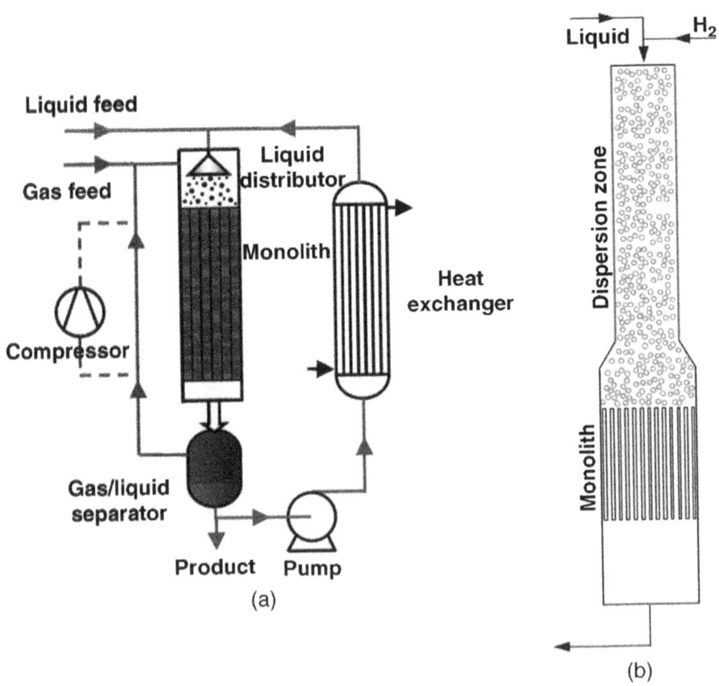

Figure 6.2 (a) Schematic diagram of the monolith loop reactor. Reprinted from Heiszwolf et al © 2001, with permission from Elsevier. (b) Diagram of the CDCR. Reprinted from Fishwick et al © 2007, with permission from Elsevier.

6.3 Hydrodynamics of Monolith Reactors

6.3.1 Flow Regimes

The performance of the monolith reactor as a high-intensity mass-transfer device is very dependent upon the flow pattern of gas and liquid within the monolith channels. Therefore, it is important to understand the conditions under which certain flow patterns occur and to be able to predict mass-transfer parameters for these flow regimes. In some earlier studies, Satterfield and Ozel [34] showed that the flow regimes are strongly influenced by the wettability between the liquid and the capillary wall and by the surface tension between the liquid and gas phases. A wealth of literature exists for gas–liquid two-phase flows in pipes and channels; for example, Taitel *et al.* [35] presented a comprehensive model for flow pattern transitions in large-diameter tubes for vertical upflow. However, such studies are not representative of the correct length scales of monolith channels with diameters of 1–2 mm. Thulasidas *et al.* [21] carried out some of the first detailed hydrodynamic studies at scales relevant to monolith channels operating in upflow mode and determined the main mass-transfer parameters, such as bubble size, shape, velocity and volume fraction of gas inside capillaries of circular or square cross-section on the basis of the superficial flow rates of gas and liquid in the feed. However, monolith reactors such as the CDCR operate in downflow mode, and later studies of downflow in monolith channels were carried out by Simmons *et al.* [36] for 2–3 mm round and square capillaries, then by Tsoligkas *et al.* [37] for 1.5–2.0 mm square glass capillaries, which represent a single channel in a monolith reactor. Using a high-speed camera, five flow patterns and three transitional flow patterns were observed, as shown in Figure 6.3. The regimes most worthy of note are (a) annular flow, (c) Taylor flow, also known as regular slug or bubble-train flow, (e) bubbly flow and (h) churn flow, with the others representing transitional regimes. Flow regime maps were determined for

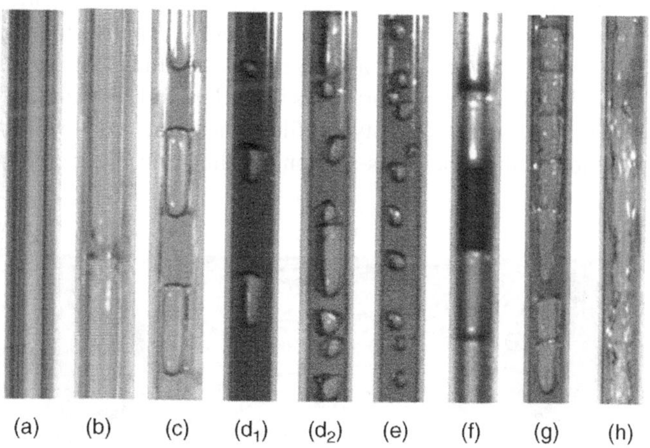

Figure 6.3 Flow patterns in a square capillary tube for vertical downflow: (a) annular flow; (b) slug–annular flow; (c) Taylor (regular slug) flow; (d) slug–bubbly flow (1 at low gas hold-up, 2 at high gas hold-up); (e) bubbly flow; (f) irregular slug flow; (g) slug–churn flow; (h) churn flow. Reprinted from Tsoligkas et al © 2007c, with permission from Elsevier.

air–water and air–water/isopropanol mixtures, which may be used to guide the selection of gas and liquid velocities in order to achieve the desired flow regime within the monolith. As mentioned in Section 6.1, Taylor flow is a desirable regime in which to operate in order to maximize the transfer of gas such as hydrogen to the catalyst surface within the three-phase reactor, owing to the thin liquid films formed at the sides of the gas bubble.

6.3.2 Mixing and Mass Transfer

In order to design and optimize the monolith reactor in three-phase operation, it is necessary to understand the different possible modes of mass transfer: (1) gas–liquid–solid (GS) via the liquid film; (2) gas–liquid (GL) via the gas bubble caps; and (3) liquid–solid (LS) within the liquid slugs (Figure 6.4). In this section, flow studies of the liquid slugs are first reviewed, followed by mass-transfer characteristics of the three-phase monolith reactor.

6.3.2.1 Recirculation in Liquid Slugs

Within the liquid slugs that occur in Taylor flow, mass-transfer processes (2) GL and (3) LS occur and it is therefore important to understand the mechanisms of mixing which assist these. Thulasidas *et al.* [21] used particle image velocimetry (PIV) to study the velocity distributions in liquid slugs in Taylor flow operated in 'upflow' mode, and showed that recirculating patterns occur with a high degree of mixing. The detailed behaviour is a function of the capillary number, where:

$$Ca = \frac{\mu V}{\sigma} \tag{6.1}$$

Depending on the capillary number of the flow, counter-rotating vortices or a complete bypass flow inside the liquid slug may be observed. Such recirculation patterns can play an important role in transporting dissolved gas from the bubble cap to the catalyst coated upon the channel walls. Tsoligkas *et al.* [37] also used PIV to study flow in liquid slugs, but in downflow mode. They found that short slugs (slug length less than the tube hydraulic diameter) led to a relatively flat velocity profile, where the axial velocity was only a function of the position in the tube cross-section, as shown in Figure 6.5. By contrast, in

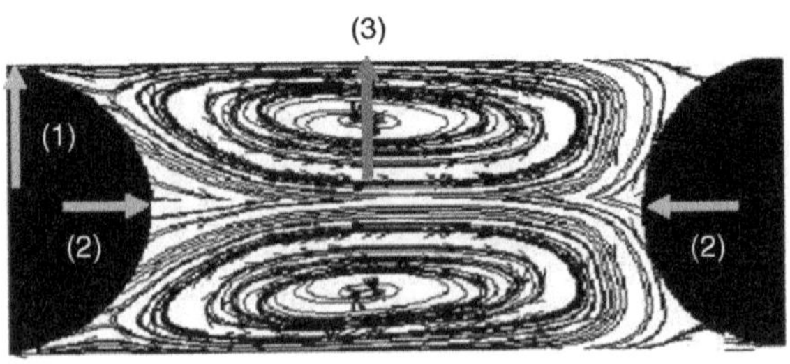

Figure 6.4 Recirculation patterns and mass-transfer processes within liquid slugs during Taylor flow: (1) gas–solid, (2) gas–liquid, (3) liquid–solid mass-transfer steps.

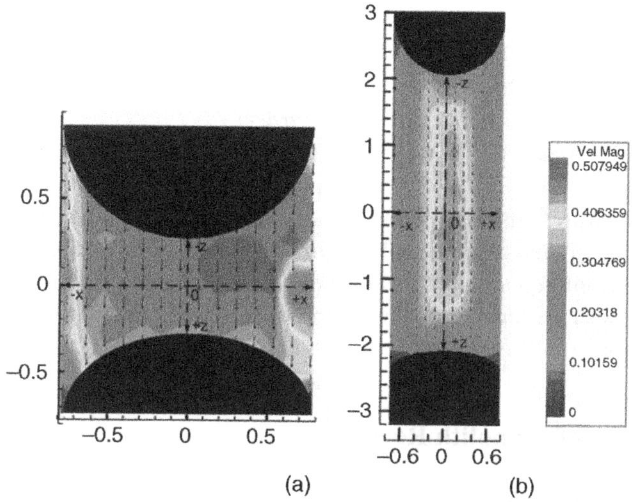

Figure 6.5 Flow fields within the liquid slugs in the $W = 1.5$ mm tube for (a) $L_S < D_c$, 30% v/v isopropanol/water–air mixture, $Ca = 0.008$, $V_{GS} = 0.384$ m/s, $V_{LS} = 0.111$ m/s, $V_B = 0.576$ m/s; (b) $L_S > D_c$, water–air mixture, $Ca = 0.0287$, $V_{GS} = 0.209$ m/s, $V_{LS} = 0.106$ m/s, $V_B = 0.343$ m/s. Units of velocity are m/s. Reprinted from Tsoligkas et al © 2007c, with permission from Elsevier.

long slugs the axial velocity component depends upon both the axial position in the tube and the tube cross-section. Parabolic velocity profiles are approximated for $V_{max}/V_b \approx 1.1$–1.7. Significant differences from upflow operation were observed, with upflow recirculation times being three times faster than downflow, which has implications for the models used to predict mass transfer and residence time distribution (RTD).

6.3.2.2 Mass-Transfer Processes

A detailed analysis of the mass-transfer steps involved in Taylor flow within the monolith reactor has been given by Kreutzer *et al.* [11] and is summarized in this section. The three mass-transfer modes of steps (1)–(3) may be combined, with the (2) GL and (3) LS steps considered as resistances in series and as in parallel with respect to (1) GS mass transfer. For the overall mass transfer, the following expression can be used:

$$k_{OV}a = k_{GS}a_{GS} + \left(\frac{1}{k_{GL}a_{GL}} + \frac{1}{k_{LS}a_{LS}}\right)^{-1} \tag{6.2}$$

The individual mass-transfer coefficients and areas are determined as:

$$k_{GS} = \frac{\mathcal{D}}{\delta} \tag{6.3}$$

$$a_{GS} = \frac{4(1-\varepsilon_L)}{D_C} \tag{6.4}$$

$$k_{GL}a_{GL} = \frac{0.133V_{TP}^2}{L_{Slug}^{0.5}} \quad (6.5)$$

$$k_{LS}a_{LS} = \frac{\mathcal{D}}{\delta}\frac{4\varepsilon_L}{D_C} \quad (6.6)$$

In Equations 6.3 and 6.6, the film thickness δ is an important parameter, which strongly influences the rate of mass transfer. The film thickness is influenced by the capillary number, and thus from Equation 6.1 it can be observed that the unit cell velocity, as well as the viscosity and surface tension of the fluid, has a strong effect upon the mass-transfer rates in the reactor.

6.4 Advantages of Monolith Reactors

6.4.1 Scale-out, Not Scale-up?

One of the advantages of monolith reactors is their relatively easy scale-up, based on the assumption that if the hydrodynamics and reaction in a single channel can be characterized, they can be assumed to be the same for all channels within the reactor. However, this assumption may only be justified under certain conditions and with a good distributor design to ensure even flow distribution of gas and liquid across all the channels in the monolith. Mantle *et al.* [38] used magnetic resonance imaging (MRI) to visualize two-phase flow in a ceramic monolith operating in Taylor flow. Special pulse sequences based on the 'single-shot RARE' sequence were developed in order to acquire data with the necessary spatial in-plane resolution of $391 \times 781\,\mu m$, with a data acquisition time of 160 ms. The technique was able to give a clear picture of the gas and liquid fractions within the channels, as shown in Figure 6.6. Bubble size and

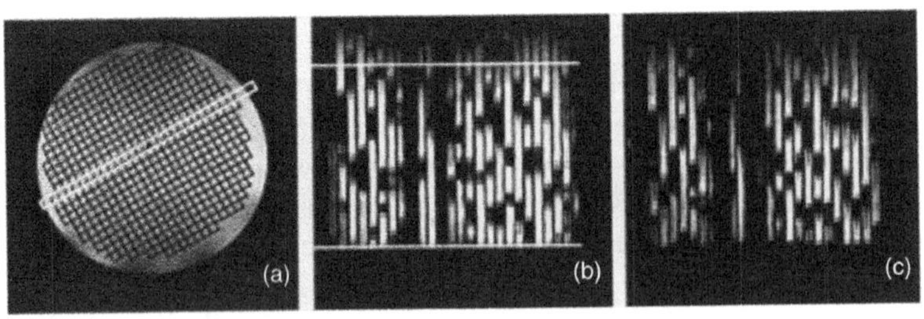

Figure 6.6 ^{1}*H magnetic resonance visualizations within the ceramic monolith; xy image of the fully water-saturated monolith. (a) shows the monolith's internal structure; each channel has a side of length 1.2 mm. It highlights the image slice position in the xz direction, for which the visualizations of gas–liquid distribution during two-phase flow are shown in (b) and (c). The data acquisition time for each image was 160 milliseconds; there was a period of 160 milliseconds between the start of each of the images shown in (b) and (c). In all images, the presence of water is indicated by high-intensity white; gas and ceramic are identified by zero-intensity black. The area within the solid white bars in (b) defines the analysis region. Reprinted from Mantle et al by permission of John Wiley & Sons. © 2002.*

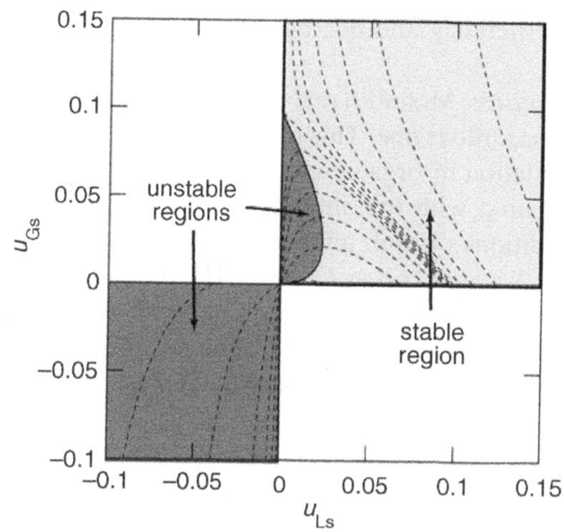

Figure 6.7 *Regions of stable and unstable operation of a monolith. Channel size d = 1.1 mm, equivalent to a monolith of 400 cpsi. The liquid properties of decane ($\mu = 0.924$ Pa.s, $\rho = 730$ kg/m^3 and $\sigma = 0.024$ N/m) and a constant slug length of four times the channel diameter were used. The pressure drop per unit length, $\Delta p/L = C$, is indicated in Pa/m. Reprinted with permission from Kreutzer et al. © 2005 American Chemical Society.*

velocity distributions were calculated from the MRI data, which confirmed that the system was operating in bubble train or Taylor flow. However, in one channel a reverse flow in the opposite direction to the superficial flow was observed, which was thought to result from the possibility of a negative pressure gradient existing at different locations across the monolith. This might lead to backmixing and deviation from plug-flow behaviour, so careful selection of the operating conditions used to achieve the desired performance is required. Kreutzer *et al.* [1] studied the scale-up of Taylor flow from capillary to monolith reactors. They demonstrated that operation in upflow mode was unstable, and they proposed a criterion for the minimal stable gas and liquid velocity for downflow mode, shown in Figure 6.7. Edvinsson Albers *et al.* [20] also found that downflow was desirable in establishing a steady Taylor flow pattern over a wide range of flowrates in scaling up a reactor to full production of hydrogen peroxide.

6.4.2 PI for Green Chemistry

The monolith has a number of advantages for greener processing, reducing the impact of the process upon the environment, which may be summarized as:

- *Improved Reaction Rate, Conversion and Selectivity*: Intensification of mass-transfer rates in the channels of the monolith may lead to the improved availability of dissolved gas at the catalyst surface. This in turn can speed up the rate of reaction, which improves the conversion of reactants and may affect the product distribution in the reactor. According to the Principles of Green Chemistry [39,40], this can lead to higher atom efficiency and decrease the production of waste byproducts for disposal.

- *Alternative Process Routes*: Monolith reactors may enable a hazardous process route to be converted to a less hazardous one. This might occur, for example, if monolith reactors make feasible the oxidation of organic chemicals using molecular oxygen under mild conditions [17], compared with the alternative route of aggressive peroxides [41] or extreme reaction conditions of high temperature or pressure.
- *Lower Running Costs, Saving Process Energy*: The pressure drop in monolith channels is lower than that for alternative catalytic reactors, such as trickle beds, which leads to lower requirement for gas compression and pumping. This can lead to savings in the electricity used for pumping. However, monolith catalysts do have a higher capital cost compared with pelleted catalysts.
- *Improved Heat Transfer*: For applications such as hydrogenation reactions, it may be necessary to cool the reaction mixture in order to keep the temperature under control. For strongly exothermic or fast reactions, an external heat exchanger can be used, located in the liquid circulation loop [24]. Alternatively, compact heat-exchanger reactors have been developed, which are a variant of monolith reactors in which alternating channels of a labyrinth design are used for the flow of the reaction mixture and the heating or cooling of fluid [42]. If the heat can be used elsewhere in the plant through process integration principles, such as the coupling together of endo- and exothermic reactions, the overall energy requirements of the plant may be decreased [43].
- *Safer Processing*: Monolith reactors present an inherently safer design than stirred tank reactors. Cybulski *et al.* [24] has summarized their advantages as: (1) lower reaction volume of hazardous material; (2) fail-safe operation, since in the event of pump failure the reaction stops automatically because the liquid drains down from the reaction zone; (3) avoidance of risky filtration operations (involving pyrophoric materials in the fine chemicals industry); and (4) improved control of temperature, preventing thermal runaway.
- *Reduction of Emissions*: Even though according to the Principles of Green Chemistry one should try to avoid the production of waste products, monoliths present a good catalyst design for end-of-pipe pollutant removal due to their low pressure drop. This makes them suitable for the removal of pollutants from a range of media, including car exhaust emissions, NO_x in stationary combustion sources [9] and oxidized pharmaceutical intermediates in wastewater [44].
- *Process Miniaturization and Local Manufacture*: The intensification of processes using monoliths may enable the development of small chemical reactors or distributed power plants [45], rather than a large centralized site such as an oil refinery or coal-fired power station. These portable plants could be operated at dispersed locations or even transported by trucks to the locations of different feedstocks. This type of localized operation could lead to lower transport distances of raw materials and products. Additionally, monoliths could be used in mobile applications, such as the cleaning of air in aircraft cabins [46] or the reforming of ethanol and dimethyl ether for onboard hydrogen production in a hydrogen vehicle [47]. The latter type of technology will help to enable the transformation of the carbon-based economy to the hydrogen economy and could help to reduce reliance on fossil fuels, depending on the method of hydrogen generation.

6.5 Applications in Green Chemistry

In this section, specific case studies are reviewed in detail with regard to applications in the chemical industry, fuel production, effluent treatment and bioreactors.

6.5.1 Chemical and Fine Chemical Industry

Monolith reactors have been studied for an increasing range of applications in three-phase reactions, and are particularly effective for hydrogenation, dehydrogenation and oxidation [15,48,49]. However, rational evaluation of the performance of such reactors must be made on the basis of comparison with other traditional reactor designs. Fishwick *et al.* [50] made such a comparison of selective hydrogenations in the monolith CDCR and stirred tank and trickle bed reactors. The selective hydrogenation of 2-butyne-1,4-diol (B3D) was used as the test reaction (shown in Scheme 6.1). This is a consecutive reaction, whose alkene intermediate, 2-butene-1,4-diol (B2D), is an important chemical in the manufacture of vitamins, pharmaceuticals and insecticides [51] and whose fully hydrogenated product, 2-butane-1,4-diol (B1D), is a raw material used in the polymer industry and the manufacture of tetrahydrofuran. Five reactors were compared, including a stirred tank reactor, trickle bed reactor, single capillary reactor, 5 cm diameter monolith CDCR and 10 cm diameter monolith CDCR. All were operated at a pressure of 200 kPa, except the single capillary reactor, which was operated at 100 kPa, and a temperature of 55 °C was used in each case. Industrial Pd/Al_2O_3 catalysts, supplied by Johnson Matthey, were tested, including a powdered 1% γ-Pd/Al_2O_3 (stirred tank), a 0.5% Pd/Al_2O_3 pellet (trickle bed) and a washcoated Pd/α-Al_2O_3 monolith. Each was typical of the industrial catalyst used in these types of reactor. In order to facilitate comparison of the different reactors, rates were reported as normalized per gram of palladium. Solvents used in the reaction included water, 2-propanol and a 30% v/v mixture of 2-propanol in water.

Table 6.1 displays the initial rate and selectivity towards B2D at B3D conversions of 90% and 100% for a range of different solvent compositions and reactor types. Selectivity,

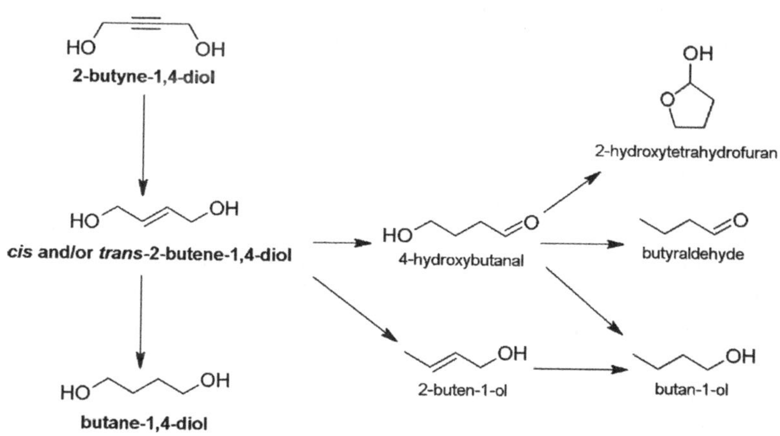

Scheme 6.1 *2-butyne-1,4-diol hydrogenation. Reprinted from Fishwick, et al © 2007, with permission from Elsevier.*

Table 6.1 *Initial reaction rates and selectivities towards cis-2-butene-1,4-diol for CDCR monoliths, stirred tank reactors and trickle bed reactors. Reprinted from Fishwick et al © 2007, with permission from Elsevier.*

Solvent	Reactor	Initial rate (mol/s/g_{Pd})	*S* (90% conversion)	*S* (100% conversion)
Water	5 cm monolith	0.186×10^{-3}	1.00	1.00
	10 cm monolith	0.186×10^{-3}	1.00	0.96
	Stirred tank reactor	0.264×10^{-3}	0.99	0.93
	Trickle bed reactor	0.033×10^{-3}	0.87	0.86
30% v/v 2-propanol in water (M)	5 cm monolith	0.206×10^{-3}	1.00	1.00
	10 cm monolith	0.206×10^{-3}	1.00	1.00
	Stirred tank reactor	0.319×10^{-3}	1.00	0.95
	Trickle bed reactor	0.330×10^{-3}	0.93	0.90
2-propanol	5 cm monolith	0.268×10^{-3}	0.99	0.90
	Stirred tank reactor	0.522×10^{-3}	0.97	0.87

S, is defined as the mole fraction of the intermediate (B2D) in the total number of moles of product. From Table 6.1 it can be observed that the B2D selectivity is generally very high, but that the structured monoliths lead to higher selectivities than both the stirred tank and trickle bed, regardless of which solvent is used. High selectivity towards B2D is thought to be associated with effective mass transfer of hydrogen to the catalyst surface. This ensures that the active sites are sufficiently populated with hydrogen for hydrogenation to occur in preference to the formation of side products by reactions that do not consume hydrogen, such as 4-hydroxybutanal, 2-buten-1-ol or the derivatives of these products shown in Scheme 6.1. However, if the concentration of hydrogen at the surface were too high, it could lead to over-hydrogenation of the alkene and the production of the alkane, and thus decrease the selectivity. Also, a close approximation to plug flow, as expected in the monoliths, could reduce the amount of time that the intermediate alkene spends in contact with the catalyst and thus decrease the extent of over-hydrogenation and formation of the alkane. In the monoliths, the mass-transfer coefficients of hydrogen at both the gas–liquid and the liquid–solid interfaces were substantially higher than in the stirred tank and trickle bed, which would be expected to lead to better availability of hydrogen at the catalyst surface and thus increase selectivity towards the hydrogenation products, rather than promoting side reactions. The performance of the monolith was also found to be dependent upon the hydrodynamic mode of operation; that is, whether the gas bubble dispersion is allowed to penetrate into the monolith channels such that Taylor flow occurs in the channels or liquid is presaturated with hydrogen by controlling the bubble dispersion to remain in the upper part of the column. Higher mass-transfer rates are observed when the monolith is operated in two-phase flow.

Referring again to Table 6.1, with the use of the mixed water/2-propanol solvent, complete selectivity towards the alkene is achieved without the need to use dopants or poison the catalyst with the addition of bases. The effect of solvent type upon the reaction behaviour can be partly explained in terms of higher hydrogen solubility in 2-propanol compared with water, leading to a higher rate of hydrogenation. Also, in mixtures of 2-propanol the bubble size is influenced by the solvent composition, which in turn affects the gas–liquid interfacial area. Of the three solvents studied, the smallest bubbles observed by a video microscope computer system occur with a mixture of 2-propanol in water, in which a higher gas–liquid interfacial area is therefore observed [52]. These studies show that suitable selection of reactor type, operating conditions and solvent can lead to improved selectivity towards the desirable product, thereby decreasing the production of undesirable waste.

In some cases, structured reactors open up new reaction routes or make feasible reactions that would otherwise have low rates and selectivities in the autoclave. The oxidation of glycerol, a byproduct of the biodiesel industry, can be used to produce a range of useful products. Glycerol is available from sustainable sources, but traditional oxidation processes used in industry sometimes involve environmentally unfriendly reagents such as dichromate, permanganate, manganese and peroxides [41,53,54]. Hutchings *et al.* [55] have shown that gold catalysts are active for selective oxidation reactions, using molecular oxygen as the oxidizing agent, and are very selective for the oxidation of glycerol to glycerate. The addition of base, such as sodium hydroxide, is required to accelerate the rate-limiting deprotonation step. Pollington *et al.* [17] studied the selective oxidation of glycerol in several multiphase reactors. The flow reactors consisted of a laboratory-scale loop system operated in batch recycle mode. The catalyst or support used included an Au/C-catalysed monolith or a cordierite monolith without metal coating, in which case the Au/C catalyst powder was placed in the feed tank with the liquid such that slurry flowed through the uncoated monolith during operation, in what is known as the 'meso-structured slurry bubble column' (MSSBC). Comparisons were made with studies of the same reaction in an autoclave. It was observed that the highest rate occurred in the MSSCB (109–201 mmol glycerol/s/m^3/bar), followed by the catalyst-coated monolith (36–65 mmol glycerol/s/m^3/bar) and finally the stirred reactor (4 mmol glycerol/s/m^3/bar). Selectivity towards glyceric acid was 100% for the monolith and autoclave, but in the MSSBC the production of dihydroxyacetone and glyceric acid was approximately equal, and thus selectivity towards glyceric acid was in the range of 47–54%. The lower selectivity in the MSSBC was attributed to better availability of oxygen than in the monolith and autoclave reactors. This study illustrated the benefits of structuring catalysts, with the monolith giving rise to enhanced reaction rates while maintaining high selectivity, which arises from enhanced mass transfer via thin liquid films when the gas–liquid mixture flows through the channels of the monolith catalyst.

6.5.2 Cleaner Production of Fuels

A major challenge facing the world today is ensuring the future supply of energy after the exhaustion of oil and gas reserves [56]. The potential use of hydrogen as an energy carrier and eventual conversion from a carbon-based to hydrogen economy is driving the development of hydrogen production and fuel cell technologies [2]. The application of monoliths to fuel production may find wider application in on-site or even portable production of hydrogen for use in fuel cells for applications in electronic device such as vehicles, portable electronics or electrical equipment in remote locations. Safety issues concerning the distribution and storage of compressed hydrogen have made on-site generation of hydrogen from hydrocarbons an attractive option. Mitchell and Kenis [2] studied the reforming of propane using SiC porous monoliths coated with ruthenium catalyst at temperatures in the range of 800–1000 °C:

$$C_3H_8 + 3\,H_2O \leftrightarrow 3CO + 7\,H_2 \tag{6.7}$$

At 900 °C and a steam-to-carbon ratio of 1.095, the observed C_3H_8 conversion was 99% and the selectivity to hydrogen was 74%, with stable performance of the catalyst over 15 thermal cycles of operation. However, the overall environmental acceptability of the process must be weighed against the energy required to generate the high reaction temperature. A lower-temperature process is the steam reforming of ethanol, which may be carried out at 400–500 °C:

$$C_2H_5OH + 3H_2O \rightarrow 6H_2 + 2\,CO_2 \tag{6.8}$$

Casanovas *et al.* [57] studied this process using conventional micro- and monolith reactors and an Si-micromonolith. The length scales of the Si-micromonolith (3.3 μm diameter channels) are considerably smaller than those of conventional monoliths (~0.9 mm channels) or 'micro'reactors (~0.7 mm channels), creating a higher surface area per unit volume and more intensive processing. An increase in specific hydrogen production of about one order of magnitude was measured when moving from cordierite monolith to microreactor and an increase of one to two orders of magnitude was measured when moving from microreactor to Si-micromonolith. Decreasing the channel size from conventional monoliths to the silicon microreactor resulted in an increase of inner surface area from 10^3 to 10^5 m^2/m^3, giving rise to a higher efficiency of hydrogen production per unit volume of reactor and the possibility of miniaturization and application in portable devices. The development of such technologies should enable the use of fuel cells in a wider range of electrical and electronic goods in future, and thus decrease reliance on batteries containing heavy metals.

6.5.3 Removal of Toxic Emissions

Monolith structures play an important role in the integrated approach to environmental protection, in which the roles of chemical reaction, separation and heat exchange can be combined. While the application of monoliths in car catalytic converters is well known, they are also finding application in gas cleaning from stationary sources – in particular the catalytic reduction of NO_x – from power stations, waste incineration plants, chemical plants, cogeneration plants [58] and the exploitation of oil sands in Canada [9]. Selective catalytic reduction (SCR) uses ammonia or urea as a reducing agent of NO_x and is one of the most widely applied post-combustion technologies in the minimization of emissions of this pollutant. Commercial SCR catalyst is usually a homogeneous mixture of titanium dioxide, tungsten trioxide and divanadium pentoxide, and is normally formed into honeycomb shapes [58]. Significant efforts have been made to develop processes for the simultaneous removal of NO_x and other pollutants, including SO_2, N_2O, VOCs and particulates. The BioDeNO_x process is an integrated biological and physicochemical process for the simultaneous removal of SO_2 and NO_x from flue gases, based upon reactive absorption, with enhancement of mass transfer through the use of monolith supports [59]. A significant challenge in the area of multiple pollutant reduction is finding a catalyst that can simultaneously reduce NO_x and N_2O. Centi *et al.* [60] found that a combination of Fe-ZSM-5 and Co-ZSM-5 is effective for reducing both pollutants by over 80%. Dhanushkodi *et al.* [9] modelled NO_x removal in a honeycomb monolith reactor by SCR with ammonia, utilizing a 2D reactor model and encompassing mass and momentum balances, with suitable inclusion of SCR reaction kinetics. This model indicated that ammonia is mainly consumed by the desired reaction at a temperature of 655 K, and under these conditions the conversion of NO is predicted to be 85%. Future developments in this area could include improved monolith preparation procedures and fluid distributor design [58].

The compact design of monolith reactors also makes them highly suitable for miniaturized applications in air purification, such as indoors or in aircraft cabins. Kalluri *et al.* [46] prepared microfibrous entrapped catalysts (MFECs) by entrapping small gamma alumina particles (150–180 μm) within sinter-locked networks of metal microfibres (8–12 μm) and formed them into thin flexible sheets (0.5–2.0 mm thickness) (Figure 6.8). Pleated structures of MFEC containing Pd/γ-Al_2O_3, conventional monoliths and packed beds were tested for the purification of air in aircraft cabins (ozone decomposition). It was found that packed beds gave too high a pressure drop, while mass transfer in monoliths was not as high as in MFEC mesh. However, Figure 6.9 shows that similar overall conversion levels of ozone were achieved in monoliths and in MFEC mesh.

An emergent area of research in monolith reactors is their application as bioreactors. Ebrahimi *et al.* [61] have asserted that as bioreactors require gas–liquid mass transfer for substrate supply, monoliths could be an attractive alternative, providing a reduction in the

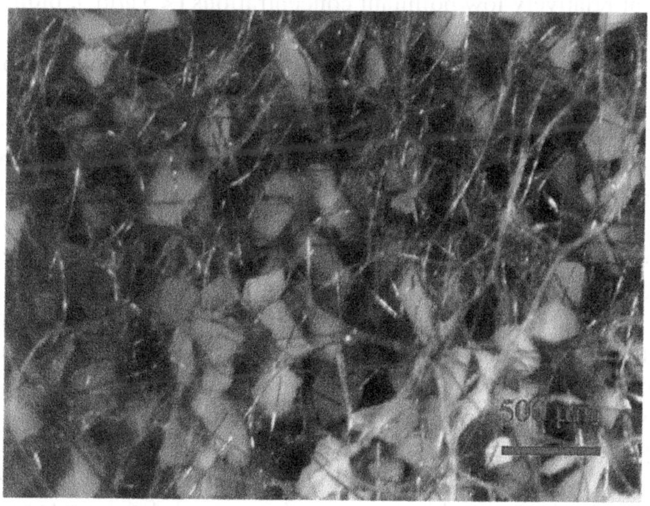

Figure 6.8 *Micrograph of 150–250 mm γ-Al_2O_3 entrapped in 4 and 8 mm (1 : 1 ratio) stainless-steel microfibres. Reprinted from Kalluri et al © 2009, with permission from Elsevier.*

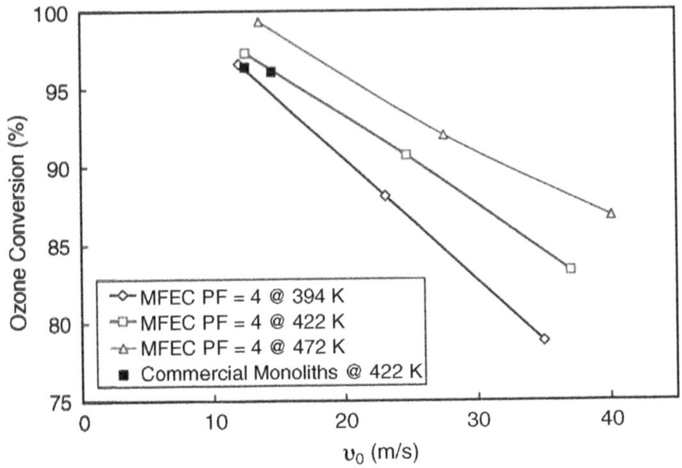

Figure 6.9 *Experimental conversion comparisons of MFEC with commercial monolith converters. Reprinted from Kalluri et al © 2009, with permission from Elsevier.*

energy dissipation that is normally required in such processes. A potential problem with the use of monoliths as bioreactors is the formation of biofilms, which could eventually lead to clogging. This can be controlled by using a high hydraulic retention time and the films can be removed by rinsing the monolith in tap water [61].

Monolith bioreactors are finding application in the removal of pollutants from gas streams, where at relatively low pollutant concentrations ($<5\ g/m^3$), biofiltration is cost effective and efficient [62] for the removal of volatile organic compounds (VOCs) from process flue gases. Rene *et al.* [62] investigated the use of a monolith bioreactor in the removal of styrene from the gas phase, using the yeast-like thermally dimorphic fungus *Sporothrix variecibatus*. The monolith reactor offers the advantages of being a support upon which to attach the biomass, while allowing liquid-phase nutrients to trickle down the channels, and of providing excellent gas–liquid-biofilm mass transfer of the pollutant to be removed from the gas phase. Rene *et al.* [62] found that up to $67.4\ g/m^3$/hour styrene could be removed at an inlet styrene load of $73.5\ g/m^3$/hour, with residence times in the empty bed varying from 19 to 77 seconds. Clogging was eventually found to be a problem, with biodegradation efficiency significantly decreasing after 92 days of operation due to excess biomass growth. Similarly, Jin *et al.* [63] studied the removal of methanol from air in a trickling monolith bioreactor. The acid-tolerant yeast *Candida boidinii* was found to be the dominant organism in the biofilm. It was able to degrade methanol at a pH as low as 2, with maximal capacity of $234\ g/m^3$/hour, more than 80% removal efficiency and complete conversion into biomass and end products, carbon dioxide and water. Again, excess biomass growth occurred after long-term operation, but this could be controlled by temporarily increasing the trickling rate. Additional studies by Jin *et al.* [64] showed that the monolith bioreactor could be used to degrade toluene in air, as a model VOC compound, and detailed studies of operational parameters of the

Figure 6.10 *Photograph of a cocurrent downflow contactor photocatalytic reactor (left) and reticulated foam monolith catalyst (right). Reprinted from Ochuma et al © 2007, with permission from Elsevier.*

process included start-up of the reactor, toluene loading rate and changes in gas and liquid flow rates.

The treatment of pollutants in liquid-phase streams can also be achieved using monolith reactors. Low concentrations of persistent pollutants, such as endocrine-disrupting chemicals or recalcitrant nitrogen-containing organic compounds which cannot be removed by biological treatment present a challenge for water-treatment companies. Processes such as supercritical water oxidation and photocatalysis have been suggested for the removal of such pollutants. Photocatalysis presents an efficient means of oxidizing pollutants using air as the oxidant, with a TiO_2 catalyst activated by ultraviolet (UV) light, but the use of slurry catalyst requires filtration of the catalyst particles from the treated water. Ochuma *et al.* [44] investigated the use of a TiO_2/alumina reticulated foam monolith for the purification of water contaminated with the recalcitrant nitrogen-containing compound 1,8-diazabicyclo [5.4.0]undec-7-ene (DBU). The reactor setup was based on the CDCR, described in Section 2.2, with the lower section of the reactor containing a 1 kW UV lamp, surrounded by an annular-shaped reticulated foam with 15 pores per inch (Figure 6.10). In contrast to the regular monolith channels, this structure consisted of random pores in sponge-like form. As shown in Figure 6.11, for the same mass concentration of catalyst the reticulate foam monolith was able to convert 100% of the DBU after 60 minutes, while with the slurry catalyst only 56% was converted. Corresponding total organic carbon (TOC) conversions were 26 and 15%, respectively. A possible limitation in the application of monoliths to photocatalytic reactions is the limited penetration of UV light into the channels, which is required for activation of the reaction. To overcome this effect, the optical-fibre monolith reactor has been designed [65,66]. This reactor is composed of a large number of parallel channels, into each of which a fibre-optic cable is inserted, which carries UV light towards the catalyst surface coated upon the channel walls. Modelling studies [66] have indicated some optimal design features that should be considered, such as the use of larger fibres and smaller fibre/channel diameter ratios for higher-efficiency pollutant degradation.

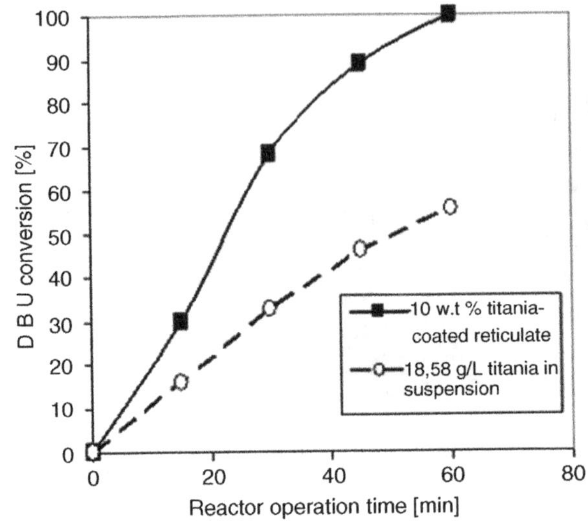

Figure 6.11 *Comparative photocatalytic conversion of DBU and TOC with TiO2 P25 (18.58 g/l suspension) and 10 wt% TiO2-coated reticulated foam monolith. Experimental conditions: 50°C, liquid circulation rate 10 l/minute, O_2 flow rate 0.1 l/minute, initial DBU concentration 0.6801×10^{-3} mol/l. Reprinted from Ochuma et al © 2007, with permission from Elsevier.*

6.6 Conclusion

This review highlighted the potential applications of monoliths in green chemistry, including in the fine chemical and chemical industries, fuel production and effluent treatment. It has been shown that the hydrodynamics of flow in the channels, particularly in slug or Taylor flow, has a significant effect upon the recirculation patterns in the liquid slugs and upon the gas–liquid mass-transfer processes occurring. The excellent gas–liquid mass-transfer capabilities of the monolith reactor and the low pressure drop are advantages that make this type of reactor attractive for the improvement of conversion, selectivity or reaction rate in a range of chemical reactions and effective for the cleaning of gas and liquid effluents. In recent years there has been a lot of research interest in monolith catalysts, but only limited use of such reactors in established industrial processes, apart from automotive applications. The cost consideration must be carefully evaluated, since monolith catalysts are more expensive to produce than slurry or pelleted catalysts, which may discourage companies from changing to a monolith reactor. However, Boger *et al.* [67] asserted that in many cases the higher costs can be offset by the reusability of monolith catalysts, together with the reduced operating costs, such as filtration and the reduction in catalyst and product losses and waste generated. Even so, the application of monoliths is likely to be more attractive in specialized applications, and particularly where miniaturization of the reactor is required. From the examples discussed, onboard hydrogen production in vehicles and air purification in aircraft are possible examples of where intensified processing will make monoliths commercially attractive.

Acknowledgment

I am grateful to EPSRC for funding and to Prof. Mike Winterbottom and Drs Mark Simmons, Reyna Natividad, Robert Fishwick, Idoko Ochuma and Andreas Tsoligkas for their contributions to the work reviewed in this chapter.

Nomenclature

a	Interfacial area, m^2/m^3
C	Pressure drop per unit length, Pa/m
Ca	Capillary number
$\mathcal{D}$	Diffusion coefficient, m^2/s
D	Diameter, m
k	Mass-transfer coefficient, m/s
L	Length, m
Δp	Pressure drop, Pa
v_0	Superficial velocity, m/s
V	Velocity, m/s

Greek Letters

δ	Film thickness, m
ε	Voidage
μ	Dynamic viscosity, Pa.s
ρ	Density, kg/m^3
σ	Surface tension, N/m

Subscripts and Superscripts

B	Bubble
C	Capillary
G	Gas
L	Liquid
OV	Overall
S	Solid
$Slug$	Slug
TP	Two-phase

References

(1) M. T. Kreutzer, J. J. W. Bakker, F. Kapteijn, J. A. Moulijn and P. J. T. Verheijen, Scaling up multiphase monolith reactors: Linking residence time distribution and feed maldistribution. Ind. Eng. Chem. Res., 44, 4898–4913 (2005).

(2) C. M. Mitchell and P. J. A Kenis, Ceramic microreactors for on-site hydrogen production from high temperature steam reforming of propane. Lab Chip, 6, 1328–1337 (2006).

(3) W. Ruettinger, O. Ilinich and R. J. Farrauto, A new generation of water gas shift catalysts for fuel cell applications. J. Power Sources, 118, 61–65 (2003).

(4) K. P. Brooks, J. Hu, H. Zhu and R. J. Kee, Methanation of carbon dioxide by hydrogen reduction using the Sabatier process in microchannel reactors. Chem. Eng. Sci., 62, 1161–1170 (2007).

(5) R. Guettel, J. Knochen, U. Kunz, M. Kassing and T. Turek, Preparation and catalytic evaluation of cobalt-based monolithic and powder catalysts for Fischer-Tropsch synthesis. Ind. Eng. Chem. Res., 47, 6589–6597 (2008).

(6) M. Sung, S. Kato, F. Kawanami and M. Sudo, Evaluation of an air-cleaning unit having photocatalytic sheets to remove acetaldehyde from indoor air. Build. Environ., 45, 2002–2007 (2010).

(7) T. van Gerven, G. Mul, J. Moulijn and A. Stankiewicz, A review of intensification of photocatalytic processes. Chem. Eng. Process., 46, 781–789 (2007).

(8) L. Lisi, R. Pirone, G. Russo and V. Stanzione, Cu-ZSM5 based monolith reactors for NO decomposition. Chem. Eng. J., 154, 341–347 (2009).

(9) S. R. Dhanushkodi, N. Mahinpey and M. Wilson, Kinetic and 2D reactor modelling for simulation of the catalytic reduction of NO_x in the monolith honeycomb reactor, Process Saf. Environ., 86, 303–309 (2008).

(10) R. K. Edvinsson and A. Cybulski, A comparison between the monolithic reactor and the trickle bed reactor for liquid phase hydrogenations, Catal. Today, 24, 173–179 (1995).

(11) M. T. Kreutzer, P. Du, J. J. Heiszwolf, F. Kapteijn and J. A. Moulijn, Mass transfer characteristics of three-phase monolith reactors. Chem. Eng. Sci., 56, 6015–6023 (2001).

(12) N. Tsoligkas, M. J. H. Simmons, J. Wood and C.G. Frost, Kinetic and selectivity studies of gas-liquid reaction under Taylor flow in a circular capillary. Catal. Today, 128, 36–46 (2007).

(13) A. N. Tsoligkas, M. J. H. Simmons and J. Wood, Two phase gas-liquid reaction studies in a circular capillary. Chem. Eng. Sci., 62, 5397–5401 (2007).

(14) A. K. Mogalicherla and D. Kunzru, Effect of gas and liquid superficial velocities on the performance of monolithic reactors. Ind. Eng. Chem. Res. 49, 1631–1641 (2010).

(15) T. A. Nijhuis, M. T. Kreutzer, A. C. J. Romijn and J. A. Moulijn, Monolithic catalysts as more efficient three-phase reactors. Catal. Today, 66, 157–165 (2001).

(16) D. I. Enache, P. Landon, C. M. Lok, S. D. Pollington and E. H. Stitt, Direct comparison of a trickle bed and a monolith for hydrogenation of pyrolysis gasoline. Ind. Eng. Chem. Res., 44, 9431–9439 (2005).

(17) S. D. Pollington, D. I. Enache, P. Landon, S. Meenakshisundaram, N Dimitratos, A. Wagland, G. J. Hutchings and E. H. Stitt, Enhanced selective glycerol oxidation in multiphase structured reactors. Catal. Today, 145, 169–175 (2009).

(18) T. A. Patrick and M. A. Abraham, Evaluation of a monolith-supported Pt/Al2O3 catalyst for wet oxidation of carbohydrate containing waste streams. Environ. Sci. Technol., 34, 3480–3488 (2000).

(19) C. P. Rodrigues, V. T. daSilva and M. Schmal, Partial oxidation of ethanol over cobalt oxide based cordierite monolith catalyst. Appl. Catal. B., 96, 1–9 (2010).

(20) R. Edvinsson Albers, M. Nyström, M. Siverström, A. Sellin, A. C. Dellve, U. Andersson and T. Berglin, Development of a monolith-based process for H2O2 production: from idea to large-scale implementation. Catal. Today, 69, 247–252 (2001).

(21) T. C. Thulasidas, M. A. Abraham and R. L. Cerro, Bubble train flow in capillaries of circular and square cross section. Chem. Eng. Sci., 50, 183–199 (1995).

(22) M. J. Stutz and D. Poulikakos, Optimum washcoat thickness of a monolith reactor for syngas production by partial oxidation of methane. Chem. Eng. Sci., 63, 1761–1770 (2008).

(23) M. T. Kreutzer, F. Kapteijn and J. A. Moulijn, Shouldn't catalysts shape up? Structured reactors in general and gas-liquid monolith reactors in particular. Catal. Today, 111, 111–118 (2006).

(24) A. Cybulski, A Stankiewicz, R. K. Edvinsson Albers and J. A. Moulijn, Monolithic reactors for fine chemicals industries: a comparative analysis of a monolithic reactor and a mechanically agitated slurry reactor. Chem. Eng. Sci., 54, 2351–2358 (1999).

(25) A. K. Heibel, J. J. Heiszwolf, F. Kapteijn and J. A. Moulijn, Influence of channel geometry on hydrodynamics and mass transfer in the monolith film flow reactor. Catal. Today., 69, 153–163 (2001).

(26) A. K. Mogalicherla and D. Kunzru, Effect of method of preparation on activity of Pd/Al_2O_3 monolith catalysts. Can. J. Chem. Eng., 9999, 367–375 (2010).

(27) D. M. Qiu, A. L. Tonkvich, M. Fanelli, S. Fitzgerald, J. Marco, R. Stevenson, M. Lamont, C. Burston, J. Lerou and L. Silva, Enhancing two-phase flow mixing and mass transfer in microchannel with surface features, in Proc. 6th Int. Conference on Nanochannels, Microchannels and Minichannels, Parts A and B, pp. 269–277 (2008).

(28) J. J. Heiszwolf, L. B. Engelvaart, M. G.van den Eijnden, M. T. Kreutzer, F. Kapteijn and J. A. Moulijn, Hydrodynamic aspects of the monolith loop reactor. Chem. Eng. Sci., 56, 805–812 (2001).

(29) H. Marwan and J. M. Winterbottom, The selective hydrogenation of butyne-1,4-diol by supported palladiums: a comparative study on slurry, fixed bed and monolith downflow bubble column reactors. Catal. Today, 97, 325–330 (2004).

(30) J. M. Winterbottom, H. Marwan, E. H. Stitt and R. Natividad, The palladium catalysed hydrogenation of 2-butyne-1,4-diol in a monolith bubble column reactor. Catal. Today, 79–80, 391–399 (2003).

(31) A. P. Boyes, A. Chugtai, Z. Khan, S. Raymahasay, A. T. Sulidis and J. M. Winterbottom, The Cocurrent Downflow Contactor (CDC) as a fixed bed and slurry reactor for catalytic hydrogenation. J. Chem. Tech. Biotechnol., 64, 55–65 (1995).

(32) K. M. de Lathouder, J. J. W. Bakker, M. T. Kreutzer, S. A. Wallin, F. Kapteijn and J. A. Moulijn, Structured reactors for enzyme immobilization – a monolithic stirrer reactor for application in organic media. Chem. Eng. Res. Des., 84, 390–398 (2006).

(33) L. L. Crynes, R. L. Cerro and M. A. Abraham, Monolith froth reactor: development of a novel three-phase catalytic system. AIChEJ, 41, 337–345 (1995).

(34) C. N. Satterfield and F. Ozel, Some characteristics of 2-phase flow in monolithic catalyst structures. Ind. Eng. Chem. Fundam., 16, 61–67 (1977).

(35) Y. Taitel, D. Bornea and A. E. Dukler, Modeling flow pattern transitions for steady upward gas-liquid flow in vertical tubes. AIChEJ, 26, 345–354 (1980).

(36) M. J. Simmons, D. C. Y. Wong, P. J. Travers and J. S. Rothwell, Bubble behaviour in three phase capillary microreactors. Int. J. Chem. React. Eng., 1, A30 (2003).

(37) A. N. Tsoligkas, M. J. H. Simmons and J. Wood, Influence of orientation upon the hydrodynamics of gas-liquid flow for square channels in monolith supports. Chem. Eng. Sci., 62, 4365–4378 (2007).

(38) M. D. Mantle, A. J. Sederman, L. F. Gladden, S. Raymahasay, J. M. Winterbottom and E. H. Stitt, Dynamic MRI visualization of two-phase flow in a ceramic monolith. AIChEJ, 48, 909–912 (2002).

(39) P. T. Anastas and J. C. Warner, Green Chemistry: Theory and Practice, Oxford University Press: New York (1998).

(40) P. Anastas and N. Eghbali, Green chemistry: principles and practice, Chem. Soc. Rev., 39, 301–312 (2010).

(41) G. B. Shul'pin, Y. N. Kozlov, L. S. Shul'pina, T. V. Strelkova and D. Mandelli, Oxidation of reactive alcohols with hydrogen peroxide catalyzed by manganese complexes. Catal. Lett, 138, 193–204 (2010).

(42) M. Grasemann, A. Renken, M. Kashid and L. Kiwi-Minsker, A novel compact reactor for three-phase hydrogenations. Chem. Eng. Sci., 65, 364–371 (2010).

(43) F. X. Yin, S. F. Ji, H. Mei, Z. L. Zhou and C. Y. Li, Coupling of highly exothermic and endothermic reactions in a metallic monolith catalyst reactor: a preliminary experimental study. Chem. Eng. J., 155, 285–291 (2009).

(44) I. J. Ochuma, O. O. Osibo, R. P. Fishwick, S. Pollington, A. Wagland, J. Wood and J. M. Winterbottom, Three-phase photocatalysis using suspended titania and titania supported on a reticulated foam monolith for water purification. Catal. Today, 128, 100–107 (2007).

(45) V. Palma, E. Palo and P. Ciambelli, Structured catalytic substrates with radial configurations for the intensification of the WGS stage in H_2 production. Catal. Today, 147, S107–S112 (2009).

(46) R. R. Kalluri, D. R. Cahela and B. J. Tatarchuk, Comparative heterogeneous contacting efficiency in fixed bed reactors: opportunities for new microstructured systems. Appl. Catal. B, 90, 507–515 (2009).

(47) C. Ledesma, M. Dominguez, N. Jimenez, E. Lopez and J. Llorca, Hydrogen generation from renewables steam reforming of ethanol and dimethyl ether over structured catalysts. Chimica Oggi-Chemistry Today, 27, 40–43 (2009).

(48) E. H. Stitt, S. D. Jackson, D. G. Shipley and F. King, Modelling propane dehydrogenation in a rotating monolith reactor. Catal. Today, 69, 217–226 (2001).

(49) R. Subramaniam, G. J. Panuccio, J. J. Krummenacher, I. C. Lee and L. D. Schmidt, Catalytic partial oxidation of higher hydrocarbons: reactivities and selectivities of mixtures. Chem. Eng. Sci., 59, 5501–5507 (2004).

(50) R. P. Fishwick, R. Natividad, R. Kulkarni, P. A. McGuire, J. Wood, J. M. Winterbottom and E. H. Stitt, Selective hydrogenation reactions: a comparative study of monolith CDC, stirred tank and trickle bed reactors. Catal. Today, 128, 108–114 (2007).

(51) R. Natividad, R. Kulkarni, K. Nuititikul, S. Raymahasay, J. Wood and J. M. Winterbottom, Analysis of the performance of single capillary and multiple capillary (monolith) reactors for the multiphase Pd-catalyzed hydrogenation of 2-butyne-1,4-diol. Chem. Eng. Sci., 59, 5431–5438 (2004).

(52) B. Hu, R. P. Fishwick, A. W. Pacek, J. M. Winterbottom, J. Wood, E. H. Stitt and A. W. Nienow, Simultaneous measurement of in situ bubble size and reaction rates with a heterogeneous catalytic hydrogenation reaction. Chem. Eng. Sci., 62, 5392–5396 (2007).

(53) R. A. Sheldon and R. S. Downing, Heterogeneous catalytic transformations for environmentally friendly production. Appl. Catal. A., 189, 163–183 (1999).

(54) W. F. Hoelderich, Environmentally benign manufacturing of fine and intermediate chemicals. Catal. Today, 62, 115–130 (2000).

(55) G. J. Hutchings, S. Carrettin, P. Landon, J. K. Edwards, D. Enache, D. W. Knight, Y. J. Xu and A. F. Carley, New approaches to designing selective oxidation catalysts: Au/C a versatile catalyst. Top. Catal., 38, 223–230 (2006).

(56) D. Abbott, Keeping the energy debate clean: how do we supply the world's energy needs? Proc. IEEE, 98, 42–66 (2010).

(57) A. Casanovas, M. Domíuez, C. Ledesma, E. López and J. Llorca, Catalytic walls and microdevices for generating hydrogen by low temperature steam reforming of methanol. Catal. Today, 143, 32–37 (2009).

(58) V. Tomašić, Application of the monoliths in $DeNO_x$ catalysis. Catal. Today, 119, 106–113 (2007).

(59) P. van der Maas, P. van den Brink, S. Utomo, B. Klapwijk and P. Lens, NO removal in continuous $BioDeNO_x$ reactors: Fe(II)EDTA(2-) regeneration, biomass growth, and EDTA degradation. Biotech. Bioeng., 94, 575–584 (2006).

(60) P. Centi, P. Ciambelli, S. Perathoner and P. Russo, Environmental catalysis: trends and outlook. Catal. Today, 75, 3–15 (2002).

(61) S. Ebrahimi, R. Kleerebezem, M. T. Kreutzer, F. Kapteijn, J. A. Moulijn, J. J. Heijnen and M. C. M.van Loosdrecht, Potential application of monolith packed columns as bioreactors, control of biofilm formation. Biotech. Bioeng., 93, 238–245 (2006).

(62) E. R. Rene, M. E. Lopez, M. C. Veiga and C. Kennes, Performance of a fungal monolith bioreactor for the removal of styrene from polluted air. Bioresour. Technol., 101, 2608–2615 (2010).

(63) Y. M. Jin, M. C. Veiga and C. Kennes, Removal of methanol from air in a low-pH trickling monolith bioreactor. Process Biochem., 43, 925–931 (2008).

(64) Y. M. Jin, M. C. Veiga and C. Kennes, Development of a novel monolith-bioreactor for the treatment of VOC-polluted air. Environ. Technol., 27, 1271–1277 (2006).

(65) H. F. Lin and K. T. Valsaraj, Development of an optical fiber monolith reactor for photocatalytic wastewater treatment. J. Appl. Electrochem., 35, 699–708 (2005).

(66) H. F. Lin and K. T. Valsaraj, An optical fiber monolith reactor for photocatalytic wastewater treatment. AIChEJ, 52, 2271–2280 (2006).

(67) T. Boger, A. K. Heibel and C. M. Sorensen, Monolithic catalysts for the chemical industry. Ind. Eng. Chem. Res., 43, 4602–4611 (2004).

7

Process Intensification and Green Processing Using Cavitational Reactors

Vijayanand Moholkar,[1] Parag Gogate[2] and Aniruddha Pandit[2]

[1]Chemical Engineering Department, Indian Institute of Technology, Guwahati, Assam, India
[2]Chemical Engineering Department, Institute of Chemical Technology, Matunga, Mumbai, India

7.1 Introduction

Cavitational reactors are a novel and promising form of multiphase reactor in which the energy is released at the location of transformation, possibly resulting in significant process intensification. The principle behind the expected beneficial effects is the phenomenon of cavitation, described as the generation, subsequent growth and collapse of cavities, which releases large magnitudes of energy over a very small volume, resulting in very high energy densities [1,2]. Cavitation occurs at millions of locations in the reactor simultaneously, and generates conditions of very high temperatures and pressures locally (a few thousand atmospheres pressure and a few thousand K temperature), with overall ambient conditions [3]. Moreover, free radicals are generated in the process due to the dissociation of vapours trapped in the cavitating bubbles, which results either in intensification of the chemical reactions or the propagation of certain reactions [4–6]. Cavitation also results in the generation of local turbulence and liquid microcirculation (acoustic streaming) in the reactor, enhancing the rates of the transport processes. These physical effects will be responsible for the intensification of physical processing applications and chemical processing applications limited by mass transfer.

Process Intensification for Green Chemistry: Engineering Solutions for Sustainable Chemical Processing, First Edition. Edited by Kamelia Boodhoo and Adam Harvey.

The energy-efficient method of producing cavities of a desired quality (type of dynamic behaviour) can be taken as the main criterion by which to distinguish among different types of cavitation: acoustic (use of sound energy), hydrodynamic (use of fluid energy and pressure/velocity alterations), optic (use of light energy) and particle (use of energy associated with movement of particles). Of these four, only acoustic and hydrodynamic cavitation generate the desired intensity for chemical or physical processing [7]. Once the cavities are generated, subsequent expansion may be achieved by reducing the ambient pressure through the use of either static or dynamic means. The surrounding fluctuating pressure field, along with the type of cavity produced (i.e. cavity containing gas, vapour or a mixture of gas and vapour), affects the final intensity of cavitation.

For cavitational reactors, two aspects of cavity dynamics are of prime importance: the maximum size reached by the cavity before its violent collapse and the cavity's lifespan. The maximum size determines the magnitude of the pressure pulse produced on collapse and hence the cavitation intensity that can be obtained in the system. The lifespan determines the distance travelled by the cavity from the point at which it is generated to the point before the collapse and hence is a measure of the active volume of the reactor in which the cavitational effects are observed. The aim of the equipment designer should be to maximize both these quantities by suitably adjusting the different parameters, including the methodology used for the generation of cavities (the type of cavity generated is a crucial parameter in deciding the intensity of the cavitation phenomena).

This chapter provides an overview of the basic aspects of cavitational reactors, the expected working mechanism for intensification, design configurations with recommendations for optimum operating parameters and some case studies related to process intensification (PI) and green chemistry.

7.2 Mechanism of Cavitation-based PI of Chemical Processing

In order to understand how cavitational collapse can affect chemical transformations, one must consider the possible effects of this collapse in different systems. In the case of homogeneous liquid-phase reactions [1,4], there are two major effects. First, the cavity that is formed is unlikely to enclose a vacuum (in the form of void) – it will almost certainly contain vapour from the liquid medium or dissolved volatile reagents. During the collapse, these vapours will be subjected to extreme conditions of high temperatures and pressures, causing molecules to fragment and generate highly reactive radical species. These radicals may then react either within the collapsing bubble or after migration into the bulk liquid. Second, the sudden collapse of the bubble also results in an inrush of the liquid to fill the void, producing fluid shear forces in the surrounding bulk liquid capable of breaking the chemical bonds of many materials dissolved in the fluid or disturb the boundary layer, facilitating the transport, in the case of heterogeneous systems.

The cavitational activation in heterogeneous systems is mainly a consequence of the physical effects of cavitation [1,4]. In a heterogeneous solid/liquid system, the collapse of the cavitation bubble results in significant structural and mechanical defects. Collapse near the solid surface produces an asymmetrical inrush of the fluid to fill the void, forming a liquid jet targeted at the surface. This effect is equivalent to high-pressure/high-velocity liquid jets and is the reason why ultrasound is used to clean solid surfaces. These jets activate the solid catalyst and increase the material mass transfer to the surface through the disruption of the interfacial boundary layers, as well as dislodging the material occupying the active sites. Collapse on the surface, particularly of powders, produces enough energy to cause fragmentation (even for finely divided metals). Thus, in this situation cavitational effects increase the surface area for a reaction and provide additional activation through efficient material mixing and enhanced mass transport.

In heterogeneous liquid/liquid reactions, cavitational collapse at or near the interface will cause disruption and mixing, resulting in the formation of very fine emulsions. When very fine emulsions are formed, the surface area available for the reaction between the two phases is significantly increased, thus increasing the rates of reaction. This is very beneficial, particularly in the case of phase-transfer-catalysed reactions or in biphasic systems.

It is very important to understand the exact mechanism of the expected intensification of chemical processing applications using the cavitational effects. When deciding on whether to select cavitational reactors as a replacement for conventional reactors in a particular chemical or physical transformation, the first question that needs to be addressed is whether the cavitational enhancement is a result of an improved mechanical process (due to enhanced mixing). If so, cavitation pre-treatment of a slurry may be all that is required before the system is subjected to a conventional transformation scheme, and the scale-up of the pre-treatment vessel will be a relatively simple task. If, however, the effect is truly based on the cavitation chemistry, cavitation must be provided during the transformation itself, either in a continuous manner or in suitable pulsed operation. The scientific database from the laboratory study provides the most effective parametric window and gives the cause–effect relationships between the operating parameters, the observed cavitational effects and the expected levels of PI. An important step in laboratory-scale investigations is to try to understand the mechanisms of interaction from the observed phenomena, so that the desired cavitation field can be created on a larger scale in order to promote similar interactions. The medium physicochemical properties are very important in deciding whether or not cavitational effects can be beneficial and determining the associated costs for the expected intensification. For example, high-viscosity media with low vapour pressure will require greater energy to generate cavitation, whereas the presence of entrained or evolved gases will facilitate cavitation, as will the presence or generation of solid particles. Overall, it is crucial that a proper analysis in terms of system parameters, the expected mechanism of the cavitational effects and the degree of PI balanced against economics is made before the selection of cavitational reactors is recommended for the specific transformation.

7.3 Reactor Configurations

7.3.1 Sonochemical Reactors

Sonochemical reactors are based on the principle of generation of cavitation by the use of ultrasound, described commonly as 'acoustic cavitation'. In the case of acoustic cavitation, the pressure variations in the liquid are effected using high-frequency sound waves, with frequencies in the range of 16 kHz–5 MHz. If a sufficiently large negative pressure is applied to the liquid such that the average distance between the molecules exceeds the critical molecular distance required to hold the liquid intact, cavities or voids are created. Subsequent compression and rarefaction cycles of the sound waves cause the cavity to expand, reach a maximum cavity size (the magnitude depends on the operating conditions) and then collapse, releasing a large amount of energy [2].

A common sonochemical reactor design is the ultrasonic horn, which is most frequently used for sonochemical research at the laboratory scale [7]. Ultrasonic horns are typically immersion-type transducers and their most important advantage is that they are capable of delivering large amounts of power directly to the reaction mixture, which can also be regulated by varying the amplitude delivered to the transducer, resulting in variable power dissipation into the system. Ultrasonic streaming from the tip of the probe may be powerful enough to provide efficient bulk mixing. Most modern units are provided with a pulse facility that allows the operator to sonicate reactions repeatedly for fractions of a second. This gives adequate time for cooling between sonic pulses. With such systems, the probe (horn) can be tuned to give optimum performance, which is important for the reproducibility of results and the optimization of power dissipation.

Apart from the conventional designs, some modifications aimed at increasing the cavitationally active volume have been reported. Horst *et al.* [8] have reported a novel modification that uses high-intensity ultrasound from a concentrator horn in combination with the conventional stirred reactor with a flow loop (Figure 7.1). It has been shown that the concept of a conical funnel fits the demands for nearly perfect radiation effectiveness and good reaction management. The approach of using a combination of sonochemical reactors with the conventional reactor might fit well in terms of easy acceptability in industrial operations. Another design, the Telsonic horn, used by Dahlem *et al.* [9], uses

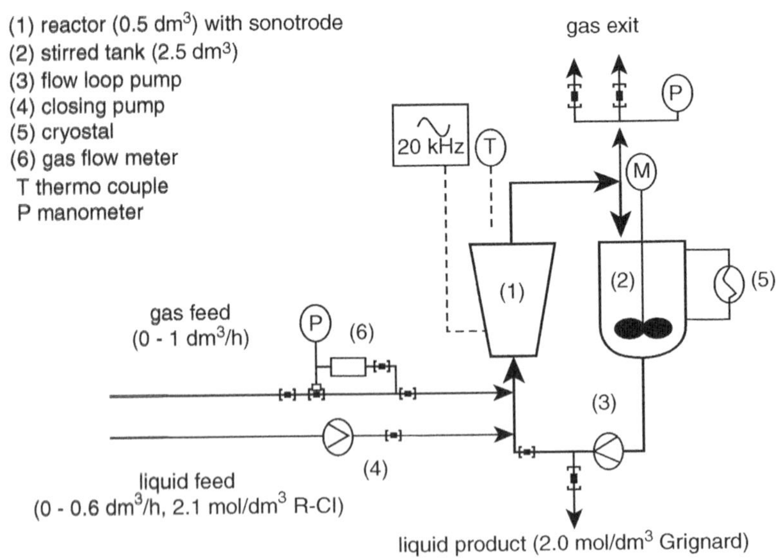

Figure 7.1 *Flow chart for the integration of conical configuration sonochemical reactors with conventional reactors. Reprinted from [ref 8] © 1996, with permission from Elsevier.*

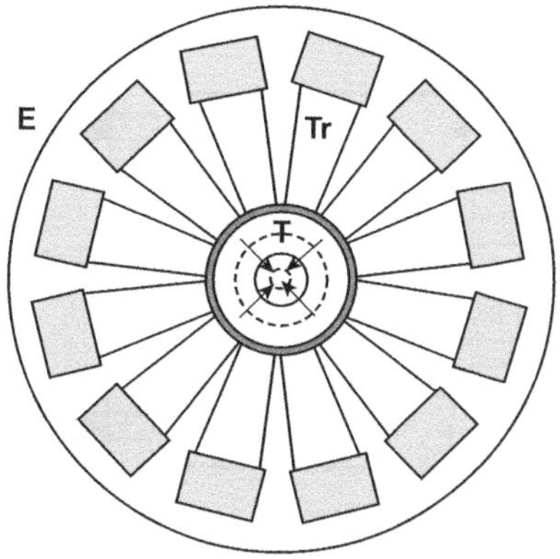

Figure 7.2 Schematic representation of the cylindrical cross-section of a sonochemical reactor. Reprinted from [ref 10] © 2009, with permission from Elsevier.

radial vibrations, as compared to the conventional longitudinal vibrations, for the immersion system, giving the twin advantages of a higher irradiating surface (lower intensity of irradiation, resulting in better yields) and good distribution of the energy in the radial direction. Moreover, even if the horn is radially vibrating, local measurements just below it also give high cavitational activity, which will again be more beneficial in enhancing the global sonochemical yields. Dion [10] has described a new continuous reactor design based on using high-power converging acoustic waves in a tube to produce a relatively large-volume confined acoustic cavitation zone in flowing liquid reagents under pressure. It has been reported that the new cylindrical sonoreactor design, depicted schematically in Figure 7.2 [10], does not contaminate the processed liquids with erosion products because the cavitation zone is maintained away from the wall of the tube. The processing capacity of the largest models may be up to several tons per hour, depending on the cavitation energy per unit volume required to produce the desired process enhancement, using an electric power input of about 50 kW.

A second sonochemical reactor design is the ultrasonic bath, which was originally manufactured and used for cleaning purposes. Typical baths have the transducers attached to the bottom; either single or multiple transducers can be used. The active zone is restricted to a vertical plane just above the transducers [11] and the area of irradiation should be increased (maximum possible) so as to get better distribution/dissipation of energy in the reactor. This has a twin advantage in terms of the decreased ultrasonic intensity (defined as power dissipation per unit area of the irradiating surface), which will increase the magnitude of the pressure pulse generated at the end of the cavitation events [12]. To increase the active zones existing in the reactor, the position of the transducers can easily be modified so that the wave patterns generated by the individual transducers will overlap, resulting in uniform and increased cavitational activity. The reaction vessel is typically immersed in the coupling fluid contained in the bath (indirect sonication), or else the bath itself can be used as the reaction vessel, but this may require additional mechanical agitation in order to achieve uniform mixing. In the case of direct sonication, the bath walls will be exposed to the reaction mixture and/or irradiation, making them susceptible to corrosion and/or erosion. When indirect sonication is used, the ultrasonic power dissipated

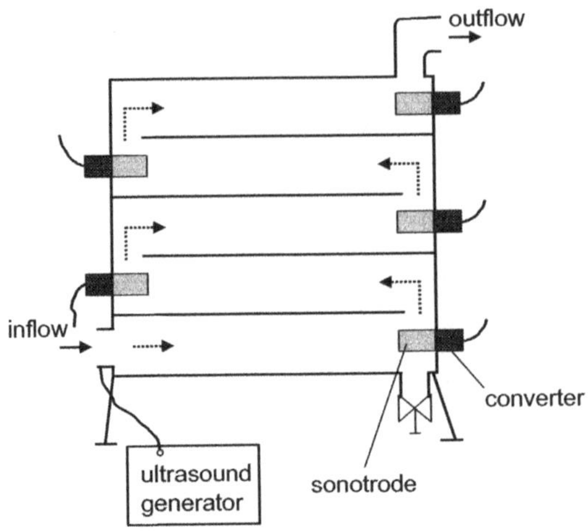

Figure 7.3 *Schematic representation of flow systems based on the ultrasonic horn. Reprinted from [ref 13] © 2007, with permission from Elsevier.*

into the system is low compared to that in other ultrasonic systems, such as a probe. Also, reproducible results may be difficult to obtain because the amount of power reaching the reaction mixture is highly dependent on the location of the reactant sample in the bath. Another disadvantage of using a bath system is that the coupling fluid surrounding the reaction vessel(s) will eventually be heated by the transfer of ultrasonic energy, making the maintenance of isothermal conditions difficult. Cooling coils can be placed within the bath, but they will have an effect of the attenuation of the sound field and may reduce the amount of power reaching the vessel containing the reactants.

Flow systems can be based on the use of either a large number probes or bath type reactors in sequential operation. The simple flow cell with an inbuilt probe system is an excellent means of processing relatively large volumes and hence is generally regarded as the best approach to industrial-scale sonochemistry. One of the simplest methods is to couple the probe transducer into a flow pipe by means of a T section [13]; a number of such transducers can be employed in this manner to give extended treatment times (a schematic representation of this assembly is given in Figure 7.3 [13]). For larger-scale operations where high-intensity ultrasound is required, the sonic horn is best utilized as part of a flow loop outside the main reactor. This permits the continuous processing of large volumes. Temperature control is provided through the circulating reaction mixture. The reactor configuration also allows the residence time of the reacting system in the flow to be varied by means of changing the flow rate. The reactor has some disadvantages, such as tip erosion, corrosion of the material and possible production of radicals, leading to side reactions. Pumping is required for circulation into the flow cell and so the system is not suitable for either very viscous reaction media or heavily particulate systems. An alternative arrangement to T sections is the use of a number of probes inserted through the walls of a pipe, allowing a much larger sonochemical treatment zone. Such a system will suffer from the problems of the individual flow cell, except that it will continue to function even if one or two probe units fail. Another disadvantage is that the sealing of reactor/probe joints,

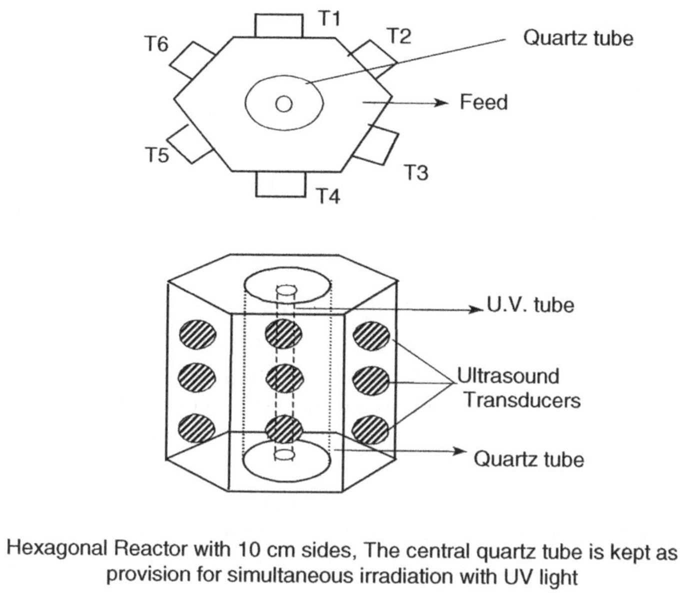

Figure 7.4 *Schematic representation of a hexagonal flow cell. Reprinted from [ref 18] by permission of John Wiley & Sons. © 2003.*

particularly for the operation of high-pressure reactions, is a crucial factor, as the small radial vibrations associated with the probe can lead to frequent leakages.

More recent developments have employed direct bonding of the transducer to the surface of the vessel. Improvements in the bonding method, and a move towards transducers with lower individual outputs, have enabled the move to systems with large numbers of transducers, giving an acoustic pattern that is uniform and noncoherent above the cavitational threshold throughout the working volume. Arrangements such as tubular reactors (either with two ends irradiated by transducers or with one end with a transducer and the other with a reflector) [14], parallel plate reactors with each plate irradiated at either the same or different frequencies [15,16] and transducers on the sides of a hexagon [17,18] can be constructed. A schematic representation of a hexagonal flow cell has been given in Figure 7.4 [18]. It is of the utmost importance to have uniform distribution of the ultrasonic activity in order to get increased cavitational effects. The use of low-output transducers gives the additional advantage of avoiding the phenomenon of cavitational blocking (acoustic decoupling), which arises where power densities close to the delivery point are very high. In addition, these multitransducer units very effectively concentrate ultrasonic intensity towards the central axis of the cylinder and away from the vessel walls, thus reducing problems of erosion and particle shedding.

7.3.2 Hydrodynamic Cavitation Reactors

Hydrodynamic cavitation can be generated simply by the passage of liquid through a constriction such as an orifice plate or a valve [19]. When the liquid passes through the constriction, the kinetic energy associated with it increases at the expense of the local pressure. If the throttling is sufficient, so that the local pressure falls below the threshold pressure for cavitation (usually the vapour pressure of the medium at the operating temperature), cavities will be generated. Subsequently, as the liquid jet expands, the pressure recovers, resulting in the collapse of the cavities. During the passage of the liquid through the constriction, boundary layer separation occurs and a substantial amount of energy is lost in the form of a permanent pressure drop. Very high-intensity turbulence is also present downstream of the constriction; the intensity depends on the magnitude of the pressure drop, which in turn depends on the geometry of the constriction and the flow conditions of the liquid.

A commonly used device based on hydrodynamic cavitation phenomena is the high-pressure homogenizer, which is in essence a high-pressure positive-displacement pump with a throttling device that operates according to the principle of a high-pressure relief valve. Liquid is pumped at very high pressure through a constriction, which converts its pressure energy into kinetic energy. At extremely high kinetic energy dissipation levels, the pressure reduces below the liquid vapour pressure, generating cavitation. A high-pressure homogenizer is designed to operate at pressures ranging from 50 to as high as 300 atm. There is a critical discharge pressure at which cavitation inception occurs, and significant cavitational yields are obtained beyond this [20]. Apart from cavitation, high-pressure homogenization also utilizes the intense shear produced due to fluid flow and impingement of high-velocity liquid on a solid wall. A general disadvantage of these types of reactor is that there is no precise control over the cavitationally active volume and the magnitude of the pressure pulses that will be generated at the end of the cavitation events (cavitational intensity), unless the valve seat designs can be substantially modified.

Hydrodynamic cavitation can also be generated in rotating equipment. When the tip speed of the rotating device (impeller) reaches a critical point, the local pressure near the periphery of the impeller drops and approaches the vapour pressure of the liquid. This results in the generation of vapourous cavities. Subsequently, as the liquid moves away from the impeller to the boundary of the tank, the liquid pressure recovers at the expense of the velocity head. This causes the cavities that have travelled with the bulk liquid to collapse. Again, as in the high-pressure homogenizer, there exists a critical rotational speed for the inception of cavitation [21]. It should be noted that the energy consumption in these types of reactor is much higher, and flexibility over the design parameters is less compared to reactors based on the use of low-pressure cavitation devices like an orifice or a venturi. These devices operate at rotating speeds of 4000–20 000 rpm. A high-speed homogenizer usually consists of a stator–rotor assembly, preferably made of stainless steel, with flexibility in the design of the stators and rotor. A plate with holes attached to the stator can be provided to insert baffles, so as to avoid vortex formation and surface aeration in the bulk liquid.

A new design of hydrodynamic cavitation reactor [7] is based on the use of a cavitation chamber in a flow loop (Figure 7.5). In this type of reactor, the flow through the main line passes through a constriction, where the local velocities suddenly increase due to a reduction in the flow area, resulting in lower pressures, which may even go below the vapour pressure of the liquid medium generating the cavities. The constriction can be a venturi, a single-hole orifice or multiple holes on an orifice plate [19]. Multiple-hole orifice plates with different combinations of numbers, diameters and shapes of holes can be constructed. Such an arrangement helps to achieve different intensities of cavitation, and generate different numbers of cavitational events. Thus, these reactors present tremendous

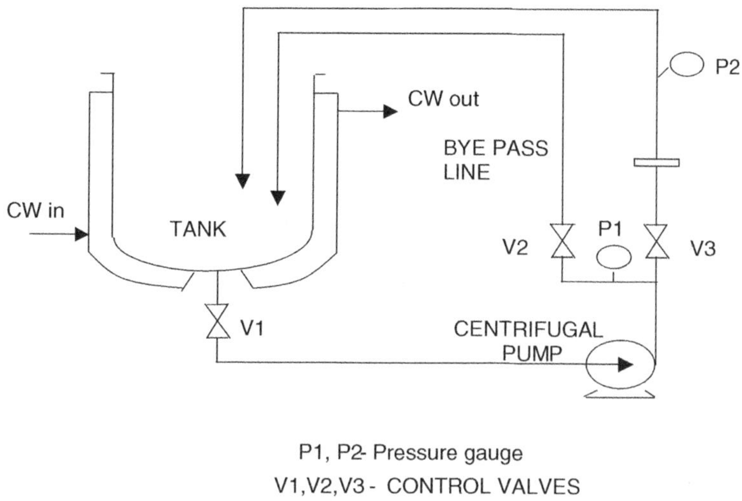

Figure 7.5 *Schematic representation of hydrodynamic cavitation setup based on a flow loop housing a cavitation chamber. Reprinted from [ref 7] © 2008, with permission from Elsevier.*

flexibility in terms of the operating (control of inlet pressure, inlet flow rate, temperature) and geometric conditions (different arrangements of holes on the orifice plates). Depending on the type of application and the requirements, geometry and operating conditions can be selected in the hydrodynamic cavitation reactor to maximize the cavitational effects at minimum possible energy consumption. Sampathkumar and Moholkar [22] recently put forth a conceptual design for a novel hydrodynamic cavitation reactor that uses a converging–diverging nozzle to create the pressure variation in the flow required to drive bubble motion, instead of orifice plates. The cavitation bubbles or nuclei are introduced in the water flow externally, upstream of the nozzle, using a sparger. Different gases can be used for the introduction of the bubbles. The size of the gas distributor (usually a glass frit), the flow rate of the gas and the pressure of gas in the reservoir (or source) from which the gas is withdrawn can be controlled to determine the initial size of the cavitational nuclei and the gaseous composition of the cavity, which significantly affects the resultant cavitational intensity. The aim should be to generate the smallest size of nuclei possible in order to maximize the intensity and hence the net cavitational effects.

7.4 Mathematical Modelling

Modelling of cavitational reactors can aid in efficient design and subsequent operation of large-scale reactors. Two aspects of modelling – the bubble dynamics and pressure field distributions in the reactor – are very important for efficient design. The bubble dynamics analysis gives an idea of the cavitational intensity, while theoretical modelling for pressure field distributions can aid in obtaining uniform cavitational activity distribution. It is not possible to cover the detailed mathematical modelling aspects given the focus of the book, but some basic discussion has been offered to aid better understanding.

The basic equation governing bubble behaviour under the action of a fluctuating sound or pressure field is the Rayleigh–Plesset equation:

$$R\left(\frac{d^2R}{dt^2}\right)+\frac{3}{2}\left(\frac{dR}{dt}\right)^2=\frac{1}{\rho_1}\left[p_i-\frac{4\mu}{R}\left(\frac{dR}{dt}\right)-\frac{2\sigma}{R}-p_\infty\right] \tag{7.1}$$

where R is the instantaneous radius of the cavity, ρ_l is the density of the cavitating medium, μ is the viscosity and σ the surface tension of the medium, p_i is the pressure inside the bubble and p_∞ is the fluctuating pressure field, which is responsible for the various stages of the cavitation phenomena. The Rayleigh–Plesset equation, which is the simplest bubble dynamics equation, has some inherent limitations in that it assumes an incompressible nature for the cavitating medium and hence underpredicts the cavitational intensity for a given set of operating parameters. Other bubble dynamics equations consider the compressibility of the liquid and the time-dependent pressure fields inside the cavitating bubble [12]. These equations are equally applicable to both sonochemical and hydrodynamic cavitation reactors, and the equations for the fluctuating pressure field will be different in these cases. In the case of sonochemical reactors, the time-varying pressure field (N/m^2) is expressed as:

$$p_\infty = p_o\text{-}p_a\sin(2\pi ft) \tag{7.2}$$

where p_o is ambient pressure, p_a is the amplitude of driving pressure $=\sqrt{2I\rho_l C}$, C is the velocity of sound in the liquid (in water $\cong$ 1500 m/s), I is the intensity of ultrasound (W/m^2) and ρ_l is the density of the liquid (kg/m^3).

In the case of hydrodynamic cavitation reactors, the time-varying pressure field is given as:

$$P_\infty = P_v + 1/2\,\rho_l(V_o^2 - V_{tn}^2) - \Delta P \tag{7.3}$$

where ΔP is the permanent pressure loss across the constriction, V_{tn} is the instantaneous turbulent velocity, V_o is the velocity at the constriction and P_v is the vapour pressure of the liquid. The instantaneous turbulent velocity (V_{tn}) can be calculated by assuming a sinusoidal velocity variation in the instantaneous local velocity with the frequency of the velocity perturbations and is given by the following equation:

$$V_{tn} = V_t + \bar{v}' \sin(2\,\pi f_T\,t) \tag{7.4}$$

where V_t is the local mean velocity and f_T is the frequency of turbulence. The details of bubble dynamics approaches to hydrodynamic cavitation reactors can be obtained from the published literature [19,20].

The solutions of these bubble dynamics equations can aid in quantifying the cavitational intensity in terms of the magnitudes of collapse pressures and temperatures, as well as the number of free radicals generated at the end of cavitational events, and also in determining the type of cavitation – stable or transient – occurring in the reactor. Typically, stable cavitation gives multiple shock effects with lower intensities and is recommended for physical processing or reactions involving lower activation energy, whereas transient cavitation results in much higher cavitational intensity and is expected to control the chemical processing applications and reactions involving stringent conditions.

Distribution of the cavitational activity in large-scale reactors, especially sonochemical reactors, is a very important issue considering the net overall effects in any processing applications. Theoretical prediction of cavitational activity in terms of the pressure field gives an efficient way of designing cavitational reactors with uniform activity. Numerical simulations can be used to characterize the ultrasonic field propagation and to obtain the spatial distribution of the mechanical effects. Based on theoretical analysis, one can obtain the pressure field distribution in any new sonochemical reactor with different geometries and operating conditions, which can aid in optimization for maximum/uniform cavitational activity. Modelling studies can be extended to the quantification of other useful parameters, such as the distribution of temperature, mass transfer coefficient and so on, which can be controlling parameters depending on the type of application. A detailed discussion of the distribution of the cavitational activity and the effects of different operating and geometric parameters, with guidelines for designing reactors based on theoretical simulations, is beyond the scope of the present work, and readers are requested to refer to existing literature on this subject [23].

7.5 Optimization of Operating Parameters in Cavitational Reactors

The intensification obtained due to the cavitation phenomena for any physical or chemical processing application are strongly dependent on the operating and geometric parameters of the equipment. The important parameters in the case of sonochemical reactors are the frequency and intensity of irradiation and the geometrical arrangement of the transducers, the geometry of the reactor and the liquid phase physicochemical properties. The important parameters in the case of hydrodynamic cavitation reactors are the inlet pressure, cavitation number, design of the cavitation chamber and liquid-phase physicochemical properties. Based on a critical overview of the literature illustrations related to modelling of cavitational reactors [24–27], some guidelines for the selection of the optimum set of operating parameters are presented in this section.

7.5.1 Sonochemical Reactors

The frequency of ultrasonic equipment is usually fixed and cannot be varied over a wide range as the maximum transfer efficiency is obtained only when the transducer is driven at its resonating frequency. Generally, it has been observed that increasing the frequency of irradiations to an optimum value (in the range of 200–500 kHz depending on the type of application) results in better conversions, especially for chemical processing applications. However, the power requirement is higher for inception of cavitation events in a high-frequency operation. The use of multiple-frequency operation can be considered an efficient alternative to the drawbacks associated with single-frequency high-power operation, especially when higher cavitational intensities are required. It has been reported that the cavitational intensity and active volume for a combination of frequencies are higher than those obtained for single-frequency operation [28–30]. It is thus recommended that the a combination of low-frequency irradiation (typically 20 kHz) with other frequencies in the range of 50–200 kHz should be used to obtain the maximum benefits from cavitational reactors in terms of achievable intensification.

The intensity of irradiation is defined as the power dissipation into the system per unit area of irradiation and hence both the actual power dissipation levels as well as the surface area of transducers play an important role in deciding the net cavitational effects. Typically, higher power dissipation levels (up to an optimum depending on the specificity of the application) with larger areas of transducers so as to give a higher cavitationally active volume are recommended for beneficial results [31].

The reactor design in terms of ratio of the diameter of the transducer to the reactor diameter, the liquid height, the position of the transducers and the geometrical configuration of the cell plays an important role in deciding the cavitational activity distribution [32]. It has been observed that higher diameters of immersion transducers relative to the reactor diameter up to an optimum ratio are beneficial. The extent of immersion of the transducer or the liquid height in the reactor which affects the extent of reflection of the incident sound waves from the liquid surface and the reactors walls, also shows an optimum value. The transducers in reactors based on the multiple-frequency–multiple-transducer arrangement should be positioned in such a way that maximum and uniform cavitational activity is obtained. Theoretical analysis of the cavitational activity distribution using wave propagation equations aids in arriving at an optimum location for the transducers [33].

The important liquid-phase physicochemical properties which affect the cavitation phenomena and hence the extent of cavitational effects for a given application include vapour pressure, viscosity and surface tension [24]. Usually, lower liquid vapour pressure is preferred, due to the fact that with an increase in the vapour pressure, the vapour content of the cavity increases, lowering the energy released during the collapse. Liquids with lower viscosities are more likely to generate cavitating conditions in the reactor. For cavitation to occur in a liquid, the natural cohesive forces present in the liquid must be overcome. Any increase in these forces will tend to increase the threshold pressure and hence the energy required to generate cavitation. In highly viscous liquids, severe attenuation of the sound intensity occurs and the active cavitating zone is substantially reduced. Similarly, liquids with the highest values of surface tension, such as water, generally result in higher cavitational intensity.

7.5.2 Hydrodynamic Cavitation Reactors

Among the different reactor configurations discussed in this chapter, the configuration based on the flow loop using a cavitation chamber in the form of a venturi or orifice seems to be the most flexible in generating varying levels of cavitational activity with better energy efficiencies. In the case of a venturi, a larger fraction of stable oscillatory radial bubble motion is obtained due to a linear pressure recovery gradient and relatively lower turbulence, whereas because of an additional oscillating pressure gradient caused by turbulent velocity fluctuation, the radial bubble motion in the case of an orifice flow results in a combination of both stable oscillatory and transient cavity behaviour. Also, the magnitude of the permanent pressure drop across the orifice is much higher than that across the venturi, resulting in a larger fraction of energy being available for cavitation. Thus, the intensity of cavitation will be higher (due to higher contribution from the transient cavitation) in the case of an orifice system as compared to the classical venturi tube.

The following important design recommendations can be made based on a detailed analysis of the existing literature [20,26,27]:

1. An orifice flow configuration is more suitable for applications requiring intense cavitational conditions, whereas for milder processes (typically requiring collapse pressure pulses between 15 and 20 bar) and for transformations based on physical effects, a venturi configuration is more suitable and more energy efficient.
2. In the case of a venturi flow, the most economical technique for increasing cavitation intensity is to reduce the length of the venturi, but for higher volumetric flow rates there might be a limitation caused by the possibility of flow instability and super-cavitation. A similar argument can be made for the enhancement in the cavitation intensity caused by reducing the venturi throat to pipe diameter ratio.
3. In the case of an orifice flow configuration, the most convenient way of controlling the cavitation intensity is to control the orifice to pipe diameter ratio and to control the cross-sectional flow area by manipulating the number and diameter of the hole on the orifice plate, although indiscriminate growth of bubbles downstream of the orifice can lead to splashing and vaporization (super-cavitation) of the flow.
4. Increasing pipe size downstream of the orifice (which offers a faster pressure recovery) is another option for intensifying cavitation effects, but using pipes of a larger size will require higher volumetric flow rates for the same cavitation number, resulting in an increase in processing costs.
5. In the case of orifice plates, it will be a good idea to have a higher number of holes on the plates for similar free-flow areas, so that the number of cavitational events and hence the cavitationally active volume is increased.
6. Higher inlet pressures into the system result in higher cavitational intensity, but optimum values should be used so as to avoid occurrence of super-cavitation, where a single large slug is formed with much lower cavitational effects.
7. The recommendations for liquid physicochemical properties are exactly similar to those discussed for sonochemical reactors.

7.6 Intensification of Cavitational Activity

At times, the net rates of chemical processing achieved using cavitational reactors may not be sufficient to prompt industrial-scale operation. This is especially important given the possibility of an even more uneven distribution of cavitational activity in large-scale reactors. It is thus important to look into supplementary strategies with the aim of intensifying the cavitational intensity. Two types of operating strategy can be recommended, depending on the type of application:

1. The use of process-intensifying parameters such as the presence of dissolved gases and/or continuous sparging of gases such as air, ozone and argon (to a limited extent), the presence of salts such as NaCl, $NaNO_2$ and $NaNO_3$ and the presence of solid particles such as TiO_2, CuO and MnO_2, which can also act as catalysts in some cases.
2. The use of a combination of cavitation and other processes with similar mechanisms such as advanced oxidation processes, including photocatalytic oxidation and Fenton

chemistry. Some recent studies also refer to combinations of cavitational reactors and microwave reactors.

7.6.1 Use of PI Parameters

The presence of gases and solid particles mainly provides additional nuclei for the cavitation phenomena and hence enhances the number of cavitation events occurring in the reactor, resulting in an enhancement in cavitational activity and hence in the net chemical/physical effects. It must be noted that the presence of these parameters also has a negative effect on the cavitational activity (aeration gives a cushioning effect and incomplete collapse, resulting in a decrease in the collapse pressure, whereas solid particles result in scattering of the sound waves in the case of sonochemical reactors, thereby decreasing the focused energy transferred into the system). The net effect of these two phenomena will be dependent on the system in question, and hence optimization is a must before operating parameters are selected for actual operation.

The properties of the dissolved gas – its polytropic constant (γ), solubility in the cavitating medium, thermal conductivity and so on – severely affects the cavitational activity due to its direct effect on the final collapse conditions. Gogate *et al.* [24] have performed simulations for various gases with different polytropic indices (in the range 1.2–1.6) and reported that the pressure developed during the collapse of the cavity increases with an increase in the polytropic constant of the gas. Thus, monoatomic gases such as argon and helium will result in maximal cavitation intensity as compared to the polyatomic gases. Moreover, monoatomic gases will offer the added advantage of lower thermal conductivity, resulting in lower heat losses and hence larger magnitudes of collapse temperatures. For those gases that have the same γ value, the cavitational effect may vary due to its dependence on gas solubility [34,35]. Usually lower gas solubility results in more violent collapse of cavities.

Considering the specific application of chemical synthesis, the presence of solid catalyst (particles/salts in a typical concentration range of 1–10% by weight of reactants; optimization is recommended in the majority of cases using laboratory-scale studies) results in intensification due to the following mechanisms:

1. Formation of more cavitation nuclei due to a greater number of discontinuities in the liquid continuum as a result of the presence of particles, giving a larger number of collapse events and resulting in an increase in the number of free radicals.
2. In a biphasic solid–liquid medium subjected to cavitating conditions, the reduction of particle sizes, leading to an increased surface area, and the formation of liquid jets at solid surfaces through the asymmetrical inrush of the fluid into the collapsing voids; these liquid jets not only provide surface cleaning but also induce pitting and surface activation effects, and increase the rate of phase mixing, mass transfer and catalyst activation.
3. Enhanced generation of free radicals due to certain catalysts, such as $FeSO_4$ and elemental iron.
4. Alteration of physical properties (vapour pressure, surface tension), facilitating the generation of cavities and results in more violent collapse of the cavities.
5. Potentially, the presence of salts, resulting in preferential accumulation of the reactants at the site of cavity collapse, and thus in an intensification of the cavitational

reactions [36]. However, it should be noted that an optimum salt concentration can exist beyond which the cavitational activity decreases. Wall *et al.* [37] have reported that the optimum concentration is usually in the range 1–2 M, and above this optimum the sonoluminescence intensity can be observed to reduce drastically.

7.6.2 Use of a Combination of Cavitation and Other Processes

Intensification can be achieved using a combination of cavitation and other processes with similar intensifying mechanisms that yield synergistic effects [38]. Combination with advanced oxidation process will be beneficial only for those chemical synthesis applications in which free radical attack is the governing mechanism. For reactions governed by a pyrolysis-type mechanism, the use of process-intensifying parameters that result in an overall increase in the cavitational intensity, such as solid particles, sparging of gases and so on, or a combination with microwave irradiations is recommended.

The observed synergistic effects of combining photocatalytic oxidation with cavitation can possibly be attributed to the mechanical effects of cavitation. The turbulence and acoustic streaming generated in the reactor increase the transport of the reactants and products at the catalyst surface and in the solution. Also, continuous cleaning and sweeping of the TiO_2 surface due to acoustic microstreaming allows for access to more active catalyst sites at any given point. There is also a possibility that the catalyst surface area is increased by fragmentation or pitting of the catalyst. Cavitation-induced turbulence enhances the rates of desorption of intermediate products from the catalyst active sites and helps in continuous cleaning of the catalyst surface. It should be noted that in situations where the adsorption of reactants at specific sites is the rate-controlling step, ultrasound will play a profound role, due to a substantial increase in the number of active sites and the increased surface area available due to fragmentation of the catalyst agglomerates under the action of turbulence generated by acoustic streaming and an increase in the diffusional rates of the reactants. Toma *et al.* [39] have given a brief overview of different studies pertaining to the effects of ultrasound on photochemical reactions, concentrating on the chemistry aspects; that is, the mechanisms and pathways of different chemical reactions. Details regarding the operating and design strategies for maximizing the synergistic effects obtained using sonophotocatalytic reactors have been described by Gogate and Pandit [40].

The combination of hydrodynamic cavitation reactors and photochemical or photocatalytic reactors has not been much investigated, though some instances of a sequential combination of hydrodynamic cavitation and photocatalytic oxidation have been reported for waste water treatment applications. A commercial process (CAV-OX process [41]) that employs ultraviolet radiation, hydrodynamic cavitation and hydrogen peroxide has been reported to oxidize organic compounds present in water at ppm-level concentrations to nondetectable levels. This system has been used with success for the effective degradation of volatile organic compounds – primarily trichloroethane, benzene, toluene, ethyl benzene and xylene. The removal efficiency of contaminants is reported to range from 20 to 99%. Hydrogen peroxide is added to the contaminated ground water, which is then pumped through a cavitation nozzle and then subjected to ultraviolet radiation. Hydrodynamic cavitation produced in a nozzle is used to generate additional hydroxyl radicals, which help in increasing the rate of degradation. Moreover, the use of cavitating conditions and the UV irradiation result in an increased number of free radicals, due to the dissociation of hydrogen peroxide.

The combined use of microwave irradiation and sonochemistry is a recent advance [42–45] that can accelerate chemical reactions and other processing applications significantly, especially in the case of heterogeneous systems. The observed dramatic acceleration effect may be attributed to a combination of enforced heat transfer due to microwave irradiation and intensive mass transfer at phase interfaces, caused by ultrasonic irradiation. In many heterogeneous reactions, the rate-determining step is the mass transfer at the interface between two (or more) phases. With sonication, the liquid jet caused by cavitation propagates across the bubble towards the phase boundary at a velocity estimated at several hundreds of metres per second and violently hits the surface. The intense agitation leads to the mutual injection of droplets of one liquid into the other, resulting in the formation of fine emulsions. These ultrasonically produced emulsions are smaller in size and more stable than those obtained conventionally and require little or no surfactant to maintain stability. Microwave irradiations complement these effects by way of enhanced instantaneous heating of the medium and increasing of the vapour pressure of the medium to give increased cavitational effects.

7.7 Case Studies: Intensification of Chemical Synthesis using Cavitation

After looking into the theoretical analysis of cavitational reactors and different reactor configurations with optimized operating parameters, we will now discuss some specific industrially important applications based on experimental investigations, focusing on the cavitation phenomena.

7.7.1 Transesterification of Vegetable Oils Using Alcohol

Various products derived from vegetable oils have been proposed as alternative fuels for diesel engines. The importance of biodiesel is on the increase due to increasing petroleum prices, limited fossil fuel reserves and their environmental benefits (decrease in acid rain and the emission of CO_2, SO_x and unburnt hydrocarbons during the combustion process). The conventional techniques for the synthesis of biodiesel refer to a catalysed chemical reaction involving vegetable oil and an alcohol to yield acid alkali esters and glycerol. They typically utilize temperatures in the range of 70–200 °C, pressures in the range of 6–10 atm and reaction times of up to 70 hours in order to achieve conversions in the range of 90–95%, based on the type of raw material used [46–48]. Cavitational reactors can be used as an alternative means of synthesising biodiesel with a significant degree of PI.

Gogate [7] has reported the application of cavitational reactors to the synthesis of biodiesel from pure vegetable oils. The hydrodynamic cavitation reactor used in this work consists of a reservoir or collecting tank with 10 l capacity connected to a multistage centrifugal pump with a power rating of 1.5 kW. The flow loop is very similar to that represented in Figure 7.3. The sonochemical reactor is a conventional cleaning tank type reactor (ultrasonic bath) equipped with three transducers at the bottom of the tank, arranged in a triangular pitch, and operates at an irradiating frequency of 20 kHz and a power dissipation of 120 W (the calorimetrically measured exact power dissipation was 45 W,

Table 7.1 Transesterfication of different vegetable oils. Reprinted from [ref 7] © 2008, with permission from Elsevier.

No.	Vegetable oil	Product	Time		Yield	
			Acoustic	Hydro	Acoustic	Hydro
1	Soyabean oil	Soyabean oil ester	15	10	97	98
2	Castor oil	Castor oil ester	10	10	99	99
3	Peanut oil	Peanut oil ester	10	10	99	90

Acoustic cavitation = vegetable oil (4 g), methanol (4 ml), NaOH (0.5%). Hydrodynamic cavitation = vegetable oil (4000 g), methanol (4000 ml), NaOH (1%), P = 1 kg/cm^2.

indicating about 37.5% energy conversion and transfer efficiency). The results obtained for a variety of oils used as starting materials [7] are shown in Table 7.1. It can be seen that cavitation can be very successfully applied to transesterification reactions, with more than 90% yield of the product (as per stoichiometry) in as little as 15 minutes of reaction time. The technique hence appears to be very effective compared to the conventional approach, as is also evident from the comparison of different techniques [7] based on quantitative criteria of energy efficiency shown in Table 7.2. It can be seen that hydrodynamic cavitation is about 40 times more efficient than acoustic cavitation and 160–400 times more efficient than the conventional agitation/heating/refluxing method. Stavarache *et al.* [49,50] have also shown a significant PI using acoustic cavitation for these transesterification reactions with pure vegetable oil as the starting raw material.

The use of pure vegetable oils for the preparation of fuels is a contentious issue and is definitely not within the scope of sustainable development. These starting raw materials can be replaced by waste vegetable oils such as fatty acids (FAs) odour cut, obtained as a byproduct from the vegetable oil refining industries or even the waste cooking oils. The feasibility of applying cavitational reactors to the synthesis of biodiesel using

Table 7.2 Comparison of energy efficiencies for different techniques. Reprinted from [ref 7] © 2008, with permission from Elsevier.

No.	Technique	Time (minutes)	Yield	Yield/kJ of energy
1	Acoustic[a]	10	99%	8.6×10^{-5}
2	Hydrodynamic[b]	15	98%	3.37×10^{-3}
3	Conventional[c]			
	I. with stirring	180	98%	2.27×10^{-5}
	II. under reflux	15	98%	7.69×10^{-6}

[a]In acoustic cavitation, 4 ml of methanol is mixed with 4 g of vegetable oil and a catalyst concentration (NaOH) of 0.5% of oil. The ultrasonic bath is the sonochemical reactor, with 20 kHz frequency and 85 W power dissipation.

[b]Operation with hydrodynamic cavitation is carried out under optimized conditions: 4 : 4 ratio (w/v) of oil to alcohol, catalyst concentration (NaOH) 1% of oil, orifice plate 1 with 16 holes of 2 mm diameter, volume of methanol 4000 ml with 4000 g of oil.

[c]For a conventional approach, 4 ml of methanol is mixed with 4 g of vegetable oil and a catalyst concentration (NaOH) of 0.5% of oil (Case I: a stirrer is used for uniform mixing, which consumes energy; Case II: a heater is used to maintain reflux conditions).

sustainable raw materials and esterification of FAs cut (C_8–C_{10}) with methanol in the presence of concentrated H_2SO_4 as a catalyst has been studied in hydrodynamic cavitation reactors and in sonochemical reactors [51]. A few experiments have also been carried out with other acid/alcohol combinations, such as coconut FAs with methanol and ethanol and FA odour cut with fatty alcohols, with the aim of investigating the efficacy of cavitation in giving the desired yields and also of quantifying the degree of PI that can be achieved using the same. Optimization in terms of the operating molar ratio of the FA to alcohol (1 : 10 ratio is the optimum in the present case) and catalyst loading (2% by weight of concentrated H_2SO_4 catalyst) results in maximum possible yields of the product. It has been observed that for all the different combinations of acid/alcohol [51], ambient operating conditions of temperature and pressure and reaction times of less than 3 hours were sufficient for more than 90% conversion. This clearly establishes the efficacy of cavitation in achieving PI of the biodiesel synthesis process. To cite a specific illustration of the degree of PI achieved in the present work, with an operating ratio of FA cut (waste FAs) to methanol of 1 : 10, 2% by weight loading of the catalyst and an operating temperature of 30 °C, 92% conversion was achieved using hydrodynamic cavitation in only 90 minutes of reaction time. By comparison, conventional methods of esterifying waste oil using methanol required about 69 hours to obtain more than 90% oil conversion to methyl esters at 65 °C operating temperature with a molar ratio of methanol to oil of 30 : 1. Comparison of the energy efficiencies of hydrodynamic cavitation reactors with sonochemical reactors indicates that hydrodynamic cavitation is more energy efficient than acoustic cavitation. Depending on the type of acid/alcohol combination used, the energy efficiency for hydrodynamic cavitation varies in the range of 1×10^{-4} to 2×10^{-4} g/J, whereas for acoustic cavitation it is an order of magnitude lower, in the range 5×10^{-6} to 2×10^{-5} g/J.

Another, cheaper sustainable raw material for biodiesel synthesis is waste frying oil. Used frying oils are generated in large quantities in food preparation, for example in fast-food networks, large restaurants, dining rooms and catering establishments. The disposal of used frying oil is problematic because disposal methods may contaminate ground water. Currently, a very small percentage of these oils have been utilized for soap production. However, due to the poor quality of the soap produced from frying oil, this is not the preferred route for disposal and most of the oil is still a waste material, and it poses significant ecological and economical problems. The production of biodiesel from used frying oil through the transesterification reaction is one of the better options for its efficient and economic utilization. Enzymatic reactions can be effectively used for this conversion [52], but the reaction times required are in the range of 24–30 hours, depending on the type of enzyme used. Hingu *et al.* [53] investigated the applicability of cavitational reactors to the synthesis of biodiesel from waste frying oils, and transesterification of used frying oil with methanol, in the presence of potassium hydroxide as a catalyst, was carried out using a low-frequency ultrasonic reactor (20 kHz). The effects of different operating parameters, such as the alcohol–oil molar ratio, catalyst concentration, temperature, power, pulse and horn position, on the extent of conversion of oil have been investigated. The optimum conditions for the transesterification process have been reported as being a molar ratio of alcohol to oil of 6 : 1, catalyst concentration of 1 wt%, temperature of 45 °C and ultrasound power of 200 W, with an irradiation time of 40 minutes. Under these optimum conditions, conversion levels of about 90% are obtained under ambient conditions in less than 1 hour.

Overall it can be said that cavitational reactors can be effectively applied to intensify the synthesis of biodiesel, based on the use of different sustainable raw materials.

7.7.2 Selective Synthesis of Sulfoxides from Sulfides Using Sonochemical Reactors

Sulfoxides are very important intermediates used in the pharmaceutical industry, foods and flavouring agents, crop protection, veterinary drugs and lubricants. They allow easy synthesis of target molecules as they can readily undergo asymmetric transformations, addition reactions and C-C bond formations. Recently they have been used widely in the synthesis of drugs such as omeprazole, lansoprazole, rabeprezole and so on, which are used as gastric proton pump inhibitors (PPIs). They are also used in the preparation of amino acid sulfoxides, which are used for the regulation of cholesterol catabolism and antibiotic activities. Sulfoxides are also used in the preparation of the potassium channel activator aprikalim and the anticancer drugs sulforaphane and spasomycin. A major problem in the synthesis of sulfoxides is that they are easily oxidized to sulfones. A proper choice of reactants/catalysts/solvents and operating conditions is required in order to obtain high sulfoxide yields by deliberately preventing the further oxidation of sulfoxides. Many efforts have been made in the past to selectively synthesize sulfoxides from sulfides. Ali and Bohnert [54] have investigated the selective synthesis of sulfoxides using silica gel supported catalysts in the presence of dichloromethane as the solvent. Ravikumar *et al.* [55] have used hexafluoro-2-propanol (HFIP) as the solvent for selective oxidation. Other reports [56–59] suggest the use of tetranitromethane, tertiarybutyl hypochlorite, periodic acid, IBX-esters and choroperbenzoic acid as oxidizing agents. These combinations of reagents and solvents on oxidation produce byproducts that are not compatible with the environment. Reports in the literature also suggest the use of complex reagents of transition metals such as vanadium, rhenium, manganese, molybdenum and titanium, but the procedures to form these reagents are generally very complex and time-consuming.

Alterations in reaction conditions brought about by cavitationally generated effects can be used for the selective synthesis of sulfoxides. It is possible that sulfone formation will be reduced if the oxidizing agent is dispersed evenly in the solvent (containing reactant) using ultrasound. The uniform distribution of hydrogen peroxide will reduce the high concentration packets in the reaction mixture, thereby inhibiting the overoxidation of sulfide present at a given location. Mahamuni *et al.* [60] have reported the efficient application of cavitational reactors for the selective synthesis of sulfoxide using methyl phenyl sulfide (MPS) as a model reactant. The reaction has been carried out in an ultrasonic bath (operating frequency 20 kHz and maximum power supply 120 W, with an actual power dissipation of 40 W, measured using calorimetric studies). Initially experiments were carried out only in the presence of hydrogen peroxide and only in the presence of ultrasound; the oxidation rate was very slow under these conditions (less than 2% conversion was observed even after 1 hour of reaction). A combined operation of hydrogen peroxide and ultrasound resulted in about 8% conversion in 1 hour. The presence of sodium tungstate as catalyst under a combined operation of hydrogen peroxide and ultrasound resulted in a dramatic increase in the rate of oxidation. It has also been reported that the periodic addition of hydrogen peroxide, instead of an initial addition (total quantity of hydrogen peroxide remains the same), gives a better selectivity for sulfoxide.

The maximum conversion achieved in the work was ~80%, with sulfone formation being only 0.3%, indicating a 99.7% selectivity for the desired product, the sulfoxides. This work clearly indicates that manipulating the operating conditions for the oxidation of sulfides using ultrasonic irradiation results in maximum yields of sulfoxides with minimum sulfone formation. However, considering the green chemistry concerns, a drawback can be considered as the use of methanol as a solvent and the fact that it is difficult to separate the catalysts used in the study from the reaction media. Green chemistry emphasizes the use of aqueous media as the solvent instead of organic media and the use of easily separable and reusable reagents. In a continuation of their earlier work, Mahamuni *et al.* [61] reported the application of sonochemical reactors to the oxidation of sulfides to sulfoxides in aqueous media. Hydrogen peroxide has been used as an oxidizing agent because of its large availability of oxygen, lower storage and transportation costs and environmentally clean nature. β-cyclodextrin is known to increase the solubilities of organic compounds in aqueous solutions through inclusion complex formation and hence has also been used for selective and partial oxidation. The reaction mixture can be easily extracted for organic compounds using ethyl acetate and separated from the aqueous phase. When the aqueous phase is filtered and cooled to 5 °C, β-cyclodextrin is obtained quantitatively and can be recycled. Thus the reported process is a green process with minimum byproduct formation and all of the reagents used in the process are environmentally benign.

The effects of various parameters, such as the presence and absence of ultrasound, the initial concentration of reactants, the reaction temperature, the molar ratio of reactant to oxidizing agent, the mode of addition of oxidizing agent, the presence and absence of β-cyclodextrin and the amount of β-cyclodextrin, have been reported [61]. The important results are given in Table 7.3. The optimum operating conditions found for the oxidation reaction at lab scale were equally effective in a scaled-up version of the reaction. The important achievements of the work can be given as follows:

1. The work has clearly established the fact that the presence of ultrasound significantly enhances the rate as well as the selectivity of the oxidation reaction as compared to the conventional technique of conducting the same reaction under stirring conditions. Also, ambient conditions of temperature and pressure are required, possibly reducing the operating costs for the process.
2. Gradual addition of hydrogen peroxide in the presence of ultrasound gives better performance (higher selectivity for sulfoxide at higher overall conversion levels of sulfide) as compared to the initial addition of the total amount of H_2O_2.
3. A molar ratio of 1 : 1 MPS : H_2O_2 is advantageous for a maximum selectivity of sulfoxide (100%) at somewhat lower conversions of sulfide (92.7%).
4. The addition of β-cyclodextrin is advantageous in increasing the selectivity of sulfoxide with higher conversion levels of sulfides.

7.8 Overview of Intensification and Green Processing Using Cavitational Reactors

Having considered the two case studies concerning PI of chemical processing applications, we will now provide an overview of the benefits that can be obtained using cavitational reactors, maintaining the focus of the book: that is, green chemistry. Initially, we highlight

Table 7.3 *Extent of conversions and yields of products in different experimental runs. Reprinted with permission from [ref 61]. © 2006 American Chemical Society.*

Sr. No.	Experiment	Time (Min)	Conversion (%)	Sulfoxide (%)	Sulfone (%)
1	1 mM MPS + Only US bath + 40 ml water	120	0	0	0
2	1 mM MPS + 1 Mm H_2O_2 + only MAG stirring + 40 ml water	120	79.75	100	0
3	1 mM MPS + 1 mM 30% H_2O_2 + only US bath + 40 ml water (32 °C)	120	92.7	100	0
4	1 mM MPS 1 mM 30% H_2O_2 + only US bath + 40 ml water + T = 15 °C	120	66.95	100	0
5	1 mM MPS + 1 mM 30% H_2O_2 + only US bath + 40 ml water + T = 55 °C	120	99.29	94.93	5.07
6	0.8 mM MPS + 1 mM 30% H_2O_2 + only US bath + 40 ml water (32 °C)	120	89.14	100	0
7	0.58 mM MPS + 1 mM 30% H_2O_2 + only US bath + 40 ml water (32 °C)	120	98.11	100	0
8	0.35 mM MPS + 1 mM 30% H_2O_2 + only US bath + 40 ml water (32 °C)	120	95.51	97.71	2.29
9	1 mM MPS + 2 mM 30% H_2O_2 + only US bath + 40 ml water (32 °C)	120	95.23	94.68	5.32
10	1 mM MPS + 4 mM 30% H_2O_2 + only US bath + 40 ml water (32 °C)	120	98.23	55	45
11	1 mM MPS + 4 mM 30% H_2O_2 + only US bath + 40 ml water + 10 wt% β-CD (32 °C)	120	99.12	84.52	15.48
12	1 mM MPS + 4 mM 30% H_2O_2 + only US bath + 40 ml water + 30 wt% β-CD (32 °C)	120	98.58	87.91	12.09
13	1 mM MPS + 4 mM 30% H_2O_2 + only US bath + 40 ml water + 50 wt% β-CD (32 °C)	120	97.08	89.67	10.33
14	1 mM MPS + 4 mM 30% H_2O_2 + only US bath + 40 ml water + 100 wt% β-CD (32 °C)	120	93.34	90.05	9.95
15	1.5 times excess MPS + periodic addition of (1 mM H_2O_2) every 10 minutes till 4 mM H_2O_2 was added + US bath + 40 ml water (32 °C)	180	100	80	20
16	1.5 times excess MPS + periodic addition of (1 mM H_2O_2) every 10 minutes till 4 mM H_2O_2 was added + US bath + 40 ml water + 100 wt% β-CD (32 °C)	180	100	96	4
17	Scaled-up reaction: MPS = 15 gm + 30% H_2O_2 = 50 ml + 11 gm β-CD + water = 3 l + US bath (32 °C)	240	100	93	7

the critical issues that should be considered when selecting possible applications of cavitating conditions for the intensification of chemical processing:

1. In the case of homogeneous reactions, only those which proceed via radical or radical–ion intermediates are usually sensitive to the cavitational effect. This means that cavitation can intensify reactions proceeding through radicals, whereas ionic reactions are not likely to be modified.
2. In the case of heterogeneous reactions, those proceeding through ionic intermediates can also be stimulated by the mechanical effects of cavitation. However, in this case appropriate balance needs to be maintained between the positive effects of cavitation and the associated costs of operation, as sometimes the same effects might be achieved by using improved agitation conditions.
3. In the case of heterogeneous reactions, where mixed mechanisms (radical and ionic) exist, cavitation can improve both. In this case, if the two mechanisms lead to the same product, an overall increase in the rate of reaction will be obtained, but if the two mechanisms lead to different product distributions, cavitational switching might occur due to enhancement in the radical pathway and the nature of the reaction products will be changed due to the cavitational effects.

In general, the different benefits that can occur due to the use of cavitational reactors in chemical processing applications (one illustrative example has been cited in each case for better understanding) are:

1. *Reaction Time Reduction*: Ultrasound can be used to substantially enhance the rates of chemical reactions. For example, Javed *et al.* [62] have shown that the Diels–Alder cyclization reaction gives 77.9% yield in 35 hours through conventional means but 97.3% in only 3.5 hours with the application of ultrasound.
2. *Increase in the Reaction Yield*: The conventional method of epoxidation of long-chain unsaturated fatty esters results in a yield of 48% in 2 hours' reaction time, whereas application of ultrasound (20 kHz frequency of irradiation) increases the yield to 92%, and importantly in just 15 minutes of reaction time [63].
3. *Possible Switching of the Reaction Pathways Resulting in Increased Selectivity*: In some cases, use of ultrasound switches the reaction pathway, thereby totally changing the product distribution. For example, with conventional reactions, stirring benzyl bromide and toluene in the presence of KCN and Al_2O_3 results in benzylation (Friedel–Crafts reaction), whereas sonication of the same components under identical conditions yields benzyl cyanide. This is due to the fact that sonication impregnates cyanide ions on Al_2O_3, thereby masking the acidic sites, and hence prevents the reaction from proceeding down the Friedel–Crafts route [64].
4. *Use of Less Forcing Conditions (Temperature and Pressure)*: Van Lersel *et al.* [65] have described a new route for the chlorination of methane using ultrasound irradiation, which allows for an intrinsically safe process at ambient pressure and temperature. Currently, the industrial chlorination of methane is performed as a noncatalytic, thermally initiated gas-phase reaction, and this route requires elevated pressures and temperatures in order to obtain a significant yield.
5. *Use of Green Solvents and Less Hazardous Reactants*: Puri *et al.* [66] have described a copper perchlorate-catalysed, highly efficient, one-pot, green protocol for the

synthesis of 2H-chromen-2-ones by the reaction of substituted phenols and β-keto esters under ultrasound irradiation in excellent yields.

6. *Increasing the Effectiveness of the Catalyst Used in the Reaction*: Cavitational reactors can be used in combination with catalysts to give beneficial results due to mechanical effects such as particle size reduction and mass-transfer facilitation through the solid–liquid film. Also, ultrasound has the ability to form fine emulsions between liquid–liquid phases, thereby increasing the interfacial area available for reaction, which can suit reactions involving phase-transfer catalysts (these work through a similar mechanism of bridging the barrier between the aqueous and organic phases, facilitating reaction between them). Hagenson *et al.* [67] reported the combination of ultrasound and phase-transfer catalyst (PTC) for the synthesis of dibenzyl sulfide from benzyl chloride and sodium sulfide. It has been reported that after approximately 4 minutes of reaction, the conversion of benzyl chloride is enhanced by a factor of 1.5 in the presence of ultrasound alone, 5.9 in the presence 0.4% w/v of the PTC tetrabutylammonium bromide (TBAB) and 6.5 in the presence of both the PTC and ultrasound. However, it should be noted that ultrasound cannot replace the PTC, and the catalyst is indeed required.

The examples mentioned here are a very few chosen from a vast number available in the literature, and several useful reviews/books are available on this subject [4–6,68–71].

7.9 The Future

It should be noted that, despite extensive research at the laboratory scale and their immense potential, there have only been a limited number of applications at the industrial scale for chemical synthesis applications. The potential major problems in the design and successful operation of cavitational reactors are related to lacunae associated with suitable design/scale-up strategies, such as the non-uniform spatial distribution of the cavitational activity, nonreproducibility of the cavitation events, lack of proper transducers capable of delivering high power densities, cavitational blocking at higher power dissipation levels and so on. Additional efforts are required in the following areas if we are to realize the dream of the application of cavitational reactors in actual practice:

1. Detailed theoretical analysis of the bubble dynamics phenomena under different operating conditions is essential to predict the cavitational intensity and its effect on the rates of chemical processing applications. Based on the detailed predictions, an optimization protocol must be established for the design and operating parameters required to maximize the cavitational activity.
2. Generalized design equations for the prediction of cavitational yields as a function of geometric and operating parameters must be developed.
3. Research into material science aspects should target the development of efficient transducers capable of delivering high amounts of power into a large volume in continuous operation without erosion of the irradiating surface.
4. In the case of hydrodynamic cavitation reactors, realistic modelling of turbulence phenomena is required in order to model the cavity/bubble dynamics, either in

isolation or in the form of cavity clusters in high-velocity flow, and their interaction with the cavity dynamics so that the cavitational intensity can be predicted.
5. Development of new geometrical configurations are required, with the objective of uniform distribution of cavitational activity, followed by an investigation of the efficacy of these reactors in different chemical processing applications.
6. The effect of PI parameters such as the presence of solid particles, dissolved salts and gases should be studied in detail. The effect of the scale of operation and of different designs of cavitational reactors on the expected intensification must also be evaluated.

7.10 Conclusion

Cavitational reactors appear to be a novel approach to PI of different chemical processing applications that might possibly lead to considerable economic savings in a sustainable and green manner. The discussion presented in the chapter has led to a detailed understanding of the different reactor configurations, with recommendations for optimization strategies for operating parameters and design configurations so that maximum benefits can be achieved. The experimental illustrations depicted have clearly demonstrated the efficacy of this novel technology. The future of sonochemical reactors lies in the design of multiple-frequency–multiple-transducer-based reactors, whereas for hydrodynamic cavitation reactors an orifice plate type configuration appears to be the most suitable. It can be said that the hydrodynamic cavitation reactors offer immediate and realistic potential for industrial-scale applications, as compared to the sonochemical reactors, and the scale-up of these reactors is comparatively easier. Overall, cavitation is a well-established technology at laboratory/pilot scale, but the combined efforts of chemists, chemical engineers and physicists are required to effectively harness the potential of this technology on an industrial scale of operation.

References

(1) T. J. Mason, Practical Sonochemistry: Users Guide in Chemistry and Chemical Engineering, Ellis Horwood Series in Organic Chemistry, Chichester, UK (1992).
(2) J. L. Luche, Synthetic Organic Chemistry, Plenum Press: New York (1999).
(3) F. R. Young, Cavitation, McGraw Hill: London (1989).
(4) J. Lindley and T. J. Mason, Use of ultrasound in chemical synthesis. Chem. Soc. Rev., 16, 275 (1987).
(5) T. J. Mason, Ultrasound in synthetic organic chemistry. Chem. Soc. Rev., 26, 443 (1997).
(6) L. H. Thompson and L. K. Doraiswamy, Sonochemistry: science and engineering. Ind. Eng. Chem. Res., 38, 1215 (1999).
(7) P. R. Gogate, Cavitational reactors for Process Intensification of chemical processing applications: a critical review. Chem. Eng. Proc., 47(4), 515–527 (2008).
(8) C. Horst, Y.-S. Chen, U. Kunz and U. Hoffmann, Design, modeling and performance of a novel sonochemical reactor for heterogeneous reactions. Chem. Eng. Sci., 51, 1837–1846 (1996).
(9) O. Dahlem, J. Reisse and V. Halloin, The radially vibrating horn: a scaling up possibility for sonochemical reactions. Chem. Eng. Sci., 54, 2829–2838 (1999).

(10) J.-L. Dion, Contamination-free high capacity converging waves sonoreactors for the chemical industry. Ultrason. Sonochem., 16, 212–220 (2009).
(11) B. Avvaru and A. B. Pandit, Oscillating bubble concentration and its size distribution using acoustic emission spectra. Ultrason. Sonochem., 16, 105–115 (2009).
(12) P. R. Gogate and A. B. Pandit, Engineering design methods for cavitation reactors I: sonochemical reactors. AIChE J., 46, 372–379 (2000).
(13) K. Nickel and U. Neis, Ultrasonic disintegration of biosolids for improved biodegradation. Ultrason. Sonochem., 14, 450–455 (2007).
(14) E. Gonze, Y. Gonthier, P. Boldo and A. Bernis, Standing waves in a high frequency sonoreactor: visualisation and effects. Chem. Eng. Sci., 53, 523–532 (1998).
(15) G. Thoma, J. Swofford, V. Popov and M. Som, Sonochemical destruction of dichloromethane and o-dichlorobenzene in aqueous solution using a nearfield acoustic processor. Adv. Env. Res., 1, 178–193 (1997).
(16) P. R. Gogate, I. Z. Shirgaonkar, M. Sivakumar, P. Senthilkumar, N. P. Vichare and A. B. Pandit, Cavitation reactors: efficiency analysis using a model reaction. AIChE J., 47, 2526–2538 (2001).
(17) M. Romdhane, C. Gourdon and G. Casamatta, Local investigation of some ultrasonic devices by means of a thermal sensor. Ultrasonics, 33, 221–227 (1995).
(18) P. R. Gogate, S. Mujumdar and A. B. Pandit, Large scale sonochemical reactors for process intensification: design and experimental validation. J. Chem. Tech. Biotech., 78, 685–693 (2003).
(19) P. R. Gogate and A. B. Pandit, Hydrodynamic cavitation reactors: a state of the art review. Rev. Chem. Eng., 17, 1–85 (2001).
(20) P. S. Kumar and A. B. Pandit, Modeling hydrodynamic cavitation. Chem. Eng. Tech., 22, 1017–1027 (1999).
(21) I. Z. Shirgaonkar, R. R. Lothe and A. B. Pandit, Comments on the mechanism of microbial cell disruption in high pressure and high speed devices. Biotech. Prog. 14, 657–660 (1998).
(22) K. Sampathkumar and V. S. Moholkar, Conceptual design of a novel hydrodynamic cavitation reactor. Chem. Eng. Sci., 62, 2698–2711 (2007).
(23) P. R. Gogate, P. A. Tatake, P. M. Kanthale and A. B. Pandit, Mapping of sonochemical reactors: review, analysis and experimental verification. AIChE J., 48(7), 1542–1560 (2002).
(24) P. R. Gogate, A. M. Wilhelm and A. B. Pandit, Some aspects of the design of sonochemical reactors. Ultrason. Sonochem., 10, 325–30 (2003).
(25) P. R. Gogate and A. B. Pandit, Sonochemical reactors: scale up aspects. Ultrason. Sonochem., 11, 105–117 (2004).
(26) V. S. Moholkar and A. B. Pandit, Modeling of hydrodynamic cavitation reactors: a unified approach. Chem. Eng. Sci., 56, 6295–6302 (2001).
(27) P. R. Gogate and A. B. Pandit, Engineering design methods for cavitation reactors II: hydrodynamic cavitation reactors. AIChE J., 46, 1641–1649 (2000).
(28) G. Servant, J. L. Laborde, A. Hita, J. P. Caltagirone and A. Gerard, On the interaction between ultrasound waves and bubble clouds in mono and dual-frequency sonoreactors. Ultrason. Sonochem., 10, 347–355 (2003).
(29) A. V. Prabhu, P. R. Gogate and A. B. Pandit, Optimisation of multi frequency sonochemical reactors. Chem. Eng. Sci., 59, 4991–4998 (2004).
(30) P. A. Tatake and A. B. Pandit, Modeling and experimental investigation into cavity dynamics and cavitational yield: influences of dual frequency ultrasound sources. Chem. Eng. Sci., 57, 4987–4995 (2002).
(31) M. Sivakumar and A. B. Pandit, Ultrasound enhanced degradation of Rhodamine B: optimization with power density. Ultrason. Sonochem., 8, 233–240 (2001).

(32) V. S. Sutkar and P. R. Gogate, Design aspects of sonochemical reactors: techniques for understanding cavitational activity distribution and effect of operating parameters. Chem. Eng. J., 155, 26–36 (2009).
(33) J. Klima, A. Frias-Ferrer, J. Gonzalez-Garcý, J. Ludvýk, V. Saez and J. Iniesta, Optimisation of 20 kHz sonoreactor geometry on the basis of numerical simulation of local ultrasonic intensity and qualitative comparison with experimental results. Ultrason. Sonochem., 14, 19–28 (2007).
(34) N. Segebarth, O. Eulaerts, Y. Kegelaers, J. Vandercammen and J. Reisse, About the Janus double horn sonicator and its use in quantitative homogenous sonochemistry. Ultrason. Sonochem., 9, 113–119 (2002).
(35) N. Shimizu, C. Ogino, M. Dadjour, K. Ninomiya, A. Fujihira and K. Sakiyama, Sonocatalytic facilitation of hydroxyl radical generation in the presence of TiO_2. Ultrason. Sonochem., 15, 988–994 (2008).
(36) J. D. Seymore and R. B. Gupta, Oxidation of aqueous pollutants using ultrasound – salt induced enhancement. Ind. Eng. Chem. Res., 36, 3453–3457 (1997).
(37) M. Wall, M. Ashokkumar, R. Tronson and F. Grieser, Multibubble sonoluminescence in aqueous salt solutions. Ultrason. Sonochem., 6, 7–14 (1999).
(38) P. R. Gogate and A. B. Pandit, A review of imperative technologies for waste water treatment II: hybrid methods. Adv. Env. Res., 8, 553–597 (2004).
(39) S. Toma, A. Gaplovsky and J.-L. Luche, The effect of ultrasound on photochemical reactions. Ultrason. Sonochem., 8, 201–207 (2001).
(40) P. R. Gogate and A. B. Pandit, Sonophotocatalytic oxidation based reactors for wastewater treatment: a critical review. AIChE J., 50, 1051–1079 (2004).
(41) Cavitation Oxidation Process (CAV-OX) Application Analysis Report, Magnum Water Technology Inc., Risk Reduction Eng. Laboratory, Office of Research and Development, USEPA, Cincinnati, USA (1994).
(42) Y. Peng and G. Song, Simultaneous microwave and ultrasound irradiation: a rapid synthesis of hydrazides. Green Chem., 3, 302–304 (2001).
(43) G. Cravotto, M. Beggiato, A. Penoni, G. Palmisano, S. Tollari, J.-M. Lévêque and W. Bonrath, High-intensity ultrasound and microwave, alone or combined, promote Pd/C-catalyzed aryl–aryl couplings. Tetrahedron Let., 46, 2267–2271 (2005).
(44) F. Trotta, K. Martina, B. Robaldo, A. Barge and G. Cravotto, Recent advances in the synthesis of cyclodextrin derivatives under microwaves and power ultrasound. J. Inclusion Phenomena Macrocyclic Chem., 57, 3–7 (2007).
(45) G. Cravotto, S.Di Carlo, M. Curini, V. Tumiatti and C. Rogerro, A new flow reactor for the treatment of polluted water with microwave and ultrasound. J. Chem. Tech. Biotech., 82, 205–208 (2007).
(46) Y. Zhang, M. A. Dube, D. D. Mclean and M. Kates, Biodiesel production from waste cooking oil: 1. process design and technological assessment. Biores. Tech., 89, 1–16 (2003).
(47) B. Freedman, R. O. Butterfield and E. H. Pryde, Transesterification kinetics of soyabean oil. J. Am. Oil. Chem. Soc., 63, 1375–1380 (1986).
(48) B. Freedman, E. H. Pryde and T. L. Mounts, Variables affecting the yields of fatty esters from transesterified vegetable oils. J. Am. Oil Chem. Soc., 61, 1638–1643 (1984).
(49) C. Stavarache, M. Vinatoru and Y. Maeda, Ultrasonic versus silent methylation of vegetable oils. Ultrason. Sonochem., 13, 401–407 (2006).
(50) C. Stavarache, M. Vinatoru and Y. Maeda, Fatty acids methyl esters from vegetable oil by means of ultrasonic energy. Ultrason. Sonochem., 12, 367–372 (2005).
(51) M. A. Kelkar, P. R. Gogate and A. B. Pandit, Intensification of esterification of acids for synthesis of biodiesel using acoustic and hydrodynamic cavitation. Ultrason. Sonochem., 15, 188–194 (2008).

(52) N. R. Sonare and V. K. Rathod, Transesterification of used sunflower oil using immobilized enzyme. J. Mol. Cat. B: Enzymatic, 66, 142–147 (2010).
(53) S. M. Hingu, P. R. Gogate and V. K. Rathod. Synthesis of biodiesel from waste cooking oil using sonochemical reactors. Ultrason. Sonochem., 17, 827–832 (2010).
(54) M. H. Ali and G. J. Bohnert, A facile and selective procedure for oxidation of sulfides with molecular bromine on hydrated silica gel in dichloromethane. Synthesis, 9, 1238–1240 (1998).
(55) K. S. Ravikumar, F. Barbier, J. P. Bégué and D. B. Delpon, Selective oxidation of sulfides: selective preparation of 1-trifluoromethyl vinyl sulfoxides. J. Fluorine Chem., 95, 123–125 (1999).
(56) D. Ramkumar and S. Sankararaman, Selective photooxidation of sulfides to sulfoxides with tetranitromethane. Synthesis, 11, 1057–1058 (1993).
(57) T. Sato and J. Otera, Oxidation of aryl sulfides to the sulfoxides with high diastereoselectivity. Cheminform, 26, doi: 10.1002/chin.199541140 (1995).
(58) S. S. Kim, K. Nehru, S. S. Kim, D. W. Kim and H. C. Jung, A mild and highly efficient oxidation of sulfides to sulfoxides with periodic acid catalyzed by $FeCl_3$. Synthesis, 17, 2484–2486 (2002).
(59) K. Y. Koposov and V. V. Zhdankin, Selective oxidation of sulfides to sulfoxides using IBX-esters. Synthesis, 1, art. no. M05204SS, 22–24 (2005).
(60) N. N. Mahamuni, P. R. Gogate and A. B. Pandit. Selective synthesis of sulfoxides from sulfides using ultrasound. Ultrason. Sonochem., 14, 135–142 (2007).
(61) N. N. Mahamuni, P. R. Gogate and A. B. Pandit, Ultrasound accelerated green and selective oxidation of sulfides to sulfoxides. Ind. Eng. Chem. Res., 45, 8829–8836 (2006).
(62) T. Javed, T. J. Mason, S. S. Phull, N. R. Baker and A. Robertson, Influence of ultrasound on the Diels-Alder cyclization reaction: synthesis of some hydroquinone derivatives and lonapalene and anti-psoriatic agent. Ultrason. Sonochem., 2, S11–S14 (1995).
(63) M. S. F. Lie Ken Jie and C. K. Lam, Ultrasound assisted epoxidation reaction of long chain unsaturated fatty esters. Ultrason. Sonochem., 2, S3–S4 (1995).
(64) T. Ando, S. Sumi, T. Kawate, J. Ichihara and H. Terukiyp, Sonochemical switching of reaction pathways in solid-liquid two phase reactions. J. Chem. Soc. Chem. Commun., 7, 439–440 (1984).
(65) M. M.van Lersel, M. A.van Schilt, N. E. Benes and J. T. F. Keurentjes, Controlled methyl chloride synthesis at mild conditions using ultrasound irradiation. Ultrason. Sonochem., 17, 315–317 (2010).
(66) S. Puri, B. Kaur, A. Parmar and H. Kumar, Ultrasound-promoted greener synthesis of 2H-chromen-2-ones catalyzed by copper perchlorate in solventless media. Ultrason. Sonochem., 16, 705–707 (2009).
(67) L. C. Hagenson, S. D. Naik and L. K. Doraiswamy, Rate enhancements in a solid-liquid reaction using PTC, microphase, ultrasound and combinations thereof. Chem. Eng. Sci., 49, 4787–4800 (1994).
(68) J. Lindley and T. J. Mason, Sonochemistry: part 2 – synthetic applications. Chem. Soc. Rev., 16, 275–311 (1987).
(69) T. J. Mason and J. P. Lorimer, Sonochemistry: Theory, Applications and Uses of Ultrasound in Chemistry, Ellis Horwood: Chichester (1998).
(70) K. S. Suslick, Organometallic sonochemistry. Adv. Organomet. Chem., 25, 73–119 (1986).
(71) Y. T. Shah, A. B. Pandit and V. S. Moholkar, Cavitation Reaction Engineering, Plenum: New York (1999).

8

Membrane Bioreactors for Green Processing in a Sustainable Production System

Rosalinda Mazzei, Emma Piacentini, Enrico Drioli and Lidietta Giorno

Institute on Membrane Technology, CNR ITM, University of Calabria, Rende, Calabria, Italy

8.1 Introduction

The exponential growth of the world population, the increase in resource use and waste production and the demand for more goods of higher quality all produce an urgent need for new, clean and safe technologies and production process. The use of efficient, selective, precise, flexible technologies, able to operate in mild conditions, such as membrane technologies, will strongly contribute to sustainable development. In particular, membrane bioreactors have the potential to play a significant role in preventing pollution and allowing resource recovery, chemical purification and energy savings. They are intensified systems in which a biochemical conversion is coupled with a membrane separation process in a single unit operation. The catalytic action of biocatalysts is extremely efficient and selective compared to conventional chemical catalysts. They can demonstrate higher reaction rates, milder reaction conditions and greater stereospecificity. Therefore, they offer unparalleled possibilities for the operation of precise (bio)chemical transformation without byproduct formation. Most of these properties come from the high molecular flexibility that biocatalysts exhibit. Transport can be appropriately tuned to control reagent

Process Intensification for Green Chemistry: Engineering Solutions for Sustainable Chemical Processing, First Edition. Edited by Kamelia Boodhoo and Adam Harvey.

supply to the catalyst and/or product removal from the reaction site. Compared to more conventional technologies, membrane bioreactors have several intrinsic advantages that make them a sustainable alternative technology, such as the possibility of reusing the biocatalyst in a continuous system (which contributes to increasing productivity and probably the economic viability of the process), the continuous selective removal of products from the reaction media and the controlled supply of the reagent to the catalytic reaction environment. Membrane bioreactors also contribute to a greener chemistry by controlling the mass transfer of reactants and products and by enhancing overall yields and conversions, minimizing solvent use and maximizing mass utilization. Some of the most important problems related to alternative energy and clean water production, together with the increasing demand for natural biofunctional products, can find their solution in the use of membrane bioreactors. These can be used in the production of:

1. Biodiesel made from natural, renewable sources, such as vegetable oil.
2. Pure water, which can be recycled from municipal or industrial waste water for reuse in a variety of places.
3. High-added-value molecules, including from vegetal waste materials.
4. Optically pure enantiomers.

The unique properties of membrane bioreactors in terms of process conditions and product quality promote their use as clean production systems of high-quality products that neither create nor discharge waste – hazardous or nonhazardous – to the environment.

8.2 Membrane Bioreactors

Based on the membrane role, membrane bioreactors can be divided into two major types (Figure 8.1): (1) where the membrane does not participate to the reaction but performs the separation process (Figure 8.1a), controls the mass transport in and out of the reactor bulk and has an indirect effect on the reaction itself (i.e. the reaction occurs somewhere other than the membrane, which affects the reaction by removing products, adding reagents, retaining the catalyst in the reaction bulk phase, etc.) and (2) where the membrane contains the biocatalyst (Figure 8.1b) and the reaction takes place at the membrane level (in which case the membrane has a direct effect on the reaction, in addition to governing the mass transport taking place through it). In the latter case, the reactor is called a biocatalytic membrane reactor.

Based on the membrane module configuration, membrane bioreactors can be divided into: (1) side stream membrane reactors, where the membrane module is placed externally to the reaction mixture (external configuration) (Figure 8.2a) and (2) submerged membrane reactors, where the membrane module is immersed within the tank containing the reaction mixture (submerged configuration) (Figure 8.2b). In both configurations, the biocatalyst can be free or immobilized.

8.2.1 Membrane Bioreactors with Biocatalyst Recycled in the Retentate Stream

When the membrane acts as a separation unit, based on where the biocatalyst is placed two main configurations can be distinguished: (1) where the biocatalyst is continuously flushed

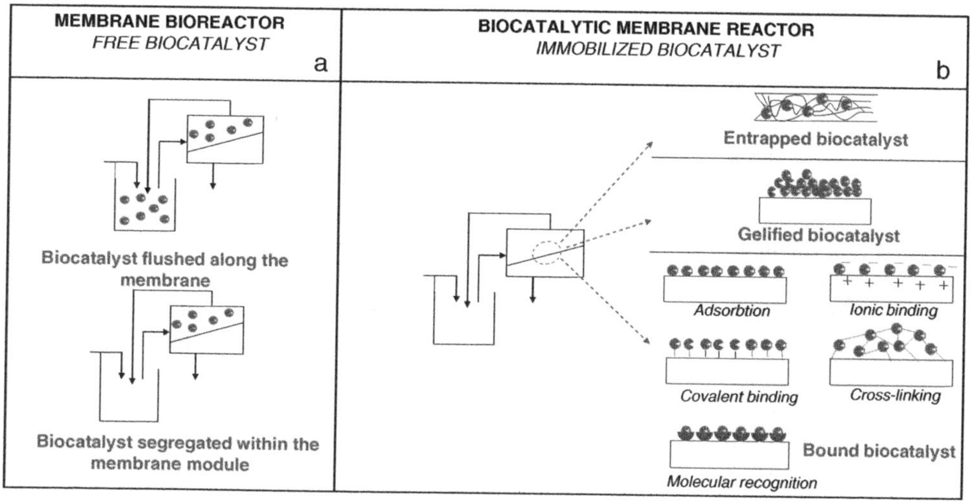

Figure 8.1 Schematic classification of (a) a membrane bioreactor and (b) a biocatalytic membrane reactor (b).

along the membrane and (2) where the biocatalyst is confined within the membrane module, in either the shell or the lumen space.

A typical example of the first configuration is where the initial solution contains both the catalyst and the substrate, while the product is separated from the feed solution by the membrane operation. This type of reactor integrates a traditional continuously stirred tank reactor (CSTR) with a separation unit (the membrane plant).

Ultrafiltration is the most common separation process used in this kind of setup. Both dead-end continuous stirred ultrafiltration cells with flat membranes and crossflow ultrafiltration through tubular membrane modules have been widely investigated.

One of the main drawbacks of this configuration is concentration polarization phenomena, which severely affect the performance of the separation process; when these occur, it is

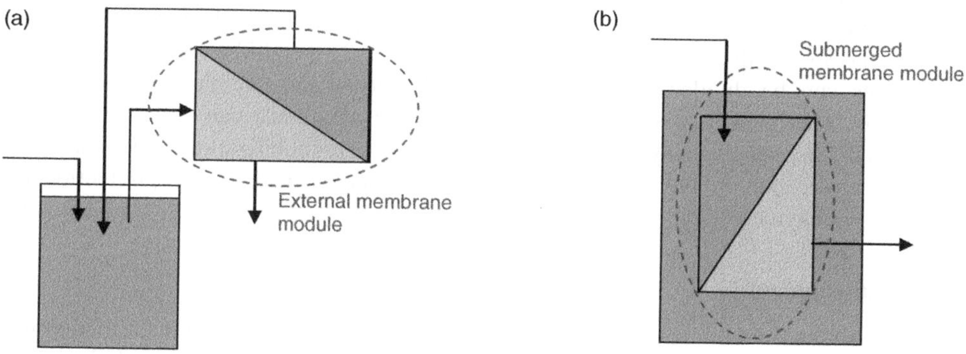

Figure 8.2 Schematic representation of (a) an external membrane reactor and (b) a submerged membrane reactor. Adapted from Giorno et al. © 2003 by permission of Wiley Periodicals, Inc. © 2003.

necessary to control the polarization layer on the membrane-pressurized side by means of reactor fluid dynamics and appropriate module design.

The biocatalyst is generally flushed along the membrane module in enzymatic systems for which a homogeneous catalyst distribution is particularly important, such as cofactor-requiring mono- and multi-enzyme systems.

Continuous membrane fermentors fall into this category. Here, microporous membranes are used to separate the fermentation broth from the product stream, thus retaining viable cells in the fermentor [1,2].

Another function that membranes can serve in membrane bioreactors is to control the supply of reagent molecules. An extractive membrane bioreactor can be used to permeate organic molecules across a membrane into an active bioreactor [3,4], and ion-exchange membrane bioreactors can be used to transport ionic compounds through a nonporous ion-exchange membrane into a biologically active compartment [5,6].

Through applying these strategies, biomass performance is not negatively affected by the waste water pH and composition.

8.2.2 Membrane Bioreactors with Biocatalyst Segregated in the Membrane Module Space

The other main configuration is where the biocatalyst is confined in a particular place within the membrane module space (in the hollow fibre lumen or in the shell surrounding the outer surface of the fibres) (Figure 8.1a). In this system the biocatalysts it is not recirculated in the effluent stream while low-molecular-weight products and inhibitors are being removed through the membrane.

One of the main applications of a biocatalyst compartmentalized in the membrane module space is in the production of bioartificial organs for therapeutic application [7,8].

8.3 Biocatalytic Membrane Reactors

In this type of system, the membrane must provide both separation and catalytic action. The different kinds of biocatalytic membrane reactor configuration depend on how the complex biocatalyst membrane is prepared. Figure 8.1b provides a schematic representation of the immobilization techniques used. In general, the different immobilization techniques can be divided into entrapment, gelification and chemical bound to the membrane. The bond between the biocatalyst and the membrane can be either random or conducted in a specific way. A novel method of enzyme immobilization, which may find a future application in the biocatalytic membrane reactor field, is the production of crosslinked enzyme crystals (CLECs) and crosslinked enzyme aggregates (CLEAs) [9].

8.3.1 Entrapment

In this membrane immobilization technique, the biocatalyst is localized within the membrane in a way that retains the protein while allowing substrate penetration. The technique can be performed during membrane preparation by phase inversion or by crossflow filtration of the biocatalyst solution through an asymmetric membrane. In the latter case, the filtration is carried out from the sponge to the thin layer of asymmetric membrane, which has a pore size smaller than the biocatalyst.

To better orientate the development of such systems, different parameters have to be considered, such as the amount of macromolecule loaded and its activity and distribution after immobilization.

Generally, the amount of immobilized biomolecule can be determined by mass balance between the initial solution and the final solutions [10]. Different studies have demonstrated that using this technique produces a uniform distribution along the membrane length and thickness [11,12]. A new integrated method [13] combining molecular biology and membrane technology allows us to determine the enzyme (β-glucosidase) spatial distribution through the membrane thickness and along the membrane module and its activity after immobilization through observation using light microscopy.

8.3.2 Gelification

The gelification of the biomolecule on the membrane is based on one of the main drawbacks of membrane processes: concentration polarization phenomena and subsequent protein precipitation upon reaching a critical concentration at the membrane interface. This type of immobilization technique is used when direct contact of the immobilized enzyme with the substrate solution cannot be otherwise achieved (i.e. the large substrate molecules cannot penetrate within the membrane structure). A drawback is biocatalyst leakage during the process.

8.3.3 Chemical Attachment

Biocatalysts can be attached to the membrane by chemical or biochemical interactions, including adsorption, covalent linking, crosslinking and ionic binding,, and molecular recognition.

8.3.3.1 Biofunctionalization of a Membrane by Adsorption

The adsorption immobilization technique is a method of preparing a biocatalytic membrane in which the interaction of the biomolecule and the membrane happens through weak and nonspecific bounds promoted by van der Waals forces and hydrogen bonds. The presence of weak interactions provides only a minimum of activation steps and avoids the use of reagents. In addition, the method is cheap, easy to carry out and less disruptive to the protein than covalent binding.

Another advantage is that the method bears the greatest similarity to the situation found in biological membranes *in vivo* and has been used to model such systems.

Protein desorption is often observed as a consequence of weak bonds when changes in temperature, pH and ionic strength are present.

8.3.3.2 Biofunctionalization of a Membrane by Covalent Linking, Crosslinking and Ionic Binding

The formation of a covalent bond between the biocatalyst and the membrane is the most stable immobilization technique. In order to carry out this kind of technique it is necessary to perform the binding reaction under conditions that do not cause protein denaturation.

In the case of biocatalysts, the groups present in the active site must not be involved in the covalent linkage to the membrane. Different strategies are applied to solve this problem, such as carrying out the immobilization in the presence of a substrate [14] or competitive inhibitor of the enzyme [15,16] that protects the active site during the immobilization.

The main advantage of this immobilization technique is that the covalent bond between the biomolecule and the membrane is very stable; therefore, no leakage of biomolecule occurs in devices prepared using this method. On the other hand, this can be a limit if the replacement of biocatalyst is required. The main application of the attachment of a biomolecule on a membrane by a covalent bond is in the production of highly selective single-use devices, such as biosensors.

Biocatalytic membranes can be also prepared by intermolecular crosslinking of the protein, either to other protein molecules or to functional groups present in an insoluble support matrix. The main drawbacks of this technique are the high cost and the inefficiency of the biocatalysts loaded, as some of the protein material will inevitably act mainly as a support.

Generally, crosslinking is used in the integration of one of the less stable immobilization methods (such as gelification, adsorption, etc.).

Another kind of membrane functionalization with biomolecules is carried out by ionic binding. This technique consists in the interaction of a charged biomolecule with the support that has opposite charge.

8.3.3.3 Site-Specific Biofunctionalization

The aim of the site-specific immobilization technique is to create the most biocompatible system, in which the interaction between the biomolecule and the membrane is directed and controlled so as not to involve the active site of the biocatalyst.

The strategies usually used are achieved by the introduction of a spacer molecule, or in the case of glycoproteins by orientating the attachment through the carbohydrate moieties [17].

Other strategies are based on gene fusion, incorporating a peptidic affinity tag at the N- or C- terminus of the biomolecule; on post-translational modification, incorporating a single biotin moiety on the protein; and on site-directed mutagenesis, introducing unique cysteine molecules to the biomolecule [18]. These can be supported by surface modification of polymeric membranes, by attaching functional groups such as sugars or polypeptides and then adsorbing the protein.

An innovative method considered to be inspired by biomimetics and shown to be efficient for biocatalyst attachment consists in using a biotin/avidin system [19,20]. The tetrameric structure of avidin permits it to interact with four different molecules of biotin at the same time. Avidin can be easily attached to membrane material, while various proteins and enzymes can easily be biotinylated. This mode of enzyme grafting has already been used in the production of electrodes and of membranes made up of conducting fibres.

8.4 Case Studies: Membrane Bioreactors

In this section some interesting green chemistry applications of membrane bioreactors will be described. These cases studies focus on the production of biofuel, waste water treatment reuse, the production of high-added-value compounds from waste and the separation of optically pure enantiomers.

8.4.1 Biofuel Production Using Enzymatic Transesterification

The increasing dependency on foreign energy supply from declining fossil fuel resources and the need to reduce global warming by lowering the emission of particles, sulfur carbon monoxides and hydrocarbons from road transport are two of the main reasons for the introduction of biodiesel as an alternative to conventional fossil-based diesel. Biodiesel can be produced from different waste materials, generally divided into: plant oil, animal fat, waste cooking and industrial waste oil. Another classification is based on the kind of generation fuel produced, which depends on the feedstock used. In Table 8.1, a comparison of petroleum refinery and first-, second- and third-generation fuels is given, together with their different feedstocks and products.

The different feedstocks used to produce biodiesel in the main biodiesel-production countries are reported in Table 8.2.The USA mainly uses genetically modified soybean oil as a raw material, while the EU and Canada mainly use rapeseed oil. Some South East Asian countries, such as Malaysia and Indonesia, are abundant in palm oil and coconuts. Although China is a large agricultural producer, the food supply system is still a big problem there. In order to follow the principle of never competing with grain land, China's recent biodiesel development has been based on waste cooking oil and linseed oil. Jatropha oil will become a good choice for China due to some unique advantages: jatropha trees can grow on waste ground with a minimal water and fertilizer demand, and the oil is not useful as food material due to the presence of antinutritional factors. Other important nonfood oils that have been successfully used as innovative feedstocks are microalgae and microbial oil, which represent the third generation of biofuel [21]. Microalgae in particular have had a lot of success because they can synthesize and accumulate large quantities of neutral lipids/oil (20–50% DCW). Their other advantages include: high growth rates (one to three doublings per day); utilization of growth nutrients such as nitrogen and phosphorus from a variety of waste water sources (e.g. agricultural runoff, industrial and municipal waste water); sequestration of CO_2 from flue gases emitted by fossil fuel-fired power plants and other sources, thus reducing emissions of a major greenhouse gas; and production of value-added coproducts and byproducts (e.g. biopolymers, proteins, polysaccharides, pigments, animal feed, fertilizer and H_2). Membrane operations such as microfiltration and utrafiltration are used to separate biomass from the culture medium, while membrane bioreactors can be used to compartmentalize the cell for microalgae cultivation.

Oils that are a mixture of free fatty acids (FFAs) and triglycerides need to be chemically altered to fatty acid alkyl esters (FAAEs) in order to be used as biodiesel fuel for our current diesel engines [22]. Different catalysts can be used for transesterification: acids, bases – both liquid and heterogeneous – and free or immobilized enzymes. The most commonly used method in industry is alkaline transesterification, in which raw material with high water or FFA content is pre-treated with an acidic catalyst in order to esterify FFAs. The pre-treatment it is carried out to reduce soap formation.

An innovative approach is to use enzymes in place of alkaline catalysts in membrane bioreactors. Enzymes do not form soaps and can promote esterification and transesterification of both FFA and triacylglycerol (TAG) in one step, avoiding the subsequent washing step. Other advantages are reported in Table 8.3.

Lipases (EC 3.1.1.3) are widely employed to catalyse the hydrolysis, alcoholysis, esterification and transesterification of carboxylic esters. These enzymes have excellent

Table 8.1 *Comparison of petroleum refinery and first-, second- and third-generation fuels.*

	Petroleum refinery	First-generation fuel	Second-generation fuel	Third-generation fuel
Feedstocks	Crude petroleum	Vegetable oils, corn sugar	Nonfood, cheap, plant-waste biomass	Microalgae
Products	Diesel, petrol, kerosene, jet fuel	Biodiesel, corn ethanol, sugar, alcohol, petrol	Hydrotreating oil, bio oil, butanol, ethanol, mixed alcohols	Biodiesel and bioethanol from microalgae and seaweeds, hydrogen from green microalgae and microbes
Disadvantages	Depletion/decline of petroleum reserve Environmental pollution Economic and ecological problems	Limited feedstock (food versus fuel) Partly blended with conventional fuel	Not competing with food Advanced technology still under development to reduce the cost of conversion	-
Advantages	—	Environmentally friendly Economics and social security	Environmentally friendly	Very short harvesting cycle Production of 15–3000 times more oil than traditional crops

Table 8.2 *Feedstock used to produce biodiesel in the main biodiesel-production countries and regions.*

Country/region	Feedstock
USA	Soybean oil
Europe	Rapeseed, sunflower
Western Canada	Rapeseed, canola oil
Africa	Jatropha
India	Jatropha
Malaysia/Indonesia	Coconut and palm oil
Philippines	Waste cooking oil
China	Linseed oil, waste cooking oil, jatropha
Greece	Cottonseed

catalytic activity and stability in nonaqueous media, which facilitates the esterification and transesterification process during biodiesel production.

As described in the literature, lipase is activated by oil-in-water interface through the phenomenon of 'interfacial activation'. This is the reason why it is particularly suitable for immobilization on membrane in the presence of oil-in-water interface. Some examples are reported in the literature where the enzyme is immobilized in a biphasic biocatalytic membrane reactor [23] for the production of optically pure isomers (Figure 8.3).

In bioediesel production, lipase is largely immobilized by adsorption, covalent-bond, crosslinking and entrapment techniques. The lipase immobilized for biodiesel production is derived from different biocatalyst sources such as *Candida antartica*, *Candida* sp. 99–125, *Pseudomonas fluorescens*, *Pseudomonas cepacia*, *Porcine pancreatic*, *Rhizomucor miehei* and *Chromobacterium viscosum* [24]. Table 8.4 gives some examples of lipases that can be immobilized on membranes for the production of biodiesel in membrane bioreactors. For large-scale application, the two kinds of lipase most frequently used are *Candida antartica*, also known as Novozym 435, and *Candida* sp. 99–125. The latter is often immobilized by adsorption on cheap textile membranes and can catalyse lard, waste oil and various vegetable oils with yields greater than 87%.

Table 8.3 *Advantages and disadvantages of the use of enzymes in free suspension in biofuel production.*

Advantages	Disadvantages
Higher compatibility with variations in the quality of raw material	Low reaction rate
Reusable	Cost
Able to produce biofuel in fewer steps, using less energy	Loss of activity after 100 days of operation
Produce less waste water	Deactivation caused by alcohol
Improve product separation	
Produce higher quality of glycerol	

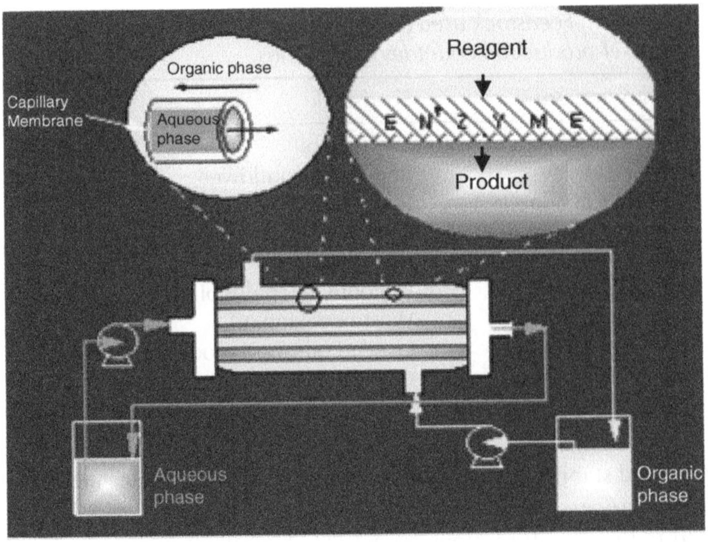

Figure 8.3 *Biphasic biocatalytic membrane reactor with immobilized lipase. Adapted from [23].*

In biodiesel biological production the acyl acceptor is generally methanol. However, the biodiesel productivity is strongly decreased in this case due to the negative effect of the alcohol on the lipase. To overcome this problem, different strategies have been applied, such as the addition of cosolvents, which must however be removed from the diesel produced by downstream processes. When methanol is replaced by methyl acetate or ethyl acetate, the lipase inactivation caused by short-chain alcohol is avoided. The only drawbacks are the low reaction rate obtained and the high cost, which limits the use of these acyl acceptors in industrial applications. Although these solvents might solve the problem of

Table 8.4 *Lipases used in membrane bioreactors for the production of biodiesel.*

Lipase	Oil/fat	Alcohol	Yield	Form	Membrane material	References
Pseudomonas cepacia	Mahua oil	Ethanol	96%	Imm.	Polypropylene	[25]
			92%	CLEA		
			99%	CLEC		
Candida spp. 99–125	Rapeseed oil	Methanol	83%	Imm.	Cotton	[24,26,27,29]
	Salad oil		95%			
	Waste oil		92%			
	Vegetable oil		96%			
	Lard, waste oil	Methanol	87%	Imm.	Textile	
Pseudomonas fluorescens	Triolein	Methanol	80%	Imm.	Polyethersulfone	[28]

CLEA, crosslinked enzyme aggregate; CLEC, crosslinked enzymes crystal; Imm, immobilized.

lipase inactivation caused by methanol, difficulties in recovering the solvent also make them less competitive at the industrial scale [24].

Supercritical fluids (SCFs) have been found to be a promising alternative to organic solvents for enzymatic reactions. Carbon dioxide (CO_2) is the most commonly used SCF, due to its various advantages. Because of the relatively low viscosity, high diffusivity and low surface tension, mass transfer is improved in supercritical CO_2 ($SCCO_2$) compared to organic solvents. Moreover, separation of CO_2 from the product can be easily achieved by reducing pressure, because the products and enzyme are not soluble in CO_2 [30].

8.4.2 Waste Water Treatment and Reuse

Waste water is water that has been used and must be treated before it can be released into another body of water, so that it does not cause pollution. Waste water comes from domestic, municipal, industrial and agricultural discharges. Today, special attention is focused on control technologies in water and waste water, involving the application of advanced techniques to minimize their environmental release and to promote water reuse and recycling.

Biological treatment technologies have been utilized in waste water treatment for over a century. The use of membranes within the treatment sequence of a water pollution control facility was initially limited to tertiary treatment and polishing, until 1969 when the application of membrane bioreactor technology in this area was reported for the first time [31]. Due to their unique advantages, membrane bioreactors are becoming increasingly popular for waste water treatment.

There are several advantages associated with the membrane bioreactor which make it a valuable alternative to other waste water treatment techniques. First, the retention of all suspended matter and most soluble compounds, including bacteria and viruses, within the bioreactor leads to excellent quality of effluents, which can be reused without additional disinfection. The oxidation of high-molecular-weight soluble compounds, which are not readily biodegradable in conventional systems, is improved due to retention and residence time increases. Membrane bioreactors eliminate process difficulties and problems associated with settling, which are usually the most troublesome parts of waste water treatment, allowing high biomass concentrations in the bioreactor. Consequently, higher-strength waste water can be treated and lower biomass yields realized. This also results in more compact systems than conventional processes, significantly reducing plant footprint, which is desirable for water recycling applications. The differences between conventional waste water treatment and membrane bioreactors are illustrated in Figure 8.4. Both aerobic and anaerobic processes can be used; the former involves the use of free or dissolved oxygen by microorganisms (aerobes) in the conversion of organic wastes to biomass and CO_2, while in the latter complex organic wastes are degraded into methane, CO_2 and H_2O through three basic steps (hydrolysis, acidogenesis including acetogenesis and methanogenesis) in the absence of oxygen. The most frequently used membrane bioreactor configuration is the aerobic, submerged membrane bioreactor [32]. Anaerobic membrane bioreactors, on the other hand, have a great potential for high-strength waste waters that are highly particulate, because the particles will be retained by the membrane. Even for low-strength waste waters, which are almost exclusively treated

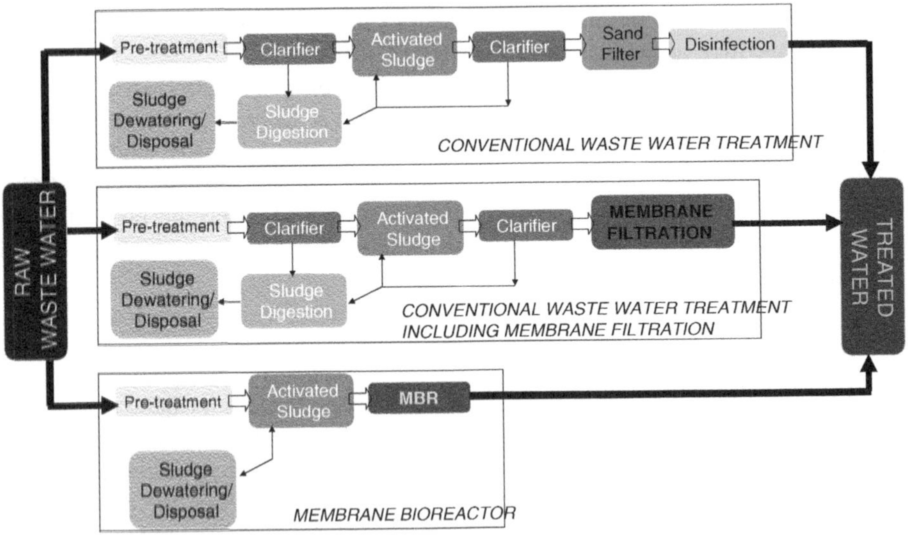

Figure 8.4 *Schematic representation of the steps required in waste water treatment by various technologies. MBR, membrane bioreactor. Adapted from Mazzei et al © 2009, with permission from Elsevier.*

aerobically, anaerobic membrane bioreactors may be a good alternative. Operational costs are lower because of the low anaerobic growth yields – there is less sludge to be disposed of [33] – and because bio gas is produced. Instead of aeration, fouling control can be achieved by sparging the bio gas produced [34]. However, since insufficient bio gas is produced, it must be recycled and compressed, so only slightly less energy than in aerobic membrane bioreactors is required [35]. A more intensive form of biodegradation can be achieved by integrating anaerobic and aerobic zones within a single bioreactor. There are four types of integrated anaerobic–aerobic bioreactor: (1) integrated membrane bioreactors with physical separation of anaerobic and aerobic zones; (2) integrated membrane bioreactors without physical separation of anaerobic and aerobic zones; (3) sequencing batch reactors (SBRs) based on temporal separation of the anaerobic and the aerobic phases; and (4) combined anaerobic–aerobic culture systems based on the principle of limited oxygen diffusion in microbial biofilms. Few examples of integrated aerobic–anaerobic membrane bioreactors without physical separation of anaerobic and aerobic zones are reported [36,37].

Several combinations of membrane bioreactors and waste water technologies (hybrid membrane bioreactors) other than the conventional activated sludge process have been proposed. The use of a hybrid membrane bioreactor is an excellent alternative for the treatment of industrial waste waters, especially when space is limited or real estate costs are high. In the biofilm membrane bioreactor (BF-MBR), biofilms grow on fluidized supports through the use of high N-removal by simultaneous biofilm denitrification [38–42]. In addition, less biomass is suspended and the circulating media generate some scrubbing action, which improves filterability despite the fact that the attached biomass has a much higher fouling potential than suspended activated sludge. Integrated operation between submerged hollow-fibre membrane filtration and a sequencing aerobic sludge blanket reactor (SASBR) leads to a novel aerobic granular sludge membrane bioreactor (AGMBR) [43]. The much better filtration characteristics of AGMBR mixed liquor are due to the low compressibility of its biomass, which is dominated by aerobic granular sludge. However, it has been reported that irreversible fouling is more severe than in conventional membrane bioreactors, which has been attributed to the colloids and solutes generated [44]. A hybrid process incorporating membrane distillation in a submerged membrane bioreactor operated at elevated temperature has been developed. Since organic particles are rejected by an 'evaporation' mechanism, the retention time of nonvolatile soluble and small organics in the submerged membrane distillation bioreactor (MDBR) is independent of the hydraulic retention time (mainly water and volatiles) [45]. A high permeate quality can be obtained in the one-step compact process. Subject to the availability of waste heat/solar thermal energy and a cooling system to exchange heat with a cooler natural resource, the MDBR may offer a low primary energy process with a very high permeate quality and a stable flux, at a practical value in a small-footprint configuration. An innovative osmotic membrane bioreactor (OMBR) has been developed, based on forward osmosis (FO) driven by an osmotic pressure difference. For the OMBR to be both technically and economically viable, the performance of the FO membranes must be sufficiently high and membrane fouling and draw solution leakage must be low. To improve the FO membrane performance for the OMBR process, the thickness and structure of the porous substructure of the FO membrane must be optimized by membrane development.

Membrane fouling and high cost are the major obstacles to the widespread use of membrane bioreactors in waste water treatment. The membrane is contaminated rapidly because of the high concentration of waste water passing through it. This decreases the membrane lifespan and increases aeration energy demands and operating costs. Different strategies are being developed to overcome these limitations. Novel nonwoven textiles may be a more economical alternative to the usual ultra- or microfiltration membranes, due to lower production costs and potentially higher permeability, which apart from smaller filtration surface areas would also lead to savings in aeration requirements. The application of nonwovens has recently been investigated in lab-scale membrane bioreactors [46], but research is at an early stage and no membrane bioreactor textiles are commercially available as yet.

Various strategies for fouling control have been developed, including:

- Feed waste water pre-treatment by physical (aquatic substance size increase), chemical (coagulants, oxidants and adsorbents) or biological (biodegradation) mechanisms.
- Biomass characteristic modification according to biomass properties.
- Optimization of operational conditions (operation below critical flux, aeration).
- Antifouling membranes, obtained by modifying hydrophobic membranes into relatively hydrophilic membranes.

8.4.3 Waste Valorization to Produce High-added-value Compounds

In this section, two case studies will be described which have gained a lot of interest in the green chemistry field: lactic acid production and phenolic compound valorization from waste materials.

Lactic acid is one of the primary platform chemicals that can be derived from renewable carbohydrates and used to make a wide variety of useful products. Lactic acid fermentation has received significant interest in recent times because it offers an alternative to the environmental pollution caused by the petrochemical industry and solves the problem of the limited supply of petrochemical resources.

Valuable compounds can also be recovered during olive oil production. In this process a considerable amount of warm water is employed in the continuous washing of the olive paste prior to separation of the oil from the paste (malaxation process). The major environmental factor here is the high concentration of polyphenols in the olive mill waste water (OMWW). These compounds present phytotoxicity, toxicity against aquatic organisms or suppression of soil microorganisms and are difficult to decompose. On the other hand, if properly processed they can be a very important category of antioxidant phytochemical. The membrane bioreactor technology used for the valorization of the phenolic compounds contained in OMWW in order to produce phytotherapic with a high antioxidant power, which are not commercially available, will be explained in Section 8.4.3.2.

The use of renewable resources such as various agrowastes opens a dual working option for value addition through an ecofriendly green technology.

8.4.3.1 Lactic Acid Production from Biomass

There is increasing interest in the recovery of fermentation products, such as organic acids, feeds or food additives and industrial chemicals, for the environmental sustainability and development of renewable resources. Currently, a great deal of attention is being paid to the biotechnological potential of agroindustrial residues such as cassava bagasse, sugarcane bagasse, sugar beet pulp, coffee husk and pulp, apple pomace, oilcakes, wheat/rice bran and so on for their use as raw materials in the production of high-added-value compounds. Lactic acid (2-hydroxypropionic acid) has many applications in the food, pharmaceutical, textile, leather and chemical industries. In particular, lactic acid can be used for products that have potential very large-volume uses in industrial applications and consumer products. The primary classes of such products are polymers for plastics and fibres, solvents for formulations and cleaning and oxygenated industrial chemicals. Environmentally friendly, 'green' solvents are another potential area for lactic acid derivatives, particularly lactate esters of low-molecular-weight alcohols such as ethyl, propyl and butyl lactate.

Although considerable pioneering effort has been made by a few companies towards the development and commercialization of lactic acid-based products, lactic acid production technologies must be further advanced and implemented before they can become technically and economically feasible and environmentally sound. Various food-processing wastes contain substrate that can be bioconverted into lactic acid. The biotechnological production of lactic acid offers several advantages over chemical synthesis, such as the low cost of substrates, low production temperature, low energy consumption and high product stereospecificity. In addition, the limited availability of petroleum resources and recent increases in the price of crude oil should be expected to render the fermentative route to lactic acid increasingly competitive. The carbon source for microbial production of lactic acid can be either sugar in its pure form, such as glucose, sucrose and lactose, or sugar-containing materials such as molasses, whey, sugarcane bagasse, cassava bagasse and starchy materials from potato, tapioca, wheat, barley and carrot. The use of alternative cheap materials is more economical than the use of purified sugars such as glucose and sucrose. Lignocellulosic resources, such as wood and crop residues, are generally considered to represent an attractive and inexpensive raw material for the microbial production of lactic acid, as they are renewable, widely distributed, abundant and cheap. On the other hand, the major technology barrier to the cost-effective production of high-purity lactic acid is its downstream separation and purification. In recent years, the possibility of the integration of highly selective membranes with conventional fermentors has opened a golden opportunity for lactic acid production. The conventional fermentors can be coupled with microfiltration [47], ultrafiltration, nanofiltration [48,49], reverse osmosis [50] and electrodialysis [51] to obtain effective separation of microbial cells, proteins, nutrients, unconverted carbon sources, water and lactic acid from the fermentation broth. The integration of such membranes with a fermentor in a single stage has failed to achieve all the goals of process intensification (PI) in lactic acid manufacturing. A judicious combination of these types of membranes and modules has the potential to achieve these goals. Efforts should be directed towards developing a fully membrane-integrated fermentative process for operation in continuous mode with continuous cell separation and recycle and continuous separation of lactic acid directly as the permeate of a properly selected nanofiltration module that overcomes the problem of membrane fouling. The integration of microfiltration membrane in the first stage, followed by a nanofiltration membrane using a flat-sheet membrane in crossflow-mode operation in the second, can help reach the target of commercial production of monomer-grade lactic acid with high productivity and concentration in an environmentally benign process [52]. The use of biocatalitic membrane reactors for lactic acid production from waste is reported as not entailing expensive raw materials, confirming the improvement of fermentation efficiency, including volumetric productivity [53,54].

8.4.3.2 Valorization of OMWW Phenolic Compounds to Produce Phytotherapic

A large variety of methods have been suggested for the treatment of wastes coming from olive oil mills, including: composting, anaerobic digestion, aerobic treatment, mixing with municipal waste water, direct land application, chemical oxidation in combination with biological treatment, adsorption and even the utilization of fungi species. However, the treatment of OMWW is extremely difficult due to its large volume and increased concentration of organic matter (biological oxygen demand (BOD) ranging between 15 and 50 g/l and chemical oxygen demand (COD) ranging as high as about 220 g/l). The major environmental factor is the phenolics compounds.

After olive oil production the phenolic compounds present in the fruit are distributed between the olive oil (2%) and waste material (98%), which is further divided into solid waste (45%) and waste water (53%), based on their different partition coefficients. The composition of the OMWW phenols is different to that of the phenols contained in the olives and in the olive oil, and is determined largely by their partition coefficient.

Oleuropein (ester 2′-(3′,4′-dihydroxypheny-1)ethanol (hydroxytyrosol)) is a hydrophilic phenol present in OMWW, in the olive tree (*Olea europaea*), in the leaves (60–90 mg/g dry weight) and in peel, pulp and seed. This compound confers resistance to pathogen attack and has been individuated as the bitter principle that must be eliminated from olives before they can be made palatable. Oleuropein is hydrolysed by the action of β-glucosidases; its hydrolysis products (elenoic acid, the isomer of oleuropein aglycon, hydroxytytrosol and dyaldeides) inhibit the activity of different pathogens such as *L. plantarum* and bacteria such as *L brevis*, *Leuconostoc mesenteroides*, *Geotricum candidum*, *Rhizopus* sp, *Phizoctonia solani* and *Pediococcus cerevisiae* involved in the lactic fermentation of green olives. The dyaldeidic form of elenoic acid linked to hydroxytyrosol, the isomer of oleuropein aglycon and hydroxytyrosol, represents the most important phenolic compound present in the oil. The former group of compounds was held to be mainly responsible for the antioxidant capacity of the olive oil [55]. Other studies demonstrated that the hydrolysis products of oleuropein have a much more marked inhibitory effect than oleuropein. In particular, the aglycon, produced by oleuropein hydrolysis, is well known as a pharmacologically active molecule. In fact, it is noted for its potential application as an antimicrobial agent in some fairly common diseases of olive trees and for its strong active antioxidant property [56–60].

Except for the isomer of oleuropein aglycon, all the oleuropein hydrolysis products are currently commercially available. The difficulty in isolating this compound is based on its low water solubility.

In order to produce the isomer of oleuropein aglycon, it is necessary to create an artificial multhipasic biocatalytic system that simulates not only the compartimentlization present *in vivo* in the olives, which separates oleuropein (substrate) from β-glucosidase (enzyme), but also the multiphasic system able to isolate the low-water-soluble compounds based on their partition coefficient. To reach this goal, β-glucosidase from almond was immobilized on polysulphone hollow-fibre membranes [61] in order to develop of a biocatalytic membrane reactor. The kinetic performance of the immobilized system was compared with a stirred tank reactor, and the results obtained demonstrated that the heterogenized enzyme has the same kinetic parameters as the batch system, maintaining the same activity for about 30 hours of continuous operation.

The biocatalytic membrane system so characterized was implemented with a membrane emulsification process (Figure 8.5). The oleuropein passed through the enzyme-loaded membrane where the reaction occurred. The water phase containing the reaction products passed through the membrane and met an organic phase recirculated into the lumen side, where water-in-oil emulsions were formed drop by drop. The droplet size and stability were adapted so that sufficient exchange area interface was achieved, and once aglycon extraction was completed the droplets broke and the phases separated spontaneously. The integrated system biocatalytic membrane reactors and membrane emulsification process (or bioreactive membrane emulsification) showed that it is possible to produce and simultaneously isolate the isomer of oleuropein aglycon in a single step.

The results obtained demonstrated that the isomer of oleuropein aglycon was produced and extracted into the organic phase of water-in-oil emulsions from OMWW with an oleuropein concentration of 0.5 mg/ml.

8.4.3.3 Production of Optically Pure Enantiomers

In the production of drugs and chemical products, the requirements for a better and more efficient catalytic methodology are concerned with process optimization, in order to obtain increased yields and energy saving, and with environmentally friendly process

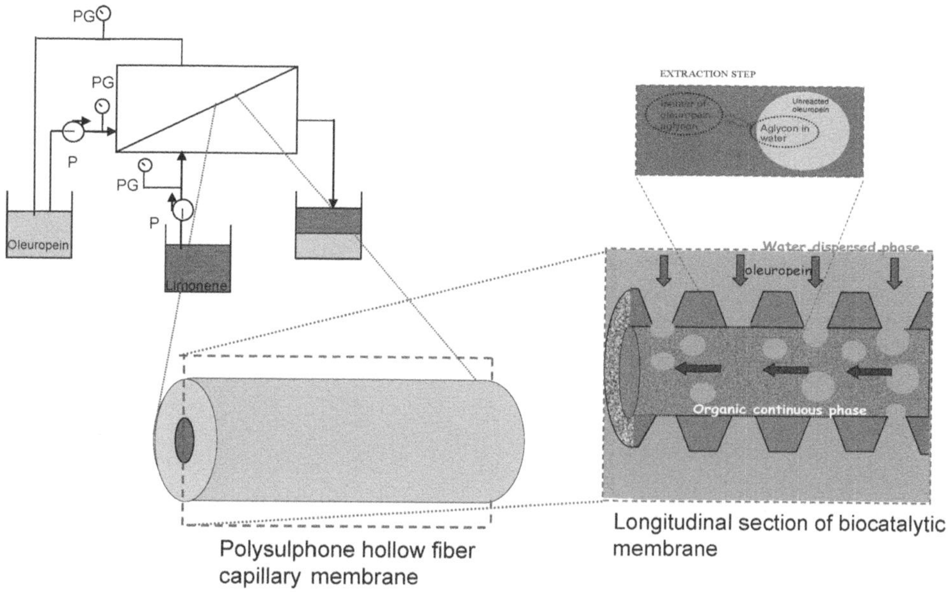

Figure 8.5 *Innovative biocatalytic membrane reactor, integrating a biocatalytic membrane reactor and a membrane emulsification process. Adapted from Mazzei et al. [62].*

development, in order to reduce waste. Compared to the classical methodologies used to separate enantiomers, such as diastereomeric and preferential crystallization, membrane bioreactor technology promotes significant advantages in terms of energy saving and environmental friendliness, in addition to greater efficiency in optically pure enantiomer production. The use of an enzymatic membrane reactor provides an innovative green technology for obtaining high-purity products without any generation of waste product.

Biocatalytic membranes using enzymes, such as lipases, in two-separate-phase systems offer a unique opportunity to produce optically pure enantiomers thanks to the enantioselective properties of the enzyme and the separation properties of the membrane.

Lipase enzymes are frequently used in biotechnology for the kinetic resolution of racemic mixtures thanks to their properties allowing discrimination between enantiomers. One of the most important aspects of lipases is their activation by organic–aqueous interface. They have been widely used in multiphase reactions (Figure 8.6a).

Giorno *et al.* [63] immobilized lipase enzyme as an oil-in-water emulsion in a two-separate-phase membrane reactor and investigated its ability to selectively convert (S)-naproxen esters from racemic mixtures. Racemic mixtures of different esters of naproxen have been used (methyl, butyl and octyl esters) to investigate the effects of –OR ester chain length on the enzyme active site enantioselectivity. The effect of the presence of emulsion on the reaction and transport rate has been also evaluated. A comparison between enzyme free in suspension and immobilized with or without emulsion has been carried out.

The two-separate-phase membrane reactor was obtained by immobilizing an oil-in-water stable emulsion, containing the enzyme at the O/W interface, within the membrane

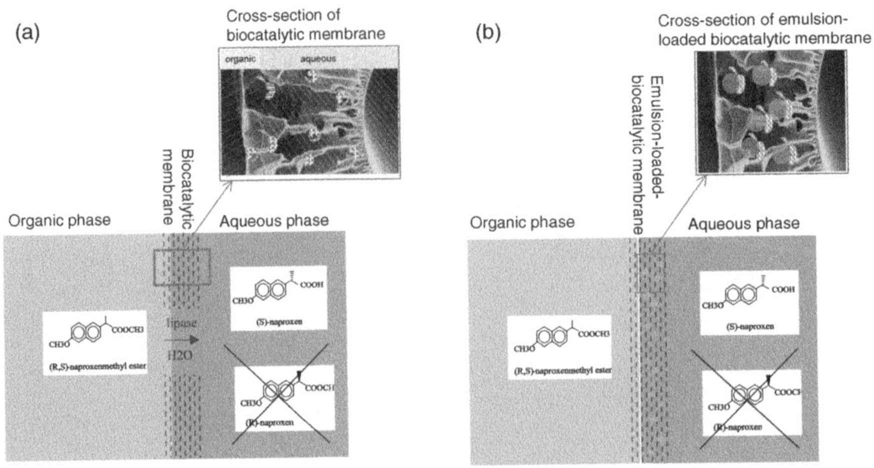

Figure 8.6 *Reaction scheme in biocatalytic membrane present in biphasic (a) and in two-separate phase membrane reactor (b). Adapted from Giorno et al. 2007.*

(Figure 8.6b). Therefore, the emulsion + enzyme-loaded membrane was used as a contactor between an organic and an aqueous phase. In this way, a configuration like organic/emulsion + enzyme-loaded membrane/aqueous phases was obtained, where the membrane contained the enzyme at the organic–aqueous interface of the immobilized oil-in-water emulsion. This condition regulates the interfacial activation of the enzyme, promotes the contact between the enzyme and the substrate, improves the transport of the organic reagent through the membrane and favours the extraction of the product into the aqueous phase as it is formed. The oil-in-water emulsion containing the enzyme was prepared by membrane emulsification [64]. The system was able to carry out enantioselective phase-transfer catalysis and simultaneous separation of the optically active reaction product, (S)-naproxen.

The reactions were carried out with the enzyme immobilized, without the presence of emulsion within the membrane pores. The use of esters with different lengths of the –OR group allowed the selectivity of the enzyme and the transport properties through the biocatalytic membrane to be investigated as a function of the reagent and micro-environment properties. The results were also compared with the enzyme used free in a stirred tank reactor.

The observed reaction rate (measured by following the product formation in the aqueous phase) was much higher for the emulsion + enzyme-loaded membrane reactor than for the enzyme-loaded membrane reactor (in absence of emulsion). The presence of the emulsion also improved the enantioselectivity of the biocatalytic membrane. For example, when using the methyl ester as a substrate, the enzyme immobilized in the presence of the emulsion showed an enantiomeric excess of 98%, while that immobilized without emulsion presented an enantiomeric excess of 75%.

The enantioselectivity of enzyme immobilized in the absence of emulsion improved with increasing –OR group length, resulting in 75, 96 and 100% for methyl, butyl and octyl ester, respectively. The enantioselectivity of enzyme free and immobilized with emulsion was maximal for all the tested esters. The productivity (mass of product/time) increased with decreasing –OR group length, as an overall effect of the chain length on the kinetic rather than the mass-transfer properties.

The stability of the immobilized enzyme was very good, showing no decay for at least 2 weeks of operation.

The overall result was that the catalytic and selective properties of the enzyme immobilized in the presence of emulsion were similar to those of the native enzyme for all of the tested substrate, and stability was much improved.

8.5 Green Processing Impact of Membrane Bioreactors

The great advantages that membrane bioreactors offer have led to great scientific interest in this research field. However, development at the industrial scale is still limited. Patent analysis (Figure 8.7) shows that the main applications are in food/biotechnology, and in particular water treatment.

The reason for the increased development in water treatment probably stems from the stringent regulation of waste water discharge in the environment, the lack of clean water and the advantages that submerged membrane bioreactors offer compared to conventional systems. As a consequence, research into and commercial application of membrane bioreactor technology are advancing rapidly around the world for both municipal and industrial waste water treatment. Market surveys of the European and the North American membrane bioreactor industries were performed by Lesjean and Huisjes [32] and Yang *et al.* [65], respectively. The detailed market analysis of the European industry shows that the industrial and municipal market has undergone a sharp increase in recent years, due to the fact that the process has become technically and economically viable with the successful introduction and commercialization of the immersed configuration [32]. The research analysis in North America shows a different behaviour to that seen in other parts of the world (Europe and Asia), with a higher ratio of industrial waste water treatment and external membrane bioreactor applications

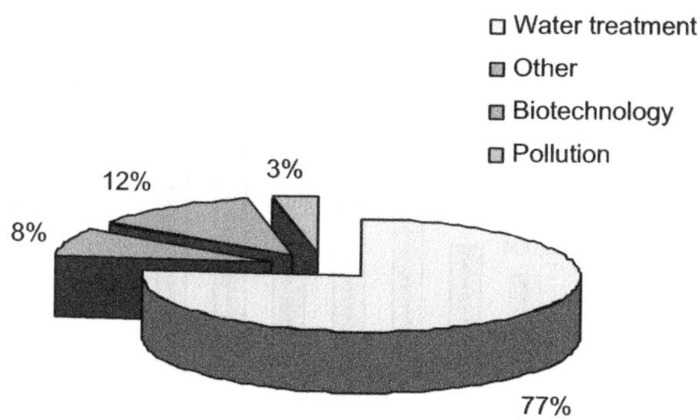

Figure 8.7 *Development of membrane bioreactor patents in different application fields.*

studied as a consequence of the more traditional and conservative North American municipalities [32].

A large number of full-scale membrane bioreactor systems are operational in various parts of the world, with about 11% located in North America. Zenon Environmental Inc. (now part of General Electric) is the major membrane bioreactor supplier in North America, while Kubota and Mitsubishi-Rayon have the most installations in other parts of the world. In Europe, Italy comes first for industrial applications, with more than 60% of installed plants, while Germany, France and the Netherlands follow with 40–50 membrane bioreactors units in total between them. The municipal market looks very different. The UK appears to be the only country with more municipal than industrial membrane bioreactor plants. In Europe, Zenon and Kubota are confirmed as the leading membrane bioreactor producer companies. The Asian markets, in particular Japan and South Korea, have embraced membrane bioreactor technology for primarily small-scale domestic applications.

The European Commission has promoted a huge research effort, investing in research and development projects focused on membranes and membrane bioreactors for waste water treatment. European companies have played a leading role in this action, in strong cooperation with the academic world. It has been estimated that about €8 million per annum has been invested in this R&D initiative, gathering about 50 companies and institutions [66], representing 10–15% of the total EU membrane bioreactor market. According to a technical market research report entitled 'Membrane Bioreactors in the Changing World Water Market', from the US-based Business Communications Co. Inc. (BCC), the global membrane bioreactor market was valued at an estimated US$216.6 million in 2006, and is rising at an average annual growth rate (AAGR) of 10.9%. This market is growing faster than the larger market for advanced waste water treatment equipment (about 5.5% AAGR) or the markets for other types of membrane system, according to the BCC study [67].

The growing number of patents published in this area confirms the fast development of membrane bioreactor technology. The number of patents increased by more than 50% from 2004 to 2010, in fact (Figure 8.8).

The demonstrated potentiality of membrane bioreactors and the urgent need to develop green technologies, supported by an increasingly restrictive legislation concerning

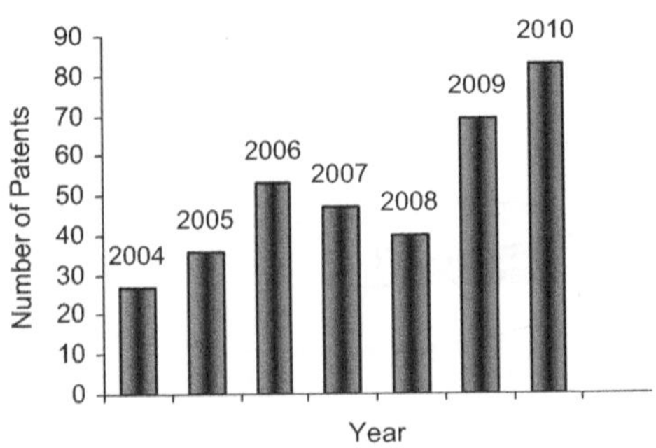

Figure 8.8 Membrane bioreactor patents filed, by year.

environmental care, will promote a future increase in the use of this technology not only for industrial waste water but also for municipal waste water. This will represent the main way to supply potable water in megacities.

8.6 Conclusion

Membrane bioreactor technology has intrinsic characteristics, such as the optimization of production processes, rational use of raw materials and energy, recycling of unavoidable waste, environmental compatibility and high selectivity, which promote its application in clean technologies.

Its main application at the industrial level is in water treatment. In the biotechnology field it is still at an emerging stage, due to some technological difficulties related to rate-limiting aspects and reproducibility at a large scale, and the lifespan, cost and the microbial contamination of the biocatalyst. The integration of advanced technologies, including molecular modelling, system engineering, molecular biology and genetics, will contribute to overcoming some crucial environmental problems related to waste water minimization and recycling, fresh water demand and alternative eco-compatible fuel production by membrane bioreactor technology.

References

(1) M. Taniguchi, N. Kotani and T. Kobayashi, J. Ferm. Technol., 65, 179 (1987).
(2) L. Giorno, E. Drioli, G. Carvoli, A. Cassano and L. Donato, Biotechnol. Bioeng., 72(1), 77–84 (2001).
(3) A. G. Livingston, J.-P. Arcangeli, A. T. Boam, S. Zhang, M. Maragon, L. M. J. Freita dos Santos, J. Membr. Sci., 151, 29–44 (1998).
(4) A. Splendiani, C. Nicolella and A. G. Livingston, Biotechnol. Bioeng., 83(1), 8–19 (2003).
(5) S. Velizarov, J. G. Crespo and M. A. Reis, Biotechnol. Prog., 18, 296–302 (2002).
(6) J. G. Crespo, S. Velizarov and M. A. Reis, Curr. Opin. Biotechnol., 15, 463–468 (2004).
(7) L. De Bartolo, S. Salerno, S. Morelli, L. Giorno, M. Rende, B. Memoli, A. Procino, V. E. Andreucci, A. Bader and E. Drioli, Long-term maintenance of human hepatocytes in oxygen-permeable membrane bioreactor. Biomaterials, 27, 4794–4803 (2006).
(8) L. De Bartolo, S. Salerno, E. Curcio, A. Piscioneri, M. Rende, S. Morelli, F. Tasselli, A. Bader and E. Drioli, Human hepatocyte functions in a crossed hollow fiber membrane bioreactor. Biomaterials, 30, 2531–2543 (2009).
(9) D. Brady and J. Jordaan, Advances in enzyme immobilization. Biotechnol. Lett., 21, 1639–1650 (2009).
(10) L. Giorno, R. Molinari, E. Drioli, D. Bianchi and P. Cesti, J. Chem. Technol. Biotechnol., 64, 345 (1995).
(11) J. P. S. G. Crespo, M. Trotin, D. Hough and J. A. Howell, Use of fluorescence labelling to monitor protein fractionation by ultrafiltration under controlled permeate flux. J. Membr. Sci., 155, 209 (1999).
(12) Z. M. Liu, J. Dubremez, V. Richard, Q. Yang, Z.K. Xu and P. Seta, Useful method for the spatial localization determination of enzyme (peroxidase) distribution on microfiltration membrane. J. Membr. Sci., 267, 2 (2005).

(13) S. Mazzuca, L. Giorno, A. Spadafora, R. Mazzei and E. Drioli, Immunolocalization of β-glucosidase immobilized within polysulphone capillary membrane and evaluation of its activity in situ. J. Membr. Sci., 285, 152–158 (2006).
(14) H. Jiang, H. Zou, H. Wang, J. Ni, Q. Zhang and Y. Zhang, J. Chromatogr. A, 903, 77–84 (2000).
(15) C. Charcosset, Biotechnol. Adv., 24, 482–492 (2006).
(16) S. Xie, F. Svec and J. M. Frechet, J. Biotechnol. Bioeng., 62, 30 (1999).
(17) J. Krenkova and J. Foret, Electrophoresis, 25, 35–50 (2004).
(18) D. A. Butterfield, D. Bhattacharyya, S. Daunert and L. Bachas, J. Membr. Sci., 181, 29–37 (2001).
(19) M. Amounas, C. Innocent, S. Cosnier and P. Seta, J. Membr. Sci., 176, 169–176 (2000).
(20) M. Amounas, V. Magne, C. Innocent, E. S. Dejean and P. Seta, Enzyme Microb. Technol., 31, 171–178 (2002).
(21) G. Dragone, B. Fernandes, A. A. Vicente and J. A. Teixeira, Third generation biofuels from microalgae. Current Research Technology and Education Topics in Applied Microbiology and Microbial Biotechnology, 1355–1366 (2010).
(22) L. Fjerbaek, K. V. Christensen and B. Norddahl, A review of the current state of biodiesel production using enzymatic transesterification. Biotechnol. Bioeng., 102, 1298–1315 (2009).
(23) L. Giorno, N. Li and E. Drioli, Use of stable emulsion to improve stability, activity, and enantioselectivity of lipase immobilized in a membrane reactor. Biotechnology and Bioengineering, 84(6), 677–685 (2003).
(24) T. Tan, J. Lu, K. Nie, L. Deng and F. Wang, Biodiesel production with immobilized lipase: a review. Biotechnology Advances, 28, 628–663 (2010).
(25) V. Kumari, S. Shah and M. N. Gupta, Preparation of biodiesel by lipase-catalyzed transesterification of high free fatty acid containing oil from *Madhuca indica*. Energ. Fuel, 21, 368–372 (2007).
(26) K. Nie, F. Xie, F. Wang and T. Tan, Lipase catalyzed methanolysis to produce biodiesel: optimization of the biodiesel production. J. Mol. Catal. B Enzym., 43, 142–147 (2006).
(27) T. Tan, K. Nie and F. Wang, Production of biodiesel by immobilized *Candida* sp. lipase at high water content. Appl Biochem Biotechnol, 128, 109–116 (2006).
(28) A. L. Machsun, M. Gozan, M. Nasikin, S. Setyahadi and Y. J. Yoo, Membrane microreactor in biocatalytic transesterification of triolein for biodiesel production. Biotechnology and Bioprocess Engineering, 15, 911–916 (2010).
(29) J. Lu, L. Deng, R. Zhao, R. Zhang, F. Wang and T. Tan, Pretreatment of immobilized *Candida* sp. 99–125 lipase to improve its methanol tolerance for biodiesel production. J. Mol. Catal. B Enzym., 62, 15–18 (2010).
(30) P. Lozano, G. Vıllora, D. Gómez, A. B. Gayo, J. A. Sánchez-Conesa, M. Rubio and J. L. Iborra, Membrane reactor with immobilized *Candida antarctica* lipase B for ester synthesis in supercritical carbon dioxide. J. of Supercritical Fluids, 29, 121–128 (2004).
(31) C. W. Smith, D. Di Gregorio and R. M. Talcott, The use of ultrafiltration membrane for activated sludge separation. Proceedings of the 24th Annual Purdue Industrial Waste Conference, Purdue University, West Lafayette, Indiana, USA, pp. 1300–1310 (1969).
(32) B. Lesjean and E. H. Huisjes, Survey of the European MBR market: trends and perspectives. Desalination, 231, 71–81 (2008).
(33) W. Fuchs, H. Binder, G. Mavrias and R. Braun, Anaerobic treatment of wastewater with high organic content using a stirred tank reactor coupled with a membrane filtration unit. Water Res., 37, 902–908 (2003).
(34) A. Akram and D. C. Stuckey, Flux and performance improvement in a submerged anaerobic membrane bioreactor (SAMBR) using powdered activated carbon (PAC). Proc. Biochem., 43, 93–102 (2008).

(35) B.-Q. Liao, J. T. Kraemer and D. M. Bagley, Anaerobic membrane bioreactors: applications and research direction, Crit. Rev. Environ., Sci. Technol., 36, 489–530 (2006).

(36) R. Del Pozo and V. Diez, Integrated anaerobic–aerobic fixed-film reactor for slaughterhouse wastewater treatment. Water Research, 39, 1114–1122 (2005).

(37) D. Zhang, P. Lu, T. Long and W. Verstraete, The integration of methanogensis with simultaneous nitrification and denitrification in a membrane bioreactor. Process Biochemistry, 40, 541–547 (2005).

(38) P. Artiga, V. Oyanedel, J. M. Garrido and R. Méndez, An innovative biofilm suspended biomass hybrid membrane bioreactor for wastewater treatment. Desalination, 179, 171–179 (2005).

(39) I. Ivanovic, T. Leiknes and H. Ødegaard, Influence of loading rates on production and characteristics of retentate from a biofilm membrane bioreactor (BF-MBR). Desalination, 199, 490–492 (2006).

(40) T. Leiknes, H. Bolt, M. Engmann and H. Ødegaard, Assessment of membrane reactor design in the performance of a hybrid biofilm membrane bioreactor (BF-MBR). Desalination, 199, 328–330 (2006).

(41) W.-N. Lee, I.-J. Kang and C. H. Lee, Factors affecting filtration characteristics in membrane-coupled moving bed biofilm reactor. Water Res., 40, 1827–1835 (2006).

(42) L. S. Downing and R. Nerenberg, Total nitrogen removal in a hybrid, membrane-aerated activated sludge process. Water Research, 42, 3697–3708 (2008).

(43) J. H. Tay, P. Yang, W. Q. Zhuang, S. T. L. Tay and Z. H. Pan, Reactor performance and membrane filtration in aerobic granular sludge membrane bioreactor. Journal of Membrane Science, 304, 24–32 (2007).

(44) J. Zhou, F. Yang, F. Meng, P. An and D. Wang, Comparison of membrane fouling during short-term filtration of aerobic granular sludge and activated sludge. Journal of Environmental Sciences, 19, 1281–1286 (2007).

(45) J. Phattaranawik, A. G. Fane, A. C. S. Pasquier, W. Bing and F. S. Wong, Experimental study and design of a submerged membrane distillation bioreactor. Chem. Eng. Technol., 32, 38–44 (2009).

(46) W.-K. Chang, A. Y.-J. Hu, R.-Y. Horng and W.-Y. Tzou, Membrane bioreactor with nonwoven fabrics as solid–liquid separation media for wastewater treatment. Desalination, 202, 122–128 (2007).

(47) J. Sikder, C. Pereira, S. Palchoudhury, K. Vohra, D. Basumatary and P. Pal, Synthesis and characterization of cellulose acetate- polysulfone blend microfiltration membrane for separation of microbial cells from lactic acid fermentation broth. Desalination, 249, 802–808 (2009).

(48) Y. Li and A. Shahbazi, Lactic acid recovery from cheese whey fermentation broth using combined ultrafiltration and nanofiltration membranes. Appl Biochem Biotechnol, 132, 985–996 (2006).

(49) M. I. Gonzalez, S. Alvarez, F. A. Riera and R. Alvareza, Lactic acid recovery from whey ultrafiltrate fermentation broths and artificial solutions by nanofiltration. Desalination, 228, 84–96 (2008).

(50) Y. Li, A. Shahbazi, K. Williams and C. Wan, Separate and concentrate lactic acid using combination of nanofiltration and reverse osmosis membranes. Appl. Biochem. Biotechnol., 147, 1–9 (2008).

(51) S. S. Yi, Y. C. Lu and G. S. Luo, Separation and concentration of lactic acid by electro-electrodialysis. Sep. Purif. Technol., 60, 308–314 (2008).

(52) P. Pal, J. Sikder, S. Roy and L. Giorno, Process intensification in lactic acid production: a review of membrane based processes. Chemical Engineering and Processing, 48, 1549–1559 (2009).

(53) R. Ohashi, T. Yamamoto and T. Suzuki, Continuous production of lactic acid from molasses by perfusion culture of lactococcus lactis using a stirred ceramic membrane reactor. Journal of Bioscience and Bioengineering, 87, 647–654 (1999).
(54) Y.-J. Wee and H.-W. Ryu, Lactic acid production by *Lactobacillus* sp. RKY2 in a cell-recycle continuous fermentation using lignocellulosic hydrolyzates as inexpensive raw materials. Bioresource Technology, 100, 4262–4270 (2009).
(55) S.-R. Cristina, C. E. Juan, and J. W. Harry, Oleuropein and related compounds. J. Sci. Food Agric., 80(7), 1013–1023 (2000).
(56) F. Visioli, G. Bellomo, G. Montedoro and C. Galli, Antioxidant and other biological activities of olive mill waste waters. Atherosclerosis, 117, 25–32 (1995).
(57) M. H. Gordon, F. Paiva-Martins and F. Almeida, Antioxidant activity of hydroxytyrosol acetate compared with that of other olive oil polyphenols. J. Agric. Food Chem., 49, 2480–2485 (2001).
(58) P. S. Rodis, V. T. Karathanos and A. Mantzavinou, Partitioning of olive oil antioxidants between oil and water phases. J. Agric. Food Chem., 50, 596–601 (2002).
(59) F. Paiva-Martins, M. H. Gordon and P. Gameiro, Activity and location of olive oil phenolic antioxidants in liposomes. Chem. Phys. Lipids, 124, 23–36 (2003).
(60) M. N. Vissers, P. L. Zock and M. B. Katan, Bioavailability and antioxidant. Effects of olive oil phenols in humans: a review. Eur. J. Clin. Nutr., 58, 955–965 (2004).
(61) R. Mazzei, L. Giorno, E. Piacentini, S. Mazzuca and E. Drioli, Kinetic study of a biocatalytic membrane reactor containing immobilized β-glucosidase for the hydrolysis of oleuropein. J. Membr. Sci., 339, 215–223 (2009).
(62) R. Mazzei, E. Drioli, L. Giorno, Biocatalytic membrane reactor and membrane emulsification concepts combined in a single unit to assist production and separation of water unstable reaction. J. Membr. Sci., 352, 166–172 (2010).
(63) L. Giorno, E. D'Amore, R. Mazzei, E. Piacentini, J. Zhang, E. Drioli, R. Cassano and N. Picci, Journal of Membrane Science, 295, 95–101 (2007).
(64) E. Piacentini, L. Giorno, A. Figoli and E. Drioli, Comprehensive Membrane Science and Engineering. 4 Cap. 3. Membrane Emulsification, Elsevier, pp. 47–78 (2010).
(65) W. Yang, N. Cicek and J. Ilg, State-of-the-art of membrane bioreactors: worldwide research and commercial applications in North America. J. Membr. Sci., 270, 201–211 (2006).
(66) B. Lesjean, T. Leiknes, R. Hochstrat, G. Schories and A. Gonzalez Gonzalez, MBR: technology gets timely EU cash boost. Filtr. Separ., November, 20–23 (2006).
(67) S. Atkinson, Research studies predict strong growth for MBR markets. Membrane Technology, February, 8–10 (2006).

9

Reactive Distillation Technology

Anton A. Kiss

Arnhem, The Netherlands

9.1 Introduction

Most chemical processes involve reaction and separation operations that are typically carried out in different sections of the plant and use various equipment types – such as continuously stirred tank reactors (CSTRs), plug flow reactors (PFRs), batch reactors and distillation columns – operated under a wide variety of conditions [1]. Recycle streams between these operating units are often used to improve conversion and selectivity, to minimize the production of undesired byproducts, to reduce energy requirements and to improve process controllability. Recent economic and environmental considerations have encouraged industry to focus on technologies based on process intensification (PI). Reactive distillation (RD) is an integrated operation that conveniently combines reaction and separation into a single unit, allowing the simultaneous production and removal of products and thus improving productivity and selectivity, reducing energy use, eliminating the need for solvents and leading to intensified, high-efficiency systems with green engineering attributes [2,3]. Some of these benefits are achieved by using reaction to improve separation (e.g. overcoming azeotropes, reacting away contaminants, avoiding or eliminating difficult separations), while others are realized by using separation to improve reaction (e.g. enhancing overall rates, overcoming reaction equilibrium limitations, improving selectivity, removing catalyst poisons) – the maximum effect being achieved when both aspects are present [2,4,5]. However, the application of RD is somewhat limited

Process Intensification for Green Chemistry: Engineering Solutions for Sustainable Chemical Processing, First Edition. Edited by Kamelia Boodhoo and Adam Harvey.

by constraints such as: common operation range for distillation and reaction (similar temperature and pressure), proper boiling point sequence (product should be the lightest or heaviest component, and side or byproducts the mid-boiling ones) and difficulties in providing proper residence time characteristics [6].

A small number of industrial applications of RD have been around for many decades [1], but even today the RD crown is carried by the Eastman process, which reportedly replaced a methyl acetate production plant with a single reactive distillation column (RDC) using 80% less energy at only 20% of the investment costs [7,8]. Nowadays, the application with the largest number of installations is methyl tertiary butyl ether (MTBE), which is used in gasoline blending. Other esters, such as ethyl tertiary butyl ether (ETBE), tert-amyl methyl ether (TAME) and fatty acid methyl ester (FAME), are also now produced by RD [4,5,9].

Considering the large extent of the topic, this chapter aims only to give a brief overview of RD. For more detailed information, the reader is kindly directed to several books and reviews published in the last decade, covering a large range of subjects concerning RD: process synthesis, conceptual design, control, operation, optimization, equipment and industrial applications [1,2,4,5,10–15].

9.2 Principles of RD

RD is especially attractive in systems where certain chemical and phase equilibrium conditions co-exist. The reaction and distillation take place in the same zone of a distillation column, the reactants being converted with simultaneous separation of the products and recycle of unused reactants. As the products must be separated from the reactants by distillation, this implies that the products should be lighter and/or heavier than the reactants. The ideal case is where one product is the lightest and the other product is the heaviest, while the reactants are the intermediate boiling components [1]. Moreover, as both operations occur simultaneously in the same unit, there must be a proper match between the temperatures and pressures required for reaction and separation [6,16] – as clearly illustrated by Figure 9.1 (left). If there is no significant overlapping of the operating conditions of reaction and separation, the combination of reaction and distillation is not possible (e.g. a high-pressure reaction cannot be combined with a vacuum distillation). Furthermore, one must also consider that working in the limited overlapping window of

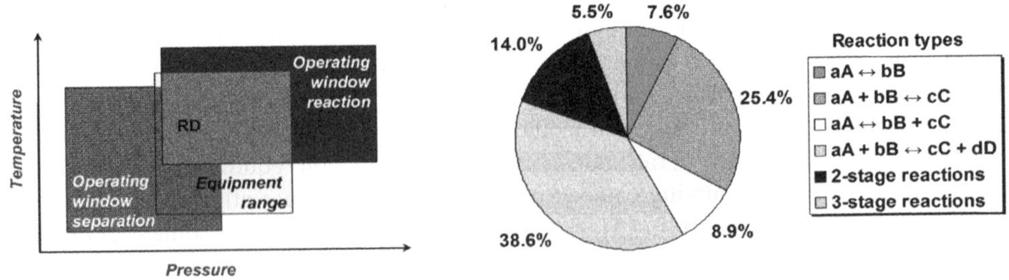

Figure 9.1 Left: overlapping of operating windows for reaction, separation and equipment. Right: reaction types used in RD [17].

operating conditions is not always the optimal solution, but merely a trade-off [16]. On the contrary, in a conventional multi-unit flow sheet, the reactors can be operated at the parameters that are most favourable for the chemical kinetics while the distillation columns can be operated at the pressures and temperatures that are most favourable for the vapour–liquid equilibrium (VLE) properties [1]. The use of residue curve maps (RCM) is recommended, as an invaluable tool for initial screening and flow sheet development of RD.

The pressure and temperature effects are much more pronounced in RD than in conventional distillation as these parameters affect both the phase equilibrium and the chemical kinetics [1,18–20]. A low temperature that gives high relative volatilities may give small reaction rates that require large amounts of catalyst or liquid holdups in order to achieve the required conversion. In contrast, a high temperature may promote undesirable side reactions or give a low equilibrium constant that makes it difficult to drive the reaction to completion [1,16].

RD is typically applied to equilibrium reactions, such as esterification, etherification, hydrolysis and alkylation. Remarkably, over 1100 articles and 800 US patents on RD were published in the past 40 years, covering in total over 235 reaction systems [1]. Figure 9.1 (right) provides a convenient overview of these systems, classified into various reaction types [17]. Most of the reactions types belong to the $a\mathrm{A} + b\mathrm{B} \rightarrow c\mathrm{C} + d\mathrm{D}$ or $a\mathrm{A} + b\mathrm{B} \rightarrow c\mathrm{C}$ or $a\mathrm{A} \rightarrow b\mathrm{B} + c\mathrm{C}$ class, with the rest falling into other categories of two- or three-stage reaction [1].

9.3 Design, Control and Applications

As compared to conventional distillation, RD sets specifications on both product compositions and reaction conversion. Consequently, the degrees of freedom (DoF) in an RDC must be adjusted to accomplish these specifications while optimizing an objective function such as the total annual cost (TAC). These design DoF include pressure–temperature, number of reactive trays, holdup of the reactive trays, location of reactant feed streams, number of stripping and rectifying trays, reflux ratio and reboiler duty [1,10].

Note that in conventional distillation the typical design specifications are the concentration of the heavy key component in the distillate and the concentration of the light key component in the bottom product. The holdup has no effect whatsoever on the steady state design of a distillation column, only on the dynamic behaviour. The column diameter is easily determined from the maximum vapour-loading correlations, after calculating the vapour rates required to achieve the desired separation. However, the holdup is very important in RD as the reaction rates depend directly on the liquid holdup and the amount of catalyst on each tray. Accordingly, RD requires an iterative design procedure, since the liquid holdup must be known before the column can be designed. This means that a tray holdup is assumed and the column is designed to achieve the desired conversion and product purities. The column diameter is then calculated, as well as the required height of liquid on the reactive trays corresponding to the assumed holdup. Note that liquid heights of over 15 cm are not recommended due to the hydraulic pressure-drop limitations. If the calculated liquid height is too large then a smaller holdup is assumed and the calculations are repeated [1].

For a long time it was assumed that reaction and distillation could be favourably combined in a column enhanced with special internals or with additional exterior volume. However,

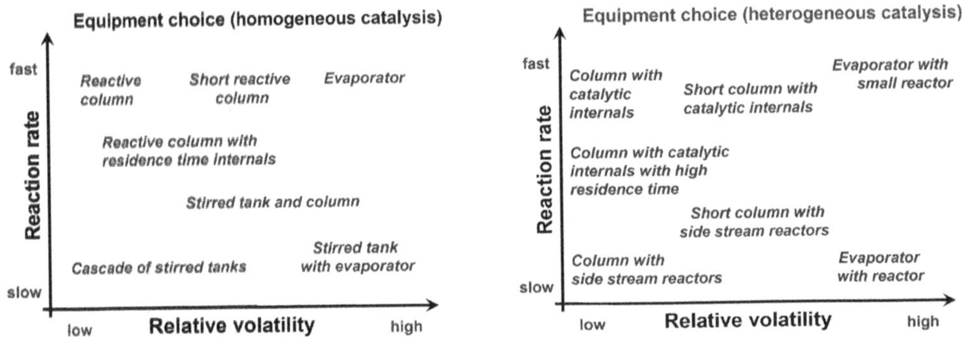

Figure 9.2 *Equipment selection in homogeneous (left) and heterogeneous (right) catalysis.*

distillation columns are an appropriate solution only for reactions that are sufficiently fast to reach high conversions within the residence time range of columns. Schoenmakers and Bessling [13] showed that two operating conditions can be distinguished: (1) distillation-controlled range, where conversion is influenced by the concentration of the components to be separated; and (2) kinetics-controlled range, where the conversion is influenced mainly by the residence time and the reaction constant. The industrial design of RD should aim at operating conditions within these two ranges, with just sufficient residence time and only the necessary expenditure for the distillation. Figure 9.2 presents the resulting principles for the choice of equipment for homo- and heterogeneous catalysis [12]. Note that in case of heterogeneous catalysis, additional separate reaction volumes are necessary in order to retain the catalyst inside the column. These volumes can be arranged either within the equipment or in a side reactor unit coupled by recycle streams.

Table 9.1 list the most important applications of RD to date [6,13,17,21]: (trans) esterification, etherification, hydrolysis, (de)hydration, alkylation, isomerization, (de) hydrogenation, amination, condensation, chlorination, nitration and so on. The most common RD configuration alternatives, from conventional RD column [12,16,22–25], to reactive dividing-wall column [26–28], and RD column combined with pre-reactor and/or side reactors [29–31], are shown in Figure 9.3 [17]. RD processes can be divided

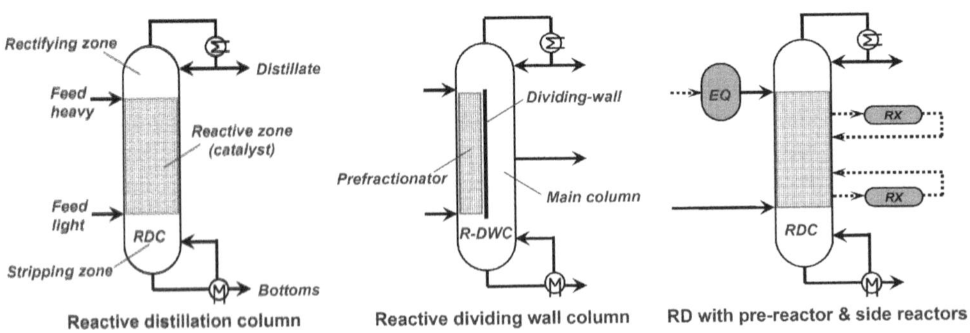

Figure 9.3 *Left: conventional RD. Middle: reactive dividing wall column. Right: RDC with pre-reactor and/or side reactors.*

Table 9.1 *Main industrial applications of RD [17].*

Reaction type	Catalyst/internals
Alkylation	
Alkyl benzene from ethylene/propylene and benzene	Zeolite β, molecular sieves
Amination	
Amines from ammonia and alcohols	H_2 and hydrogenation catalyst
Carbonylation	
Acetic acid from CO and methanol/dimethyl ether	Homogeneous
Condensation	
Diacetone alcohol from acetone	Heterogeneous
Bisphenol-A from phenol and acetone	N/A
Trioxane from formaldehyde	Strong acid catalyst, zeolite ZSM-5
Esterification	
Methyl acetate from methanol and acetic acid	H_2SO_4, dowex 50, amberlyst-15
Ethyl acetate from ethanol and acetic acid	N/A
2-methyl propyl acetate from 2-methyl propanol and acid	Katapak-S
Butyl acetate from butanol and acetic acid	Cation exchange resin
Fatty acid methyl esters from fatty acids and methanol	H_2SO_4, amberlyst-15, metal oxides
Fatty acid alkyl esters from fatty acids and alkyl alcohols	H_2SO_4, amberlyst-15, metal oxides
Cyclohexyl carboxylate from cyclohexene and acids	Ion-exchange resin bags
Etherification	
MTBE from isobutene and methanol	Amberlyst-15
ETBE from isobutene and ethanol	Amberlyst-15/pellets, structured
TAME from isoamylene and methanol	Ion-exchange resin
DIPE from isopropanol and propylene	ZSM 12, amberlyst-36, zeolite
Hydration/dehydration	
Mono ethylene glycol from ethylene oxide and water	Homogeneous
Hydrogenation/dehydrogenation	
Cyclohexane from benzene	Alumina supported Ni catalyst
MIBK from benzene	Cation-exchange resin with Pd/Ni
Hydrolysis	
Acetic acid and methanol from methyl acetate and water	Ion-exchange resin bags
Acrylamide from acrylonitrile	Cation exchanger, copper oxide
Isomerization	
Iso-parafins from n-parafins	Chlorinated alumina and H_2
Nitration	
4-nitrochlorobenzene from chlorobenzene and nitric acid	Azeotropic removal of water
Transesterification	
Ethyl acetate from ethanol and butyl acetate	Homogeneous
Diethyl carbonate from ethanol and dimethyl carbonate	Heterogeneous
Vinyl acetate from vinyl stearate and acetic acid	N/A
Unclassified reactions	
Monosilane from trichlorsilane	Heterogeneous
Methanol from syngas	$Cu/Zn/Al_2O_3$ and inert solvent
DEA from monoethanolamine and ethylene oxide	N/A

MTBE, methyl tertiary butyl ether; ETBE, ethyl tertiary butyl ether; TAME, tert-amyl methyl ether; DIPE, diisopropyl ether; MIBK, methyl isobutyl ketone; ZSM-5, Zeolite Socony Mobil 5.

into homogeneous (autocatalysed or homogeneously catalysed) and heterogeneous processes catalysed by a solid catalyst (catalytic distillation). Note that the rate of autocatalytic reactions can only be influenced by the temperature or pressure of the RD equipment. Homogeneous catalysis allows influence of the reaction rate by changing the concentration of the catalyst, enabling it to be adapted over a wide range to the needs of the RD equipment. While homogeneous catalysis is more flexible, it requires additional expensive separation steps for catalyst recovery. Heterogeneous catalysis requires a special construction (catalyst packed in 'tea bags' on trays, or sandwiched in structured packing such as Sulzer Katapak) to fix the catalytic particles in the reactive zone, thus limiting the catalyst concentration that can be achieved. The reaction rate can be enhanced only to the limit set by the attainable concentration range. Heterogeneous catalysis is simpler in principle, but it requires more equipment volume and suffers from an increased operating temperature and limited catalyst lifetime [13].

RD provides an excellent example of the ever-present interaction between design and control. Both steady state and dynamic aspects must be considered at all stages of the development and commercialization of a chemical process: laboratory, pilot plant and production [1,32]. Typically, the controllability of a RDC is improved by adding more reactive trays, while taking into account the conflict between the steady state design and dynamic controllability [1]. In neat operation mode, the reactants are fed according to the stoichiometric ratio. Consequently, the control system must be able to detect any imbalance, which will inevitably result in a gradual accumulation of one of the reactants and a loss of conversion and product purities [33]. An alternative that may be more convenient, is the operation with an excess of one reactant. This makes the control of the RDC easier, but it requires the recovery and recycle of the reactant that is in excess [1].

9.4 Modelling RD

RD is a multicomponent process and hence it is qualitatively more complex than similar binary processes. Thermodynamic and diffusional coupling in the phases and at the interface are accompanied by complex chemical reactions [6]. Consequently, RD models must take into consideration the column hydrodynamics, mass-transfer resistances and reaction kinetics. The complexity of an RD model depends of its application and the problem to be solved: plant design, model-based control and real-time optimization. A process model consists of submodels for mass transfer, reaction and hydrodynamics of various complexities. The mass transfer between the vapour and liquid phase can be described by the rigorous rate-based approach (Maxwell–Stefan diffusion equations) or can be accounted for by the simple equilibrium (EQ) stage model, assuming thermodynamic equilibrium between the vapour and liquid phases [6].

Detailed modelling of RD has been covered by several studies [34–41]. Figure 9.4 illustrates both the equilibrium and the non-equilibrium stage models [42]. The equilibrium stage model assumes that each vapour stream leaving a tray or a packing segment is in thermodynamic equilibrium with the correspondent liquid stream leaving the same tray or segment. In the case of RD, the chemical reaction is additionally considered via reaction equilibrium equations or via rate expressions integrated into the mass and energy balances [6,42]. The use of the dimensionless Hatta number – which represents the

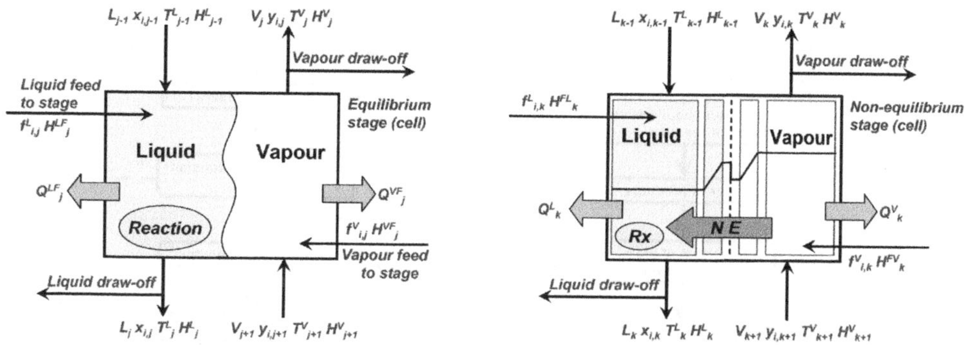

Figure 9.4 *Modelling RD: equilibrium (left) and non-equilibrium (right) stage.*

reaction rate in reference to that of the mass transfer rate – helps to discriminate between very fast, fast, average and slow chemical reactions. If a fast reaction system is considered, RD can be satisfactorily described assuming reaction equilibrium. A more physically consistent way of describing a column stage is the rate-based approach, which takes directly into account the actual rates of multicomponent mass and heat transfer and of chemical reactions [6,26,35,39,41,42]. Both approaches can be conveniently simulated in commercial process simulators, such as Aspen Plus (e.g. RADFRAC distillation unit with RateSep model), Aspen Custom Modeler [43], ChemCAD, HYSYS, ProSim, Pro II and gPROMS.

It is also important to note that the introduction of an *in situ* separation function within the reaction zone leads to complex interactions between VLE, vapour–liquid mass transfer, catalyst diffusion and chemical kinetics. Such interactions have been shown to lead to the phenomenon of multiple steady states and complex dynamics, which has been verified in experimental laboratory and pilot plant units [42,44,45].

9.5 Feasibility and Technical Evaluation

9.5.1 Feasibility Evaluation

The development of a realistic process model is the key step in evaluating the feasibility of an RD process. Moreover, a successful modelling effort requires a complete understanding of both the equilibrium and the kinetic limits of the process. Thery *et al.* [16] give an overview of the state-of-the-art methods available in the literature for the feasibility analysis and design of RD processes. A recent approach to the feasibility evaluation of RD, proposed by Kiss *et al.* [46], is based on fundamental explanations and industrial applications of RD reported in the scientific literature (Figure 9.5). In order to perform this evaluation, some basic information on the chemical process is required, such as: VLE, stoichiometry of reactions, kinetics and enthalpy of reactions. Note that the boundary values mentioned in this framework should only be taken as guideline figures and should not be interpreted as very strict limits of the RD process.

The first step is to check the number of products and the reaction type. If there is only one product present, the last step of the main reaction is irreversible, and when there are no side

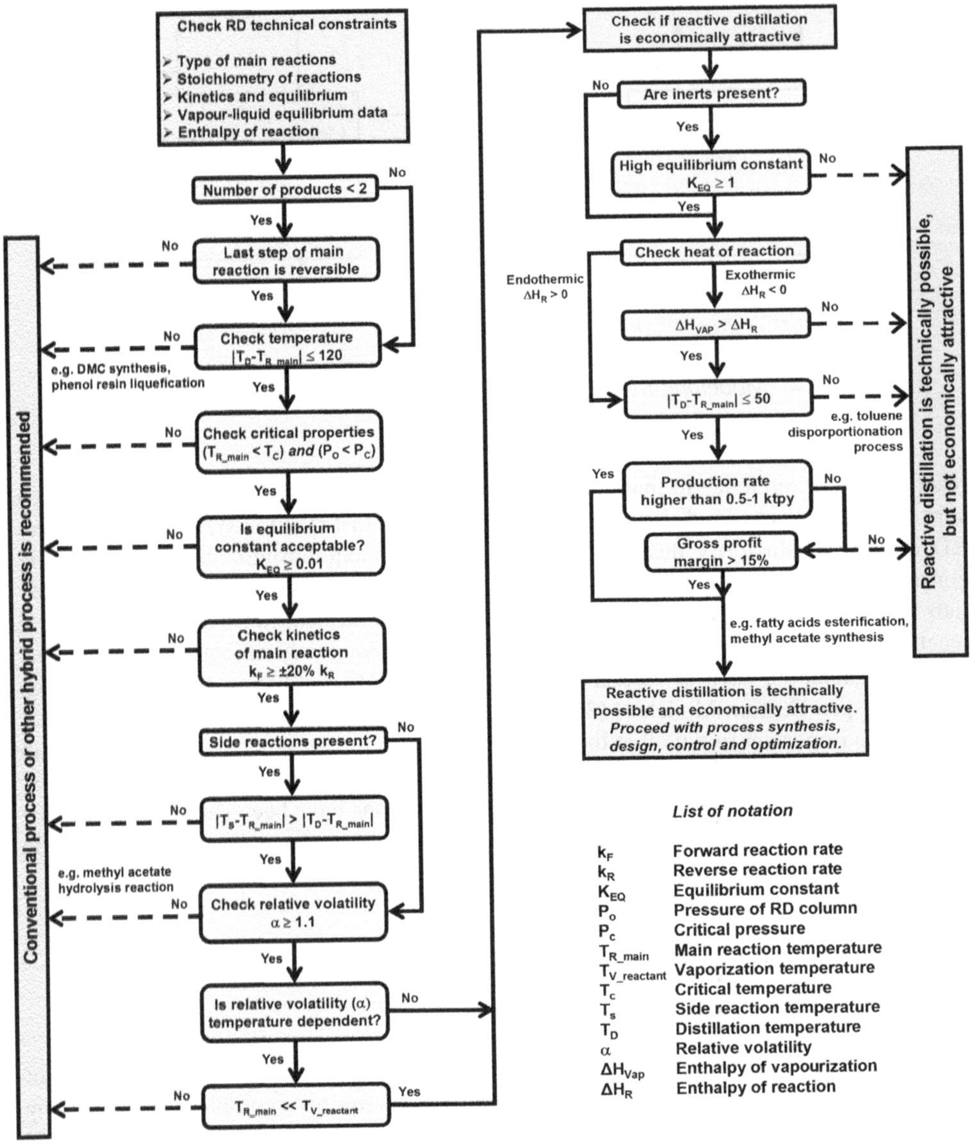

Figure 9.5 *Framework for checking whether RD is an attractive option. Reprinted from Kiss et al. © 2011, with permission from Elsevier.*

reactions there is no advantage to using RD over a simple reactor [37]. One must not forget also that the main advantages of RD rely on overcoming the equilibrium limitations and enhancing selectivity towards the desired product [4,42].

As both operations occur simultaneously in the same unit, there must be a proper and significant match between the temperatures required for reaction and separation [42], or else the combination of reaction and distillation is not possible (e.g. a high-pressure reaction cannot be combined with a vacuum distillation). Moreover, one must also keep in mind that working in the limited overlapping window of operating conditions is not always the optimal solution, but merely a trade-off. For a feasible RD process the temperature difference between separation and main reaction should be less than about 50 °C. In the case of the conventional hydro-dealkylation (HDA) process the temperature difference between the reaction and the separation process is 120 °C. Consequently, although RD has been found to be technically applicable, it is not sufficiently attractive from an economical viewpoint [46]. Moreover, the operating pressure and temperature should not be close to the critical region of the key components, since this might potentially lead to one supercritical phase [47]. If the column operates at the critical pressure of key components, these will be present in the vapour – in the vast majority of RD processes, the reaction takes place in liquid phase [4].

The temperature dependence of the vapour pressure of individual components can result in decreased relative volatility as temperature increases in multicomponent systems. This can create a mismatch of favourable temperature for kinetics and relative volatilities which can make the RD process unattractive [1]. A minimum relative volatility of 1.1 was chosen here, as this is the typical minimum value for distillation processes [1,10]. For instance, in the hydrolysis reaction of methyl-acetate, the reactant (MeAc) is the lightest component and it is difficult to keep it in the reactive zone.

The chemical equilibrium constant influences the reactant flow and residence time in an RDC. A low equilibrium constant will require a significant excess of reactants and a higher number of reactive trays for the RDC [47], thus resulting in an increased cost of the RD process. Consequently, a value of $K_{EQ} > 0.01$ (equivalent to an equilibrium conversion of less than 10%) is recommended in order to be economically attractive. Otherwise, the cost of recycling the excess reactant will outweigh the benefits of integrating reaction and separation.

For various reasons, inerts are sometime present in many chemical processes, as well as in some RD commercial systems, where the lighter reactants are fed together with an inert. However, the presence of inerts reduces the concentration of reactants and results in lower reaction rates as well as reduced K_{EQ}. Nevertheless, certain amount of inerts can be beneficial for optimum conversion – for example, in MTBE production, n-butene serves as a coolant for the reactive zone, thereby keeping the temperature of the reactive zone at a level where the equilibrium is favourable for MTBE conversion. Note that in RD processes, the specific reaction rate for the main reaction cannot be too low as this will require large liquid holdups, huge amounts of catalyst on each reactive tray and eventually a larger column [47]. Therefore, the reaction rate for the main reaction should be higher than 10^{-5} kmol/kgcat.sec, as for example in the hydrolysis of methyl-acetate. In RD processes, the desired column temperature should be selected such that the secondary reactions are minimized while the productivity is maintained sufficiently high. Moreover, the heat of the reaction should be lower than the heat of vaporization of key components. A higher reaction heat results in the drying out of trays and reduced conversion.

The last criterion is checking of the production rate – if it is above 0.5–1.0 ktonnes/year then the proposed RD process becomes economically attractive. In the case of lower production rates, it is important to also evaluate the gross profit of the process. A gross profit higher than 15% for small-scale production is also suitable for RD process application (e.g. in the pharmaceuticals industry). Ultimately, only if all conditions are fulfilled can the RD process be considered feasible and possible economically attractive – such as fatty acids esterification [33,48,49,59] – and the procedure continue with the detailed technical evaluation.

9.5.2 Technical Evaluation

The model of an RD process must account for the phase equilibrium, the rate of reaction and mass transfer in both the bulk and the film. The equilibrium behaviour of most RD systems varies between two boundaries, corresponding to phase equilibrium and chemical equilibrium control. An RCM derived from the chemical and phase equilibrium for a reactive and nonreactive section of the RDC represents variations of the stationary points – corresponding to the pure components and azeotropes – between equilibrium boundaries for a given temperature and pressure. Variations in these singular points can be represented by the Damkohler number (*Da*), defined as:

$$Da = \frac{H_0 k_f}{V} \tag{9.1}$$

where H_0 is the liquid holdup (mol), k_f is a pseudo first-order rate constant (1/s) and V is the vapour rate (mol/s). The Damkohler number is the ratio of characteristic residence time (H_0/V) to characteristic reaction time ($1/k_f$). For low values of Damkohler number ($Da \leq 0.1$) the reaction rate at each stage is relatively slow compared to the residence time available, and the system is dominated by phase equilibrium. For large values of Damkohler number ($Da > 10$) the reaction rate is fast and chemical equilibrium is approached at the reactive stages. If the Damkohler number does not lie between these values then neither the phase equilibrium nor the chemical equilibrium is controlling – hence the process is in fact kinetically controlled [50].

Moreover, a combination of the Damkohler number and the chemical equilibrium constant can be used to perform a preliminary screening of the suitability of the process for an RD application. RD is significantly beneficial when the process demonstrates combination of low *Da* ($Da \leq 0.1$) and high K_{EQ} ($K_{EQ} > 1$) or high *Da* ($Da > 0.1$) and low K_{EQ} ($K_{EQ} \leq 1$). The combination of low *Da* and high K_{EQ} represents a slow forward reaction, but a slower reverse reaction leads to high product formation. RD offers benefits as long as the required holdup is not too large. The combination of high *Da* and low K_{EQ} represents a fast product formation, but a fast reverse reaction also leads to little product formation. RD is beneficial because the product can be removed quickly from the reactive zone, shifting the equilibrium to the product side. Since the product removal rate solely depends on the rate of mass transfer between phases, the process must not be mass transfer controlled.

The combination of high *Da* and high K_{EQ} represents a fast forward reaction and a slower reverse reaction, leading to a situation of instantaneous irreversible reaction, and therefore a simple reactor is sufficient to carry out the process. For this class of process, RD can only be useful compared to a simple reactor when the higher selectivity of the main reaction over the side reaction is noticed. The reactive stage can be assumed to be at chemical equilibrium, and such a reactive stage can be modelled as a VLE stirred tank reactor (STR). RD is not beneficial when the process demonstrates a combination of low

Da and low K_{EQ}; this represents a slow forward reaction and fast reverse reaction, which essentially leads to no product formation. This class of process requires an optimally designed reactor with a large holdup.

This technical feasibility analysis of RD processes and process limitations based on *Da* and the chemical equilibrium constant is represented in Figure 9.6 [50].

The working regime of the process must be identified in order to confirm whether the process is mass transfer or kinetically controlled and whether the reaction takes place only in the bulk or also in the film. This allows both the requirements of internals for the RDC and the modelling approach that must be applied in order to design the RD process to be established. Identification of the working regime can be performed by using the Hatta number (*Ha*), pseudo first-order rate constant (k_f) and product of mass-transfer coefficient and interfacial area ($k_L a$). The Hatta number is the ratio of the maximum possible conversion in the film to the maximum diffusion transport through the film. For higher-order reactions of two components, the Hatta number is defined as:

$$Ha = \sqrt{\frac{\frac{2}{n+1} k_f C_A^{n-1} C_B D_A}{k_L^2}} \tag{9.2}$$

where n is the order of reaction (−), k_f is the forward reaction rate constant (1/s), C is the concentration (mol/m^3), D is the diffusivity (m^2/s) and k_L is the mass-transfer coefficient (m/s). A Hatta number less than one represents a slow RD process where the reactions take place in the bulk, while a Hatta number greater than one represents a fast RD process where the reaction taking place in both the bulk and the film. This identification of the working regime based on the Hatta number is illustrated in Figure 9.6. The pseudo first-order rate constant (k_f) and the product of the mass-transfer coefficient and interfacial area ($k_L a$) particularly determine whether the process operates under the slow kinetic regime ($k_f < k_L a$), slow diffusion regime ($k_f > k_L a$) or slow mixed regime ($k_f = k_L a$). Depending on the working regime, a proper selection of the internals can be conveniently performed [50]. Figure 9.6 also summarizes the internal and model requirements for different working regimes. As clearly shown in this section, the technical evaluation of any RD process can be quickly but systematically performed based only on the Damkohler number, the chemical equilibrium constant, the Hatta number, the pseudo first-order rate constant and the product of mass-transfer coefficient and interfacial area [50].

9.6 Case Studies: RD

9.6.1 Biodiesel Production by Heat-Integrated RD

Biodiesel is an alternative, renewable and biodegradable fuel produced mainly from green sources such as vegetable oils, animal fat or even waste cooking oils from the food industry. However, waste raw materials can contain a substantial amount of free fatty acid (FFA), up to 100%. Accordingly, the development of an efficient continuous biodiesel process is required, in which the use of a solid catalyst is especially desirable in order to suppress the costly processing and waste treatment steps. Note that the common problem of all

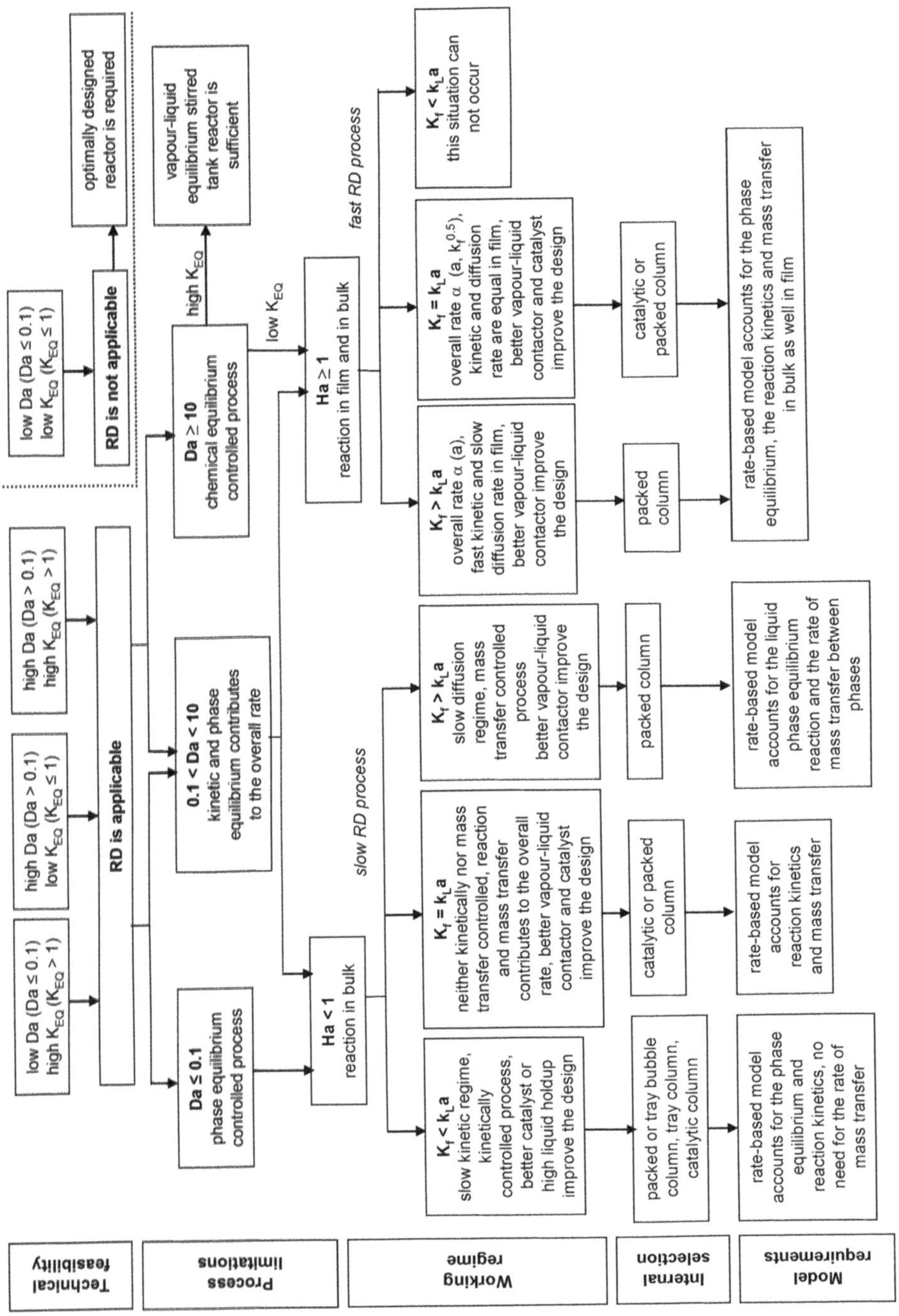

Figure 9.6 *Systematic framework for the technical evaluation of RD processes. Reprinted from Kiss © 2012, with permission from Elsevier.*

conventional processes is the use of liquid catalysts that require neutralization and an expensive multistep separation which generates salt waste streams.

To solve these problems, we use solid acids applied in an esterification process based on catalytic RD. Such an integrated process is able to shift the chemical equilibrium to completion and preserve the solid catalyst activity by continuously removing the products, while also leading to lower investment and operating costs [21,33,49,51,59].

The integrated RD process was designed according to previously reported process synthesis methods for reactive separations [6,52]. Rigorous simulations embedding experimental results were performed with Aspen Plus [53]. The RDC was simulated using the rigorous RADFRAC unit with the RateSep (rate-based) model and considering three phase balances.

The physical properties required for the simulation and binary interaction parameters for the methanol–water and acid–ester pairs were available in the Aspen Plus database of pure components, while the other interaction parameters were estimated using the UNIFAC-DMD group contribution method. The fatty components were conveniently lumped into one fatty acid and its FAME according to the reaction $R\text{-}COOH + CH_3OH \leftrightarrow R\text{-}COO\text{-}CH_3 + H_2O$. Dodecanoic (lauric) acid/ester was selected as the lumped component due to the availability of experimental results, kinetics and vapor–liquid–liquid equilibrium (VLLE) parameters for this system [48,54]. The lumping of components is very reasonable because fatty acids and their corresponding esters have similar properties. This approach has already been reported to have been successfully used to simulate other fatty ester production processes [21,27,48].

In this case study, sulfated zirconia is considered as the solid acid catalyst due to its good performance and available kinetics [21,48]. The esterification reaction is a second-order reversible reaction, so the reaction rate accounts for both direct and reverse reactions.

$$r = (k_1 W_{cat})\, C_{Acid}\, C_{Alcohol} - (k_2 W_{cat})\, C_{Ester}\, C_{Water} \qquad (9.3)$$

k_1 and k_2 are the kinetic constants for the direct (esterification) and reverse (hydrolysis) reactions, W_{cat} is the amount of catalyst and $C_{component}$ is the molar concentration of the components present in the system. As water is continuously removed from the system, the reverse hydrolysis reaction is extremely slow and hence the second term of the reaction rate can be practically neglected.

The conceptual design of the process is based on an RDC that integrates the reaction and separation steps into a single operating unit capable of shifting the reaction equilibrium towards product formation by the continuous removal of reaction products, instead of using an excess of reactant. An additional flash and a decanter are used to guarantee the high purity of the products. An RDC consists of a core reactive zone completed by rectifying and stripping separation sections, whose extent depends on the separation behaviour of the reaction mixture. Since methanol and water are much more volatile than the fatty ester and acid, these will separate easily in the top. Figure 9.7 (top) presents the flow sheet of this biodiesel process based on conventional RD, as reported by Kiss et al. [21].

The reference flow sheet presented in Figure 9.7 (top) is relatively simple, with just a few operating units, two cold streams that must be preheated (fatty acid and alcohol) and two hot streams that must be cooled down (top water and bottom fatty esters). Therefore, the heat integration is performed by applying previously reported heuristic

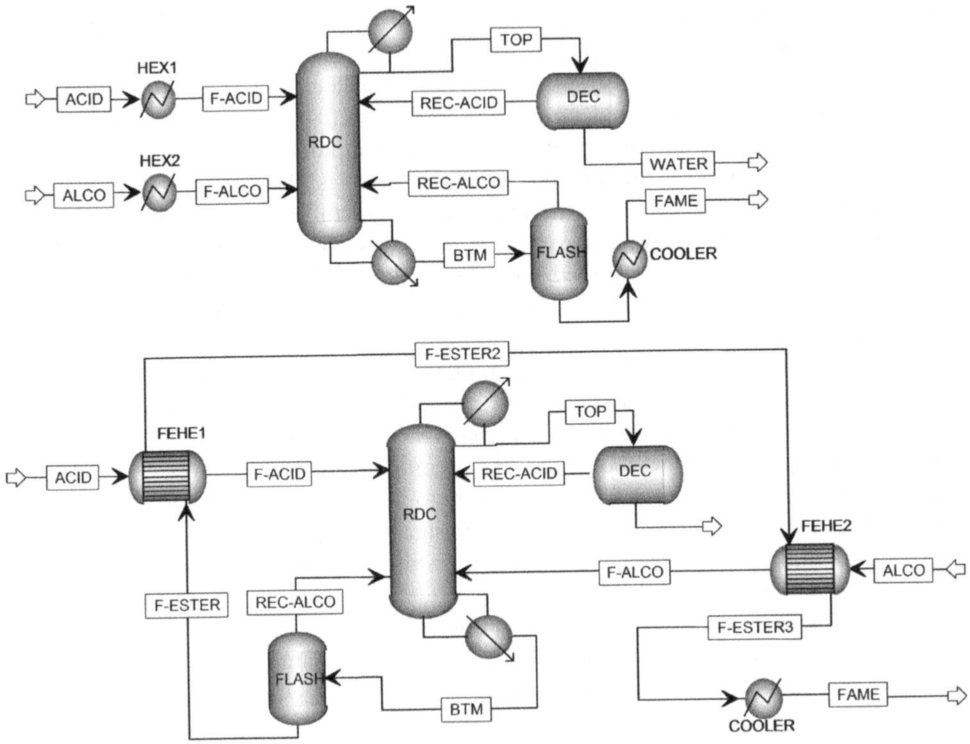

Figure 9.7 Synthesis of fatty esters by RD: base case flow sheet (top) and heat-integrated RD flow sheet (bottom). Reprinted from Kiss © 2011, with permission from Elsevier.

rules [55,56]. A feed–effluent heat exchanger (FEHE) should therefore replace each of the two heat exchangers, HEX1 and HEX2. Figure 9.7 (bottom) illustrates the improved process design, including heat integration around the RDC [33]. The hot bottom product of the column (FAME) is used to preheat both reactants: the fatty acid and alcohol feed streams. Remarkable, there is no longer any need for an external hot utility to preheat the reactant feed streams, and no additional heat exchanger is required by this heat-integrated setup.

The main design parameters are listed in Table 9.2 [33]. High conversion of the reactants is achieved, with the productivity of the RD unit exceeding 20 kg fatty esters per kg catalyst hour. and the purity specifications over 99.9 wt% for the final biodiesel product (FAME stream). Figure 9.8 (top) shows the liquid and vapour composition, as well as the reaction rate and temperature profiles along the RDC. The RDC is operated in the temperature range 70–210 °C, at ambient pressure. As the reaction takes place mainly in the reactive zone, the reaction rate exhibits a maximum in the middle of the column. The concentration of water increases from the bottom to the top of the column, while the concentration of fatty ester increases from the top to the bottom. Therefore, in the top of the reactive separation column there is mainly water with negligible amounts of fatty acids, while in the bottom there is liquid fatty ester product (biodiesel) with a very limited amount of methanol.

Table 9.2 Design parameters simulating the heat-integrated RDC. Reprinted from Kiss © 2011, with permission from Elsevier.

Parameter	Value	Units/Remarks
Total number of theoretical stages	15	Reactive from 3 to 12
Column diameter	0.4	m
HETP	0.5	m
Valid phases	VLL	—
Volume liquid holdup per stage	18	l
Mass catalyst per stage	6.1	kg
Catalyst bulk density	1050	kg/m^3
Fatty acid conversion	>99.99	%
Fatty acid feed, on stage 3 (liquid, at 145 °C)	1167	kg/hour
Methanol feed, on stage 10 (liquid, at 65.4 °C)	188	kg/hour
Reboiler duty	136	kW
Condenser duty	−72	kW
Reflux ratio (mass ratio R/D)	0.10	kg/kg
Boil-up ratio (mass ratio V/B)	0.12	kg/kg
Production of biodiesel (FAME)	1250	kg/hour
RD column productivity	20.4	kg FAME/kg cat hour
Specific energy requirements	108.8	kW hour/ton ester

FAME, fatty acid methyl ester; HETP, height equivalent to a theoretical plate; VLL, Vapor–Liquid–Liquid.

Although, heat-integrated RD offers major advantages, such as reduced capital investment and operating costs, no catalyst-related waste streams and no soap formation. The controllability of the process is just as important as the savings in capital and operating costs. In processes based on reactive separations, feeding the reactants according to their stoichiometric ratio is essential to achieving high product purity [48,57]. This constraint must be fulfilled not only during normal operation but also during the transitory regimes that arise due to planned production rate changes or unexpected disturbances. In spite of the high degree of integration, the heat-integrated RD process can well be controlled – a key result being an efficient control structure that can ensure the reactants ratio required for the total conversion of fatty acids and the prevention of difficult separations (for details, see [33]).

Figure 9.8 (bottom) depicts the dynamic simulation results [33]. The simulation starts from the steady state. At time $t = 1$ hour, the acid flow rate is increased by 10%, from 1168.0 kg/hour to 1284.4 kg/hour. Then at time $t = 5$ hours the acid flow rate is decreased to 1051.2 kg/hour, representing a 10% decrease with respect to the nominal value. The new production rate is achieved in about 2 hours. The purity of FAME remains practically constant throughout the dynamic regime, with the main impurity being methanol. Notably, the acid concentration stays below the 2000 ppm requirement of the ASTM D6751-08 standard (i.e. acid number < 0.50 mg KOH/g biodiesel).

It is worth noting that both design alternatives (base case and heat-integrated flow sheet) presented here are suitable for a large range of fatty acids and alcohol feedstocks. These processes based on RD have no additional separation steps and produce no waste salt streams, as water is the only byproduct.

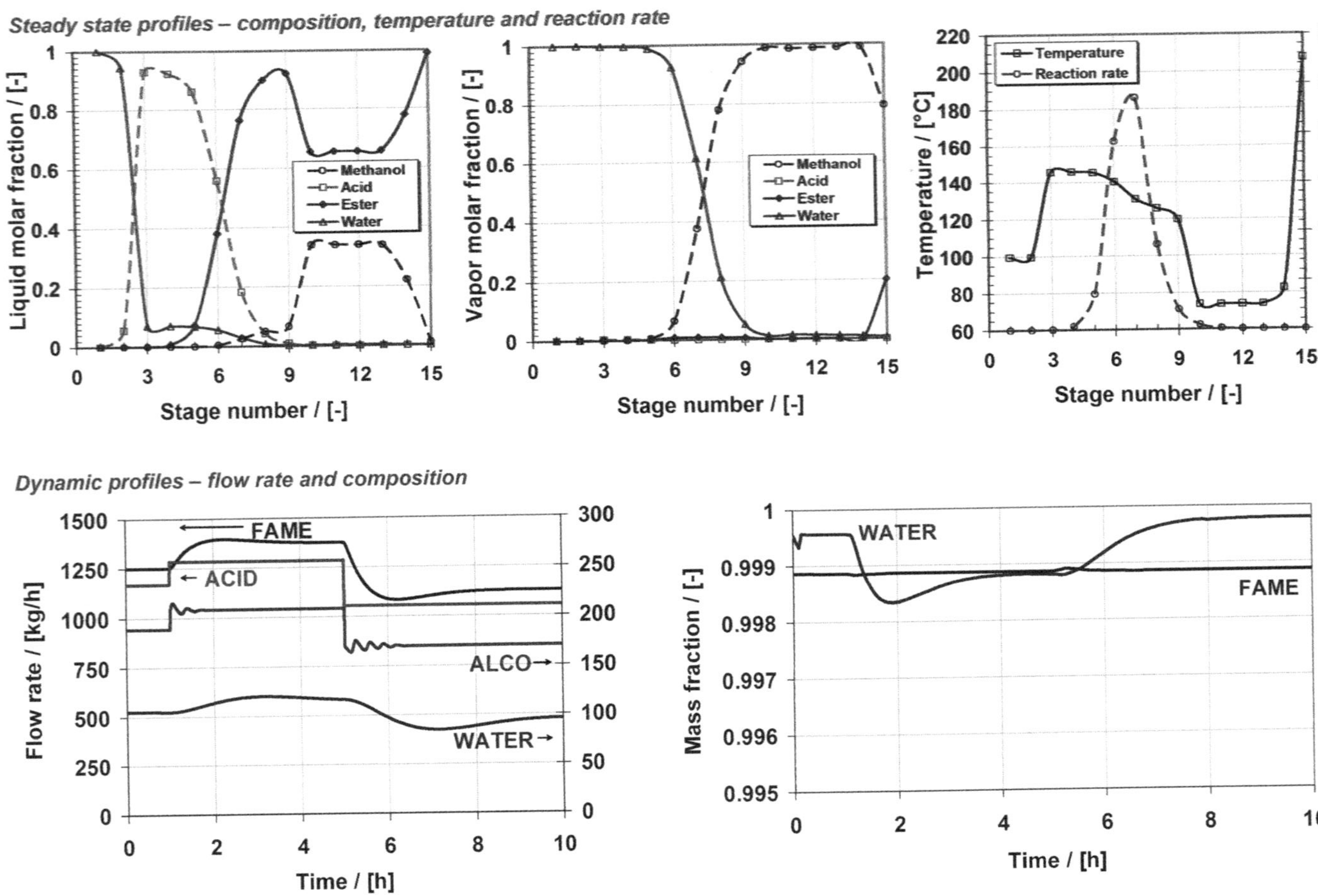

Figure 9.8 *Top: steady state profiles – liquid/vapour molar composition, temperature and reaction rate profiles along the reactive column. Bottom: dynamic simulation results – acid flow rate disturbance of +10% at 1 hour and −10% at 5 hours. Reprinted from Kiss © 2011, with permission from Elsevier.*

9.6.2 Fatty Esters Synthesis by Dual RD

Fatty esters are key products of the chemical processing industry, being incorporated in a wide variety of high-added-value products, from cosmetics to plasticizers to biodetergents. An important problem in the synthesis of fatty esters by RD is effective water removal with a view to protecting the solid catalyst and avoiding costly recovery of the alcohol excess. This case study presents a novel approach based on dual esterification of fatty acid with light and heavy alcohols, namely methanol and 2-ethylhexanol. These two complementary reactants have an equivalent reactive function but synergistic thermodynamic features. The setup behaves rather like a reactive absorption, combined with reactive azeotropic distillation using heavy alcohol as co-reactant and water-separation agent. A super acid solid catalyst based on sulfated zirconia, whose activity is comparable for the two alcohols, can be used at temperatures of 130–200 °C and moderated pressure. The control of the inventory of alcohols is realized by fixing the reflux of heavy alcohol and the light alcohol column inflow. This strategy allows both stoichiometric reactant feed rate and large flexibility in ester production to be achieved. The distillation column used to recover light alcohol from water is no longer necessary. The result is a compact, efficient and easy-to-control multiproduct reactive setup, as clearly illustrated in Figure 9.9 [48]. The other design parameters of the RDC are conveniently provided in Table 9.3 [48].

In this control structure, the reactants are fed into the process in a ratio that satisfies the overall mass balance imposed by the reaction stoichiometry and the phase equilibrium at

Table 9.3 *Design parameters simulating the dual RDC. Reprinted from Dimian* et al. *© 2009, with permission from Elsevier.*

Parameter	Value	Units/Remarks
Number of theoretical stages	25	Reactive from 5 to 24
Lauric acid feed (on stage 5, at 3.5 bar, 150 °C)	100	kmol/hour
Methanol feed (directly in reboiler, at 4 bar, 100 °C)	130	kmol/hour
2-ethyl hexanol (fed in decanter, at 3.5 bar, 130 °C)	13.48	kmol/hour
Catalyst bulk density	1050	kg/m^3
Volume holdup per stage	0.050	m^3
Mass catalyst per stage	55	kg
Reflux flow rate	2500	kg/hour
Column diameter	1.2	m
HETP	0.5	m
Fatty acid conversion	>99.99	%
Reboiler duty	1750	kW
Condenser duty	−1492	kW
Production rate	22 333	kg ester/hour
Productivity of RD column	20.3	kg ester/kg cat/hour
Bottom product composition (mass fraction)	650 ppb acid	kg/kg
	11 ppm water	kg/kg
	0.058 methanol	kg/kg
	0.788 methyl ester	kg/kg
	0.174 2EH ester	kg/kg
Specific energy requirements	166.8	kW/hour/ton ester

HETP, height equivalent to a theoretical plate.

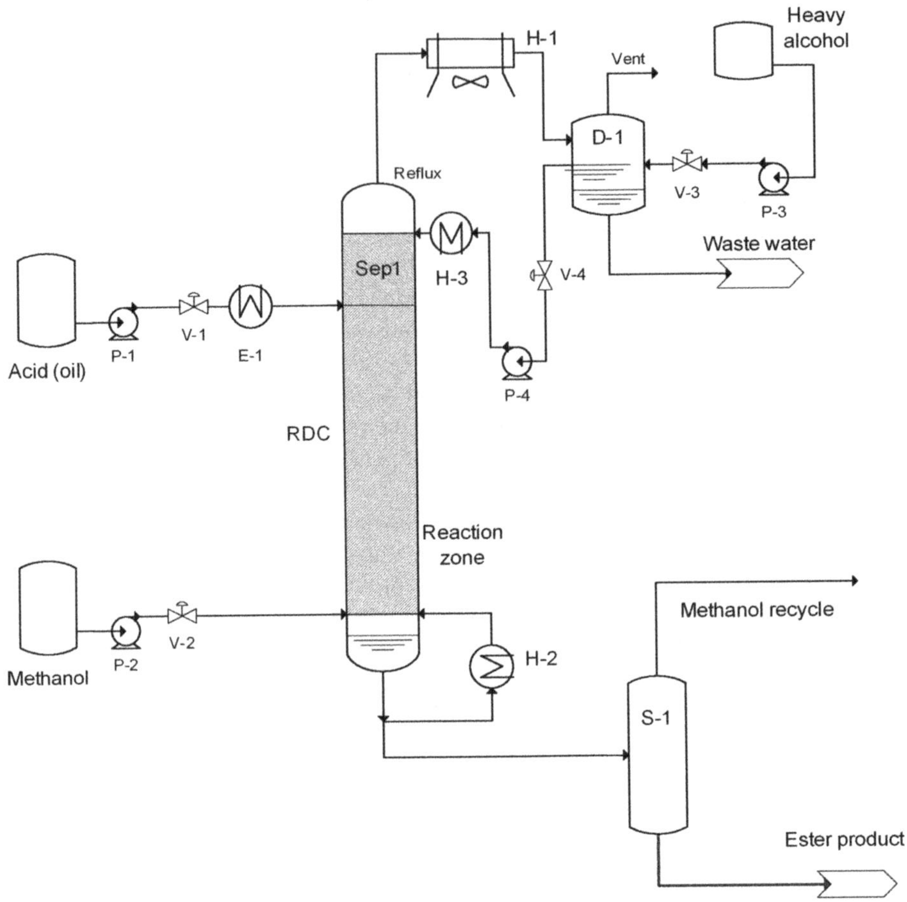

Figure 9.9 *Flow sheet configuration and control strategy of the dual RD setup. Reprinted from Dimian* et al. *© 2009, with permission from Elsevier.*

the top and the bottom of the RDC. In contrast, control structures fixing the feed rates of all reactants (acid, light and heavy alcohol) will not work in the presence of small control implementation errors, the failure manifesting in the accumulation or depletion of one reactant [58].

Figure 9.10 (top) compares the temperature profiles for the base case and for a 10% increase of the lauric acid flow rate with and without temperature control [48]. Accurate control of the lauric acid concentration in the bottom stream is achieved by using a concentration controller that prescribes the set point of the temperature controller in a cascade structure. For this control configuration, the change in the lauric acid feed flow rate leads to a change in the methyl-ester production rate. In contrast, when both flow rate ratios (i.e. lauric acid feed/methanol entering the column and lauric acid feed/heavy alcohol reflux) are constant, the change in the lauric acid feed flow rate leads to changes in both the methyl-ester and ethyl-hexyl ester production rates.

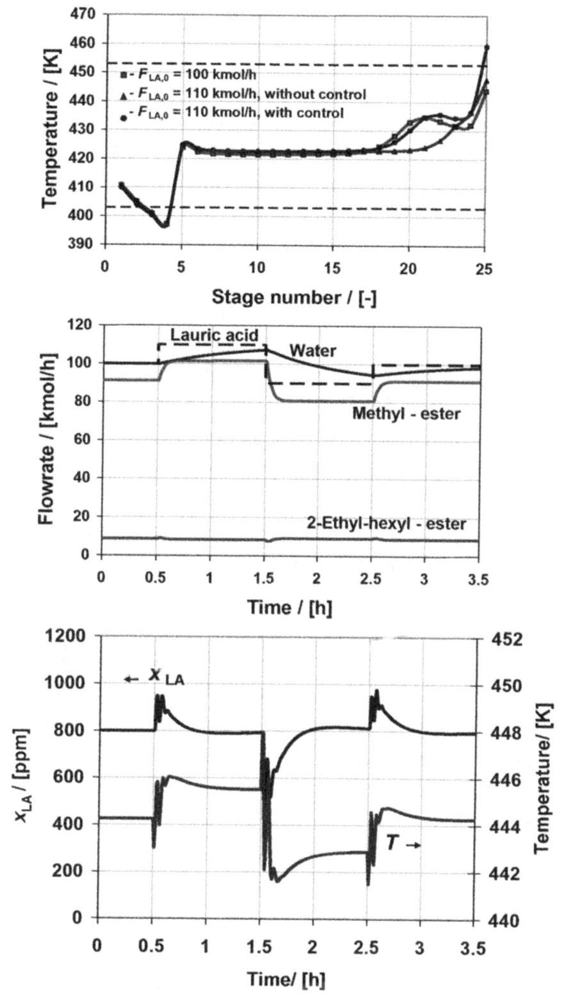

Figure 9.10 *Top: temperature profiles along the RDC. Middle: dynamic simulation results showing the increase and decrease of the production rate. Bottom: temperature and concentration of lauric acid in the bottom stream during production rate changes. Reprinted from Dimian* et al. *© 2009, with permission from Elsevier.*

Figure 9.10 (middle) presents the performance of the control system for the following scenario: the simulation starts from the steady state (feed rate of lauric acid: 100 kmol/hour), which is maintained for 0.5 hours. Then the feed rate of lauric acid is increased to 110 kmol/hour. After 1 hour it is decreased to 90 kmol/hour. Finally, the initial flow rate of 100 kmol/h is restored. The change in the acid feed flow rate leads to a change of the same magnitude in the light ester production rate, while the production rate of the heavy ester remains constant. The dynamics are fast, with only 20 minutes being necessary to achieve the new production rate. The amount of water obtained at the top of the column reflects the amount of ester formed. During the entire transient period, the concentration of water on the reactive trays remains below the 2 wt% limit [48].

Figure 9.10 (bottom) presents the temperature and the concentration of lauric acid in the bottom stream for the same scenario. Both variables remain very close to the nominal values. Notably, the tuning of the controllers is not critical with respect to the performance of the control system. In this case study, the parameters of the controllers were set as follows. The range of the controlled variable was set to the nominal value ±10 °C for the temperature control loops, and the nominal value ±50% for the level control loops. For all loops, the range for the manipulated variable was set to twice the nominal value. The gain of the feedback controllers was set to 1%/%. An integration time of 20 minutes was selected for the temperature controllers. In conclusion, the control structure presented here achieves stable operation and it is able to modify the throughput while keeping the characteristics of the products at their design values [48].

At optimal operation, the highest yield and purity can be achieved by using stoichiometric feeds in the desired ratio of fatty esters. At this point the amount of methanol lost in top is practically negligible. The heavy ester plays the role of a solvent and prevents escaping methanol in the top product. On the top stages, the heavy alcohol enhances water concentration in the vapour phase, from which it is separated by condensation and decanting, while heavy ester is produced in an amount proportional to the reflux flow rate. Optimal operation is based on controlling the inventory of reactants by using the principle of fixed recycle flows of co-reactants: in this case the reflux of the organic phase and the methanol inflow to the RDC. This strategy allows large changes in the production rate. The control strategy is generic and can be employed for esterification involving the formation of azeotropes, as for ethanol and (iso)propanol. The overall result when integrating design and process control is a compact, efficient and easy-to-control multiproduct reactive setup.

9.7 Green Processing Impact of RD

The main business drivers for RD applications are economical (over 20% reduction of variable cost, capital expenditure and energy use), environmental (lower CO_2 emissions, no or reduced salt waste) and social (improved health and safety due to reduced footprint, lower reactive content and runaway sensitivity). The industrial-scale implementation of RD by cooperation of partners in research, scale-up, design and operation can be considered a model for the rapid implementation of other PI techniques in the chemical processing industry [4]. Taking into account the remarkable progress made in hardware development, modelling for design and simulation, control strategies and real-time optimization, and considering the pace at which new applications are being explored, RD is certainly one of the most important tools for PI using green chemistry and engineering [4,5,10].

In spite of the more complex design, control and equipment, RD is a PI technology that fulfils the principles of green engineering – prevention instead of treatment, design for separation, maximization of efficiency, output-pulled versus input-pushed, meeting of needs and minimization of excess, integratation of local material and energy flows, design for a commercial afterlife, renewable versus depleting – through unique features such as a reduced number of processing units, enhanced overall rates, overcome-unfavourable equilibrium, avoidance of difficult separations, improved selectivity, reduced energy requirements and low or no solvent use [3,10].

Remarkably, for the production of biodiesel and fatty esters, the heat-integrated RD process described here eliminates all conventional catalyst-related operations, efficiently uses the raw materials and equipment and considerably reduces energy requirements and CO_2 emissions. Moreover, despite the high degree of integration, the novel process for fatty ester synthesis can be very well controlled, as demonstrated by the results of rigorous dynamic simulations [33,48].

9.8 Conclusion

RD is today an established unit operation in chemical processing technology, as well as the front-runner in the field of PI [4,5]. At present, a variety of models are available in the literature for the screening, analysis, design and optimization of RD processes: RCMs are invaluable for initial screening and flow sheet development, equilibrium models have their place in the creation of preliminary designs and non-equilibrium models are used for final designs, development of control strategies and commercial RD plant design and simulation [4,13,42].

A major focus of research and development in the future should be a methodology for choosing the most suitable equipment and for the scale-up of equipment, since great deficiencies still lie in this area. Moreover, considering the green opportunities created by the use of reactive separations, a future aim will be to spread the use of integrated RD technologies through the production of bulk or specialty chemicals, such as ethers, esters, polyesters and biofuels – as for example in the manufacture of biodiesel by the green processing of waste materials with high FFA contents [33,50].

References

(1) W. L. Luyben and C. C. Yu, Reactive Distillation Design and Control, Wiley-AIChE (2008).

(2) M. F. Malone and M. F. Doherty, Reactive distillation. Industrial & Engineering Chemistry Research, 39, 3953–3957 (2000).

(3) M. F. Malone, R. S. Huss and M. F. Doherty, Green chemical engineering aspects of reactive distillation. Environmental Science & Technology, 37, 5325–5329 (2003).

(4) G. J. Harmsen, Reactive distillation: the front-runner of industrial process intensification: a full review of commercial applications, research, scale-up, design and operation. Chemical Engineering and Processing, 46, 774–780 (2007).

(5) G. J. Harmsen, Process intensification in the petrochemicals industry: drivers and hurdles for commercial implementation. Chemical Engineering and Processing, 49, 70–73 (2010).

(6) C. Noeres, E. Y. Kenig and A. Gorak, Modelling of reactive separation processes: reactive absorption and reactive distillation. Chemical Engineering & Processing, 42, 157–178 (2003).

(7) V. H. Agreda, L. R. Partin and W. H. Heise, High-purity methyl acetate via reactive distillation. Chemical Engineering and Processing, 86, 40–46 (1990).

(8) J. J. Siirola, Industrial applications of chemical process synthesis. Advances in Chemical Engineering, 23, 1–62 (1996).

(9) A. A. Kiss and C. S. Bildea, A review on biodiesel production by integrated reactive separation technologies, Journal of Chemical Technology and Biotechnology, 87, 861–879 (2012).
(10) K. Sundmacher and A. Kienle, Reactive Distillation: Status and Future Directions, Wiley-VCH: Weinheim (2003).
(11) K. Sundmacher, A. Kienle and A. Seidel-Morgenstern, Integrated Chemical Processes: Synthesis, Operation, Analysis, and Control, Wiley-VCH: Weinheim (2005).
(12) A. Tuchlenski, A. Beckmann, D. Reusch, R. Dussel, U. Weidlich and R. Janowsky, Reactive distillation – industrial applications, process design & scale-up. Chemical Engineering Science, 56, 387–394 (2001).
(13) H. G. Schoenmakers and B. Bessling, Reactive and catalytic distillation from an industrial perspective. Chemical Engineering and Processing, 42, 145–155 (2003).
(14) A. Stankiewicz, Reactive separations for process intensification: an industrial perspective. Chemical Engineering and Processing, 42, 137–144 (2003).
(15) R. S. Hiwale, N. V. Bhate, Y. S. Mahajan and S. M. Mahajani, Industrial applications of reactive distillation: recent trends. International Journal of Chemical Reactor Engineering, 2, R1 (2004).
(16) R. Thery, X. M. Meyer and X. Joulia, Analysis of the feasibility, synthesis and conception of processes of reactive distillation: state of the art and critical analysis. Canadian Journal of Chemical Engineering, 83, 242–266 (2005).
(17) A. A. Kiss, Applying reactive distillation, NPT Procestechnologie, 19(1), 22–24 (2012).
(18) J. G. Stichlmair and J. R. Fair, Distillation, Principles and Practice, Wiley-VCH: New York (1998).
(19) M. F. Doherty and M. F. Malone, Conceptual Design of Distillation Systems, McGraw-Hill: New York (2001).
(20) H. Schmidt-Traub and A. Gorak, Integrated Reaction and Separation Operations, Springer: New York (2006).
(21) A. A. Kiss, A. C. Dimian and G. Rothenberg, Biodiesel by reactive distillation powered by metal oxides. Energy & Fuels, 22, 598–604 (2008).
(22) L. A. Serafimov, Y. A. Pisarenko and N. N. Kulov, Coupling chemical reaction with distillation: thermodynamic analysis and practical applications. Chemical Engineering Science, 54, 1383–1388 (1999).
(23) H. Subawalla and J. R. Fair, Design guidelines for solid-catalyzed reactive distillation systems. Industrial & Engineering Chemistry Research, 38, 3696–3709 (1999).
(24) O. S. Daza, E. S. Perez-Cisneros, E. Bek-Pedersen and R. Gani, Graphical and stage-to-stage methods for reactive distillation column design. AIChE Journal, 49, 2822–2841 (2003).
(25) T. Damartzis and P. Seferlis, Optimal design of staged three-phase reactive distillation columns using non-equilibrium and orthogonal collocation models. Industrial & Engineering Chemistry Research, 49, 3275–3285 (2010).
(26) I. Mueller and E. Y. Kenig, Reactive distillation in a Dividing Wall Column – rate-based modeling and simulation. Industrial & Engineering Chemistry Research, 46, 3709–3719 (2007).
(27) A. A. Kiss, H. Pragt and C. van Strien, Reactive Dividing-Wall Columns – how to get more with less resources? Chemical Engineering Communications, 196, 1366–1374 (2009).
(28) S. Hernandez, R. Sandoval-Vergara, F. O. Barroso-Munoz, R. Murrieta-Duenasa, H. Hernandez-Escoto, J. G. Segovia-Hernandez and V. Rico-Ramirez, Reactive dividing wall distillation columns: simulation and implementation in a pilot plant. Chemical Engineering & Processing, 48, 250–258 (2009).
(29) M. Klöker, E. Y. Kenig, M. Schmitt, K. Althaus, H. Schoenmakers, A. P. Markusse and G. Kwant, Influence of operating conditions and column configuration on the performance of

reactive distillation columns with liquid-liquid separators. Canadian Journal of Chemical Engineering, 81, 725–732 (2003).
(30) R. Baur and R. Krishna, Distillation column with reactive pump arounds: an alternative to reactive distillation. Chemical Engineering and Processing, 43, 435–445 (2004).
(31) D. B. Kaymak and W. L. Luyben, Optimum design of a column – side reactor process. Industrial & Engineering Chemistry Research, 46, 5175–5185 (2007).
(32) N. Sharma and K. Singh, Control of reactive distillation column – a review. International Journal of Chemical Reactor Engineering, 8, R5 (2010).
(33) A. A. Kiss, Heat-integrated reactive distillation process for synthesis of fatty esters. Fuel Processing Technology, 92, 1288–1296 (2011).
(34) J. H. Grosser, M. F. Doherty and M. F. Malone, Modeling of reactive distillation systems. Industrial & Engineering Chemistry Research, 26, 983–989 (1987).
(35) R. Taylor and R. Krishna, Multicomponent Mass Transfer, John Wiley & Sons: New York (1993).
(36) J.-H. Lee and M. P. Dudukovic, A comparison of the equilibrium and nonequilibrium models for a multicomponent reactive distillation column. Computers & Chemical Engineering, 23, 159–172 (1998).
(37) L. U. Kreul, A. Górak and P. I. Barton, Modeling of homogeneous reactive separation processes in packed columns. Chemical Engineering Science, 54, 19–34 (1999).
(38) R. Baur, A. P. Higler, R. Taylor and R. Krishna, Comparison of equilibrium stage and nonequilibrium stage models for reactive distillation. Chemical Engineering Journal, 76, 33–47 (2000).
(39) E. Y. Kenig, A. Gorak, A. Pyhalahti, K. Jakobsson, J. Aittamaa and K. Sundmacher, Advanced rate-based simulation tool for reactive distillation. AIChE Journal, 50, 322–342 (2004).
(40) Y. Egorov, F. Menter, M. Klöker and E. Y. Kenig, On the combination of CFD and rate-based modelling in the simulation of reactive separation processes. Chemical Engineering and Processing, 44, 631–644 (2005).
(41) I. Mueller, C. Pech, D. Bhatia and E. Y. Kenig, Rate-based analysis of reactive distillation sequences with different degrees of integration. Chemical Engineering Science, 62, 7327–7335 (2007).
(42) R. Taylor and R. Krishna, Modelling reactive distillation. Chemical Engineering Science, 55, 5183–5229 (2000).
(43) Aspen Technology, Aspen Dynamics – Reference & User Guide (2009).
(44) T. E. Güttinger and M. Morari, Predicting multiple steady states in distillation: singularity analysis and reactive systems. Computers & Chemical Engineering, 21, S995–S1000 (1997).
(45) G. Venimadhavan, M. F. Malone and M. F. Doherty, Bifurcation study of kinetic effects in reactive distillation. AIChE Journal, 45, 546–556 (1999).
(46) A. A. Kiss, P. Singh and C. J. G. van Strien, A systematic approach towards applicability of reactive distillation. Computer Aided Chemical Engineering, 29A, 191–195 (2011).
(47) D. B. Kaymak and W. L. Luyben, Effect of the chemical equilibrium constant on the design of reactive distillation columns. Industrial & Engineering Chemistry Research, 43, 3666–3671 (2004).
(48) A. C. Dimian, C. S. Bildea, F. Omota and A. A. Kiss, Innovative process for fatty acid esters by dual reactive distillation. Computers & Chemical Engineering, 33, 743–750 (2009).
(49) A. A. Kiss, Separative reactors for integrated production of bioethanol and biodiesel. Computers and Chemical Engineering, 34, 812–820 (2010).
(50) M. Shah, A. A. Kiss, E. Zondervan and A. B. de Haan, A systematic framework for the feasibility and technical evaluation of reactive distillation processes. Chemical Engineering and Processing: Process Intensification, 60, 55–64 (2012).

(51) A. A. Kiss, A. C. Dimian and G. Rothenberg, Solid acid catalysts for biodiesel production – towards sustainable energy.. Advanced Synthesis & Catalysis, 348, 75–81 (2006).
(52) G. Schembecker and S. Tlatlik, Process synthesis for reactive separations, Chemical Engineering & Processing, 42, 179–189 (2003).
(53) Aspen Technology, Aspen Plus: User Guide, volumes 1 & 2 (2009).
(54) A. A. Kiss, G. Rothenberg, A. C. Dimian and F. Omota, The heterogeneous advantage: biodiesel by catalytic reactive distillation.. Topics in Catalysis, 40, 141–150 (2006).
(55) Y. H. Chen and C. C. Yu, Design and control of heat-integrated reactors. Industrial & Engineering Chemistry Research, 42, 2791–2808 (2003).
(56) A. C. Dimian and C. S. Bildea, Chemical Process Design – Computer-aided Case Studies, Wiley-VCH: Weinheim (2008).
(57) C. S. Bildea and A. A. Kiss, Dynamics and control of a biodiesel process by reactive absorption, Chemical Engineering Research and Design, 89, 187–196 (2011).
(58) A. A. Kiss, C. S. Bildea and A. C. Dimian, Design and control of recycle systems by non-linear analysis. Computers & Chemical Engineering, 31, 601–611 (2007).

10

Reactive Extraction Technology

Keat T. Lee and Steven Lim

School of Chemical Engineering, Universiti Sains Malaysia, Engineering Campus, Pulau Pinang, Malaysia

10.1 Introduction

10.1.1 Definition and Description

'Reactive extraction' is generally defined as a chemical process that involves reaction and separation processes which occur simultaneously in a single processing unit. Recently, this simultaneous process has received a lot of attention as one of the more viable ways to save considerable investment and energy costs by reducing the need for separate equipment to cater for separate processes. Phase separation in a reactive system can either occur naturally, depending on the characteristics of the reaction, or be forced to occur by the deliberate introduction of a suitable solvent into the reaction [1]. Reactive extraction can be categorized into two primary categories according to the types of separation involved: liquid–liquid extraction and solid–liquid extraction. In liquid–liquid reactive extraction, a second liquid phase or solvent is usually added into the reaction system. This normally exhibits high miscibility and selectivity with the intermediates or products of the reaction, so that they can be continuously extracted from the reaction as soon as they have been produced. It must also be chemically inert and highly immiscible towards the reactants of the reaction in order to avoid the occurrence of any undesired side reactions. The removal of intermediates or products can help to prevent further reaction, which otherwise might reduce the yield of desired products [2]. The extraction can also be employed to remove

Process Intensification for Green Chemistry: Engineering Solutions for Sustainable Chemical Processing, First Edition. Edited by Kamelia Boodhoo and Adam Harvey.

unwanted impurities from the product stream, which will maintain the final product quality and eliminate the need for additional purification in the downstream process.

On the other hand, solid–liquid reactive extraction involves the extraction of materials from the solid into the liquid phase, where they undergo reaction and produce the required products. Solid biomass normally contains a large number of useful raw materials such as lipids, hydrocarbon, fatty acids and proteins. After being extracted, these can be further processed into highly valuable products in the energy and pharmaceuticals sectors. By combining the extraction from solid- and liquid-phase reaction in a single reactive extraction unit, a huge reduction in processing cost and time can be produced. Currently, solid–liquid reactive extraction is under investigation for the production of biodiesel from oil seeds biomass. Sometimes also termed '*in situ* transesterification', this places the oil-bearing solid biomass in direct contact with short-chain alcohol instead of pre-extracted oil. Extraction and transesterification proceed simultaneously, with alcohol acting as both an extraction solvent and a transesterification reagent [3]. The kinetics of solid–liquid reactive extraction can be increased through the addition of an appropriate catalyst or cosolvent, or by subjecting the reaction to supercritical conditions.

10.1.2 Literature Review

The integration of reaction and liquid–liquid extraction was first proposed in the 1960s by Piret *et al.* [4]. A reactive extraction design method was developed for both dilute and concentration systems which shows an enhancement in rate of reaction, volumetric efficiency and reactant conversion. Anderson and Veysoglu [5] employed dichloromethane as the hydrocarbon solvent to continuously extract the epoxide products from 6-methylhept-5-en-2-one as the reaction proceeded, in order to prevent them from degrading to undesirable products. Reactive extraction carried out for the hydrolysis of formate esters by King *et al.* [6] has also proven to give higher productivity and yield and reduced waste compared to conventional hydrolysis. Many bioprocessing operations, such as fermentations, which are often inhibited by high concentrations of product have taken advantage of the reactive extraction techniques to overcome their limitations [7,8]. Application of reactive extraction for the intensification of biodiesel production can help to increase product selectivity, conversion and purity through the synergistic effect between transesterification reaction and liquid–liquid extraction. Two immiscible liquid phases (biodiesel-enriched and glycerin-enriched) that are formed during the transesterification reaction can be removed separately during the process as extract and raffinate [9,10].

The idea of using solid–liquid reactive extraction in transesterification was first adopted in analytical studies in order to detect the fatty acid compositions of materials from different organisms [11]. Its application in biodiesel production was introduced by Harrington and D'Arcy-Evans [12] in their work on the *in situ* acid-catalysed esterification of sunflower seed to produce biodiesel. It was reported that the yield of fatty acid methyl ester (FAME) produced by this method was greater than that in the conventional transesterification process. This was attributed to the capability of *in situ* transesterification (using an acid catalyst) to react with lipid materials that were not extracted from the seed by hexane. Qian *et al.* [13] have conducted an experiment on the *in situ* transesterification of cottonseed oil using homogeneous alkaline catalyst. In this study, cottonseed was reacted with methanol instead of pre-extracted cottonseed oil and 98% conversion of oil was obtained under the following conditions: <2% moisture content in cottonseed flours, 0.300–0.355 mm particle size, 0.1 M NaOH concentration in methanol, methanol to oil ratio 135 : 1, 40 °C reaction temperature and 3 hours' reaction time. In addition, petroleum ether was used as cosolvent in order to speed up the extraction rate of oil from cottonseed.

10.2 Case Studies: Reactive Extraction Technology

10.2.1 Reactive Extraction for the Synthesis of FAME from *Jatropha curcas* L. Seeds

Biodiesel from *Jatropha curcas* L. seed is conventionally produced via a two-step method: extraction of oil and subsequent esterification/transesterification to FAME. We have examined a single-step *in situ* extraction: esterification and transesterification (or collectively known as reactive extraction) of *Jatropha curcas* L. seed to biodiesel. Preliminary studies on the effects of different seed sizes and moisture contents on reactive extraction were carried out. The catalyst used was sulfuric acid, H_2SO_4. Since *Jatropha* seeds were used directly as the oil source in the reactive extraction, instead of preextracted oil, it was very important to investigate the effect of seed size on the reaction in order to eliminate mass-transfer limitation before further experimental work was carried out. In addition, the effect of the seed moisture content was also studied at the beginning of the research in order to reduce irregular feedstock properties.

10.2.1.1 Material and Method

Reactor System. A transesterification reaction using reactive extraction technology was carried out in a batch-type reactor system, which consisted of a dual-neck, 500 ml, round-bottomed flask equipped with a reflux system, magnetic stirrer, thermometer and heater. The heater used was a temperature-controlled heater equipped with magnetic stirring. The thermometer was used to monitor the temperature of the reaction mixture. The reflux system consisted of a condenser with cooling water as the cooling medium. Figure 10.1 shows a schematic diagram for the batch-type reactor system setup used.

Determination of Moisture and Oil Content in Jatropha Seeds. Initially, fresh jatropha seeds were blended and sieved to less than 1 mm. They were then repeatedly weighed and dried in an oven at 76 °C until a constant weight was achieved for moisture content measurement. The dried seeds were sieved again to obtain fine particles $\leq$0.355 mm in size. In order to determine the maximum amount of oil that could be extracted from the seeds using the conventional method, a Soxhlet extractor with excess n-hexane as the solvent was utilized for 12 hours. After the extraction process, the hexane was removed using a rotary evaporator and the weight of the extracted oil was measured.

Effect of Different Seed Sizes. The blended jatropha seeds were sieved into the desired sizes: $\leq$0.250 mm, 0.250 mm $< x \leq$ 0.355 mm, 0.355 mm $< x \leq$ 1 mm and >1 mm. Then the seeds of different sizes were subjected to a drying process until they were completely

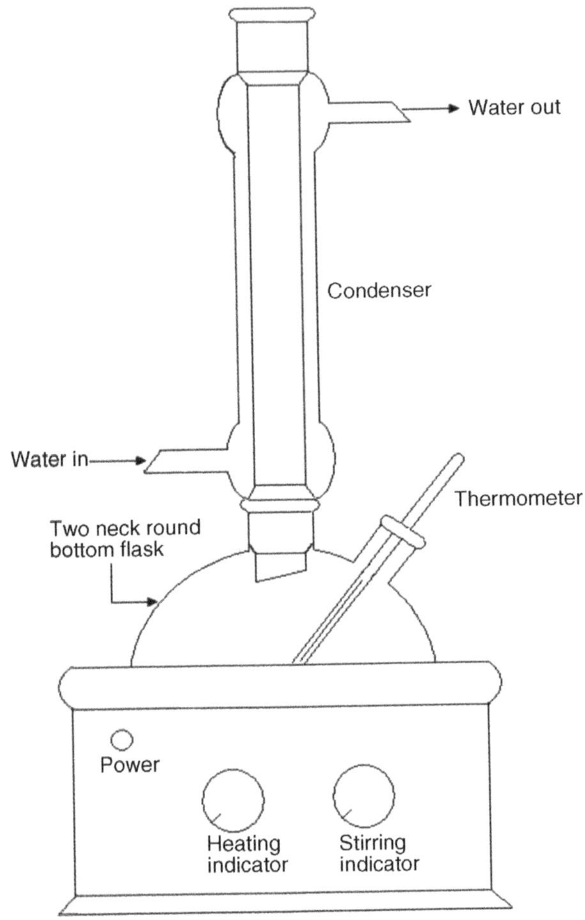

Figure 10.1 *Schematic diagram of the reactor system for an acid-catalysed reactive extraction reaction.*

dry. Next, 20 g of jatropha seeds of the desired size were loaded into the round-bottommed flask. Then 150 ml of methanol (solvent-to-seed ratio was fixed at 7.5 ml/g) and 1.65 ml of concentrated sulfuric acid, H_2SO_4 (15 wt% of seed) were added. The amount of chemicals used in this study was based on literature reported elsewhere [14,15]. After that, the reaction mixture was heated to 60 °C for different reaction periods: 0, 4, 8, 12 and 24 hours. Upon completion, the mixture was cooled to room temperature. The cooled reaction mixture was filtered before being further purified by repeated washing with solvent (depending on the type of solvent used during the reaction) and the excess solvent in the filtrate was recovered using a rotary evaporator. The yield of FAME was calculated based on the analysis result from gas chromatography (GC) for the upper layer of the samples (the layer of biodiesel). The samples were diluted with methyl heptadecanoate, which served as the internal standard (IS), before being analysed using GC (PerkinElmer, claurus 500) equipped with a flamed ionized detector (FID) and Nukol capillary column (15 m × 0.53 mm; 0.5 μm film). N-hexane was used as the solvent, while helium was used as the carrier gas. The oven temperature was set at 110 °C and then increased to 220 °C at a rate of 10 °C/minute. The temperatures of the detector and injector were set at 220 and 250 °C, respectively. The yields of FAME were then calculated using the formula in Equation 10.1:

$$\text{FAME yield (wt\%)} = \frac{Weight\ of\ FAME\ (g)}{Weight\ of\ oil\ in\ 20g\ of\ seeds\ (g)} \times 100(\%) \qquad (10.1)$$

'Oil extraction efficiency' is defined as the quantity of oil that can be extracted from the blended and sieved jatropha seeds during reactive extraction. The volume of the upper layer, formed after evaporation of the reaction filtrate, was measured and recorded to calculate oil extraction efficiency, as shown in Equation 10.2. The densities of the jatropha oil and jatropha methyl esters were considered to be equal because the difference was less than 5% at room temperature.

$$\text{Extraction efficiency (\%)} = \frac{Final\ volume\ of\ collected\ sample}{Total\ volume\ of\ oil\ in\ original\ sample} \times 100(\%) \qquad (10.2)$$

10.2.1.2 Result and Discussion

The average moisture content of *Jatropha curcas* L. seeds was found to be 5.39%. This was in agreement with those reported in the literature, with values ranging from 4.75 to 19.57% [16]. The average oil content for jatropha seeds was 54.4%, and this value also falls within the range reported in the literature [17], which states that jatropha seeds normally contain 40–60% oil, depending on variety and species.

Effect of Different Seed Sizes. For solid–liquid reactive extraction, seed size is an important factor that must be investigated in order to show that mass transfer is not a problem for FAME production using this method. The reaction was performed using 20 g of blended and dry jatropha seeds in different sizes: $\leq$0.250 mm, 0.250 mm $< x \leq$ 0.355 mm, 0.355 mm $< x \leq$ 1 mm and $>$1 mm. The reaction was performed under the following reaction conditions: reaction temperature of 60 °C, methanol to seed ratio of 7.5 ml/g, H_2SO_4 as catalyst at 15 wt% and different reaction periods of 0, 4, 8, 12 and 24 hours. The amounts of chemicals used in this study were based on numbers reported in the literature [14,15]. Figure 10.2 shows the extraction efficiency of the reactive extraction process at different reaction periods and for different ranges of seed particle size, while Figure 10.3 shows the yield of FAME. As shown in Figure 10.2, it was observed that the oil extraction efficiency gradually increased with increasing reaction period, and within 8 hours almost 50% of the oil content in the seeds had already been extracted, even for seeds with larger particle sizes. Initially, all ranges of particle size exhibited similar trends for extraction efficiency up to a reaction period of 8 hours. Beyond that, for particles with a size of either 0.355 mm $< x \leq$ 1 mm or $>$1 mm, the extraction efficiency remained almost constant, while for particles of less than

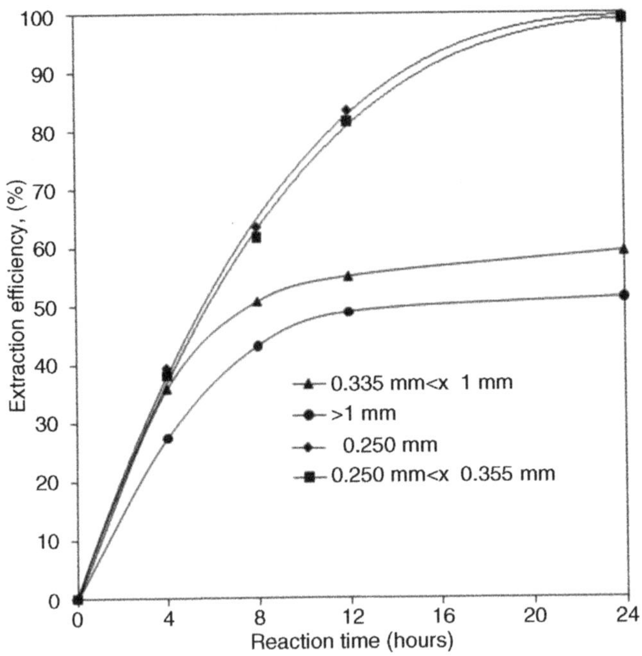

Figure 10.2 *Oil extraction efficiency for acid-catalysed reactive extraction of* Jatropha curcas *L. seeds versus reaction period at different ranges of seed size. Reprinted from Shuit et al © 2010, with permission from Elsevier.*

0.355 mm, the extraction efficiency kept on increasing, until it reached almost 100% after a reaction period of 24 hours. The likely explanation for this phenomenon is that particles with a smaller size have a larger surface area, which favours the extraction of oil from the seeds. For seeds with a larger particle size, there will be a stage at which a mass-transfer limitation stops the oil trap deep in the core from being extracted, at the point at which the reaction period is 8 hours.

Figure 10.3 shows the yield of FAME for the reactive extraction process at different reaction periods and at different ranges of seed particle size. The results showed that the yield of FAME exhibited similar trends to the extraction efficiency for all ranges of seed particle size. In reactive extraction, the yield of FAME is very much dependent on the rate of dissolution (extraction) and alcoholysis (transesterification) [13]. It is noted here that the total weight of oil in the sample used during reactive extraction is the same for all reaction periods. Therefore, since both yield of FAME and extraction efficiency exhibited a similar trend, it can be concluded that in reactive extraction, once oil is extracted from the seed, it is instantaneously converted to FAME. Therefore, the limiting factor for reactive extraction is the rate of extraction, rather than the rate of trans-esterification. The results in Figure 10.3 also show that, using seeds of size 0.250 mm $< x \leq 0.355$ mm, the yield of FAME can reach almost 100% after a reaction period of 24 hours. However, with a further decease in seed particle size there is no significant improvement in the extraction efficiency and yield of FAME.

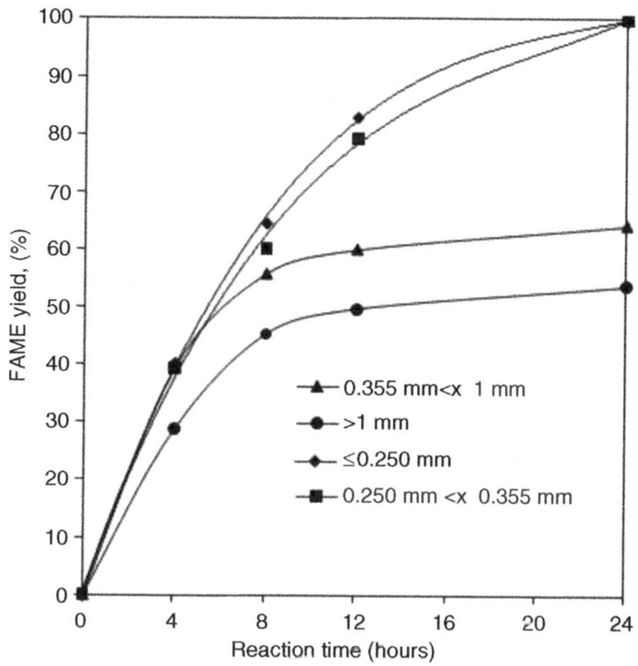

Figure 10.3 *FAME yield for acid-catalysed reactive extraction of* Jatropha curcas *L. seeds versus reaction period at different ranges of particle size. Reprinted from Shuit et al © 2010, with permission from Elsevier.*

10.2.1.3 Conclusion

In this study, acid-catalysed reactive extraction has been proven to be a promising technology for the production of biodiesel from feedstock that contains high free fatty acid (FFA), such as *Jatropha curcas* L. seeds. It has the potential to reduce the cost of production compared to conventional biodiesel production technology.

10.2.2 Supercritical Reactive Extraction for FAME Synthesis from *Jatropha curcas* L. Seeds

Noncatalytic supercritical methanol for biodiesel production had also been proven to be superior in terms of reaction time, product separation, FAME yield and process complicity compared to conventional biodiesel processing [18]. Fluid in a supercritical phase can be considered an intermediate between liquid and gas. This special state has been attributed to several distinctive characteristics, such as low viscosity, high diffusion coefficients, variation of density and dielectric constant as a function of pressure. Consequently, supercritical fluids (SCFs) are an excellent extraction solvent as well as chemical reaction reagent. As an extension to the acid-catalysed reactive extraction, the main objective of this research was to determine the feasibility of noncatalytic supercritical reactive extraction using methanol to produce biodiesel from *Jatropha curcas* L. seeds in a high-pressure batch reactor with n-hexane as cosolvent. The effects of the

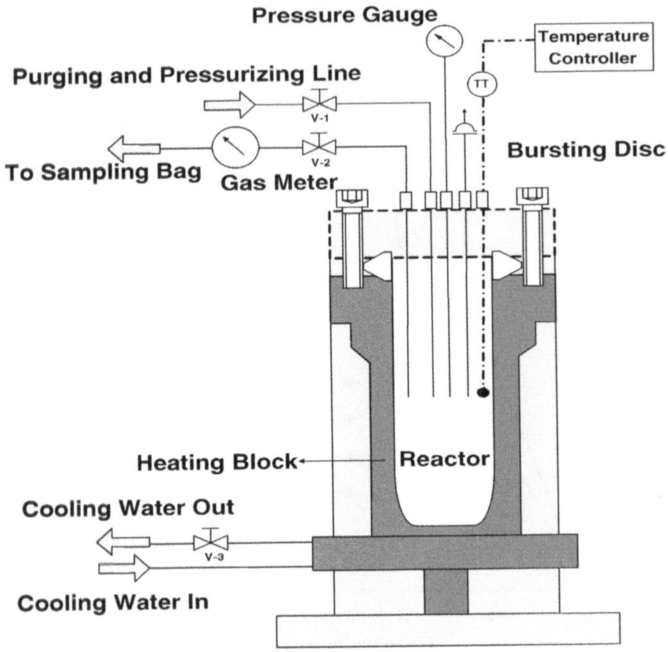

Figure 10.4 *Schematic diagram of the experimental rig for SCF reactive extraction.*

solid particle size of the jatropha seeds and the reaction temperatures on the FAME yield were studied to obtain a clearer picture of its suitability and future prospects.

10.2.2.1 Materials and Methods

Reactor System. All the experimental runs were carried out in a high-pressure 'Limbo Li' reactor manufactured by BuchiGlasUster, Switzerland, as shown in Figure 10.4. In a typical run, 20 g of blended and sieved jatropha seeds were loaded into the 450 ml reactor with 200 ml of methanol (solvent to seed ratio fixed at 10 ml/g) and 50 ml of n-hexane (2.5 ml/g cosolvent to seed ratio), according to the literature [19].

Effect of Solid Particle Size and Reaction Temperature. Methanol was to act both as an extraction agent and transesterification reagent, while the addition of n-hexane was required to increase the oil solubility in the reaction mixture and to separate the FAME from the glycerol in the later process. The reaction mixture was first subjected to mechanical agitation at 400 rpm for 30 minutes under room conditions before being heated to the desired reaction temperature. The operating pressure and reaction time, which were not studied independently, were not controlled but correlated by the reaction temperature. The heating rate was adjusted so that the final temperature was always reached in the shortest possible time (40–75 minutes), after which it was immediately quenched with cooling water to room temperature. The reaction mixture was then filtered and the solid residue was washed thrice with 30 ml of recycled n-hexane. In order to separate the FAME from glycerol and any n-hexane-insoluble byproduct from the SCF reactive extraction, the filtrate was stirred with 100 ml recycled n-hexane for 30 minutes before being transferred to a separating funnel. The upper dark yellow-coloured layer was decanted after 1 hour and excess n-hexane was evaporated to recover the FAME. The separation procedure was repeated twice for the lower dark brown-coloured layer to ensure complete separation of FAME from glycerol. The volume of the collected FAME sample was then measured and recorded for calculation using Equation 10.1.

10.2.2.2 Result and Discussion

Figure 10.5 shows the variation of temperature and pressure towards FAME yield for supercritical reactive extraction with different solid particle sizes. At lower temperatures (≤240 °C), FAME yield was extremely low, as predicted, due to the poor miscibility of methanol with the extracted oil without the addition of catalyst or supercritical conditions (above 240 °C and 8.1 MPa) [18,19]. Oil conversion to FAME was found to be directly proportional to the reaction temperature, particularly beyond the supercritical condition, where the conversion rate was intensified. Smaller particle sizes, which led to higher extraction efficiency, were favourable for higher FAME yields (with the only exception at particle size ≤0.5 mm; maximum 71.4% w/w at 300 °C), since the smaller particles, which are stickier, tend to agglomerate and thus limit FAME conversion due to their lower surface area. It can therefore be concluded that the rate of oil extraction is higher than the rate of transesterification at low temperature ranges due to the significant extraction efficiency of n-hexane. The kinetics of the transesterification process, which was low before attaining supercritical conditions, becomes the limiting factor. However, at higher temperature and pressure ranges, supercritical conditions enable the transesterification to proceed at a much higher pace than the extraction of oil [18]. Thus, jatorpha oil will be converted to FAME as soon as it is extracted from the seeds, and the kinetics of extraction becomes the limiting factor. The 103.5% FAME yield, exceeding the theoretical 100.0% yield for particle size ≤1 mm at 300 °C, was due to the higher pressure exerted under supercritical conditions, which was able to

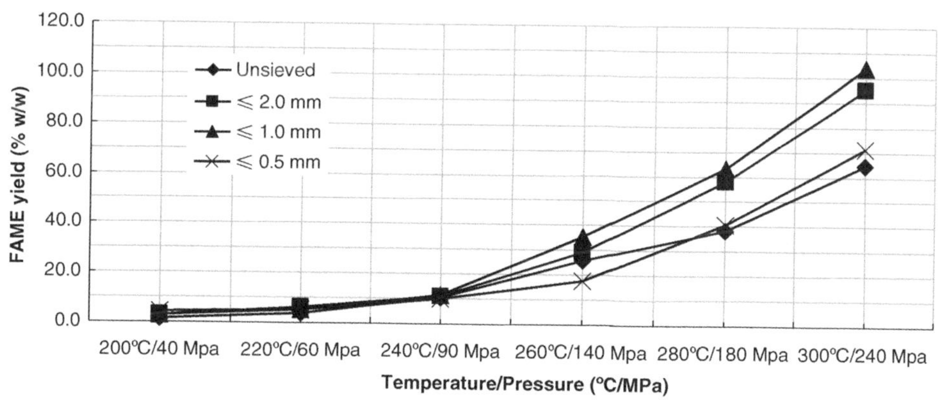

Figure 10.5 *Effect of varying temperatures and pressures to biodiesel (FAME) yield with different particle size for SCF reactive extraction. Reprinted from Lim et al © 2010, with permission from Elsevier.*

extract more oil trapped deep inside the core of the seeds than conventional 24-hour Soxhlet extraction using n-hexane. Another significant finding was that the unsieved jatropha oil seeds, which contained different solid particle sizes, tended to inhibit FAME conversion in the supercritical reactive extraction process. The varying particle sizes were expected to render the reaction rate inconsistent, while larger solid particle sizes, which were easier to settle, reduced the oil–methanol contact area for the transesterification process. Consequently, screening of blended *Jatropha curcas* L. seeds to a smaller particle size through sieving is important to maximizing the FAME or biodiesel output.

10.2.2.3 Conclusion

This experimental study has confirmed the feasibility of supercritical reactive extraction technology for the commercial production of biodiesel from oil seeds. Almost 100.0% FAME yield can be achieved in a relatively short time, even when skipping the conventional oil extraction stage, which might take up to 24 hours. At higher temperatures and pressures (300 °C and 240 MPa), the extraction of oil from seeds has a higher efficiency than conventional oil extraction, whether by chemical solvent or by mechanical pressing. Moreover, no addition of catalyst is required, which will greatly simplify downstream processes such as catalyst separation and washing. The solid particle size has been found to be a significant factor, with the optimum at $\leq$1.0 mm.

10.3 Impact on Green Processing and Process Intensification

As shown in Table 10.1, solid–liquid or liquid–liquid reactive extraction is an attractive method for the production of biodiesel since it is a simple process with fewer processing steps than other processes and provides a similarly high FAME yield and quality. Consequently, reactive extraction has significant revolutionary potential for the process and plant design of green processing technology and process intensification (PI). First, this novel technology provides a more efficient extraction method, especially for solid biomaterials. It can effectively reduce the mass-transfer limitation inherent in conventional bioprocessing technology by a large margin. This can increase the final product yield at a lower cost. Moreover, combining the extraction and reaction processes in a single processing unit can reduce the number of waste streams generated, which will reduce the extra cost for transportation to centralized waste treatment. The final products produced from reactive extraction have also been proven to contain fewer impurities. This will reduce the separation intensity in the downstream processes and allow higher-quality products to be obtained. A reduction of the processing unit and volume will directly help to lower the capital and production costs in conventional green processing plants. Indirect reduction costs will include lower energy usage due to less operating equipment and inherently safer operation due to a decrease in the use of hazardous chemicals such as hexane. The total reduction in cost will essentially render products of conventional bioprocessing more economically competitive than their counterparts in the market. This in turn will garner greater corporate interest in providing monetary capital to drive its development.

Table 10.1 Comparison of different biodiesel production methods.

	Conventional transesterification	Liquid–liquid reactive extraction	Solid–liquid reactive extraction
Description	Preextracted oil will react with transesterification reagent to produce biodiesel and glycerin	Preextracted oil will react with transesterification reagent to produce biodiesel and glycerin	Oil-bearing materials will be in direct contact with the transesterification reagent without preextraction to produce biodiesel and glycerin Transesterification reagent will also act as an extraction solvent
Additional solvent	Not required	Suitable solvent will be added to increase the rate of separation	Appropriate cosolvent can be added to increase the extraction of oil
Product separation and purification	Product separation and purification will occur after the reaction in the product stream	Product separation will occur simultaneously with the reaction	Product separation and purification will occur after the reaction in the product stream
Products	Low yield and purity	Moderate yield and higher purity	Higher yield and moderate purity
Cost	High cost	Moderate cost	Low cost
Advantages	More established and easily optimized	Higher product yield, selectivity and purity Reduced recycle and waste streams	Reduced processing steps Faster processing time
Disadvantages	High processing cost	More difficult to optimize Requires use of additional solvent	Requires higher amount of solvent and catalyst

10.4 Conclusion

As already stated, advances in reactive extraction technology will be beneficial to green processing technology, especially for the production of bio-energy such as biodiesel. However, as a relatively new technology, extraction technology still needs to overcome several challenges in order to unleash its whole potential. Process optimization will be more difficult to perform as it is an intensive processing technology with a high mixing rate and simultaneous reactions. Conventional processing technology cannot be fully integrated with reactive extraction until more studies and more detailed information at a larger scale become available. Regardless, reactive extraction has been proven to be very promising, both in accelerating the development of green processing technology and in incorporating PI into it. Therefore, this novel technology will definitely warrant more thorough research and investigation in the future.

Acknowledgement

The authors are grateful for a Postgraduate Research Grant (USM-RU-PRGS) and Research University Grant (814062) from Universiti Sains Malaysia, which has fully supported the write-up of this chapter and USM Vice-Chancellors Award of a student scholarship to Steven Lim.

References

(1) K. D. Samant and K. M. Ng, Synthesis of extractive reaction processes. AIChE Journal, 44(6), 1363–1378 (1998).
(2) K. J. Dussan, C. A. Cardona, O. H. Giraldo, L. F. Gutiérrez and V. H. Pérez, Analysis of a reactive extraction process for biodiesel production using a lipase immobilized on magnetic nanostructures. Bioresource Technology, 101(24), 9542–9549 (2010).
(3) G. Kildiran, S. Yücel and S. Türkay, In-situ alcoholysis of soybean oil. Journal of the American Oil Chemists' Society, 73(2), 225–228 (1996).
(4) E. L. Piret, W. H. Penney and P. J. Trambouze, Extractive reaction: batch and continuous-flow chemical reaction systems dilute case. AIChE Journal, 6(3), 394–400 (1960).
(5) W. K. Anderson and T. Veysoglu, Simple procedure for the epoxidation of acid sensitive olefinic compounds with m-chloroperbenzoic acid in an alkaline biphasic solvent system. The Journal of Organic Chemistry, 38(12), 2267–2268 (1973).
(6) M. L. King, A. L. Forman, C. Orella and S. H. Pines, extractive hydrolysis for pharmaceuticals. Chemical Engineering Progress, 81(5), 36–39 (1985).
(7) A. Freeman, J. M. Woodley and M. D. Lilly, In situ product removal as a tool for bioprocessing. Nat Biotech, 11(9), 1007–1012 (1993).
(8) R. P. Chauhan and J. M. Woodley, Increasing the productivity of bioconversion processes. CHEMTECH, 27(6), 26–30 (1997).
(9) L. F. Gutiérrez, Ó. J. Sánchez and C. A. Cardona, Process integration possibilities for biodiesel production from palm oil using ethanol obtained from lignocellulosic residues of oil palm industry. Bioresource Technology, 100(3), 1227–1237 (2009).
(10) Ó. J. Sánchez, L. F. Gutiérrez and C. A. Cardona, Production of biodiesel from palm oil by extractive reaction. Chemical Engineering Transactions, 21, 1231–1236 (2010).

(11) K. Abel, H. deSchmertzing and J. I. Peterson, Classification of microorganism by analysis of chemical composition. 1. Feasibility of utilizing gas chromatography. J. Bacteriol., 85, 1039 (1963).
(12) K. J. Harrington and C. D'Arcy-Evans, Transesterification in situ of sunflower seed oil. Industrial & Engineering Chemistry Product Research and Development, 24(2), 314–318 (1985).
(13) J. Qian, F. Wang, S. Liu and Z. Yun, In situ alkaline transesterification of cottonseed oil for production of biodiesel and nontoxic cottonseed meal. Bioresource Technology, 99, 9009–9012 (2008).
(14) A. Mondala, K. Liang, H. Toghiani, R. Hernandez and T. French, Biodiesel production by in situ transesterification of municipal primary and secondary sludges. Bioresource Technology, 100, 1203–1210 (2009).
(15) E. Su, W. Xu, K. Gao, Y. Zheng and D. Wei, Lipase-catalyzed in situ reactive extraction of oilseeds with short-chained alkyl acetates for fatty acid ester production. Journal of Molecular Catalysis B: Enzymatic, 48, 28–32 (2007).
(16) D. K. Garnayak, R. C. Pradhan, S. N. Naik and N. Bhatnagar, Moisture-dependent physical properties of jatropha seed (*Jatropha curcas* L.). Industrial Crops and Products, 27, 123–129 (2008).
(17) A. Kumar and S. Sharma, An evaluation of multipurpose oil seed crop for industrial uses (*Jatropha curcas* L.): a review. Industrial Crops and Products, 28(1), 1–10 (2008).
(18) S. Hawash, N. Kamal, F. Zaher, O. Kenawi and G. E. Diwani, Biodiesel fuel from jatropha oil via non-catalytic supercritical methanol transesterification. Fuel, 88, 579–582 (2009).
(19) S. H. Shuit, K. T. Lee, A. H. Kamaruddin and S. Yusup, Reactive extraction and in situ esterification of *Jatropha curcas* L. seeds for the production of biodiesel. Fuel, 89, 527–530 (2010).
(20) S. Lim, S. S. Hoong, L. K. Teong and S. Bhatia, Supercritical fluid reactive extraction of *Jatropha curcas* L. seeds with methanol: a novel biodiesel production method. Bioresource Technology, 101(18), 7169–7172 (2010).

11

Reactive Absorption Technology

Anton A. Kiss

Arnhem, The Netherlands

11.1 Introduction

Current legislation imposes tight restrictions to reduce the impact of the chemical processing industry on the environment. As a consequence, much research effort is focused on developing sustainable processes that allow the green processing of waste or the production of valuable chemicals.

During the last few decades, the chemical processing industry has shown increasing interest in the development of reactive separation processes that combine reaction and separation into a single, integrated unit. Compared to conventional processes, reactive separations bring several important benefits, such as: increased selectivity and reaction yield, overcoming of thermodynamic limitations, lower energy requirements and a significant reduction in water and solvent consumption [1–3]. The most important and well-known examples of reactive separations are reactive distillation (RD) and reactive absorption (RA). By integrating absorption and chemical reactions into a single operating unit with enhanced performance, RA qualifies as a great process intensification (PI) technique.

RA is essentially a mature process and has been known since the foundation of the modern chemical industry. In recent years its role as a core environmental protection process has grown significantly, and today RA is the most widely applied reactive separation process.

As an industrial process, the most commonly encountered use of RA is in the separation and/or purification of a gas mixture by the absorption of one of its parts (e.g. CO_2, H_2S,

Process Intensification for Green Chemistry: Engineering Solutions for Sustainable Chemical Processing, First Edition. Edited by Kamelia Boodhoo and Adam Harvey.

NOx or SOx) in a solvent that is afterwards regenerated. However, apart from gas cleaning, RA is also applied in the production of bulk chemicals, such as nitric and sulfuric acid. Recently, reactive separations (RD and RA) using green catalysts have offered great opportunities for the manufacture of fatty esters, which are involved in specialty chemical and biodiesel production [2]. Integrating reaction and separation into one production unit brings key benefits, such as: simplified operation, elimination of waste, reduced capital investment and low operating costs [4–9].

This chapter introduces the basics of RA (theory, models, equipment, operation and control) and briefly presents several industrial applications in gas purification (removal of CO_2, H_2S, NOx and SOx) and the production of various bulk chemicals, such as sulfuric acid, nitric acid and fatty acid methyl esters (FAME, biodiesel).

11.2 Theory and Models

Absorption may be either a physical or a chemical process and it is defined as a process by which a substance included in one state is transferred into another in a different state – typically a gas is absorbed into a liquid. This should not be confused with adsorption, which is the physical adherence/bonding of a chemical on the surface of another substance. Physical absorption of a gas or part of a gas mixture in a liquid solvent involves the mass transfer that occurs at the interface between the gas and the liquid and the rate at which the gas diffuses into the liquid depends on the solubility of the gas and the working conditions (pressure and temperature). A classic example is the absorption of carbon dioxide (CO_2) into water (H_2O) – usual in the beverage industry.

Chemical absorption, or reactive absorption, involves a chemical reaction between the substance being absorbed and the bulk liquid, and depends upon the stoichiometry of the reaction and the concentration of the reactants. An illustrative example of chemical absorption is the purification of natural gas by passing it through an aqueous solution of monoethanolamine (MEA). The acid gases (e.g. H_2S, CO_2) are removed by reaction with MEA. RA may be reversible or irreversible, depending on the reaction type (e.g. equilibrium or irreversible).

The rigorous modelling of RA has received considerable attention, reflected in quite a number of recent publications [1,10–12]. The optimal design of a RA process requires adequate models covering column hydrodynamics, mass and heat transfer and reaction kinetics. Large-scale applications are modelled by dividing the columns into smaller segments, called stages. Each stage corresponds to a single tray or a segment of packed column. A general overview of RA modelling approaches is given by Kenig and Gorak [11]. Several models have been developed throughout the history of RA – the main ones are described in this section.

11.2.1 Equilibrium Stage Model

Basically, the equilibrium stage model assumes that the gas and liquid streams leaving a stage are in thermodynamic equilibrium. Due to its simplicity, this model has been used for a variety of applications in the past few decades, especially for nonreactive systems. Reactive systems require extensions before a proper description can be produced. The chemical reactions must be taken into consideration, using a source term in mass and energy balances. Usually, their effect is beneficial only in regions of low concentration within the gas phase, due to limitations of the reaction stoichiometry or equilibrium. Yet, in real absorption processes, the thermodynamic equilibrium is usually not attained within a stage; for this reason, tray efficiencies or HETP (height equivalent of a theoretical plate) values are introduced in order to build a link to real columns. However, this method often fails for RA processes, due to the specific features of multicomponent mixtures [13].

11.2.2 HTU/NTU Concepts and Enhancement Factors

A simple approach to determining transfer units from the mass-transfer kinetics was developed early on by Chilton and Colburn [14]. The column height is determined as a product of the number and height of the transfer units. The height of a theoretical unit (HTU) is estimated via empirical Sherwood correlations, whereas the number of theoretical units (NTU) is obtained by numerical or analytical integration of the inverse of the driving force over the column height. Chemical reactions are considered via enhancement factors, which are defined as the quotient of mass-transfer rate with and without reaction [15]. Since the complexity of the reactions is described using a single parameter, this method often leads to inaccurate results.

11.2.3 Rate-based Stage Model

The rate-based approach takes direct account of the multicomponent mass and heat transfer and the chemical reaction. The mass transfer between the phases can be described by various different theories, such as the two-film model or the penetration/surface renewal theory. The corresponding model parameters are determined using empirical correlations. For many applications, the two-film model parameters can be found in the literature, and therefore this method is often preferred [10]. Figure 11.1 illustrates the modelling of RA using the two-film model, looking at the reaction of *A* (from gas) with *B* (from liquid) [16].

Despite relentless research efforts, the behaviour of RA is still not properly understood. This can be attributed to various factors, such as: intricate mass transfer–reaction coupling; complex combinations of thermodynamics and kinetics, with reaction rates varying over a wide range; and complicated reaction schemes that typically also include ionic species.

11.3 Equipment, Operation and Control

RA can be performed in a variety of equipment types that provide a continuous flow of both contacting phases. Just as RD, RA is typically carried out in tray or packed columns. However, RA requires no reboiler or condenser [7,12,16]. The RA process is characterized by independent flow of both gas and liquid phases, and allows co-current (downflow and upflow) and more typical counter-current operating regimes. RA operating units can be conveniently classified according to which phase is in a continuous or a disperse form. Using this criterion, the classification of the RA equipment can be summarized as [17]:

(1) *Both phases in a continuous form:* Packed columns, thin-film contactors, wetted-wall columns, contactors with flat surfaces, laminar jet absorber, disc (sphere) columns.

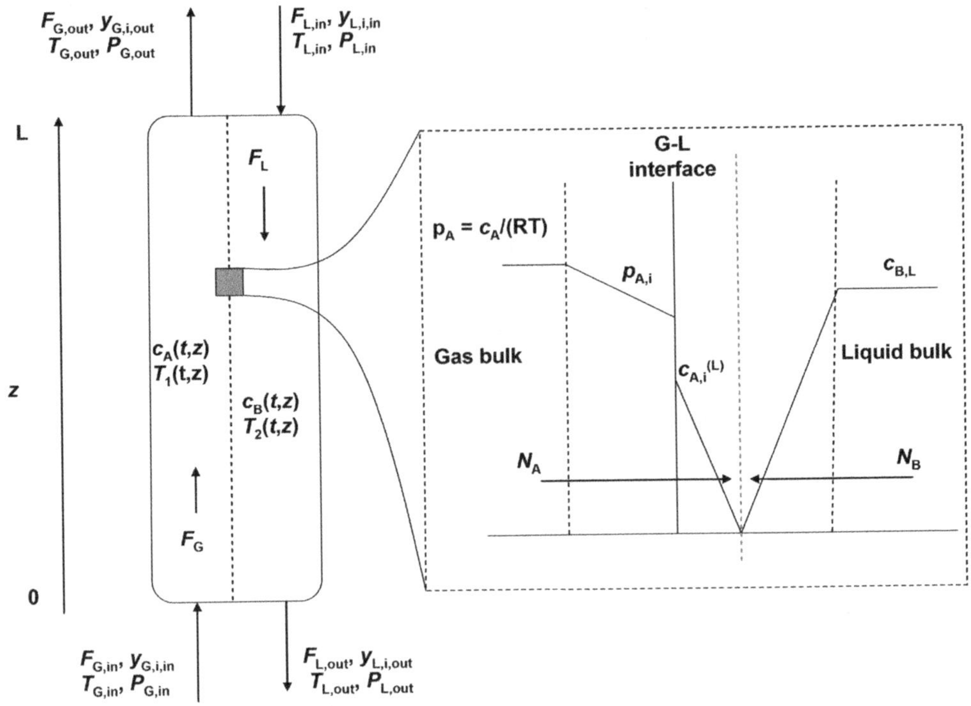

Figure 11.1 *RA modelling using the two-film model. Reprinted from Kiss* et al. *© 2010, with permission from Elsevier.*

(2) *A disperse gas phase and a continuous liquid phase:* Plate columns, plate columns with packing, bubble columns, packed bubble columns, mechanically agitated columns, jet absorbers.
(3) *A disperse liquid phase and a continuous gas phase:* Spray columns, Venturi scrubbers.

In integrated processes based on RD or RA, feeding the reactants according to their stoichiometric ratio is essential to achieving high product purity [18,19]. Therefore, the inlet liquid flow rate is controlled depending on the inlet gas flow rate. A level controller is also applied and the pressure difference in the column is measured in order to ensure a uniform load distribution with respect to the column height and to avoid overload or flooding. Since RA applications are manifold, the key controlled parameters may differ from process to process. For CO_2 absorption by MEA, the most effective parameters are the flow rate of the circulating liquid stream and the energy required in the desorption step. For the removal of NOx by aqueous nitric acid, the level of the liquid hold-up at the bottom of the column and the temperature along the column are the major parameters. In particular, the temperature is controlled several times along the column due to strong temperature-dependent reactions [20].

Moreover, the choice of the optimal control method is important. Although RA processes are nonlinear, linear control methods can be applied with satisfying results in the required range of operating conditions [21]. Commonly, proportional–integral–derivative (PID) controllers are applied for this purpose. However, over the whole operating region, nonlinear controllers are necessary to compensate for process disturbances [21]. Kershenbaum [22] investigated the importance of carrying out both dynamic simulations and experiments, and showed that these methods are complementary.

11.4 Applications in Gas Purification

The main goals of absorption processes are the removal or separation of specific chemicals from the gas phase and the production of bulk chemicals in the liquid phase. In the case of gas purification applications, the typical industrial absorption operations are usually accomplished by combining absorption and desorption units, for the capture and subsequent recovery of certain chemicals.

The current market for RA processes is growing, mainly due to extensive efforts in the area of CO_2 capture and storage [23], but also to novel proposed applications such as biofuel production [2,7,9]. With new, more restrictive legislation, industrial needs are also shifting towards more efficient, less expensive, sustainable and eco-friendly reactive separation processes. The main industrial applications remain focused on the purification of gas streams and the production of chemicals [11]:

- Removal of harmful substances (e.g. coke oven gas purification, CO_2 and NOx removal).
- Retrieval of valuable substances or nonreacted reactants (e.g. solvent regeneration).
- Production of chemical products (e.g. sulfuric and nitric acid, formaldehyde synthesis).
- Water removal (e.g. from natural gas, air drying).
- Conditioning of gas streams (e.g. synthesis gas conditioning).
- Separation of substances (e.g. olefin/paraffin separations).

11.4.1 Carbon Dioxide Capture

Carbon dioxide is mainly produced by gasification, reformation or in power plants from major sources such as coal, oil, natural gas and biomass. The application of absorption processes occurs either post-combustion (in power plants) or pre-combustion (reformation or gasification of natural feed stocks). These industrial applications are illustrated in Figure 11.2 (left) [24].

The vast majority of these processes belong to fossil fuel combustion. Depending on the type of power plant, the concentration of CO_2 in the gas phase can vary from 3–5 mol% (gas-fired power plant) to 12–16 mol% (coal-fired power plant). Various solvents are available for the capture of CO_2 from gas mixtures. The solvent chosen must be selective and should not absorb other gaseous components or form corrosive salts with nitrogen or sulfuric oxides.

Typical capture of CO_2 by RA employs the use of alkanol-amines, such as MEA, diethanolamine (DEA), tri-ethanolamine (TEA) and methyl-di-ethanolamine (MDEA). The primary amines (MEA) have a high enthalpy of reaction and vaporization, high reaction rate and high corosivity, while the tertiary amines (TEA, MDEA) have a high loading capacity. The reaction of CO_2 with primary or secondary amines can be described using a zwitterion mechanism [15]. In the first step, a zwitterion is produced by CO_2 and the amine

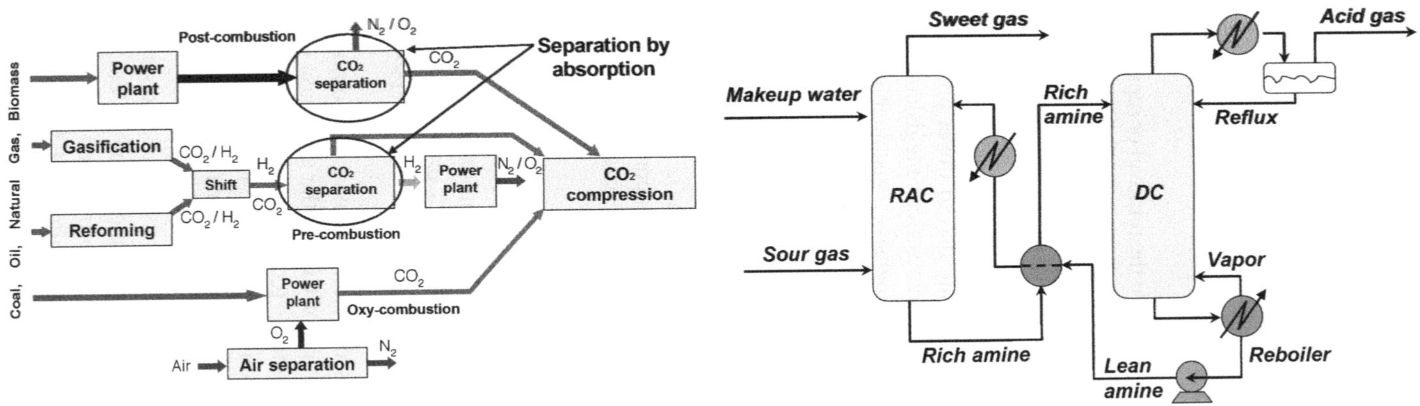

Figure 11.2 Left CO_2 sources and process flow. Right: process flow diagram of a typical amine treatment process used in industrial plants for CO_2 and H_2S capture [24].

(indicated with AmH). The zwitterion is then deprotonated by a base (B), which results in the formation of a carbamate [25]. On the other hand, the absorption of CO_2 with tertiary amines can be described using the based-catalysed hydration mechanism [25]. The tertiary amines (R_3N) cannot react directly with CO_2, but it is worth noting that these amines can have a base catalytic effect on the hydration of CO_2.

$$CO_2 + AmH \underset{k_1}{\overset{k_1}{\rightleftharpoons}} AmH^+COO^- \tag{11.1}$$

$$AmH^+COO^- + B \xrightarrow{k_B} AmCOO^- + BH^+ \tag{11.2}$$

The typical flow sheet of an RA process consists of an absorption column, which captures CO_2 from the feed gas (acid or sour gas) in an amine solvent, thus leaving the outlet gas free of CO_2 (sweet gas), followed by a stripping column, which performs the desorption of CO_2 and the recovery of the amine used as a recyclable solvent (Figure 11.2, right). CO_2 can then be conveniently stored or converted into valuable chemicals, such as CO and oxygen, synthetic fuels (alcohols, ethers, hydrocarbons), carboxylic acids, monomers for polymers (e.g. polycarbonates), fertilizers and food.

RA of CO_2 using MEA is a well-established technology. However, its application is limited due to the corrosive salts formed by MEA and the SO_x present in process streams. Other drawbacks are the relatively high enthalpy of reaction and evaporation, resulting in high energy requirements for the recovery of the solvent. Nowadays, the Econamine FG (Fluor Daniel) is one of the most widely used processes in industry. It utilizes a 30 wt% MEA solution containing corrosion inhibitors that can recover 85–95% of the CO_2 in flue gases, producing high-purity CO_2 (over 99%). Non-inhibiting solutions can also be used. The ABB Lummus Crest process works with low MEA concentrations (15–20 wt%) that do not require the addition of inhibitors to avoid corrosion [24].

Sterically hindered amines have been developed as MEA alternatives. These are less corrosive, have higher capacity and regeneration rates (1.0 instead of 0.5 mol/mol amine) and require less energy for regeneration. Kansai Electric Power and Mitsubishi Heavy Industries have developed a process that uses hindered amines (e.g. KS1, AMP). A comparison of these commercially available processes, based on a number of key performance parameters, was reported by Bailey and Feron [26].

CO_2 removal is also carried out by using hot potassium carbonate in many places, such as ammonia plants and natural gas treatment. The most frequently applied technology is Universal Oil Products (UOP's) Benfield process, which makes use of DEA as an activator. The Catacarb process (Eickmeyer & Associates) also uses hot potassium carbonate, which is applied for the removal of CO_2 from ethylene oxide recycle gas. The Exxon Flexsorb HP process uses a hindered amine activator, whereas the Giammarco–Vetrocoke uses an organic activator [27].

The design and the selection of an optimal RA process for CO_2 capture depend on several factors. The CO_2 concentration can vary depending on the origin of the flue gas, which influences the carbon capture process performance. Moreover, the solvent type and its capacity and ability to absorb CO_2 are important factors. The size of the equipment and the energy requirement are determined by the solvent recycle flow rate. In addition to the thermal energy required for solvent regeneration, electrical energy is required to operate liquid pumps, the flue gas fan and the CO_2 compressor.

11.4.2 Sour Gas Treatment

Natural gas and other industrial gases containing hydrogen sulfide (H_2S) are called sour gases. As H_2S is a known poison that leads to catalyst deactivation, its removal from sour gases is crucial to natural gas processing and to gas cleaning for the synthesis of ammonia and methanol. Petroleum refining is a major source of sour gas with low concentrations of CO_2. Another key application of the sour gas treatment is in coke oven gas purification. The gas, produced during the carbonization of coal to form coke, consists mainly of CO, H_2, CH_4, H_2S and CO_2 – but it is also contaminated with other components that must be removed. Sour gases are industrially cleaned using a similar amine scrubbing process, as previously described (Figure 11.2, right).

The most widely used amines for sour gas treatment are MEA (10–20% concentration), DEA (10–30%), di-isopropanolamine (DIPA) (30–50%), N-MDEA (30–50%) and diglycol amine (DGA) (40–70%) [28]. Sour gas removal can also be carried out using hot potassium carbonate or the Benfield process [27]. In the purification of coke oven gas contaminated with various substances, an aqueous ammonia solution is usually used as a solvent. 'Ammonia hydrogen sulfide circulation scrubbing' is also applied in industry for coke oven gas purification [24].

11.4.3 Removal of Nitrogen Oxides

Nitrogen oxide is a highly reactive gas that contains varying amounts of nitrogen and oxygen. NO_x is formed during the combustion of fuel at high temperatures. According to Chung *et al.* [29], the primary sources of nitrogen oxides are transportation (61%), the energy industry (18%), the chemical processing industry (14%) and others (7%). The applied removal processes can be subdivided into dry processes (catalytic or noncatalytic denitrification) and wet processes (which make use of water or aqueous nitric acid for NOx scrubbing in RA columns). Note that the wet process based on RA is also used for the production of nitric acid, described in a later section. During NOx absorption, the following main reactions in gas- and liquid-phase occur [24]:

$$2NO + O_2 \rightarrow 2\,NO_2 \quad \text{(gas phase)} \tag{11.3}$$

$$2NO_2 \rightarrow N_2O_4 \quad \text{(gas phase)} \tag{11.4}$$

$$NO + NO_2 \rightarrow N_2O_3 \quad \text{(gas phase)} \tag{11.5}$$

$$2\,NO_2 + H_2O \rightarrow HNO_3 + HNO_2 \quad \text{(liquid phase)} \tag{11.6}$$

$$N_2O_4 + H_2O \rightarrow HNO_3 + HNO_2 \quad \text{(liquid phase)} \tag{11.7}$$

$$N_2O_3 + H_2O \rightarrow 2\,HNO_2 \quad \text{(liquid phase)} \tag{11.8}$$

$$3\,HNO_2 \rightarrow HNO_3 + H_2O + 2\,NO \quad \text{(liquid phase)} \tag{11.9}$$

$$2\,NO + 3\,H_2O_2 \rightarrow 2\,HNO_3 + 2\,H_2O \quad \text{(liquid phase)} \tag{11.10}$$

$$HNO_2 + H_2O_2 \rightarrow HNO_3 + H_2O \quad \text{(liquid phase)} \tag{11.11}$$

Absorption of NOx can be further optimized by the addition of hydrogen peroxide (H_2O_2), which destroys HNO_2 and NO in the liquid phase and hence extends nitric acid production [24,30]. Equations 11.10 and 11.11 describe the additional reactions that take place.

Moreover, quite a number of research activities have been dedicated to aqueous alkaline solutions (e.g. NaOH), since valuable byproducts can be obtained from these, such as nitric acid, nitrates and nitrites [31]. This can be improved by the addition of H_2O_2 and the use of structured packings as column internals [30].

11.4.4 Desulfurization

Sulfur compounds (e.g. SO_x) are produced in several industrial processes, such as combustion of organic fuels or the reforming of oil. There is a wide range of industrial processes for the removal of sulfur oxides from gas streams. The applications can be classified into [32]:

(1) *Wet processes:* Limestone gypsum, sea water washing (SWW), ammonia scrubbing, Wellman–Lord.
(2) *Semi-dry processes:* Circulating fluidized bed, spray drying, duct spray drying.
(3) *Dry processes:* Furnace sorbent injection, sodium bicarbonate injection.

Of these, only the wet processes are performed by RA. The limestone gypsum process is the most common flue gas desulfurization (FGD) process and is applied in different variants, depending on the supplier and client's requirements. The typical limestone gypsum process is characterized by an open spray tower [32]. The flue gas passes a precipitator that removes the fly ashes and then is cooled and passed into the absorber. Here, the gas is contacted with limestone slurry, which removes SO_2 and almost all HCl. The gas leaving the top of the RA column is heated in the reheater before being exhausted. In the absorber, SO_2 is dissolved and neutralized in water, which reacts ultimately with the $CaCO_3$ to produce gypsum. This process is well understood and is applied by many companies, such as Marsulex, ABB, Lurgi Lentjes Bischoff (LLB), Babcock Borsig, Kawasaki and IHI, who operate and sell this technology. Babcock and Wilcox (B&W) and Babcock Hitachi have a very similar design, while Mitsubishi Heavy Industries (MHI) uses a slightly different one, without an open spray tower.

Flue gas can be also cleaned using sea water, which has natural alkalinity. This technology is produced by two suppliers: ABB and LLB. The SWW schematic is shown in Figure 11.3 (left) [24]. Flue gas enters the reheater, where it is cooled, gas and then is scrubbed with sea water to remove up to 99% SO_2 and almost all HCl [32]. This technology is expanding very quickly due to its major advantages: simple plant design with no solid product. However, it is limited to coastal sites due to the sea water requirements.

Ammonia scrubbing technology – offered by LLB and Marsulex – uses NH_3 as an absorbent that reacts with SO_2 to produce ammonium sulfate, which can be used in the fertilizer industry [32]. Figure 11.3 (right) depicts the self-explanatory ammonia scrubbing process [24].

The Wellman–Lord process uses an aqueous sodium sulfite solution, which leads to valuable products such as elementary sulfur, sulfuric acid and liquid SO_2. The main advantages of this process are the minor amount of waste produced and the low quantities of absorbent required.

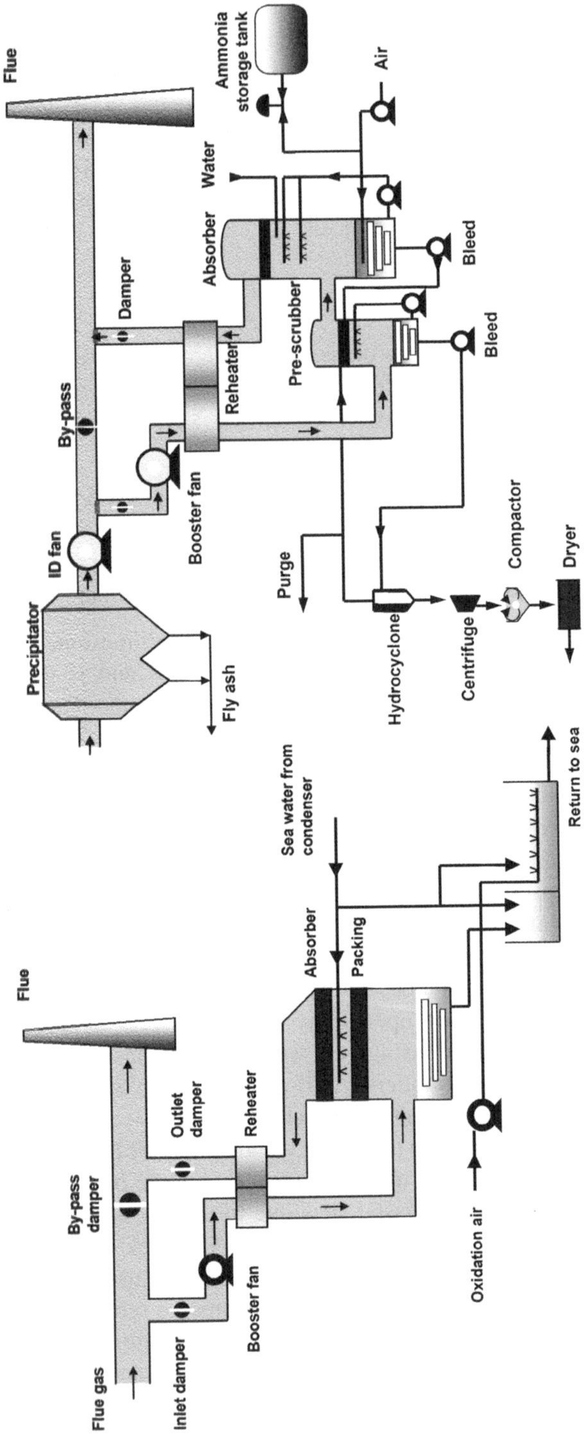

Figure 11.3 *SWW process (left). Ammonia scrubbing process (right) [24].*

11.5 Applications to the Production of Chemicals

RA is also applied as an intensified green processing technology for the production of chemicals, such as sulfuric and nitric acid [33,34], and more recently for the production of biodiesel and fatty esters [2,7,9]. Remarkably, sulfuric acid is the most produced chemical substance worldwide – with a production rate of up to 150 million tonnes/year – while production of biodiesel as a renewable fuel exceeds 36 million tonnes/year.

11.5.1 Sulfuric Acid Production

Sulfuric acid is largely used as an ingredient. The fertilizer industry is its largest single consumer, mostly for the manufacture of phosphate fertilizers, but the (petro)chemical and oil refining industries also make use of it, for example as an acidic dehydrating agent.

There are two major processes used in sulfuric acid production: the old lead chamber process and the current contact process [33]. The main steps in the latter consist of burning sulfur (S) in air to form sulfur dioxide (SO_2), converting SO_2 to sulfur trioxide (SO_3) using oxygen (O_2) from air and then absorbing SO_3 in water (H_2O) or a diluted solution of sulfuric acid (H_2SO_4) to form a concentrated solution of acid (>96%). The exothermic reactions that take place are [16]:

$$S(s) + O_2(g) \rightleftharpoons SO_2(g) \qquad \Delta H_R = -296.8\ \mathrm{kJ/mol} \qquad (11.12)$$

$$SO_2(g) + {}^1/_2\, O_2(g) \rightleftharpoons SO_3(g) \qquad \Delta H_R = -96.2\ \mathrm{kJ/mol} \qquad (11.13)$$

$$SO_3(g) + H_2O(lq) \rightleftharpoons H_2SO_4(lq) \qquad \Delta H_R = -132\ \mathrm{kJ/mol} \qquad (11.14)$$

Modern production plants are using the double absorption technique, which can lead to overall conversion rates higher than 99.85% [16]. The simplified flow sheet of such an industrial process consists of a sulfur burner, a multipass converter, heat exchangers and absorbers (Figure 11.4). SO_2 conversion is further improved and tail gas emissions are reduced through an intermediate SO_3 absorption step (Abs1), in which the production of sulfuric acid takes place. This absorption step is strategically placed after the fourth bed of catalyst in order to change the gas composition by removing most of the SO_3 and thus shifting the equilibrium curve to higher conversions. The absorption of SO_3 is finalized in the second absorber (Abs2), which also purifies the outlet gas emissions. Figure 11.5 illustrates the typical temperature and composition profiles along the two counter-current operated absorbers [16]. SO_3 concentration is reduced from 9.81 to 0.01% in the intermediate absorber, and from 0.44% to ppm levels in the final step. The dimensionless length is actually the ratio between any given length and the total absorber length.

11.5.2 Nitric Acid Production

Nitric acid is one of the most important inorganic acids, being used mainly in the fertilizer industry. High-strength nitric acid (98–99%) is produced by dehydration followed by bleaching, condensation and absorption in a weak nitric acid, while commercial-grade nitric acid solutions (52 and 68%) are produced via the Ostwald process. Weak nitric acid (30–70%) is produced by catalytic oxidation of ammonia, followed by condensation and absorption in water. For the production of weak nitric

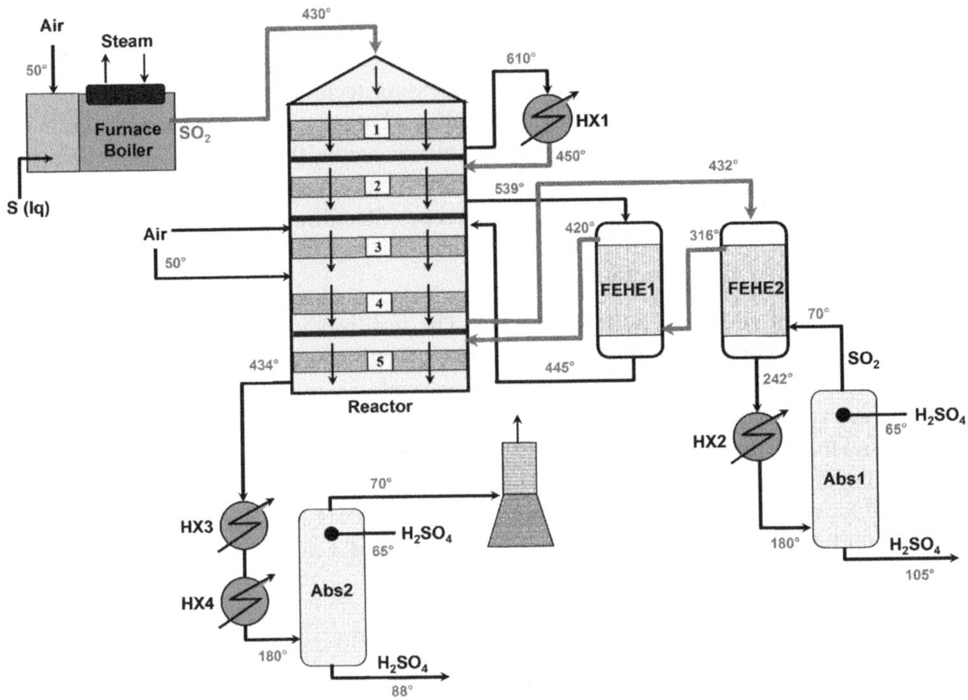

Figure 11.4 *Flow sheet of a modern sulfuric acid production plant [16]. Reprinted from Kiss et al. © 2010, with permission from Elsevier.*

acid, either a single-pressure process (1–14 bar for oxidation and absorption) or a dual-pressure process (1.0–6.5 bar for oxidation and 6.5–13.0 bar for absorption) is applied. Most new plants work with dual-pressure processes based on the oxidation of ammonia-as illustrated in Figure 11.6 [24]. An air–NH_3 mixture of 9 : 1 is catalytically converted mainly to NO, which is afterwards oxidized and reacted with residual O_2 to form NO_2

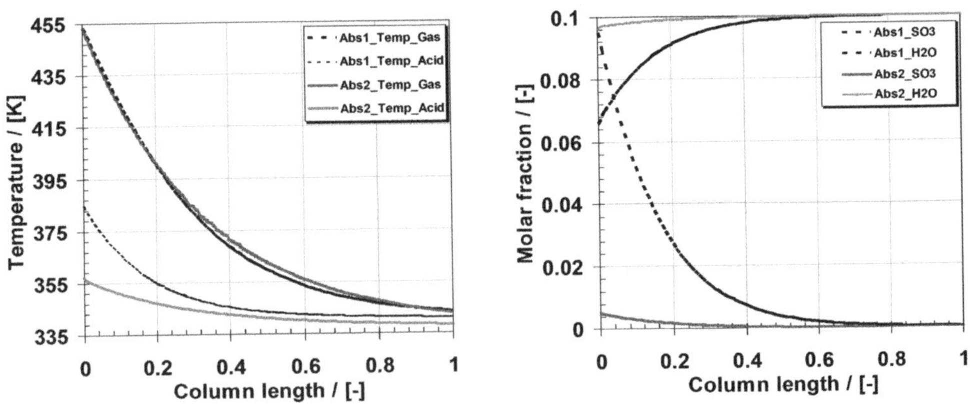

Figure 11.5 *Temperature and composition profiles along RA columns. Reprinted from Kiss et al. © 2010, with permission from Elsevier.*

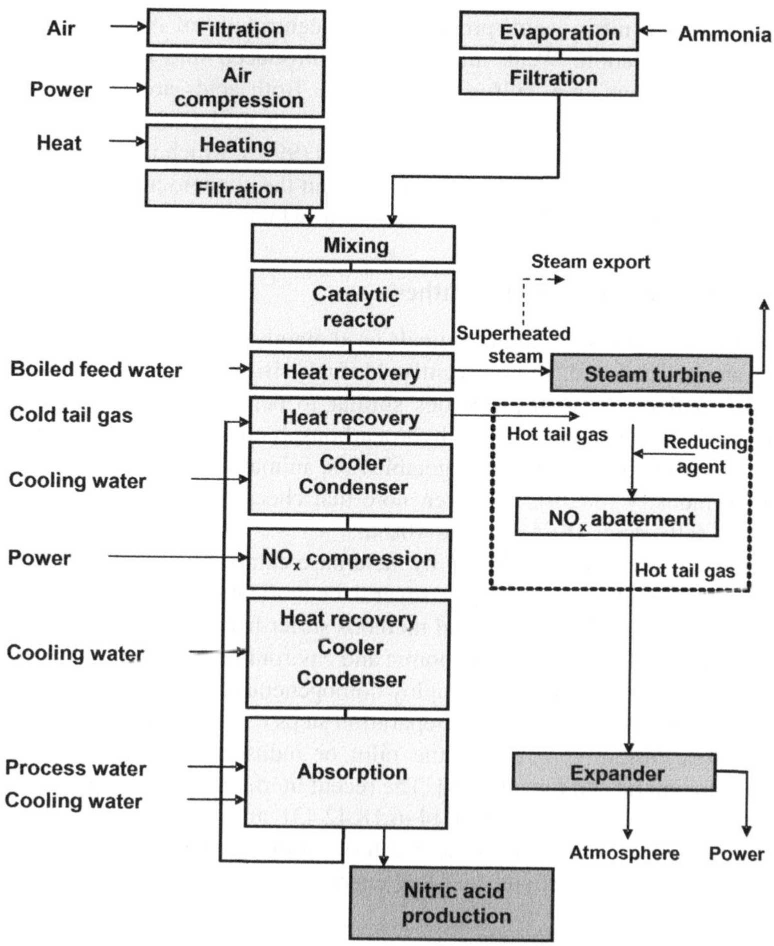

Figure 11.6 *Flow chart of a dual-pressure plant for nitric acid production.*

and NO_2-dimer. The mixture of NO_2/dimer is introduced into the bottom of the absorption tower, in counter-current with deionized process water. Another air stream is introduced into the column to reoxidize the NO that is formed in the reaction. The nitric acid is then withdrawn from the bottom of the column. Oxidation takes place in the free space between the trays, while absorption occurs on the trays – usually sieve or bubble cap trays. The following exothermic reactions occur [24,34]:

$$2\,NH_3(g) + 5\,O_2(g) \rightarrow 4\,NO + 6\,H_2O(g) \qquad \Delta H_R = -226\,kJ/mol \qquad (11.15)$$

$$2\,NO(g) + O_2(g) \rightarrow 2\,NO_2(g) \qquad \Delta H_R = -114\,kJ/mol \qquad (11.16)$$

$$3\,NO_2(g) + H_2O(l) \rightarrow 2\,HNO_3(aq) + NO(g) \qquad \Delta H_R = -117\,kJ/mol \qquad (11.17)$$

$$4\,NO_2(g) + 2\,H_2O(l) + O_2(g) \rightarrow 4\,HNO_3(aq) \qquad \text{only when air is present} \qquad (11.18)$$

In the high-strength nitric acid process, a concentration of 98–99% is obtained using extractive distillation. Weak nitric acid is introduced into a distillation column with a dehydrating agent (e.g. sulfuric acid, 60%). Both acids are fed into the top of the column, at atmospheric pressure, in counter-current to the ascending vapours. The vapour leaving the column is concentrated nitric acid (99%), which is passed to a bleacher and condenser. Concentrated nitric acid is obtained in the condenser. The byproducts (O_2 and NO) are introduced into an absorption column, which produces weak nitric acid.

11.5.3 Biodiesel and Fatty Esters Synthesis

Fatty esters are important specialty chemicals used mainly in cosmetics, pharmaceuticals, cleaning products, food and, more recently, biofuels. Biodiesel is a biodegradable and renewable alternative fuel with properties similar to petroleum diesel [35–37]. Unlike petroleum diesel, which is a mixture of hydrocarbons, biodiesel consists of FAME that is produced from green sources such as vegetable oils, animal fat or even waste cooking oils from the food industry [38,39]. However, note that cheap waste oils have a substantial amount of free fatty acids (FFAs), up to 100%.

Fatty esters are currently produced by acid/base-catalysed trans-esterification with methanol or ethanol [40], followed by several neutralization and product purification steps [41]. Nevertheless, all conventional methods suffer from problems related to the use of liquid catalysts, leading to severe economic and environmental penalties. At present, the most common biodiesel technologies employ homogeneous catalysts, in batch or continuous processes, where both reaction and separation steps can create bottlenecks. There are several processes currently in use at the pilot or industrial scale: batch, continuous, supercritical, enzymatic and two-step [9]. The recent literature is quite abundant in studies on integrated processes such as RD [4–6,18,42,43] and RA [7,9]. Several reactive separation processes based on fatty acid esterification have been reported, aiming at high performance and productivity and low energy requirements [2]:

- *RD:* 191.2 kW.hour/ton biodiesel [6];
- *Dual RD:* 166.8 kW.hour/ton biodiesel [18];
- *RA:* 138.4 kW.hour/ton biodiesel [7];
- *Heat-integrated RD:* 108.8 kW.hour/ton biodiesel [6];
- *Heat-integrated RA:* 21.6 kW.hour/ton biodiesel [9].

Figure 11.7 shows a comparison of the energy requirements for a conventional two-step process – acid and base catalysis [44] – versus recently reported reactive separation processes [2,6,7,9,18]. It is worth noting here the difference in scale.

This section presents a novel energy-efficient RA process for biodiesel production that is very easily controlled despite the high degree of integration. The integration of reaction and separation into one unit, corroborated with the use of a heterogeneous catalyst, offers major advantages, such as: reduced capital investment, low operating costs, simplified downstream processing steps and zero catalyst-related waste streams and soap formation. Rigorous process simulations have been used to design and control a plant producing 10 ktpy biodiesel (1250 kg/hour) by esterification of methanol with FFA, using sulfated zirconia as a solid acid catalyst [7,9].

The conceptual design of the process is based on an RA column that integrates the reaction and separation steps into one operating unit. The chemical equilibrium is shifted

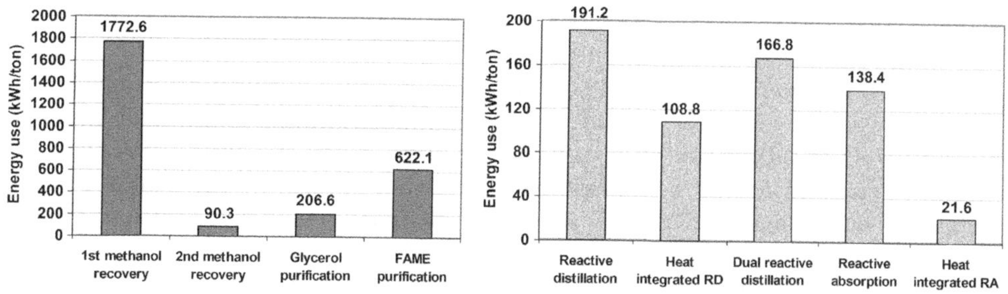

Figure 11.7 *Energy requirements for conventional process (left) and reactive separations (right).*

towards product formation by continuous removal of the reaction products, instead of using an excess of a reactant – typically the alcohol. An additional flash vessel and a decanter are used to guarantee the high purity of both products. Since methanol and water are much more volatile than the fatty ester or fatty acid, these will separate easily in the top of the column. Figure 11.8 (left) presents the flow sheet for this process, based on conventional RA, as reported by Kiss [5]. This reference flow sheet is relatively simple, with just a few operating units, two cold streams to be preheated (fatty acid and alcohol) and two hot streams to be cooled (top water and bottom fatty esters). Therefore, heat integration is performed by the application of previously reported heuristic rules [45–46]. Two feed–effluent heat exchangers (FEHEs) should therefore partially or totally replace each of the two heat exchangers HEX1 and HEX2. Figure 11.8 (right) illustrates the improved process, including heat integration around the RA column [9]. The hot liquid product of the FLASH, a mixture of fatty esters (biodiesel), is used to preheat and vaporize the alcohol feed stream. If production changes are expected, the

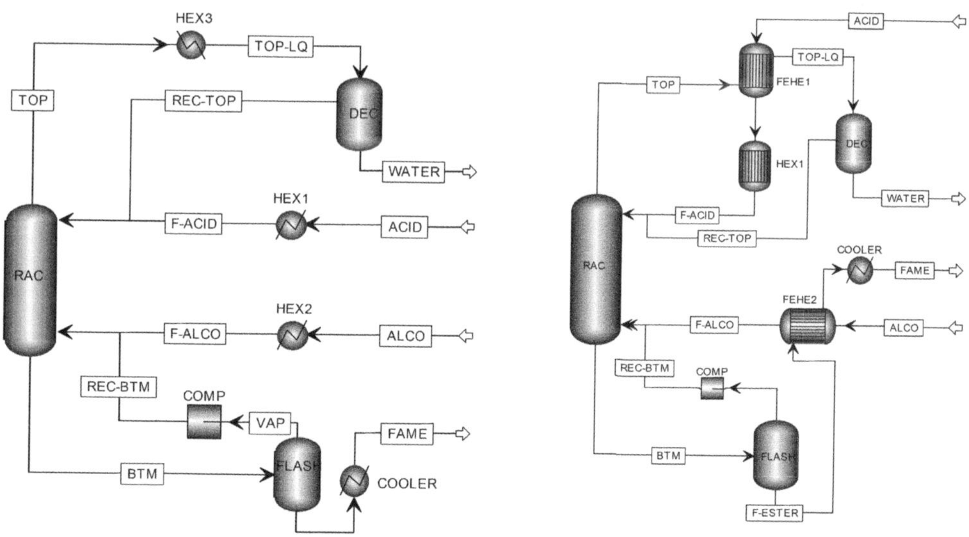

Figure 11.8 *Synthesis of fatty esters by RA base case flow sheet (left) and heat-integrated process (right). Reprinted from Kiss and Bildea © 2011, with permission from Elsevier.*

nominal design should include a bypass of the hot stream (dashed line in Figure 11.8, right), which can be used for control objectives.

Process simulations that include experimental results were performed using Aspen Plus [47]. The RA column was simulated using the rigorous RADFRAC unit with the RateSep (rate-based) model, and considering three phase balances. The physical properties required for the simulation and the binary interaction parameters for the methanol–water and acid–ester pairs were available in the Aspen Plus database of pure components [47]. The fatty components were conveniently lumped into one fatty acid and its fatty ester, according to the reaction: R-COOH + $CH_3OH \leftrightarrow$ R-COO-CH_3 + H_2O. Dodecanoic (lauric) acid/ester was selected as the lumped component due to the availability of experimental results, kinetics and vapor–liquid–liquid equilibrium (VLLE) parameters for this system [4,5,7,9,18].

Table 11.1 shows the complete mass balance of the process, while Table 11.2 lists the main design parameters, such as column size, catalyst loading and feed condition [9]. The composition, temperature and rate profiles are plotted in Figure 11.9 [9].

High conversion of the reactants is achieved, with the productivity of the RA unit exceeding 19 kg fatty ester/kg catalyst/hour. The purity specification is higher than 99.9 %wt for the final biodiesel product (FAME stream). Note that the total of the optional recycle streams (REC-BTM) is not significant, representing less than 0.9% of the total biodiesel production rate.

Table 11.1 *Mass balance of a 10 ktpy FAME process based on integrated RA. Reprinted from Kiss and Bildea © 2011, with permission from Elsevier.*

	F-ACID	F-ALCO	BTM	REC-BTM	REC-TOP	TOP	WATER	FAME
Temperature (°C)	160	65.4	136.2	146.2	51.8	162.1	51.8	30
Pressure (bar)	1.05	1.05	1.03	1.216	1	1	1	0.203
Vapour fraction	0	1	0	1	0	1	0	0
Mole flow (kmol/hour)	5.824	5.876	6.125	0.252	0.059	5.886	5.828	5.873
Mass flow (kg/hour)	1166.7	188.3	1261.3	11.3	9.369	114.4	105.06	1250
Mass flow (kg/hour)								
Methanol	0	188.3	9.125	7.544	0.002	0.103	0.101	1.581
Acid	116.74	0	Trace	Trace	9.218	9.233	0.016	Trace
Water	0	0	Trace	Trace	0.24	105.2	104.93	Trace
Fame	0	0	1252.2	3.764	0.846	0.846	Trace	1248.4
Mass fraction								
Methanol	0	1	0.007	0.667	172 ppm	894 ppm	965 ppm	0.001
Acid	1	0	Trace	Trace	0.894	0.08	148 ppm	Trace
Water	0	0	Trace	10 ppb	0.023	0.912	0.999	Trace
Fame	0	0	0.993	0.333	0.082	0.007	513 ppb	0.999
Mole fraction								
Methanol	0	1	0.046	0.931	873 ppm	546 ppm	0.001	0.008
Acid	1	0	Trace	Trace	0.726	0.008	13 ppm	Trace
Water	0	0	Trace	26 ppb	0.211	0.992	0.999	Trace
Fame	0	0	0.954	0.069	0.062	670 ppm	43 ppb	0.992

Table 11.2 *Design parameters for simulation of the RA column. Reprinted from Kiss and Bildea © 2011, with permission from Elsevier.*

Parameter	Value	Units
Total number of theoretical stages	15	–
Number of reactive stages	10 (from 3 to 12)	–
Column diameter	0.4	m
HETP	0.6	m
Valid phases	VLL	–
Volume liquid holdup per stage	18	l
Mass catalyst per stage	6.5	kg
Catalyst bulk density	1050	kg/m^3
Fatty acid conversion	>99.99	%
Fatty acid feed (liquid, at 160 °C)	1167	kg/hour
Methanol feed (vapour, at 65 °C)	188	kg/hour
Production of biodiesel (FAME)	1250	kg/hour
RA column productivity	19.2	kg FAME / kg cat hour

HETP, height equivalent of a theoretical plate; VLL, vapor–liquid–liquid; FAME, fatty acid methyl ester.

Kiss and Bildea [9] used sensitivity analysis to evaluate the range of the operating parameters: reactants ratio, temperature of feed streams, decanting temperature, flashing pressure and recycle rate. The optimal molar ratio of the reactants (alcohol: acid) is very close to the stoichiometric value of one [7]. In practice, using a very small excess of methanol (up to 1%) or an efficient control structure is sufficient for the complete conversion of the FFAs.

Table 11.3 shows a head-to-head comparison of this novel heat-integrated RA process [9] with previously reported reference RD and RA processes [6,7]. These heating and cooling requirements are figures that ultimately translate into equipment size and cost. Remarkably, the energy demand is less than 22 kW.hour/ton biodiesel (34 kg steam/ton biodiesel). Also, the heating and cooling requirements are significantly reduced as compared to the reference RA base case [5], by 85% and 95%, respectively.

Heat-integrated RA offers significant advantages, but the controllability of a process is just as important as capital and operating cost savings. In processes based on RD or RA, feeding the reactants according to their stoichiometric ratio is essential to achieving high

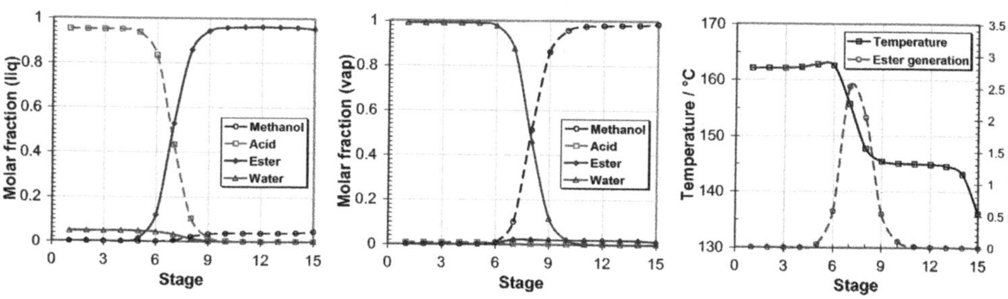

Figure 11.9 *Liquid–vapour composition, temperature and reaction-rate profiles along the column. Reprinted from Kiss © 2011 and Kiss and Bildea © 2011, with permission from Elsevier.*

Table 11.3 Comparison of integrated RA and RD processes (at a production rate of 1250 kg/hour fatty esters). Reprinted from Kiss and Bildea © 2011, with permission from Elsevier.

Equipment/parameter/units	RD	HI-RD	RA	HI-RA
Reactive column – reboiler duty (heater), kW	136	136	n/a	n/a
HEX-1 heat duty (fatty acid heater), kW	95	0	108	27
HEX-2 heat duty (methanol heater), kW	8	0	65	0
Reactive column – condenser duty (cooler), kW	−72	−72	n/a	n/a
HEX-3 water cooler/decanter, kW	−6	−6	−77	0
COOLER heat duty (biodiesel cooler), kW	−141	−38	−78	−14
FLASH heat duty (methanol recovery), kW	0	0	0	0
Compressor power (electricity), kW	0.6	0.6	0.6	0.6
Reactive column, number of reactive stages	10	10	10	10
Feed stage number, for acid/alcohol streams	3/10	3/10	1/15	1/15
Reactive column diameter, m	0.4	0.4	0.4	0.4
Reflux ratio (mass ratio R/D), kg/kg	0.10	0.10	n/a	n/a
Boil-up ratio (mass ratio V/B), kg/kg	0.12	0.12	n/a	n/a
Productivity, kg ester/kg catalyst/hour	20.4	20.4	19.2	19.2
Energy requirements per ton biodiesel, kW/hour/ton FAME	191.2	108.8	138.4	21.6
Steam consumption, kg steam/ton FAME	295	168	214	34

product purity [9,18]. However, the integrated biodiesel processes based on RA have fewer degrees of freedom as compared to RD. This makes it challenging to correctly set the reactant feed ratio and consequently avoid impurities in the products. Kiss and Bildea [9] proposed an efficient plantwide control structure that can ensure the stoichiometric ratio of reactants and which fulfils the excess-of-methanol operating constraint, sufficient for the total conversion of the fatty acids.

Figure 11.10 depicts the dynamic simulation results for the flow sheet presented in Figure 11.8 (right) for the recycle of methanol vapours [9]. Production rate changes are easily achieved and the product purity is maintained at high values, with the acid concentration in FAME below the 2000 ppm requirement of the ASTM D6751-08 standard.

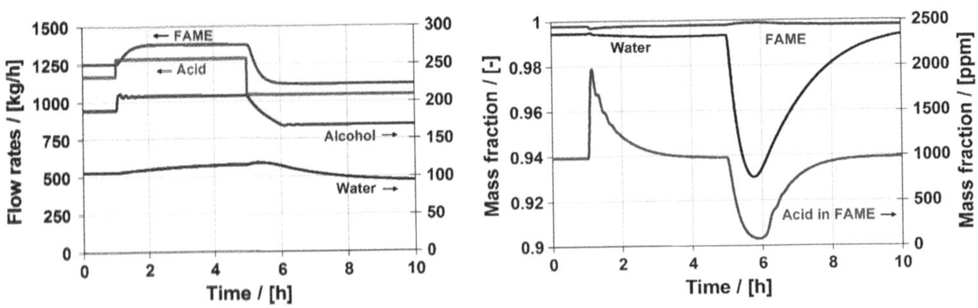

Figure 11.10 Dynamic simulation results for the flow sheet with methanol recycling: acid flow rate disturbance of +10% at 1 hour and −10% at 5 hours. Reprinted from Kiss and Bildea © 2011, with permission from Elsevier.

11.6 Green Processing Impact of RA

RA is essentially an established technology, known and used since the foundation of the chemical industry. Recently, the role of RA as a core environmental protection process has grown significantly, and today RA is the most widely applied reactive separation process. As illustrated by the case studies presented in this chapter, RA can lead to significantly lower investment and operating costs as compared to conventional processes. Moreover, it can bring significant green advantages such as decreased waste production, lower energy requirements, cleaner gas emissions and drastically reduced greenhouse gas (GHG) pollution.

Remarkable, for biodiesel production the novel heat-integrated RA process described here eliminates all conventional catalyst-related operations, uses raw materials and equipment efficiently and considerably reduces energy requirements – 85% lower as compared to the base case. Moreover, in spite of the high degree of integration, the process is easy to control, as demonstrated by the results from rigorous dynamic simulations.

11.7 Challenges and Future Prospects

In terms of industrial challenges, these depend mainly on the nature of the applied process [24]. For NO_x removal, the latest research activities have been dedicated to aqueous alkaline solutions, since valuable byproducts can be obtained (e.g. nitric acid, nitrates and nitrites). The recent research on SO_x removal focuses on wet processes, and significant progress has been made so far (e.g. limestone technology, Wellman–Lord process, ammonia or sea water scrubbing).

The current research into CO_2 removal from gaseous mixtures is closely following industrial requirements, focusing on the development of new solvents (e.g. amine mixtures or hindered amines) with high capacities and easy regeneration properties. The main area for future improvement of industrially used processes is in the regeneration of the solvent, which is responsible for 70–80% of the operating costs. Either the existing processes, including the equipment, must be improved or different solvents will have to be considered. In addition, the absorption could be integrated into the total plant, which could significantly reduce the total energy requirements.

Considering current tighter legislation, the future outlook is focused on spreading the use of integrated RA technologies to the carbon capture, the eco-efficient removal of contaminats, biogas purification and the production of biofuels – as for example in the manufacturing of biodiesel by the green processing of waste materials with high FFA contents.

References

(1) C. Noeres, E. Y. Kenig and A. Gorak, Modelling of reactive separation processes: reactive absorption and reactive distillation. Chemical Engineering & Processing, 42, 157–178 (2003).

(2) A. A. Kiss and C. S. Bildea, A review on biodiesel production by integrated reactive separation technologies, Journal of Chemical Technology and Biotechnology, 87, 861–879 (2012).

(3) A. A. Kiss, Applying reactive distillation, NPT Procestechnologie, 19(1), 22–24 (2012).

(4) A. A. Kiss, G. Rothenberg, A. C. Dimian and F. Omota, The heterogeneous advantage: biodiesel by catalytic reactive distillation. Topics in Catalysis, 40, 141–150 (2006).
(5) A. A. Kiss, A. C. Dimian and G. Rothenberg, Solid acid catalysts for biodiesel production – towards sustainable energy. Advanced Synthesis & Catalysis, 348, 75–81 (2006).
(6) A. A. Kiss, A. C. Dimian and G. Rothenberg, Biodiesel by reactive distillation powered by metal oxides. Energy & Fuels, 22, 598–604 (2008).
(7) A. A. Kiss, Novel process for biodiesel by reactive absorption. Separation & Purification Technology, 69, 280–287 (2009).
(8) A. A. Kiss, Heat-integrated reactive distillation process for synthesis of fatty esters. Fuel Processing Technology, 92(7), 1288–1296 (2011).
(9) A. A. Kiss and C. S. Bildea, Integrated reactive absorption process for synthesis of fatty esters. Bioresource Technology, 102, 490–498 (2011).
(10) E. Y. Kenig, L. Kucka and A. Gorak, Rigorous modeling of reactive absorption processes. Chemical Engineering & Technology, 26, 631–646 (2003).
(11) E. Y. Kenig and A. Gorak, Reactive absorption, in Integrated Chemical Processes – Synthesis, Operation, Analysis, and Control, K. Sundmacher, A. Kienle and A. Seidel-Morgenstern (Eds.), Wiley-VCH: Weinheim (2005).
(12) E. Y. Kenig and P. Seferlis, Modeling reactive absorption. Chemical Engineering Progress, January, 65–73 (2009).
(13) R. Taylor and R. Krishna, Multicomponent Mass Transfer, John Wiley & Sons: New York (1993).
(14) T. H. Chilton and A. P. Colburn, Mass transfer (absorption) coefficients prediction from data on heat transfer and fluid friction. Industrial & Engineering Chemistry, 26, 1183–1187 (1934).
(15) P. V. Danckwerts, Gas Liquid Reactions, McGraw-Hill: New York (1970).
(16) A. A. Kiss, C. S. Bildea and J. Grievink, Dynamic modeling and process optimization of an industrial sulfuric acid plant. Chemical Engineering Journal, 158, 241–249 (2010).
(17) R. Zarzycki and A. Chacuk, Absorption: Fundamentals and Applications, Pergamon Press: Oxford (1993).
(18) A. C. Dimian, C. S. Bildea, F. Omota and A. A. Kiss, Innovative process for fatty acid esters by dual reactive distillation. Computers & Chemical Engineering, 33, 743–750 (2009).
(19) C. S. Bildea and A. A. Kiss, Dynamics and control of a biodiesel process by reactive absorption. Chemical Engineering Research and Design, 89, 187–196 (2011).
(20) N. Dalaouti and P. Seferlis, Design sensitivity of reactive absorption units for improved dynamic performance and cleaner production: the NOx removal process. Journal of Cleaner Production, 13, 1461–1470 (2005).
(21) S. D. Dore, J. D. Perkins and L. S. Kershenbaum, Application of geometric nonlinear control in the process industries – a case study. Control Engineering Practice, 3, 397–402 (1995).
(22) L. Kershenbaum, Experimental testing of advanced algorithms for process control: when is it worth the effort? Chemical Engineering Research and Design, 78, 509–521 (2000).
(23) M. R. Rahimpour and A. Z. Kashkooli, Enhanced carbon dioxide removal by promoted hot potassium carbonate in a split-flow absorber. Chemical Engineering and Processing, 43, 857–865 (2004).
(24) O. Yildirim, A. A. Kiss, N. Huser, K. Lessmann, E. Y. Kenig, Reactive absorption in chemical process industry: A review on current activities, Chemical Engineering Journal, 213, 371–391 (2012).
(25) P. D. Vaidya and E. Y. Kenig, Gas-liquid reaction kinetics: a review of determination methods. Chemical Engineering Communications, 194, 1543–1565 (2007).
(26) D. W. Bailey and P. H. M. Feron, Post-combustion decarbonisation processes. Oil & Gas Science and Technology – Revue de l'Institut Francais du Petrole, 60, 461–474 (2005).

(27) A. Aboudheir, P. Tontiwachwuthikul and R. Idem, Rigorous model for predicting the behavior of CO2 absorption into AMP in packed-bed absorption columns. Industrial & Engineering Chemistry Research, 45, 2553–2557 (2006).
(28) B. Mandal and S. S. Bandyopadhyay, Simultaneous absorption of CO_2 and H_2S into aqueous blends of N-methyldiethanolamine and dietha-nolamine. Environmental Science & Technology, 40, 6076–6084 (2006).
(29) S. J. Chung, K. C. Pillai and I. S. Moon, A sustainable environmentally friendly NO_x removal process using Ag(II)/Ag(I)-mediated electrochemical oxidation. Separation and Purification Technology, 65, 156–163 (2009).
(30) J. L. Paiva and G. C. Kachan, Absorption of nitrogen oxides in aqueous solutions in a structured packing pilot column. Chemical Engineering and Processing, 43, 941–948 (2004).
(31) J. A. Patwardhan and J. B. Joshi, Unified model for NOX absorption in aqueous alkaline and dilute acidic solutions. AIChE Journal, 49, 2728–2748 (2003).
(32) R. K. Srivastava, W. Jozewicz and C. Singer, SO2 scrubbing technologies: a review. Environmental Progress, 20, 219–227 (2001).
(33) T. L. Muller, Sulfuric acid and sulfur trioxide, in Kirk-Othmer Encyclopedia of Chemical Technology, John Wiley & Sons (2006).
(34) S. I. Clarke and W. J. Mazzafro, Nitric acid, in Kirk-Othmer Encyclopedia of Chemical Technology, John Wiley & Sons (2005).
(35) M. Bowman, D. Hilligoss, S. Rasmussen and R. Thomas, Biodiesel: a renewable and biodegradable fuel. Hydrocarbon Processing, 85, 103–106 (2006).
(36) M. Balat and H. Balat, A critical review of bio-diesel as a vehicular fuel. Journal of Energy Conversion and Management, 49, 2727-2741 (2008).
(37) G. Knothe, Biodiesel: current trends and properties. Topics in Catalysis, 53, 714–720 (2010).
(38) J. M. Encinar, J. F. Gonzalez and A. Rodriguez-Reinares, Biodiesel from used frying oil. Variables affecting the yields and characteristics of the biodiesel. Industrial & Engineering Chemistry Research, 44, 5491–5499 (2005).
(39) M. G. Kulkarni and A. K. Dalai, Waste cooking oil-an economical source for biodiesel: a review. Industrial & Engineering Chemistry Research, 45, 2901-2913 (2006).
(40) A. A. Kiss, Separative reactors for integrated production of bioethanol and biodiesel. Computers and Chemical Engineering, 34, 812–820 (2010).
(41) M. A. Hanna, L. Isom and J. Campbell, Biodiesel: current perspectives and future. Journal of Scientific & Industrial Research, 64, 854–857 (2005).
(42) B. B. He, A. P. Singh and J. C. Thompson, A novel continuous-flow reactor using reactive distillation for biodiesel production. Transactions of the ASAE, 49, 107–112 (2006).
(43) K. Suwannakarn, E. Lotero, K. Ngaosuwan and J. G. Goodwin, Simultaneous free fatty acid esterification and triglyceride transesterification using a solid acid catalyst with in situ removal of water and unreacted methanol. Industrial & Engineering Chemistry Research, 48, 2810–2818 (2009).
(44) E. Vlad, C. S. Bildea, V. Plesu, G. Marton and G. Bozga, Design of biodiesel production process from rapeseed oil. Revista de Chimie, 61, 595–603 (2010).
(45) Y. H. Chen and C. C. Yu, Design and control of heat-integrated reactors. Industrial & Engineering Chemistry Research, 42, 2791–2808 (2003).
(46) A. C. Dimian and C. S. Bildea, Chemical Process Design – Computer-aided Case Studies, Wiley-VCH: Weinheim (2008).
(47) Aspen, Technology , Aspen Plus: User Guide, volumes 1 & 2 (2009).

12
Membrane Separations for Green Chemistry

Rosalinda Mazzei, Emma Piacentini, Enrico Drioli and Lidietta Giorno

Institute on Membrane Technology, CNR-ITM, University of Calabria, Rende, Calabria, Italy

12.1 Introduction

Resource shortages and environmental concern have prompted the development of clean technologies aiming at a rational use of raw materials, water and energy, prevention or minimization of waste and management of waste through recovery and recycling. Optimizing production processes, reducing equipment size, boosting plant efficiency, reducing capital costs, minimizing environmental impact and increasing safety, remote control and automation are among the challenges the green strategy must face. Process intensification (PI), a design approach to achieving real benefits in manufacturing and processing, responds well to this strategy.

Separation processes play a key role in green chemistry. Membrane operations are among the most suitable clean and energy-saving separation processes. They can promote separation at the molecular level on the basis of physical–chemical properties, including molecular size, electrostatic interactions, interfacial tension and solubility. They work in mild temperature and pressure conditions and do not need chemical additives.

Membrane technology has the potential to replace conventional energy-intensive techniques by permitting the selective and efficient transport of specific components,

Process Intensification for Green Chemistry: Engineering Solutions for Sustainable Chemical Processing, First Edition. Edited by Kamelia Boodhoo and Adam Harvey.

improving the performance of reactive processes and providing reliable options for sustainable industrial growth.

Membrane processes are actually the most effective separation processes and their increasing development promotes new and alternative strategies for their application in green technology. Their intrinsic characteristics of efficiency, selectivity, modularity, reduced energy consumption and waste production, decreased equipment size/productivity ratio, increased safety, remote control and automation fit the requirements of green and intensified processes well. In the last 45 years, membranes have been the subject of increasing interest at the industrial level, passing from lab to large scale in different application fields, with high technical and economical impact as clean, selective and efficient technologies.

In the first part of this chapter, some basic aspects of membranes and membrane processes are briefly described. In the second part, some relevant case studies are illustrated. Finally, water purification from heavy metal ions, hydrocarbons and organic and inorganic acids will be considered together with some examples of integrated membrane processes in water desalination and fruit juice concentration.

12.2 Membranes and Membrane Processes

In the most general sense, a membrane can be defined as a discontinuous phase that separates and/or contacts two adjacent compartments/phases and promotes exchange of matter, energy and information between the compartments under the action of a driving force (Figure 12.1).

Membranes show a great variety in their physical structures and the materials they are made from. In Table 12.1 a membrane classification according to materials, structures and configurations is reported.

Separation in membrane processes is the result of the different transport rates of different components through the membrane. Transport rates of individual chemical components through the membrane are determined by the driving force acting on them, their mobility and their concentration within the membrane.

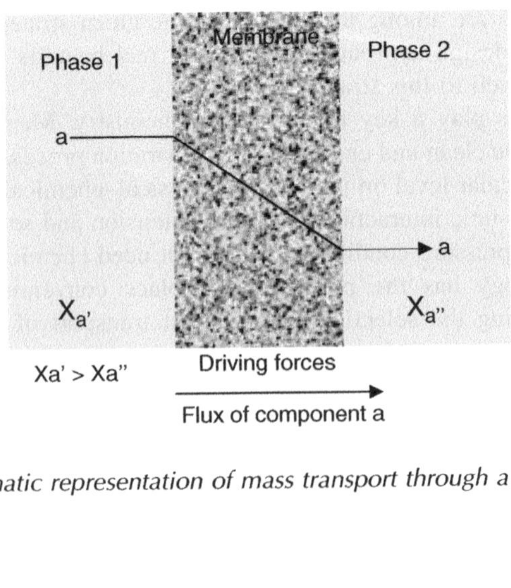

Figure 12.1 *Schematic representation of mass transport through a synthetic membrane.*

Table 12.1 Materials, structures and configurations of technically relevant membranes.

Membrane materials	Membrane structures		Membrane configuration
	Symmetric	***Asymmetric***	
Polymers	Homogeneous film	Integral asymmetric	Flat-sheet
Ceramics	Cylindrical pores	Composite structure	Spiral wound
Glass	Sponge-type structures		Tubular
Metals			Capillary
Liquids			Hollow fiber

The driving force is the difference in energy of a component in the two phases separated by the membrane. Therefore, in a very general form, the driving force for the transport of a component through a membrane can be expressed by its energy gradient; that is, the gradient in its electrochemical potential, which is the result of hydrostatic pressure, concentration, electrical potential and the temperature difference between the two phases separated by the membrane. The mobility of a component in the membrane is primarily determined by its size and the physical structure of the membrane material. The concentration of a component in the membrane is primarily determined by its chemical compatibility with the membrane material.

Transport processes are conventionally described by well-known phenomenological equations that relate fluxes of components to corresponding driving forces in the form of linear relations:

$$J = -P\frac{dX}{dz} \tag{12.1}$$

For example, Darcy's law describes the relation between a volume flux and a hydrostatic pressure difference; Fick's law describes the relation between the flux of individual components and a concentration gradient; Ohm's law describes the relation between an electrical current and an electrical potential gradient; and Fourier's law describes the relation between heat transport and a temperature gradient.

Microfiltration, ultrafiltration, nanofiltration and reverse osmosis are typical pressure-driven membrane processes. In these processes a feed solution is separated into a permeate (i.e. the collected fraction that passes through the membrane) and a retentate (i.e. the fraction that contains particles or molecules retained by the membrane), as illustrated in Figure 12.2. Some specific properties, together with the driving forces and fields of application, of pressure-driven membrane processes are summarized in Table 12.2.

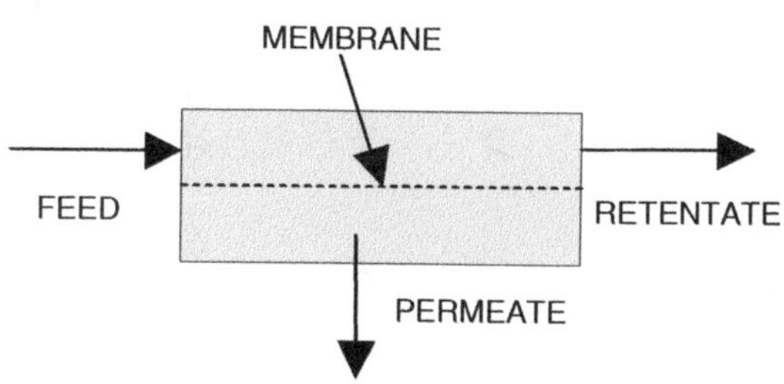

Figure 12.2 Schematic representation of pressure-driven membrane processes.

Table 12.2 *Principles and applications of common pressure-driven membrane processes.*

Process	Driving force	Separation mechanism	Transport mode	Membrane	Pressure (bar)	Permeability ($l/h/m^2/bar$)	Applications
Microfiltration	Δp	Molecular sieving	Convection	Macro porous 0.1–10.0 μm	0.1–2.0	>1000	Clarification, pre-treatment, removal of bacteria
Ultrafiltration	Δp	Molecular sieving	Convection	Meso porous 2–10 nm	0.1–5.0	10–1000	Removal of macromolecules, bacteria, viruses
Nanofiltration	Δp ($\Delta\mu_i$)	Molecular sieving Charge effect	Diffusion Donnan-exclusion	Micro porous 0.5–2.0 nm	3–20	1.5–30.0	Removal of (multivalent) ions and relatively small organics
Reverse osmosis	Δp ($\Delta\mu_i$)	Solution–diffusion	Diffusion	Dense and micro porous < 0.5 nm	5–120	0.05–1.5	Removal of ions, ultrapure water production and sea water desalination

Membrane separation processes can differ greatly with regard to membranes, driving forces, areas of application and industrial or economic relevance. The term 'microfiltration' is used when particles with a diameter of 0.1–10.0 μm are separated from a solvent or other low-molecular component. Particles are separated according to their dimensions, and the separation mechanism is based on the molecular sieving effect. In ultrafiltration the components to be retained by the membrane, mainly of asymmetric structure, are macromolecules or submicron particles. The membranes used for nanofiltration are mostly asymmetric and the components to be separated are molecular mixtures and ions. The mechanisms involved in the separation process are molecular size exclusion, diffusion and Donnan dialysis. Reverse osmosis is a membrane process that permits separation of particles, macromolecules and low-molecular-mass compounds (salts, sugars, etc.) from a solvent.

Feed solutions, therefore, often have a significant osmotic pressure, which must be overcome by the applied hydrostatic pressure.

Gas separation and pervaporation are the other two important membrane processes used in many applications. Some of these, together with their principles, are reported in Table 12.3.

In gas separation the membranes used are both porous and dense. With porous membranes the transport is based on the so-called 'Knudsen diffusion', while with dense solid membranes the gas transport is based on a solution–diffusion mechanism. Both Knudsen diffusion and solution–diffusion transport can result in a selective transport of gases and thus in a separation of gases. However, the extent of the separation – that is, the separation factor – is much higher in a solution–diffusion transport than in Knudsen diffusion.

Pervaporation is a process in which the liquid mixture to be separated is in direct contact with one side of a dense membrane and the permeated product is removed as vapour from the other side by the application of a lower pressure, inert gas or freezing. The principle behind this process is selective sorption, diffusion of the selected components from the liquid solution through the membrane and then evaporation from the permeate side of the membrane at the downstream side with a low partial pressure. In this case the driving force is the chemical potential gradient.

Another important family of membrane operations is the membrane contactors. A membrane contactor is a device that achieves gas/liquid or liquid/liquid mass transfer without dispersion of one phase within another. The interface is established at the membrane mouth and the transport is guided by simple diffusion through the membrane pores. Membrane contactors can be used to separate both immiscible and miscible liquids, depending on their type (Table 12.4). Examples of membrane contactors in which the two solvents are miscible are: membrane-distillation, osmotic-distillation and membrane crystallization. In a membrane distillation process, a porous hydrophobic membrane is in contact with aqueous solutions of different temperatures on either side of it. The high interfacial tension between the hydrophobic membrane surface and the aqueous phase prevents the passage of the aqueous phase through the membrane as a liquid phase. The temperature gradient promotes evaporation, forming a vapour–liquid interface at the pore entrance. At the interface, the more volatile compound evaporates, diffuses and/or convects across the membrane and is condensed and/or removed at the other side of the system.

Table 12.3 *Principles and applications of gas separation and pervaporation.*

Process	Driving force	Separation mechanism	Transport mode	Membrane	Pressure (bar)	Permeability ($l/h/m^2/bar$)	Applications
Gas separation	Δp_i	Solution–diffusion Diffusion	Diffusion Surface diffusion, capillary condensation, molecular sieving, Knudsen diffusion	Dense Macro porous <1 μm	0.1–30.0	1–5000	Gas separation
Pervaporation	ΔC	Solution–diffusion	Diffusion	Dense	0.001–10.0	06–60	Solvent dehydration, removal of organic solvents from industrial waste waters, concentration of hydrophobic flavour compounds in aqueous solutions, breaking of azeotropes

Table 12.4 *Types of membrane contactor.*

Membrane contactor	Contacting phases	Phases properties	Applied driving forces	Type of membrane	
Membrane distillation	Phase 1 Liquid	Phase 2 Liquid	Miscibile	Partial pressure gradient	Hydrophobic porous
Osmotic distillation	Phase 1 Liquid	Phase 2 Liquid		Partial pressure gradient	
Membrane crystallization	Phase 1 Liquid Solid	Phase 2 Liquid		Temperature gradient/partial pressure gradient	
Membrane-based solvent extraction	Phase 1 Liquid	Phase 2 Liquid	Immiscibile	Concentration gradient	Hydrophilic or hydrophobic, mesoporous
Membrane stripper/scrubber	Phase 1 Gas Liquid	Phase 2 Liquid		Concentration gradient	Hydrophilic or hydrophobic, porous
Phase-transfer membrane reactor	Phase 1 Liquid	Phase 2 Liquid		Concentration gradient	Hydrophilic or hydrophobic, mesoporous

Osmotic distillation is a concentration technique for aqueous solutions in which the driving force is promoted by a hypertonic solution used as a strip phase.

Membrane crystallizers are membrane contactors in which solvents (usually water) evaporate through microporous hydrophobic membranes in order to concentrate feed solution above their saturation limit. The driving force is a partial pressure gradient, while the supersaturated environment permits the nucleation and growth of crystals.

Examples of membrane contactors in which the two phases are immiscible are: membrane-based solvent extraction, membrane strippers/scrubbers, membrane emulsifiers and phase-transfer membrane reactors.

Another membrane process mentioned in the case studies (Section 12.3) is electrodialysis, one of the first membrane processes used in desalination and brackish water for the production of potable water. In this process, cation-exchange membranes carrying fixed negative charges and anion-exchange membranes carrying positive fixed charges are placed in alternating series between two electrodes. The cells, compartmentalized by the different membranes, are filled with an electrolyte solution and contain the feed solution. When an electrical potential is established between the electrodes, the cations migrate towards the negatively charged cathode, while the anions migrate towards the positively charged anode.

12.3 Case Studies: Membrane Operations in Green Processes

Membranes and membrane operation systems find a huge number of applications in green chemistry. Table 12.5 shows the main applications in industry. Some selected applications are discussed in this section. In particular, attention is focused on: metal ion removal from waste water, acid separation, hydrocarbon separation and optically pure enantiomer production. Furthermore, the concept of integrated membrane operation for innovative system design and processing is highlighted through a description of water desalination and fruit juice processing.

12.3.1 Membrane Technology in Metal Ion Removal from Waste Water

Toxic heavy metals such as zinc, copper, nickel, mercury, cadmium, chromium and lead are elements with atomic weights between 64 and 200 u.m.a[1]. With the fast development of industry in developing countries, they are increasingly directly or indirectly discharged into the environment. They are not biodegradable and if not recovered they can become very dangerous to living organisms, in which they can accumulate, causing toxic and carcinogenic problems [5] (see Table 12.6).

Many methods are currently used to remove these toxic compounds from waste water, such as chemical precipitation, ion exchange, adsorption and electrochemical treatment technologies. In Table 12.7 the methodological principles underlying the different traditional techniques are reported, together with their advantages and disadvantages.

Membrane processes are very promising systems for heavy metal removal [16] (Table 12.8). The membrane processes most frequently used to remove metal ions are: ultrafiltration for pre-treatment, reverse osmosis, nanofiltration and electrodialysis.

[1] Atomic mass unit

Table 12.5 *Major membrane process applications in industry (for more details on the various applications see [1–4]).*

Field	Application	Membrane processes	Advantages compared to conventional processes
Water industry	Ultapure water production	MF and UF are used for the pre-treatment and removal of microorganisms. NF, RO and MD are used to remove contaminants, including ions. Membrane contactors are used for water degassing. Electrically enhanced membrane processes are used for ion removal. MD is used in desalination to produce ultrapure water. Membrane reactors are used in photoxidation processes.	Simplicity of layout and lower energy compared to thermal desalination processes. Modular design that permits easy reproducibility at large scale.
	Water treatment Industrial, municipal and agricultural waste water	Integrated membrane systems are most frequently used. MF and UF are used in pre-treatment for particle and microbe removal, and in conjunction with adsorbent and coagulant for organic compound removal. NF and RO are used in ground water treatment for salt and natural organic matter removal. For water reuse and surface water applications, integrated membrane systems are most frequently used (such as biophenols from vegetation water). Membrane bioreactors are used in municipal and agricultural waste water treatment.	Minimization or even elimination of the need for dosing chemicals and reduction of the demand for disinfectants. Production of integrated systems able to integrate biodegradation with the clarification step.

(*Continued*)

Table 12.5 *(Continued)*

Field	Application	Membrane processes	Advantages compared to conventional processes
	Water softening	UF, NF and RO.	High product quality compared to lime treatment.
	Water sanitation	MF and UF are used for preliminary disinfection. Complete virus rejection is obtained by using a charged membrane relevant to the electric charge of the viruses. The smallest viruses can be rejected by NF.	Reducing demand of disinfectants and physical removal of microorganisms.
	Water denitrification	ED, NF and RO.	Higher nitrate recovery. Removal of both mixed-liquor suspended solids and other contaminants such as the residues of the carbon source.
	Sulfate removal from water	NF.	Very high retention of sulphates.
	Phosphorous removal from water	RO and ED. MF/UF with adsorbers.	Complete phosphorous removal.
	Fluoride removal	RO (not used in drinking water production), NF, ED and Donnan dyalisis.	Alternative technique.
	Removal of organic pollutants from water	NF and RO. UF (with or without additives) to remove macromolecular organic matter.	Effective and economic control of THMs.
	Biocide removal	NF and RO.	Alternative to very expensive conventional methods (e.g. activated carbon in powder form, PAC).
	Waste water treatment and reuse	Integrated membrane systems. Membrane bioreactors. MF, UF and NF used to pre-treat (also to disinfect for MF and UF) water. RO used for dissolved solids reduction.	Waste water with low solid content, disinfected, with low sludge production and compact plant configuration.

	Water desalination	**1.** Desalination: RO. **2.** Surface water recovery: MF and UF for the removal of particles, turbidity, bacteria and disinfection; MF and UF integrated with PAC, NF and RO for water with high water content feed. **3.** Ground water recovery: NF.	Reduction of energy consumption, reduction of treated volume .
	Remediation of contaminated soil	RO, RO and membrane bioreactor, extractive membrane bioreactor. Reverse osmosi-NF-cristallization. Polymer-enhanced ultrafiltration.	Full and selective purification.
Paper industry		Membrane processes used for concentration of pulping wastes. UF used to treat paper effluents for the removal of suspended solids, colloidal materials, anionic trash, bacteria, colours etc. NF used for multivalent ion retention. Integrated UF–NF for paper machine circulation, water reuse and the recovery of bleaching effluents. Integrated MF–UF–NF for white water recovery.	Specific removal of compounds that affect paper quality (chelatin agents, e.g. citrate, tartarate, EDTA etc.)
Textile industry		MF used to remove colloidal dyes or as a pre-treatment to the NF step. UF used to recover dyeing effluents for particles and macromolecules removal. NF used for the removal of low-molecular-weight organic compounds and divalent salts. RO suitable for the removal of ions and larger species from the dye bath. Integrated systems.	Complete removal of colour, suspended dissolved solids and refractory chemicals.

(Continued)

Table 12.5 *(Continued)*

Field	Application	Membrane processes	Advantages compared to conventional processes
Tannery industry		MF used for the reduction of total suspended solids. UF used for macromolecules and colloids. NF used for multivalent salt removal. RO used for water reclamation. Pervaporation can remove and concentrate VOCs in pure form.	Recovery and reuse of primary resources, reduction of environmental impact, easy reuse of sludge, saving of chemicals and water etc.
Metal plating industry		RO used to reclaim metals from electroplating industry. RO ion exchange and electrodyalisis used for water recycling. NF and integrated systems of NF and MF used to recycle spent final rinse water from electroless plating.	More selective separation and cheaper method.
Electronic industry		Cross-flow electrofiltration/electrodyalisis process. RO in semiconductor industry.	Less sensitive to fouling phenomena compared to other conventional filtration process. Lower operating expenses compared to distillation and ion exchange for organic pollutant diminution in the semiconductor industry.
Pharmaceuticals industry	Antibiotic production, concentration and removal from waste water	UF used to separate antibiotics from cells. NF used to concentrate antibiotics. Facilitated liquid membrane and membrane contactors used for amino acid separation from fermentation broth and for the purification of amino acids and antibiotics. UF and RO used for antibiotic recovery from waste water.	Improved efficiency of antibiotic separation from fermentation broth and cost reduction.

Pharmaceutical production	Membrane bioreactors.	Continuous systems.
Vaccine production	MF used for the purification of protein polysaccharides conjugate vaccine.	Possibility of carrying out the process in sterile conditions.
Peptide production	UF, ED with UF, supported liquid membranes.	Clean and concentrated samples.
Protein fractionation	UF (separation of proteins and nucleic acids). High-performance tangential flow filtration.	High throughput and easy scale-up compared to electrophoresis and chromatographic separation.
Plasma purification	Affinity membranes.	Alternative technique to the expensive process of centrifugation.
Enzyme purification		No intrapartcle diffusion, short axial-diffusion path, low pressure drop, no bed compaction. Mild operation conditions that do not denaturate the proteins.
Grow factor purification		Mild operation conditions that do not alter macromolecules structures.
Monoclonal antibody purification	Ion exchange membrane.	Binding capacity independent of flow rate.
Virus removal	MF and UF.	Treatment of large quantities with a recovery of about 90% of the colony-forming vector particles in the retentate.
Endocrine disruptor removal	Filtration through hydrophobic membranes and pervaporation.	Selective and efficient removal.
Estrogenic hormone removal	NF and RO.	Alternative techniques.
Endotixin removal	UF.	Alternative techniques.
Enantiomer separation	Affinity membranes, membrane (bio) reactors, dyalisis membranes, liquid membranes.	Easy scale-up, energy saving.

(*Continued*)

Table 12.5 *(Continued)*

Field	Application	Membrane processes	Advantages compared to conventional processes
Food industry	Fruit juice and pulp treatment	Integrated membrane process of MF/UF and enzymatic treatment to concentrate fruit pulp. Fruit juice concentration carried out by integrated systems of RO, NF, UF, MD and OD. Fruit juice clarification carried out by MF and UF.	High-quality fruit juice preserving nutritional, sensory and storage characteristics. High productivity and cost reduction through yield improvement, high-quality product, low-energy operating cost, decreased time and lower waste disposal.
	Aroma processing	Pervaporation, vapour permeation and nondispersive extraction through membrane contactors used in recovery concentration and removal of natural aromas. Other processes used are RO, NF and membrane (bio)reactors.	High selectivity, low energy consumption, physical separation techniques, moderate operating temperature and no requirement for additives.
	Wine processing	Wine stabilization carried out by electrodialysis. Wine clarification and sterilization carried out by MF. Grape and must concentration carried out by RO to increase the sugar content in the must through partial water removal. Pervaporation used for odorous compound analysis in wine.	Avoidance of generation of hazardous waste (e.g. DE filtration) during wine clarification. Precise control of the final crystalline salts of potassium tartarate (KHT) during wine stabilization.

Beverage processing	UF and RO extensively used to treat waste water from the brewing industry. MF used in beer clarification for the removal of yeast cells, flocks formed by the coagulation of polyphenolic and proteinaceous compounds. MF also used for pasteurization and sterilization. OD.	Integration of clarification/stabilization/sterile filtration in one single continuous operation.
Edible oil production	MF and UF used in waste water pre-treatment to remove suspended solids and macromolecules. Integrated systems of MF/UF/NF/RO most frequently used to recover water and isolate bioactive molecules.	Less energy intensive than evaporation and incineration.
Potato and starch industry	Membrane processes used in waste water treatment. UF used to recover proteins and RO used to recover process water in potato-starch factories.	Cheap method supported, able to treat water and recover valuable compounds.
Sugar processing	MF, UF, NF, ion exchange and ED used for decolourization and removal of impurities in sugar cane, to reduce viscosity, to reduce calcium oxide content and to improve crystallization.	Significant improvement of sugar quality and yield.
Meat technologies	Treatment of waste water from abattoirs by recovery of proteins from animal blood. MF used as a pre-treatment step. UF mainly used to fractionate and isolate the proteins. UF and RO used to pre-concentrate the plasma.	Less expensive compared to freeze drying gel filtration. The recovered proteins can be directly used as food ingredients.

(*Continued*)

Table 12.5 *(Continued)*

Field	Application	Membrane processes	Advantages compared to conventional processes
	Milk and dairy industry	Membrane processes used to clarify concentrate and recover and fractionate. MF used in milk partial sterilization and in the production of casein-enriched cheese milk. UF used for protein fractionation and recovery in cheese manufacturing, milk standardization, whey treatment and fractionation of milk proteins (e.g. a-lactalbumin, b-lactoglobulin). NF can be used to produce milk with low lactose content. NF and RO used in the treatment of effluents coming from the dairy industry, and for the recovery of valuable compounds.	Less expensive processes.
	Food additive production	Removal of polyphenols from fruit juice waste water to produce food additives. Enzymatic oxidations using a membrane bioreactor used for bioremediation. Integrated membrane processes used for polyphenol isolation and concentration.	Possibility of purifying water and recovering high-added-value compounds in integrated systems.

MF, microfiltration; UF, ultrafiltration; NF, nanofiltration; RO, reverse osmosis; MD, membrane distillation; ED, electrodyalisis; THM, trihalomethae; OD, osmotic distillation; PAC, powdered activated carbon; EDTA, Ethylenediaminetetraacetic acid; VOC, volatile organic compounds.

Table 12.6 *Effects of heavy metals on human health.*

Heavy metal	Area of negative effect	Reference
Zinc	Stomach, skin	[6]
Nickel	Lung and kidney, skin; cancer	[7]
Mercury	Central nervous system, lung and kidney; chest pain, dyspnoea	[8]
Cadmium	Kidney; carcinogenic	[5]
Lead	Central nervous system, kidney, liver, reproductive system	[9]
Chromium	Skin, lung; cancer	[10]

Table 12.7 *Traditional techniques used to remove heavy metals from waste water.*

Technique	Advantages	Disadvantages	Methodological principle	Reference
Chemical precipitation	Simple, inexpensive	Sludge to be processed	Reaction of chemicals with the metal to form insoluble precipitates	[11]
Ion exchange	High treatment capacity, high removal efficiency, fast kinetics	Heavy metal uptake affected by pH, temperature, initial metal concentration and contact time	Ion-exchange resins exchange cations with metals	[12]
Adsorption	Inexpensive, flexibility in design and operation, high-quality treated effluent, easy regeneration of adsorbents	Easy desorption of heavy metal	Interaction of adsorbant and heavy metal through weak bonds	[5]
Coagulation	Very efficient	Act only on insoluble substances, requiring other techniques to treat heavy metals (filtration)	Destabilization of colloids by coagulant	[13]
Flocculation	Inexpensive, flexibility in design and operation, high-quality treated effluent	Needs other techniques to treat heavy metals (filtration)	Use of polymer to form bridges between the flocs and formation of agglomerates	[5]

(*Continued*)

Table 12.7 (*Continued*)

Technique	Advantages	Disadvantages	Methodological principle	Reference
Flotation	Efficient	Sludge to be processed	Separation of heavy metal from a liquid phase by bubble attachment from mineral processing	[14]
Electrochemical treatment	Very efficient	Large capital investment, expensive electricity supply	Plating out of metal ions on a cathode surface	[15]

Ultrafiltration pores (2–10 nm) are generally too large compared to heavy metal ions. To obtain high removal efficiency, two main techniques are used: micellar-enhanced ultrafiltration (MEUF) and polymer-enhanced ultrafiltration (PEUF). The principle of the former is the formation of micelles, which capture the ions, producing a complex metal structure. The micelles are formed by adding surfactants, such as sodium dodecyl sulphate (SDS), to water. The binding complex is formed when the surfactant reaches the critical micelle concentration (CMC). The micelle/metal ion complexes are easily retained by the ultrafiltration process due to their large size. Different parameters can influence the efficiency of the separation process, such as surfactant and metal ion concentration, pH, ionic strength and so on [17,18]. The membranes generally used in this process are both inorganic (ceramic) and organic (polysulphone, regenerated cellulose, polycarbonate).

In the PEUF technique, the system used to complex the metal ions is based on their interaction with polymers. This complex forms a macromolecular structure with a molecular weight higher than the molecular weight cut-off of the membrane. The retained species needs an additional treatment to remove ions and to recover the starting polymer. One of the key studies related to this technique is to find suitable polymers that can achieve a good complexation with metal ions. The complexing agents generally used are polyacrilic acid (PAA), polyethyleneimine (PEI), diethylaminoethyl cellulose and humic acid, while the membrane materials are both inorganic (ceramic) and organic (polyethersulphone, polysulphone, etc.).

Reverse osmosis is increasingly applied to heavy metal removal and waste water treatment. Different examples of heavy metal removal are reported in the literature [19–22], but these have yet to be widely applied. A pilot plant bioreactor with a reverse osmosis unit that produced high removal of heavy metal has also been developed [5,23].

Figoli *et al.* [24] analysed the removal of pentavalent arsenic by two commercial nanofiltration membranes. One of the main results of this study was the increase in metal ion removal produced by changing operating parameters such as temperature, pH and feed concentration. There is again a great deal of literature describing the efficiency of nanofiltration for the removal of heavy metal [25,26]. Liu *et al.* [26] studied the treatment

Table 12.8 *Membrane processes used in heavy metal removal.*

Heavy metal	MF/flocculation	UF	NF	RO	ED	Extraction by liquid membrane	Adsorption by ion-exchange membrane	Extraction by supported liquid membrane
Arsenic	+	+ with flocculation	+	+				
Boron				+	+			
Cadmium		+ PEUF	+ charged membrane			+	+	+
Cesium								+
Chromium		+ integrated with NF and RO, and with PEUF	+	+			+	+
Cobalt		+ MEUF, adsorption						+
Copper		+ with complexing agent				+		
Gold								+
Indium			+					
Lead		+ PEUF	+		+			
Manganese	+					+		
Mercury		+ PEUF						
Molybdenum						+		
Nickel		+ PEUF	+		+	+		
Palladium						+		
Plutonium								+
Selenium			+					
Silver								+
Thorium						+		
Uranium			+			+		
Zinc				+	+			+

MF, microfiltration; UF, ultrafiltration; RO, reverse osmosis; ED, electrodyalisis; PEUF, polymer-enhanced ultrafiltration; MEUF, micellar-enhanced ultrafiltration.

of waste water from the metallurgical industry for the purification of water containing heavy metals. They concluded that water purified by both nanofiltration and reverse osmosis satisfied the state reutilization qualification, but nanofiltration is more suitable for development at the industrial scale.

Electrodyalisis is another membrane process that appears very promising in heavy metal waste water treatment. Generally this process is used for the production of drinking water and to process brackish water and sea water, treat industrial effluents, recover useful materials from effluents and produce salt. An electrodyalisis pilot plant comprising a set of ion-exchange membranes was developed by Nataraj *et al.* [27] for the removal of hexavalent chromium. The performance of this process was also evaluated, with success, for the separation of Cu and Fe from solutions in copper electrowinning operations. Lambert *et al.* [28] studied the separation of Cr (III) by electrodyalisis using modified cation-exchange membranes, while Mohammadi *et al.* [29] studied Pb^{2+} separation from waste water. In the latter case, increasing voltage and temperature improved cell performance, while an increase of flow rate decreased the separation percentage.

12.3.2 Membrane Operations in Acid Separation from Waste Water

Membrane processes also find application in inorganic and organic acid separation. Inorganic acids are commonly used in numerous industrial processes, such as steel treatment, surface treatment, metal refining, the electronics and glass industries and pigment and chemical production, while organic acids are increasingly used in the food, pharmaceuticals and packaging industries. Recovery of inorganic acids provides economic and environmental benefits as they are both valuable materials and hazardous waste. Organic acids are frequently used as food or pharmaceutical additives, as active ingredients in some care products and as monomer units in biodegradable polymer synthesis. The interest that exists in the recovery of organic acids is because:

- Some low-molecular-weight carboxylic acids appear as waste materials in different industrial waste waters (petrochemical, pharmaceutical, polymer, etc.).
- Organic acids are produced by fermentation and they need to be extracted from a complex mixture in order to obtain a pure product.

Table 12.9 illustrates some interesting examples of inorganic and organic acids used for different applications in industry and the sources from which they can be extracted.

The disposal of inorganic acids is usually carried out by a neutralization–precipitation–discharge sequence that uses solvent extraction, stripping, precipitation and evaporation. This makes the process more complicated and increases the cost of operation and the accumulation of toxic solid waste.

The recovery of inorganic acids through simultaneous concentration and purification has the advantages of waste disposal cost savings and the recovery of valuable materials that can be reused in the production line. Membrane processes, such as electrodialysis and diffusion dialysis, can be used to separate inorganic acids from industrial waste water, while nanofiltration, reverse osmosis and membrane distillation can purify acids such as phosphoric acid of cationic impurities. Innovative production schemes based on membrane operations may be applied during organic acid production in fermenters and by conversion

Table 12.9 *Examples of inorganic and organic acids used in industry.*

	Acid	Source	Industrial use
Inorganic	HCl	Pickling bath	Carbon steel pickling
	H_3PO_4	Agricultural and industrial waste water	Production of detergents, food products, toothpastes and fertilizers
	HF	Pickling bath Etching solutions Glass production washing water	Electronic element pickling Silicate glass etching
	H_2SO_4	Pickling bath	Titanium pickling
	HNO_3	Industrial waste water	Production of fertilizers Explosives Etching and dissolution of metals
Organic	Acetic acid	Chemical industry waste water	Organic intermediate in the syntheses of terephthalic acid, vinyl acetate, acetic anhydride and cellulose esters
	Amino acid	Fermentation broth media	Nutraceutical additives Sweeteners for hypocaloric beverages Enrichment of human and animal diet Cell culture
	Citric acid		Active ingredient in some cleaning solutions Pharmaceutical or food additive
	Gluconic acid		Food additive Active ingredient in some cleaning solutions
	Lactic acid		Biodegradable polymer synthesis Food and pharmaceutical additive
	Valeric acid	Chemical industry waste water	Polymer manufacture

of organic salts. Membrane extraction, through membrane contactors, pervaporation, electrodialysis or nanofiltration/ultrafiltration, represents a suitable choice for organic acid purification.

A few case studies will be illustrated that have application for the separation of inorganic and organic acids.

12.3.2.1 Inorganic Acid Separation

In the field of inorganic acid separation, some species need to be recovered as valuable waste materials from industrial waste water, such as HCl, HNO_3 or H_2SO_4. They are used to clean the metal surface as pickling liquors in order to remove surface oxides. Pickling rinsing water contains acid that can be reused. As an example, membrane technology plays a particular role in hydrochloric acid recovery, which can be carried out by:

- electrodialysis [30–32];
- diffusion dialysis [33,34];
- nanofiltration [35];
- membrane distillation [36,37].

Electrodialysis allows the acid to be simultaneously reconcentrated and purified. Diffusion dialysis is an attractive tool for acid recovery due to its energy efficiency as the operation only needs electrical energy in order to pump the liquids into the overhead tank. Few examples are reported in the literature in which nanofiltration is used for this application. The concentration and recovery of HCl by membrane distillation has been performed using polypropylene membranes with successful results for industrial application. Another interesting application of membrane processes in the separation of inorganic acids is the purification of inorganic acids, such as H_3PO_5, after chemical production. In this case, reverse osmosis and nanofiltration are used as efficient, alternative methodologies to the conventional solvent extraction method.

12.3.2.2 Organic Acid Separation

One interesting example of organic acid separation is the extraction of lactic acid using membrane processes. Lactic acid is widely used in the food industry for acidulants, preservatives and flavour enhancers, as well as in the pharmaceuticals, cosmetics, textile and leather industries. In addition, polylactic acid can be used to produce biodegradable polymers. The biological process that permits the production of a stereoisomer of lactic acid requires the use of an efficient and economic downstream process to recover lactic acid and isolate it from various impurities in the fermentation broth. Moreover, separation and purification operations often represent the majority of the production cost.

Microfiltration and ultrafiltration are efficient operations for clarifying such complex fermentation media. Clarifying fermentation broths by filtration is thus becoming more and more attractive in the present context of industrial waste reduction. Microfiltration is used to separate microbial cells for their subsequent recycling to the bioreactor, ensuring high cell concentration and increasing productivity, while ultrafiltration can also retain proteins. Carrère [38] has compared different operating modes of microfiltration in terms of productivity and fouling rate. The constant transmembrane pressure mode was found to be more convenient than the constant permeate flux mode because of its higher productivity levels and moderate fouling rates. The reactive liquid–liquid extraction method using a microporous hollow-fibre membrane module may potentially overcome many of the disadvantages found in the extraction of lactic acid, such as high cost, large volumes of waste, multiple steps and low-efficiency operation with under practical conditions [39–41]. Reverse osmosis and nanofiltration can be used to separate nutrients, salts and unconverted carbon sources from lactic acid. For industrial application, an organic phase with sunflower oil is attractive because of its cost nontoxicity and environmental benefits. It has proven to be effective at operating conditions (pH and temperature) similar to those of fermentation and it shows stability with the commercially available large processing device [42]. One problem related to fermentation processes is the production of lactate salts as a consequence of the addition of alkali to neutralize the pH and increase the fermentation productivity. Electrodialysis represents one of the most important methods for desalting solutions. Two steps can be indentified: 1) separation and concentration of lactate salts by conventional electrodialysis and 2) lactate salts conversion into lactic acid by bipolar electrodialysis. Due to the poor conductivity of the organic phase, power consumption in electrodialysis is always high and electrodialysis fermentation for lactic acid has barely been commercialized.

The successful separation of fermentation components, integrating membrane-based processes with a fermentor, has been demonstrated by experimental study, but a judicious combination of the types of membranes and modules is necessary in order to develop a continuous cost-effective fermentative process using a cheap, renewable carbon source. Integration of the microfiltration membrane in the first stage, followed by the nanofiltration membrane in a flat-sheet crossflow module in the second stage, can help reach the targets of commercial production of pure lactic acid with high productivity and concentration in an environmentally sustainable process.

12.3.3 Membrane Operation for Hydrocarbon Separation from Waste Water

The most important groups of organic substances affecting the environment are listed in Table 12.10. Nonhalogenated and halogenated volatile organic substances can be removed from water using membrane processes such as pervaporation, vapour permeation, membrane distillation, reverse osmosis, nanofiltration and hybrid systems. Pervaporation and vapour permeation have been applied to waste waters and waste gases containing nonhalogenated volatile organic substances from the contaminated streams to promote their recovery or elimination [43–45]. Hydrophobic polydimethyl siloxane (PDMS) membranes are used to remove a wide variety of nonhalogenated volatile substances from aqueous solutions because they have high permselectivity, permeability and diffusivity [46]. Crosslinking of PDMS membranes permits an improvement in mechanical strength [47]. Commercial reverse osmosis and nanofiltration membranes can be used for the separation of organic pollutants [48]. Retention is related to membrane type, solute size and solute–membrane interaction. In particular, solutes with larger widths and lengths and higher logarithms of ocatnol–water partition coefficients are reported to have higher retentions [49]. Hybrid systems based on pervaporation–bioconversion are also used for wastewater treatment [50,51].

The request for more efficient and less costly technologies for the recovery of organic solvents from air has led to an increasing interest in the use of vapour permeation and pervaporation [52,53].

Separation of olefins–olefins or olefins–paraffin is another field in which membrane systems represent an attractive alternative to conventional methods, with the potential for retrofit and energy saving. Olefins can be produced by catalytic cracking in the conversion of crude oil into gasoline and must be separated from the purge gas in order to meet regulations on atmospheric emissions. Facilitated transport membranes can be used as an alternative to conventional distillation in olefin separations or else can be incorporated into hybrid processes with distillation or adsorption in order to enhance the process efficiency [54]. Facilitated transport membranes using carriers are more selective than other membrane processes. The carrier agent acts as a shuttle to selectively transport one component from the feed to the product side of the membrane. The bonds formed by the chemical complexation are stronger than those by van der Waals forces alone, so it is possible to achieve high selectivity and high

Table 12.10 *Organic substances with negative effects on the environment.*

Organic substance	Common compounds	Use	Common sources	Dangerous properties
Nonhalogenated volatile	1-butanol, acetone, cyclohexanone, ethanol, styrene	Chemical industry (washing solvents, polymer synthesis), cosmetic and technical applications (aereosols), fuel additives, insecticides, paint thinners	Municipal waste, disposal areas, solid waste, agricultural runoff containing herbicides and pesticides	Carcinogens, reproductive hazards, neurotoxins, endocrine disruptors
Nonhalogenated semivolatile	Nitrophenols, phthalates, polyaromatic hydrocarbons	Plastics, chemical manufacturing, coke production, coal burning, motor vehicles	Incomplete combustions of organic matter (wood and fossil fuels), uncombusted coal and oil	Carcinogens, effects on respiratory function, endocrine disruptors
Halogenated volatile	Trihalomethanes, dichloromethane, 1,2-dichloroethane, trichloroethylene, 1,1,2- trichloroethane, tetrachloroethylene	Chlorination of drinking water, solvents in industrial plants, plastics and vinyl products, removal of grease from metal parts	Chemical manufacturing plants or disposal areas, contaminated marine sediments, disposal wells and leach fields, electroplating/metal finishing shops, firefighting training areas, leak collection and system sanitary lines, leak storage tanks, paint stripping and spray booth areas, pesticide/herbicide mixing areas, vehicle maintenance areas	Carcinogens

Halogenated semivolatile	1,1-bis (4-chlorophenyl)-2,2,2-trichloroethane, 2-chlorophenol, hexachlorobenezene, quintozene	Pesticides	Agrochemical industries, agricultural waste, pesticide/herbicide mixing areas	Carcinogens, endocrine disruptors
Fuels	Cumene, cyclohexane, ethylbenzene, benzene, toluene, xylenes	Manufacture of plastics, detergents, pesticides, chemicals, solvents in paint, printing, rubber	Aircraft areas, surface impoundments, chemical disposal areas and wells, vehicle maintenance areas	Carcinogens, skin irritation, swallowing or vomiting, fetotoxicity

capacity for the component to be bound, though they are still weak enough to be broken by simple engineering operations such as raising the temperature or decreasing the pressure.

However, the water required for transport must be removed, increasing the cost of process. An alternative approach is to use solvent-swollen ion-exchange membrane supports [55]. Various petrochemical streams contain a mixture of olefins and saturated hydrocarbons. The recovery of high-purity olefin gases from paraffins in a petrochemical stream is very important, since the aliphatic unsaturated hydrocarbons are generally used as intermediate reactants for industrial chemical syntheses such as the production of polymers.

Another important application is the separation of molecules dissolved in organic liquids by nanofiltration [56]. Common purification processes (such as distillation, extraction, chromatography, adsorption and crystallization) determine the energy or materials consumption and, in some cases, the efficiency. Organic solvent nanofiltration uses solvent-stable membranes for the separation of molecules of different sizes in solution. This method is applied to product purification, monomer/dimer separation, molecular fractionation, room-temperature solvent exchange, catalyst recovery and recycling, decolouration and solvent recycling. The main advantages are related to an increase product value, reduced operating cost, reduced processing time and environmental friendliness.

12.3.4 Membrane Operations for the Production of Optically Pure Enantiomers

The importance of enantiopure materials in the pharmaceuticals industry has increased as technologies for measuring and making enantiopure materials and the production of enantiomers have become commonplace, with many of the top-selling drugs in the world now being sold in enantiopure form. Membrane processes are emerging technologies for the resolution of enantiomers that have recently become widely used in providing rapid access to enantiopure materials, in order to support pharmaceutical development. Green aspects of membrane technology include waste reduction, safer solvents and energy efficiency. Elimination of waste is always a key green chemistry concern and is an important factor in any separation of enantiomers. The racemization and recycling of the undesired enantiomer will permit higher yields to be obtained and waste to be reduced. The possibility of combining membrane operations in the same production system or with traditional techniques permits the use of process integration capable of pursuing this aim. The green chemistry principle of a reduction in solvent utilization can be reached using membrane technology due to the compactness of the plants, control of fluid-dynamic conditions and recycling of the solvent. Energy efficiency is another advantage of considerable importance.

In this section, enantiocatalytic membrane reactors and enantioselective membrane processes are described in terms of enantiomer production/separation, a very interesting field for green chemistry process development in the pharmaceuticals industry. Enantiomers are chemical compounds that are identical to each other in all physical and chemical properties except optical rotation; therefore, their separation is one of the most complicated problems in chemical technology.

The chiral nature of living systems has important implications for biologically active compounds interacting with them. On a molecular level, chirality is an intrinsic property of fundamental compounds such as amino acids and sugars, and consequently of peptides, proteins and polysaccarides. Processes mediated by biological systems are very sensitive to stereochemistry, and in fact different responses can be observed when the activities of a pair of enantiomers are studied.

As a consequence, it is crucial to consider stereochemistry when xenobiotics are under study.

In the last 2 decades this field has gained a lot of interest for a variety of scientific and economic reasons, with the pharmaceuticals industry being the main contributor and driving force. Single-enantiomer drug sales show a continuous growth worldwide and many of the top-selling drugs are marketed as single enantiomers (269 of the top 500 drugs). In addition, a large number of drugs used for the treatment of illness have shifted from acute to chronic (Alzheimer's, cancer, obesity, asthma, arthritis and neurological disease). These drugs contain one or more stereogenic centre.

The broad range of examples in which stereisomers of drugs have differences in terms of bioavailability (some examples are reported in Table 12.11) have guided scientific research towards the development of alternative technologies by which to produce optically pure isomers and/or resolve racemic mixture of enantiomers. The technologies actually available for the production of pure enantiomers are mainly divided into physical methods, asymmetric syntheses and biological methods (Table 12.12).

In classical optical resolution, a racemic mixture of enantiomers (ER, ES) reacts with an enantiomer (R′) to give two diastereisomers (ER·R′ and ES R′). The mixture can then be separated by physical methods through crystallization. Afterwards, the two diastereoisomers react with another component (Z) to regenerate each enantiomer (ER and ES) and the new component R′Z. Conventional methods of optical resolution have the common drawback that only a small amount of material can be treated in one operation; this is the reason why other technologies must be used in the development of large-scale separation.

A stereoselective synthesis has been described and applied for the large-scale preparation of single enantiomeric substances, but the process is both expensive and time-consuming. Preparative techniques for the separation therefore have interesting potential.

Membrane processes are an emerging and promising alternative to the conventional resolution methods. There are basically two ways to use membrane technology to produce enantiomers (Figure 12.3). In one case, a kinetic resolution using an enantiospecific biocatalyst is combined with a membrane separation process; the membrane separates the product and the substrate based on their relative chemical properties, such as solubility. In the other case, the membrane is intrinsically enantioselective: the membrane is the chiral system which selectively separates the desired isomers on the basis of its special conformation.

In the final method, the main membrane techniques used are affinity ultrafiltration, dialysis, membrane extraction and liquid membranes. Where the enantiomer resolution is carried out by an enantiocatalytic membrane reactor, the membrane separates the product from the reactant. It has enantiocatalytic properties, thanks to the presence of biocatalysts able to discriminate between the two enantiomers. Stereoselective hydrolases (lipase in particular) are widely used and are very attractive in productive-scale production. This is mainly due to the fact that these enzymes are not expensive and do not require cofactors.

A specific example of an enantiocatalytic membrane reactor using lipase is the multiphase/extractive membrane reactor developed for the production of the diltiazem chiral intermediate on a commercial scale [57]. In this system, the reaction is carried out in a two-separate-phase reactor. The role of the membrane in this case is to compartmentalize the enzyme and to keep the two phases in contact (organic/aqueous) while maintaining them in

Table 12.11 *Examples of differences between (S)-enantiomers and (R)-enantiomers (adapted from [67]).*

Component	Enantiomer differences
Dopa	(S)-enantiomer: anti-Parkinson (R)-enantiomer: serious side effects
Penicillamine	(S)-enantiomer: antiarthritic (R)-enantiomer: mutagen
Tahlidomide	(S)-enantiomer: teratogen (R)-enantiomer: sedative
Ketamine	(S)-enantiomer: anaesthetic (R)-enantiomer: hallucinogen
Ethambutol	(S)-enantiomer: (S,S)-tuberculostatic (R)-enantiomer: (R,R)-; blindness
Novrad	(S)-enantiomer: (2R, 3S); analgesic (R)-enantiomer: (2S, 3R)-(-); antitussive
Indacrinone	(S)-enantiomer: uricosuric (R)-enantiomer: diuretic
Plactobutrazol	(S)-enantiomer: (2R, 3S); plant growth regulation (R)-enantiomer: (2R, 3S); fungicide
Asparagine	(S)-enantiomer: bitter (R)-enantiomer: sweet
Aspartame	(S)-enantiomer: (S,S); sweet (R)-enantiomer: (R,R); bitter
Limonene	(S)-enantiomer: lemon odour (R)-enantiomer: orange odour
Metadone	L-methadone: antiallodynic D-methadone: inactive
Citalopram	(S)-citalopram is over 100-fold more potent an inhibitor of the serotonin reuptake transporter than (R)-citalopram
Verapamil	The bioavailability of (R)-verapamil is more than double that of (S)-verapamil, due to reduced hepatic first-pass metabolism
Ibuprofen	The (S)-enantiomer is over 100-fold more potent as an inhibitor of cyclo-oxygenase than (R)-ibuprofen

Table 12.12 *Technologies actually available for the production enantiomers.*

Physical methods	Classical resolution (diasteromeric crystallization) Membrane separation Chromatography Capillary electrophoresis Sensors Liquid–liquid extraction
Asymmetric synthesis	Using chemical and biochemical catalysts
Biological methods	Biotransformation-based processes (e.g. biocatalytic membrane reactor)

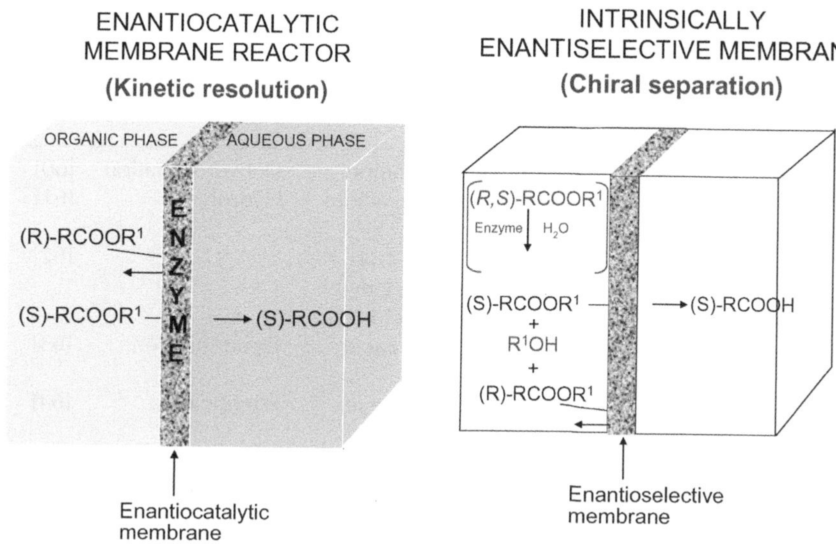

Figure 12.3 *Schematic configuration of membrane technologies used to produce enantiomers.*

two different compartments. The biocatalyst (lipase) is heterogenized on polyacrilonitrile (PAN) hollow-fibre membranes through entrapment. The reactant is fed into the solvent, while the product is extracted into the water from the other side of the membrane.

An improvement on the multiphase membrane reactor system using heterogenized lipase for enantiomer separation is the emulsion enzyme membrane reactor. In this system the organic water interface at the membrane level is obtained by immobilizing an oil-in-water emulsion produced through membrane emulsification on the biocatalytic membrane. In this way, each pore works as a microreactor containing immobilized enzyme [58]. By this process the selectivity and productivity of the biocatalytic system are significantly improved.

A further improvement was developed by Giorno *et al.* [59], in which enantioselective particles were produced through a membrane emulsification process. In these particles, oil-in-water emulsions were stabilized using lipase heterogenized at the interface between the oil and water. The biocatalyst in the original application had the action of both the emulsion stabilizer and the enantioselective catalyst.

Other examples of membrane bioreactors used for enantiomer separation are given in Table 12.13.

As previously described, the other method for enantiomer separation is based on a physical stereoselectivity by an enantioselective membrane. In order to carry out optical resolution, the membrane must contain a chiral environment in its structure. The production of a polymeric membrane with intrinsic enantioselective properties can be prepared for by using chiral polymers [1] or by chiral modification of achiral porous membranes using chiral selectors of natural (proteins, oligosaccarides, polysaccarides, antibiotics, etc.), semisynthetic (modified oligosaccarides, modified polysaccarides, polysaccaharide sulfates, etc.) or synthetic (helical synthetic polymers, etc.) origin [70]. Some examples of recent and innovative works on the development of enantioselective membranes are given in Table 12.14.

Table 12.13 *Examples of enantiomer production and separation using membrane bioreactors.*

Enantioselective biocatalyst	Membrane reactor	Isolated enantiomer	Reaction	References
Dehydrogenase	MBR	(L)-phenylalanine	Dehydrogenation	[60]
Lipase from *C. Cylindracea*	BMR	(S)-ibuprofen acid	Hydrolysis	[61]
Lipase from *Serratia marcescens*		(+)-(2S, 3R)-3-(4-methoxyphenyl) glycidic acid		[62]
Acinobacter sp. AK 226	MBR	(S)-ibuprofen acid	Denitrification	[63]
Lipase from *Rhizomucor miehei*	Byphasic-BMR	(S)-bultil ibuprofen ester	Esterification	[64]
α-chymotrypsin		(L)-phenylalanine	Hydrolysis	[65]
Lipase from *Pseudomonas* sp.	BMR	(1R;2R)-2-methyl-1-cyclohexanol	Transesterification	[66]
Lipase from *Candida rugosa*	Byphasic-BMR	(S)-naproxen acid	Hydrolysis	[58,59,67]
Lipase from *Candida rugosa*	Emulsion-BMR			[68]
Baker's yeast	Byphasic-MBR	R-citronellol	Reduction	[69]

MBR, membrane bioreactor; BMR, biocatalytic membrane reactor.

Table 12.14 *Some examples of enantioselective membranes.*

Membrane material	Chirality	Compound separated	References
Poly[(acrylonitrile)-co-(acrylic acid)]	Imprinted polymer with amino acids	Amino acids	[71]
Chiral helical poly[*p*-(oligopinanylsiloxanyl) phenylacetylene]s	Chiral polymer Phenylacetylenes containing pinanyl groups		[72]
Polyamide	Chiral polymer		[73]
Polyamide	Chiral polymer synthesized with Schiff's base complexes of Cu (II) and Zn (II)		[74]
Nitrocellulose	Chiral polymer		[75]
Composite membrane (polysulphone polymerized with l-lysine and piperazine monomers with trimesoyl chloride)	Chiral polymer Phenylacetylenes containing pinanyl groups		[76]

Table 12.14 *(Continued)*

Membrane material	Chirality	Compound separated	References
Polysulphone membranes	Chiral polymer (glucopyranose units that contain a large amount chirally active carbons)	Alcohol intermediate reagent in drug preparation	[77]
Polypropilene	*N*-3,5-dinitrobenzoyl-l-alanine-1-octylester	Acid	[78]
Cholesteric liquid crystalline	Mixtures of 4-pentyl-4-cyanobiphenyl (5CB) and cholesteryl oleyl carbonate (COC)		[79]
Polysulfone	*N*-hexadecyl-l-hydroxyproline	Drug	[80]
Polypropylene	Imprinted polymer with (S)-naproxen		[81]
Composite microsphere porous cellulose	Imprinted polymer nanoparticles with S-propanolol	Alcohol	[82]
Cellulose acetate butyrate	Chiral polymer		[83]
Chitosan membranes			[84]
Ethycellulose			[77]
Cholesteric liquid crystalline			[85]
Cellulose acetate butyrate		Oxide compound intermediate in drug preparation	[16]
Cellulose acetate			[86]

MEUF is a particular kind of enantioselective membrane in which the chiral selector is anchored to non-ionic micelles [87]. During ultrafiltration, the micelles are retained by the membrane, including the bound enantiomer, while the other enantiomer passes through the membrane. A specific example has been carried out by Overdevest *et al.* [88], who developed a multistage system using MEUF for the separation of *D,L*-phenylalanine enantiomers using cholesteryl-L-glutamete as the chiral selector, anchored in non-ionic micelles.

Membrane materials used for membrane preparation or modification are chiral substances and as a rule are obtained through multistage chemical transformation. They are expensive. The development of new approaches to obtaining enantioselective membranes is thus directed towards alternative and more economic strategies. Molecular imprinting is an alternative technique used to produce enantioselective membranes. In this process, the enantioselective surrounding is created by crosslinking a functional monomer or monomers in the presence of a template substance by radical polimerization. After polymerization, the template molecules are washed out, leaving the selective cavities free to interact in a specific manner with the target molecule.

In order to provide a high selectivity, the imprinted polymer should possess the following characteristics: stiffness of the polymer structure, enabling the cavities to retain the shape corresponding to the template; good accessibility to the specific cavities; and a high number and strength of interactions between complementary functional groups of the template and the polymer [89].

Several research papers on the development of chirally imprinted membranes have focused on different strategies. An alternative method for the production of imprinted polymers was developed by Yoshikawa *et al.* [90]. They preformed noncrosslinked polymers, which served simultaneously as functional recognition elements for the templates and as supporting matrices for the future binding sites. Imprinted membranes are obtained by simple casting of the template-containing polymer solutions using cross-linking polymerization reactions, avoiding the unfavourable reaction of thermal and/or photochemical stress on templates and pre-polymerization complexes. A variety of chiral and achiral polymers have been evaluated for alternative molecular imprinting. Examples include blends of peptide-grafted polystyrene–polyacrylonitrile [90], cellulose acetate [91], carboxylated polysulfones [92] and polyamides [93,94].

12.4 Integrated Membrane Processes

The possibility of integrating various membrane operations in the same production line or in combination with conventional separation units allows, in many cases, better performance in terms of product quality, plant compactness, environmental impact and energy use. These characteristics are fundamental to the development of a green process. In this section, some important examples of integrated membrane processes will be presented, beginning with water desalination.

12.4.1 Integrated Membrane Processes for Water Desalination

'Desalination' refers to processes that remove salts and other minerals from water in order to produce fresh water suitable for human consumption or for irrigation in countries where the availability of water is scarce.

The principal desalination technologies are divided into thermal and membrane. With the former, it is possible to separate water from salts by evaporation and condensation, while with the latter, as already explained in Section 12.1, a semipermeable membrane is used. The main technologies used in thermal desalination are divided into multistage flash (MSF) and multi-effect distillation (MED). The main membrane desalination technologies are reverse osmosis, nanofiltration and electrodialysis.

Membrane processes are already recognized as the best technology for water desalination, since thermal desalination techniques are about 10 times less energy efficient. The production capacities of membrane and thermal systems for water desalination are shared equally, with reverse osmosis dominating the membrane processes and MSF the thermal processes. When the number of plants is taken into consideration, a different picture is obtained: membrane-based systems are the most widely used, with their installation representing about 80% of the worldwide total [95,96].

Saudi Arabia is the world leader in desalination processes, with about the 26% of the global production capacity. The USA is second with 17%, mainly achieved using membrane processes and particularly reverse osmosis integrated with nanofiltration. The desalination

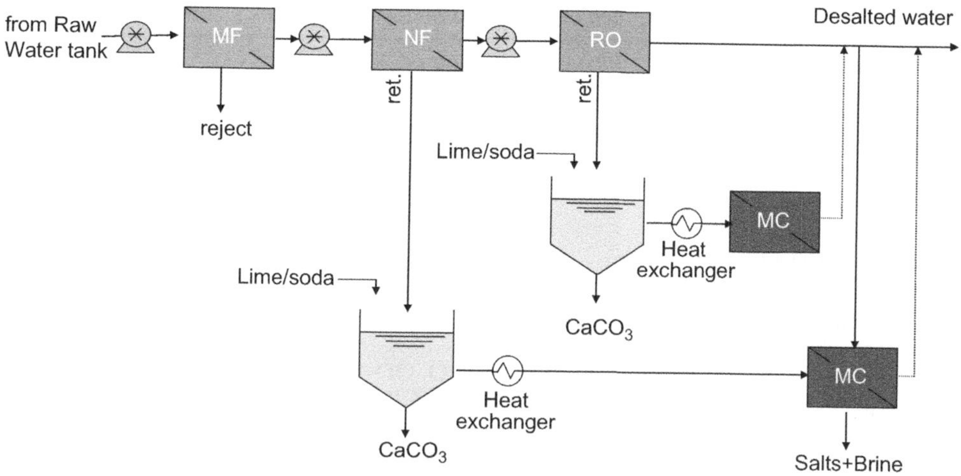

Figure 12.4 *Schematic representation of the integrated system developed in the MEDINA project (after [94]).*

capacity in Europe is largely distributed between Spain and Italy. In Spain, more than 69% of desalination plants use reverse osmosis, while in Italy the figure is about 20%.

In Asia the market for desalination plants is growing fast, due to the enormous populations and economic growth. The wide use of reverse osmosis in desalination processes is guided by the fast development of reverse osmosis membranes. In this process, a suitable pre-treatment is indispensable for efficient plant operation. Ultrafiltration, microfiltration and nanofiltration are frequently used to pre-treat a solution before reverse osmosis as they provide many advantages over conventional processes. Another membrane process that uses reverse osmosis pre-treatment is the membrane bioreactor, most frequently used for waste water treatment. An advanced integration of different membrane technologies for desalination, which can also be used for the valorization of brine, is illustrated in Figure 12.4. In particular, this introduces the use of a membrane contactor to crystallize salts at different steps in the purification (retentate of nanofiltration and reverse osmosis). This strategy has been developed in the MEDINA project, funded by the European Commission within the 6th Framework Program.

One of the main problems with sea water desalination is brine disposal. An interesting solution is offered by membrane contactors, such as membrane distillation and membrane crystallization [98–100]. Both of these are thermally driven, and they are not limited by concentration polarization phenomena. Therefore, nanofiltration and/or reverse osmosis retentate streams can be treated with these techniques to recover water and salts contained within them.

12.4.2 Integrated Membrane Processes for the Fruit Juice Industry

The use of integrated systems is very attractive for the production of high-quality concentrated juices fruit. The high percentage of pectins and solids creates a very viscous flow, which gives low permeate flux when processed by reverse osmosis and osmotic

Table 12.15 *Examples of integrated membrane processes for fruit juice concentration.*

Application	Integrated unit	Reference
Passion fruit juice	UF, EV	[103]
Apple juice concentration	51 RO stages	[104]
Kiwi fruit concentration	UF, MD	[105]
Orange juice concentration	UF, EV	[106]
Fruit juice concentration	RO, NF	[107]
Apple juice production and aroma concentration	BMR, RO, PV, EV	[108]
Blood orange and kiwi fruit juice concentration	UF, RO, OD	[109]
Blood orange juice concentration	UF, RO, OD–UF, OD	[101]
Orange juice concentration	MF/UF, followed by two RO stages	[101]
Black currant juice	MF, RO, OD	[110]

UF, ultrafiltration; EV, evaporation; RO, reverse osmosis; MD, membrane distillation; NF, nanofiltration; BMR, biocatalytic membrane reactor; PV, pervaporation; MF, microfiltration

distillation. When processes such as microfiltration and ultrafiltration are used in combination with reverse osmosis to remove the suspended solids, the reverse osmosis feed viscosity decreases and the flux increases [101]. In addition, the use of integrated membrane systems is very helpful in volatile (alcohols, aldehydes, esters) retention, generally associated with the serum and considered very important to the production of high-quality products.

Different systems have been reported in the literature in which microfiltration or ultrafiltration is used in combination with reverse osmosis to produce fruit juice of the same quality as fresh juice. One of the first works in this field covered the use of ultrafiltration membranes prior to heat treatment to minimize flavour loss and deterioration. The serum containing the flavours was concentrated by reverse osmosis above 42 ° Brix [102].

Some years later, Separasystems LP developed a combined membrane process, called FreshNote, which was able to concentrate orange juice to 60 ° Brix, retaining almost fresh juice flavours. The integrated process was made by a plurality of two-stage reverse osmosis systems following microfiltration and ultrafiltration processes. Some other examples are reported in Table 12.15, in which different integrated systems for various applications in the fruit juice industry are described.

12.5 Green Processing Impact of Membrane Processes

Membrane technology is one of the disciplines at the centre of the technological innovation necessary to face the problems caused by a growing global population, water stress, environmental pollution and increasing CO_2 emission into the atmosphere. Green chemistry, the principles of which have been discussed in depth in Chapter 2, has a role to play here. The ways in which membrane processes fulfil some of these principles are highlighted in Table 12.16.

Table 12.16 *"Green" properties of membrane processes.*

Green chemistry fundamentals	Membrane processes properties
Prevention: It is better to prevent waste than to treat or clean up waste after it has been created.	Conservation and rational use of raw materials by recovery, reuse and recycling of unreacted substrates, water and production media such as catalyst, solvents, surfactants, adsorbents, cooling agents etc. Products and co-products of high quality without waste generation. Selective use of (bio)catalyst in membrane reactors which prevent side-product generation.
Atom economy: Methods of synthesis must be designed so to maximize the incorporation of all materials used in the process into the final product.	
Less hazardous chemical synthesis: Wherever practicable, synthetic methodologies should be designed to use and generate substances that possess little or no toxicity to human health and the environment.	
Designing safer chemicals: Chemical products should be designed to effect their desired function while minimizing their toxicity.	
Reduce derivatives: Unnecessary derivatization (blocking group, protection/deprotection, temporary modification) should be avoided whenever possible.	Selective and controlled production and/or separation of products of interest with high purity avoiding generation of undersidered chemical compunds.
Design for degradation: Chemical products should be designed so that at the end of their function they do not persist in the environment and break down into innocuous degradation products and do not persist in the environment.	
Safer Sovents and Auxiliaries: The use of auxiliary substances (e.g. solvents, separation agents and so forth) should be made unnecessary wherever possible and innocuous when used.	Selective compartmentalization that permit to recycle, recovery and reuse auxiliary substances and compactness of plants able to reduce needed volume.
Design for Energy efficinecy: Energy requirements should be recognized for their environmental and economic impacts and should be minimized. Synthetic methods should be conducted at ambient temperature and pressure.	Low energy consuption due to mild operative conditions (temperature, pressure), combination of operations in the same system or with traditional techniques and modularity.
Use of renovable feedstocks: A raw material of feedstock should be revewable rather than depleting whenever technically and economically	Ability to treat complex solution such as waste and ra materilas to obtain high added value products and coproducts with high purity.

(*Continued*)

Table 12.16 *(Continued)*

Green chemistry fundamentals	Membrane processes properties
Catalysis: Catalytic reagents (as selective as possible) are superior to stoichiometric reagents.	Use of heterogenized or free (bio)catalyst in membrane reactor with improved selectivity and productivity.
Real-time analysis for pollution prevention: Analytical methodologies need to be further developed to allow for real-time, in-process monitoring and control prior to the formation of hazardous substances.	Possibility to produce mebrane (bio)sensors to monitor and control hazardous compounds.
Ineretly safer chemistry for accident prevention: Substances and the form of a substance used in a chemical process should be chosen to minimize potential for chemical accidents, including releases, explosions, and fires.	Alternative processes able to substitute conventinal methodologies (e.g. membrane (bio)reactor) using rationalized solvet feed, selective removal of products, controlled supply of one reagent in order to keep low concentration, mass transport controll and multistep reaction.

The strategy of PI is one of the key aspects of green chemistry. A European Roadmap has been prepared by an international team of experts to identify the potential benefits of PI and illustrate the actions needed to accelerate PI implementation in the process industry [111]. In this roadmap, membrane technologies are indicated as being one of the interesting options in the PI strategy. The benefits and state of development of membrane technologies, according to the criteria established by European experts in this roadmap, are shown in Table 12.15. Among the membrane processes considered, membrane-assisted reactive distillation is deemed the most efficient in terms of energy saving, CO_2 impact and high-quality compound production. These properties, together with the economic feasibility, paint a future in which this methodology will replace traditional options. The other membrane technologies analysed (membrane crystallization, membrane distillation, membrane reactor) show great potential in the same direction.

Despite membrane operations offering important process potentials, some standards for fundamental/strategic research are needed in order to promote industrial-scale development. At the moment, microfiltration, ultrafiltration, reverse osmosis and electrodialysis are well-established unit operations, while nanofiltration, gas separation and pervaporation are at a developing level. Membrane contactors remain at an emerging step of development and major problems need to be solved before industrial systems will be installed on a large scale. Such problems include additional resistance to mass transfer offered by the membrane, limited operating pressure below the breakthrough value, relatively low membrane lifetime and high replacement costs.

The robustness and suitability of membrane technology are proven in various fields at commercial scales. Interesting examples are in sea water desalination (El Paso desalination plant, Texas, 104 m^3/day), in waste water treatment and reuse (submerged membrane system from Zenon) and in the treatment of gas streams.

12.6 Conclusion

Membrane separation technologies offer new options for designing, rationalizing and optimizing innovative production cycles that respond well to green technology requirements as intensified, efficient and clean processes.

Membrane separation process have been successfully used in various fields, such as metal ion, hydrocarbon and organic and inorganic acid removal from waste water, in which a green approach is needed to prevent environmental damage and protect health. Integrated membrane processes have overcome problems related to water desalination in the production of potable water and in concentrated fruit juice production with high purity and improved quality. Further advances, including membrane contactors, should be developed in the future by the interchange of various areas of research: advanced membrane materials, module design and economics, membrane morphology supramolecular engineering, massive production and process control.

References

(1) E. Drioli and L. Giorno, Biocatalytic Membrane Reactors: Application in Biotechnology and the Pharmaceutical Industry, Taylor & Francis: Padstow (1999).

(2) E. Drioli and L. Giorno, Comprehensive Membrane Science Technology, Elsevier (2010).

(3) E. Drioli and L. Giorno, Membrane Operation, Wiley-VCH (2009).

(4) B. Koltuniewicz and E. Drioli, Membranes in Clean Technologies, Wiley-VCH (2008).

(5) F. Fu and Q. Wang, Removal of heavy metal ions from wastewater: a review. Journal of Environmental Management, 92, 407–418 (2011).

(6) N. Oyaro, O. Juddy, E. N. M. Murago and E. Gitonga, The contents of Pb, Cu, Zn and Cd in meat in Nairobi, Kenya. Int. J. Food Agric. Environ., 5, 119–121 (2007).

(7) C. E. Borba, R. Guirardello, E. A. Silva, M. T. Veit and C. R. G. Tavares, Removal of nickel(II) ions from aqueous solution by biosorption in a fixed bed column: experimental and theoretical breakthrough curves. Biochem. Eng. J., 30, 184–191 (2006).

(8) C. Namasivayam and K. Kadirvelu, Uptake of mercury (II) from wastewater by activated carbon from unwanted agricultural solid by-product: coirpith. Carbon, 37, 79–84 (1999).

(9) R. Naseem and S. S. Tahir, Removal of Pb(II) from aqueous solution by using bentonite as an adsorbent. Water Res., 35, 3982–3986 (2001).

(10) L. Khezami and R. Capart, Removal of chromium(VI) from aqueous solution by activated carbons: kinetic and equilibrium studies. J. Hazard. Mater., 123, 223–231 (2005).

(11) Y. Ku and I. L. Jung, Photocatalytic reduction of Cr(VI) in aqueous solutions by UV irradiation with the presence of titanium dioxide. Water Res., 35, 135–142 (2001).

(12) B. Alyüz and S. Veli, Kinetics and equilibrium studies for the removal of nickel and zinc from aqueous solutions by ion exchange resins. J. Hazard. Mater., 167, 482–488 (2009).

(13) A. G. El Samrani, B. S. Lartiges and F. Villiéras, Chemical coagulation of combined sewer overflow: heavy metal removal and treatment optimization. Water Res., 42, 951–960 (2008).

(14) M. Lundh, L. Jönsson and J. Dahlquist, Experimental studies of the fluid dynamics in the separation zone in dissolved air flotation. Water Res., 34, 21–30 (2000).

(15) G. H. Chen, Electrochemical technologies in wastewater treatment. Sep. Purif. Technol., 38, 11–41 (2004).

(16) W. Wang, W. Xiong, M. Zhao, W. Sun, F. Li and L. Yuan, Chiral separation of trans-stilbene oxide through cellulose acetate butyrate membrane. Tetrahedron: Asymmetry, 20, 1052–1056 (2009).
(17) J. Landaburu-Aguirre, V. García, E. Pongrácz and R. L. Keiski, The removal of zinc from synthetic wastewaters by micellar-enhanced ultrafiltration: statistical design of experiments. Desalination, 240, 262–269 (2009).
(18) E. Sampera, M. Rodrígueza, M. A.de la Rubia and D. Prats, Removal of metal ions at low concentration by micellar-enhanced ultrafiltration (MEUF) using sodium dodecyl sulfate (SDS) and linear alkylbenzene sulfonate (LAS). Sep. Purif. Technol., 65, 337–342 (2009).
(19) M. Mohsen-Nia, P. Montazeri and H. Modarress, Removal of Cu2þ and Ni2þ from wastewater with a chelating agent and reverse osmosis processes. Desalination, 217, 276–281 (2007).
(20) L. N. Zhang, Y. J. Wu, X. Y. Qu, Z. S. Li and J. R. Ni, Mechanism of combination membrane and electro-winning process on treatment and remediation of Cu2þ polluted water body. J. Environ. Sci., 21, 764–769 (2009).
(21) B. K. C. Chan and A. W. L. Dudeney, Reverse osmosis removal of arsenic residues from bioleaching of refractory gold concentrates. Miner. Eng., 21, 272–278 (2008).
(22) U. Ipek, Removal of Ni(II) and Zn(II) from an aqueous solution by reverse osmosis. Desalination, 174, 161–169 (2005).
(23) E. Dialynas and E. Diamadopoulos, Integration of a membrane bioreactor coupled with reverse osmosis for advanced treatment of municipal wastewater. Desalination, 238, 302–311 (2009).
(24) A. Figoli, A. Cassano, A. Criscuoli, M. S. I. Mozumder, M. T. Uddin, M. A. Islam and E. Drioli, Influence of operating parameters on the arsenic removal by nanofiltration. Water Res., 44, 97–104 (2010).
(25) E. Cséfalvay, V. Pauer and P. Mizsey, Recovery of copper from process waters by nanofiltration and reverse osmosis. Desalination, 240, 132–142 (2009).
(26) F. N. Liu, G. L. Zhang, Q. Meng and H. Z. Zhang, Performance of nanofiltration and reverse osmosis membranes in metal effluent treatment. Chin. J. Chem. Eng., 16, 441–445 (2008).
(27) S. K. Nataraj, K. M. Hosamani and T. M. Aminabhavi, Potential application of an electrodialysis pilot plant containing ion-exchange membranes in chromium removal. Desalination, 217, 181–190 (2007).
(28) J. Lambert, M. Avila-Rodriguez, G. Durand and M. Rakib, Separation of sodium ions from trivalent chromium by electrodialysis using monovalent cation selective membranes. J. Membr. Sci., 280, 219–225 (2006).
(29) T. Mohammadi, A. Razmi and M. Sadrzadeh, Effect of operating parameters on Pb2þ separation from wastewater using electrodialysis. Desalination, 167, 379–385 (2004).
(30) E. Paquaya, A.-M. Clarinvalb, A. Delvauxc, M. Degrezd and H. D. Hurwitz, Applications of acid picking wastewater treatment. Chem. Eng. J., 79, 197–201 (2000).
(31) J. Pierard, E. Paquaya and M. Degrezd, Recycling by electrodialysis: from lab to industrial applications. Desalination, 149, 393–398 (2002).
(32) O. D. Linnikov, E. A. Anokhina and V. E. Scherbakov, Investigation on purification of hydrochloric acid by membrane method. Desalination, 132, 299–306 (2000).
(33) J. Xu, S. Lu and D. Fu, Recovery of hydrochloric acid from the waste acid solution by diffusion dialysis. Journal of Hazardous Materials, 165, 832–837 (2009).
(34) M.-S. Kang, K. Yoo, S.-J. Oh and S.-H. Moon, A lumped parameter model to predict hydrochloric acid recovery in diffusion dialysis. Journal Membrane Science, 188, 61–70 (2001).

(35) R. Weber, H. Chemiela and V. Mavrov, Characteristic and application of new ceramic nano filtration membranes. Desalination, 157, 113–125 (2003).

(36) M. Tomaszewska, M. Gryta and A. W. Morawski, Recovery of hydrochloric acid from metal pickling solutions by membrane distillation. Separation and Purification Technology, 22-23, 591–600 (2001).

(37) M. Tomaszewska, M. Gryta and A. W. Morawski, The influence of salt in solutions on hydrochloric acid recovery by membrane distillation. Separation and Purification Technology, 14, 183–188 (1998).

(38) H. Carrère and F. Blaszkow, Comparison of operating modes for clarifying lactic acid fermentation broths by batch cross-flow microfiltration. Process Biochemistry, 36, 751–756 (2001).

(39) H. Huang, S. T. Yang and D. E. Ramey, A hollow-fiber membrane extraction process for recovery and separation of lactic acid from aqueous solution. Appl. Biochem Biotechnol., 113-116, 671–688 (2004).

(40) H. G. Joglekar, I. Rahman, S. Babu, B. D. Kulkarni and A. Joshi, Comparative assessment of downstream processing options for lactic acid. Sep. Purif. Technol., 52, 1–17 (2006).

(41) R. Datta and M. Henry, Lactic acid: recent advances in products, processes and technologies – a review. J. Chem. Technol. Biotechnol., 81, 1119–1129 (2006).

(42) M. M. Hossain and J. L. Maisuria, Effects of organic phase, fermentation media, and operating conditions on lactic acid extraction. Biotechnol. Prog., 24, 757–765 (2008).

(43) W. W. Y. Lau, J. Finlayson, J. M. Dickson, J. Jiang and M. A. Brook, Pervaporation performance of oligosilylstyrene-polydimethylsiloxane membrane for separation of organics from water. J. Membr. Sci., 134, 209–217 (1997).

(44) D. Yang, S. Majumdar, S. Kovenklioglu and K. K. Sirkar, Hollow fiber contained liquid membrane pervaporation system for the removal of toxic volatile organics from wastewater. J. Membr. Sci., 103, 195–210 (1995).

(45) J. S. Cha, V. Malik, D. Bhaumik, R. Li and K. K. Sirkar, Removal of VOCs from waste gas streams by permeation in a hollow fiber permeate. J. Membr. Sci., 128, 195 (1997).

(46) C. K. Yeom, H. K. Kim and J. W. Rhim, Removal of trace VOCs from water through PDMS membranes and analysis of their permeation behaviours. J. Appl. Polym. Sci., 73, 601–611 (1999).

(47) K. Inui, H. Okumura, T. Miyata and T. Uragami, Permeation and separatiom of benzene/ cyclohexane mixtures through cross-linked poly(alkylmetacrylate) membranes. J. Membr. Sci., 132, 193–202 (1997).

(48) K. Ksutic and B. Kunst, Removal of organics from aqueous solutions by commercial RO and NF membranes of characterized porosities. Desalination, 142, 47–56 (2002).

(49) K. O. Agenson, J. Oh and T. Urase, Retention of a wide variety of organic pollutants by different nanofiltration/reverse osmosis membranes: controlling parameters of process. J. Membr. Sci., 225, 91–103 (2003).

(50) T. A. C. Oliveira, J. T. Scarpello and A. G. Livingston, Pervaporation-biological oxidation hybrid process for removal of volatile organic compounds from wastewaters. J. Membr. Sci., 195, 75–88 (2002).

(51) L. M. F. dos Santos and A. G. Livingston, Novel membrane bioreactor for detoxification of Voc wastewaters: biodegradation of 12-dichloroethane. Water. Res., 29, 179–194 (1995).

(52) W. Kujawski, Application of pervaporation and vapor permeation in environmental protection. Polish Journal of Environmental Studies, 9(1), 13–26 (2000).

(53) A. Jonquières, R. Clément, P. Lochon, J. Néel, M. Dresch and B. Chrétien, Industrial state-of-the-art of pervaporation and vapour permeation in the western countries. Journal of Membrane Science, 206, 87–117 (2002).

(54) J. C. Davis, R. J. Valus, R. Eshraghi and A. E. Velikoff, Facilitated transport membrane hybrid systems for olefin purification. Sep. Sci. Technol., 28, 463 (1993).
(55) P. M. Thoen, R. D. Noble and C. A. Koval, Unexpectedly large selectivity for olefin separations utilizing silver ion-exchange membranes. Chem. Mat., 6, 947 (1994).
(56) L. G. Peeva, S. Malladi and A. G. Livingston, Nanofiltration operations in nonaqueous systems, in Comprehensive Membrane Science and Engineering, E. Drioli and L. Giorno (Eds.), Elsevier, Academic Press: Oxford (2010).
(57) J. L. Lopez and S. Matson, A multiphase/extractive enzyme membrane reactor for production of diltiazem chiral intermediate. Journal of Membrane Science, 125, 189–211 (1997).
(58) L. Giorno, E. D'Amore, E. Drioli, R. Cassano and N. Picci, Influence of -OR ester group length on the catalytic activity and enantioselectivity of free lipase and immobilized in membrane used for the kinetic resolution of naproxen esters. Journal of Catalysis, 247, 194–200 (2007).
(59) L. Giorno, E. D'Amore, R. Mazzei, E. Piacentini, J. Zhang, E. Drioli, R. Cassano and N. Picci, An innovative approach to improve the performance of a two separate phase enzyme membrane reactor by immobilizing lipase in presence of emulsion. Journal of Membrane Science, 295, 95–101 (2007).
(60) E. Schmidt, D. Vasić-Rački and C. Wandrey, Enzymatic production of l-phenylalanine from the racemic mixture of d,l-phenyllactate. Applied Microbiology and Biotechnology, 26, 42–48 (1987).
(61) Z. Xiao-Ming and I. W. Wainer, On-line determination of lipase activity and enantioselectivity using an immobilized enzyme reactor coupled to a chiral stationary phase. Tetrahedron Letter, 34, 4731–4734 (1993).
(62) H. Matsumae, M. Furui and T. Shibatani, Lipase-catalyzed asymmetric hydrolysis of 3-phenylglycidic acid ester, the key intermediate in the synthesis of diltiazem hydrochloride. Journal of Fermentation and Bioengineering, 75, 93–98 (1993).
(63) M. Takagi, K. Oishi, F. Ishimura and I. Fujimatsu, Production of S-(+)-ibuprofen from a nitrile compound by enzymatic reaction combined with ultrafiltration. Journal of Fermentation and Bioengineering, 78, 54–58 (1994).
(64) M. T. Lopez-Belmonte, A. R. Alcantara and J. V. Sinisterra, Enantioselective esterification of 2-arylpropionic acids catalyzed by immobilized *Rhizomucor miehei* lipase. Journal of Organic Chemistry, 62, 1831–1840 (1997).
(65) K. Abe, M. Goto and F. Nakashio, Novel optical resolution of phenylalanine racemate utilizing enzyme reaction and membrane extraction. Separation Science and Technology, 32, 1921–1935 (1997).
(66) J. Ceynowa and M. Rauchfleisz, Resolution of racemic trans-2-methyl-1-cyclohexanol by lipase-catalysed transesterification in a membrane reactor. Journal of Molecular Catalysis B: Enzymatic, 15, 71–80 (2001).
(67) K. Sakaki, L. Giorno and E. Drioli, Lipase-catalyzed optical resolution of racemic naproxen in biphasic enzyme membrane reactors. Journal of Membrane Science, 184, 27–38 (2001).
(68) L. Giorno, N. Li and E. Drioli, Use of stable emulsion to improve stability, activity, and enantioselectivity of lipase immobilized in a membrane reactor. Biotechnology and Bioengineering, 84, 677–685 (2003).
(69) R. Valadez-Blanco, F. C. Ferreira, R. F. Jorge and A. G. Livingston, A membrane bioreactor for biotransformations of hydrophobic molecules using organic solvent nanofiltration (OSN) membranes. Journal of Membrane Science, 317, 50–64 (2008).
(70) N. M. Maier, P. Franco and W. Lindner, Separation of enantiomers: needs, challenges, perspectives. J. Chromatography A, 906, 3–36 (2001).

(71) N. Ul-Haq, T. Khan and J. K. Park, Enantioseparation with D-Phe- and L-Phe-imprinted PAN-based membranes by ultrafiltration. J. Chem. Technol. Biotechnol., 83, 524–533 (2008).
(72) A. Toshiki, F. Tomoyuki, S. Ken-ichi, K. Takashi, T. Masahiro and Y. Masayuki, Synthesis of chiral helical poly(p-(oligopinanylsiloxanyl)phenylacetylene)s and enantioselective permeability of their membranes. Journal of Polymer Science Part A: Polymer Chemistry, 42(18), 4502–4517 (2004).
(73) M. Nakagawa, Y. Ikeuchi and M. Yoshikawa, Chiral separation of racemic amino acids with novel polyamides having N-a-acetyl-L-glutamyl residue as a diacid component. Polymer, 49, 4612–4619 (2008).
(74) T. Hashimoto and M. Yoshikawa, Chiral separation of racemic mixtures with chiral polyamide membranes containing aspartyl residues in their main chains. Current Nanoscience, 7, 915–924 (2011).
(75) Y. Jiang, J.-H. Zhang, S.-M. Xie, Y.-C. Lv, M. Zhang, C. Ma and L.-M. Yuan, Chiral separation of D, L-tyrosine through nitrocellulose membrane. Journal of Applied Polymer Science, 1–7 (2011).
(76) K. Singh, P. G. Ingole, J. Chaudhari, H. Bhrambhatt, A. Bhattacharya, H. C. Bajaj, Resolution of racemic mixture of α-amino acid derivative through composite membrane. Journal of Membrane Science, 378, 531–540 (2011).
(77) M. Yang, M. Zhao, S.-M. Xie and L.-M. Yuan, Optical resolution of (R, S)-2-Phenyl-1-propanol through enantioselective ethycellulose membranes. Journal of Applied Polymer Science, 112, 2516–2521 (2009).
(78) P. Hadika, L.-P. Szabo, E. Nagya, and Zs. Farkas, Enantioseparation of d,l-lactic acid by membrane techniques. Journal of Membrane Science, 251, 223–232 (2005).
(79) S. Han and S. M. Martin, Enantioselective cholesteric liquid crystalline membranes characterized using nonchiral HPLC with circular dichroism detection. Journal of Membrane Science, 367, 1–6 (2011).
(80) T. Gumí, M. Valiente and C. Palet, Elucidation of SR-propranolol transport rate and enantioselectivity through chiral activated membranes. Journal of Membrane Science, 256, 150–157 (2005).
(81) L. Donato, A. Figoli and E. Drioli, Novel composite poly(4-vinylpyridine)/polypropylene membranes with recognition properties for (S)-naproxen. Journal of Pharmaceutical and Biomedical Analysis, 37, 1003–1008 (2005).
(82) C. Jantarat, N. Tangthong, S. Songkro, G. P. Martin and R. Suedee, S-propranolol imprinted polymer nanoparticle-on-microsphere composite porous cellulose membrane for the enantioselectively controlled delivery of racemic propranolol. International Journal of Pharmaceutics, 349, 212–225 (2008).
(83) S. Xie, W.-F. Wang, P. Ai, M. Yang and L.-M. Yuan, Chiral separation of (R, S)-2-phenyl-1-propanol through cellulose acetate butyrate membranes. Journal of Membrane Science, 321, 293–298 (2008).
(84) W. Xiong, W. Wang, L. Zhao, Q. Song and L.-M. Yuan, Chiral separation of (R, S)-2-phenyl-1-propanol through glutaraldehyde-crosslinked chitosan membranes. Journal of Membrane Science, 328, 268–272 (2009).
(85) S. Han and S. M. Martin, Enantioselective cholesteric liquid crystalline membranes characterized using nonchiral HPLC with circular dichroism detection. Journal of Membrane Science, 367, 1–6 (2011).
(86) M. Zhao, X. Xu, Y. Jiang, W. Sun, W. Wang and L. Yuan, Enantioseparation of trans-stilbene oxide using a cellulose acetate membrane. Journal of Membrane Science, 336, 149–153 (2009).
(87) T. J. M. De Bruin, A. T. M. Marcelis, H. Zuilhof, L. M. Rodenburg, H. A. G. Niederländer, A. Koudijs, P. E. M. Overdevest and E. J. R. Sudhölter, Separation of amino acid enantiomers by micelle-enhanced ultrafiltration. Chirality, 12(8), 627–636 (2000).

(88) P. E. M. Overdevest, A.Van Der Padt, J. T. F. Keurentjes and K. Van't Riet, Langmuir isotherms for enantioselective complexation of (D/L)- phenylalanine by cholesteryl-L-glutamate in nonionic micelles. Colloids and Surfaces A: Physicochemical and Engineering Aspects, 163(2-3), 209–224 (2000).
(89) G. Wulff, Molecular imprinting in cross-linked materials with aid of molecular template. Chem. Int. Ed. Engl., 34, 1812–1832 (1995).
(90) M. Yoshikawa, T. Fujisawa and J. Izumi, Molecularly imprinted polymeric membranes having EFF derivatives as a chiral recognition site. Macromol Chem Physic, 200(6), 1458–1465 (1999).
(91) M. Yoshikawa, T. Ooi and J. I. Izumi, Alternative molecularly imprinted membranes from a derivative of natural polymer, cellulose acetate. J. Appl. Polym. Sci., 72, 493–499 (1999).
(92) M. Yoshikawa, J. I. Izumi, T. Ooi, T. Kitao, M. D. Guiver and G. P. Robertson, Carboxylated polysulfone membranes having a chiral recognition site induced by an alternative molecular imprinting technique. Polym. Bull., 40, 517–524 (1998).
(93) Y. Kondo, O. M. Yoshikawa and H. Okushita, Molecularly imprinted polyamide membranes for chiral recognition. Polym. Bull., 44, 517–524 (2000).
(94) P. S. Reddy, T. Kobayashi and N. Fujii, Molecular imprinting in hydrogen bonding networks of polyamide nylon for recognition of amino acids. Chem. Lett. 293–294 (1999).
(95) V. Frenkel, Water reuse. Int. Desalin., 17, 47–50 (2008).
(96) F. Macedonio and E. Drioli, Membrane engineering progresses in desalination and water reuse. Memb. Water Treat., 1, 75–81 (2010).
(97) E. Drioli, A. Criscuoli and F. Macedonio, Membrane Based Desalination: An Integrated Approach (European Water Research), IWA Publishing: London (2011).
(98) E. Drioli, E. Curcio, A. Criscuoli and G. Di Profio, Integrated system for recovery of $CaCO_3$, NaCl and $MgSO_4{\cdot}7H_2O$ from nanofiltration ritentate. J. Memb. Sci., 239, 27–38 (2004).
(99) F. Macedonio, E. Drioli, E. Curcio and G. Di Profio, Experimental and economical evaluation of a membrane crystallizer plant. Desalin. Water Treat., 9, 49–53 (2009).
(100) F. Macedonio and E. Drioli, in Comprehensive Membrane Science and Engineering, E. Drioli and L. Giorno (Eds.), Elsevier: Oxford (2010).
(101) B. Jiao, A. Cassano and E. Drioli, Recent advances on membrane processes for the concentration of fruit juices: a review. Journal of Food Engineering, 63(3), 303–324 (2003).
(102) J. T. Lawhon and E. W. Lusas, Method of producing sterile and concentrated juices with improved flavor and reduced acid. US Patent 4,643,902 (1987).
(103) Z. R. Yu and B. H. Chiang, Passionfruit juice concentration by ultrafiltration and evaporation. Journal of Food Science, 51(6), 1501–1505 (1986).
(104) J. Mans, The champagne of apple juice concentrates. Prepared Foods, 161, 70 (1992).
(105) B. Jiao, R. Molinari, V. Calabro and E. Drioli, Application of membrane operations in concentrated citrus juice processing. Agro-Industry Hi-tech, 3(1), 19–27 (1992).
(106) J. R. Johnson, R. J. Braddock and C. S. Chen, Kinetics of ascorbic acid loss and nonenzymatic browning in orange juice serum: experimental rate constants. Journal of Food Science, 60(3), 502–505 (1995).
(107) H. Nabetani, Development of a membrane system for highly concentrated fruit juice. Journal of Membrane (Japanese), 21(2), 102–108 (1996).
(108) S. Alvarez, F. A. Riera, R. Alvarez, J. Coca, F. P. Cuperus, S. Bouwer, G. Boswinkel, R. W. van Gemert, J. W. Veldsink, L. Giorno, L. Donato, S. Todisco, E. Drioli, J. Olsson, G. Tragardh, S. N. Gaeta and L. Panyor, A new integrated membrane process for producing clarified apple juice and apple juice aroma concentrate. Journal of Food Engineering, 46(2), 109–125 (2000).

(109) A. Cassano, E. Drioli, G. Galaverna, R. Marchelli, G.Di Silvestro and P. Cagnasso, Clarification and concentration of citrus and carrot juice by integrated membrane process. Journal of Food Engineering, 57, 153–163 (2003).

(110) A. Kozak, S. Banvolgyi, I. Vincze, I. Kiss, E. Bekassy-Molnar and G. Vatai, Comparison of integrated large scale and laboratory scale membrane processes for the production of black currant juice concentrate. Chemical Engineering and Processing, 47, 1171–1177 (2008).

(111) Creative Energy, European Roadmap for Process Intensification, http://ispt.eu/cusimages/Report%20European%20Roadmap%20for%20Process%20Intensification%20(December%202007).pdf, last accessed 14 September 2012.

13

Process Intensification in a Business Context: General Considerations

Dag Eimer[1] and Nils Eldrup[2]

[1]*D-IDE AS, Teknologisenteret, Porsgrunn, Norway*
[2]*Sivilingeniør Eldrup AS, Teknologisenteret, Porsgrunn, Norway*

13.1 Introduction

Process intensification (PI) has been reviewed by others in the past [1]. Here the emphasis is on using PI to drive the innovation process, the challenges encountered and the selection of problems to solve. PI is traditionally defined as the art of reducing process equipment by two orders of magnitude. This will not be challenged here, but the present approach is partially different, although equally radical.

PI *is* about radical improvements to the processing of chemicals. It is about creating exciting new ideas. Sometimes somebody comes up with a stroke of genius, but more often improvements come about in vague, roundabout ways. (There are, however, techniques for making a search for ideas more efficient [2]). In the beginning there is an idea, probably a weakly defined idea that needs a lot of nurturing before it can bloom. Most ideas that can be classified as PI are likely to have had a difficult start. Being radical, they will have been opposed by critics. Some of this criticism will have been merited, because the idea in its original form was likely raw and unready. It may indeed have been a mere whimsical thought. Any such weakly formulated idea, or beginning of an idea, needs strengthening and development.

Process Intensification for Green Chemistry: Engineering Solutions for Sustainable Chemical Processing, First Edition. Edited by Kamelia Boodhoo and Adam Harvey.

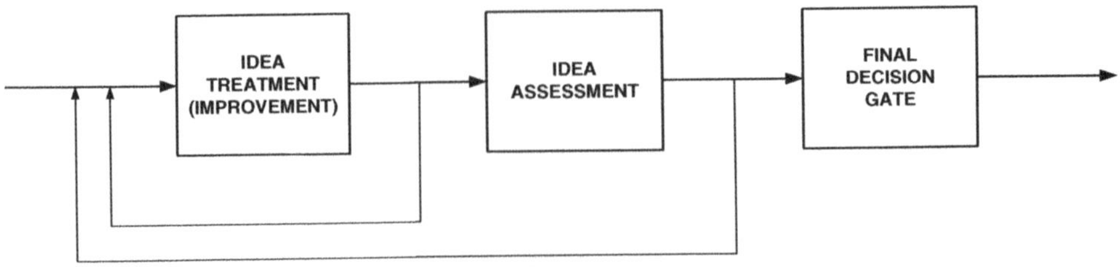

Figure 13.1 *Progress of idea development.*

There are formal techniques available to make this idea development work more efficient [2]. Figure 13.1 illustrates the progression of ideas and the need to reiterate development steps. The application of such techniques to technological problems and idea development is described by Eimer *et al.* [3] for a variety of radical ideas within the field of CO_2 capture. In this context, the idea assessment stage plays a very important role. When assessing technical ideas, it is prudent to do a few simple cost estimates, which point the finger in the direction, or directions, in which the idea might prove to be interesting. Some introductory engineering considerations will also be helpful towards this end. There are probably many ideas that have been prematurely, and perhaps wrongly, dismissed because they were not improved and strengthened beyond their early descriptions. It certainly makes sense to go back for further improvement after carrying out an initial assessment.

13.2 The Industrial Setting

Chemicals produced in bulk normally come from big capacity plants that are kept onstream for more than 8000 hours per year. A well-operated plant would these days be expected to be onstream close to 100% of the time, except for planned maintenance stops. Such stops are to a large degree dictated by key equipment in the process. There might be 2–3 years between planned shutdowns, although brief stops may have to be carried out to deal with problems as they arise. Any PI equipment that is introduced in the process train must live up to this standard. Pumps built to American Petroleum Institute (API) standard are supposed to run 25 000 hours between maintenance stops. Compressors and turbines usually have shorter periods specified by the manufacturer.

In a large chemical train process items with moving parts, such as compressors and pumps, require regular maintenance. Heat exchangers may need to be cleaned and repaired as leaks arise. Catalysts are changed at regular intervals due to gradual deactivation. Such stops provide the window of opportunity to maintain PI equipment as well, if special maintenance is needed. However, stops are kept as short as possible, and new maintenance tasks during a turnaround are undesirable. There must be a perceivable benefit to the operation of the plant to merit putting new technology into use, particularly if it involves a foreseeable need for periodic maintenance.

Operation of large plants is usually associated with a risk-aversion philosophy. There is a large investment to be protected. Profit margins are not necessarily very large, and may often be marginal. Risk has to be minimized in order to keep a high onstream factor. There is limited opportunity to stop and rebuild or 'make good'. The major opening in this barrier against new technology arises for so-called 'enabling technology', which opens up the manufacture of new products, allows a major reduction in investment and operating costs or facilitates a superior product specification. An example of such a PI application is Dow's process for producing low-chlorides hypochlorous acid (HOCl) [4]. Another compelling reason to accept a PI item is debottlenecking of a plant. Size may then be decisive: it is obviously an attractive project to debottleneck a 200 million GBP plant by say 5–10% for a marginal investment, but the situation might be different if the process plant is small and batch-orientated.

A shutdown in a batch plant is less dramatic as it will be partially stopped and started at short intervals anyway. There are also often many parallel units to keep other plant items in continuous operation. In such a setting, it is easier to accept new equipment, possibly with moving parts, even when it has a risk level that is not proven by previous long operation.

In general, there has to be a significant upside in order for investors to accept PI technology. Savings on the cost of an individual piece of equipment are very marginal. It has to be remembered that the equipment cost in a plant is in the order of 25% of the plant's erected cost. Hence any saving in a process item is only a saving in a quarter of the total cost. There are much bigger savings to be had if the PI technology proposed also reduces the cost of the remaining 75%, for example by reducing the support structure needed, avoiding piling operations, combining two process functions in the same apparatus and so on. Even so, it might be difficult to argue the point in a multi-unit process train. If a new unit were to enable a noticeable energy saving, it might be easier to convince investors. Such an energy saving could for instance come about if a reactor were to produce a product with fewer byproducts, thereby simplifying the ensuing distillation train. Again, the power of enabling new products or superior specifications must be emphasized.

Time to market in a batch plant that produces high-value chemicals or pharmaceuticals has been discussed in PI meetings in the past. Clearly an increase in the time available to exploit a product advantage before patent expiration is very valuable, and bringing the cash flow forward is not negative either.

Small equipment items may in themselves represent enabling technology by allowing retrofitting of new process functions or debottlenecking when space is a challenge. Space is usually a problem in any existing plant. For example, it might be prohibitive to add a gas-treating function to an oil and gas platform using conventional technology, but a small unit might be squeezed in.

HiGee has been a spectacular example and probably the flagship of the PI community. It has, however, not been the huge commercial success that was visualized 25 years ago. There are successful applications about, such as Dow's [4] and de-aeration plants in China [5], but the major breakthrough has not come, and presently there is no active marketing of the unit. In the main, it could be said that the battle against risk aversion has been lost. The applications attempted have certainly been associated with conservative process environments [6,7]: these were demonstration projects in which HiGee was used to treat natural gas. The first demonstration was for selective absorption of H_2S over CO_2, where a short residence time is important. The second was of CO_2 capture and water removal from natural gas. Both demonstration projects were claimed to be successful. Removal targets were met, and HiGee operated successfully.

PI is much more than HiGee and the other items usually talked about in the PI community, however. One example not usually put forward is the transport of powder. In earlier days this was dominated by belt conveyers. Today, more often than not the powder is fluidized and conveyed in a pipeline. There must have been a compelling economic incentive for this change. The picture is complex, but it includes maintenance, energy, spillage and manpower.

Another PI-based gain that has generally not seen action is the end to the practice of 'piped-in spare pumps'. This seems trivial, but there are considerable sums of money attached to it. Fifty years ago it was taught in chemical engineering that pumps should always be doubled as they were unreliable and broke down. The pump norm from API says that pumps shall be designed and built for 25 000 hours' maintenance-free operation. It would seem that there is a case for reconsidering this practice. Colleagues in operations that the authors have spoken to seem to confirm this notion.

In the oil industry, the phase separator that processes the stream from the well into gas, oil and a water phase does not look the same today as it did 30–40 years ago. The development has been evolutionary, and the term 'PI' has not been used, although it might be applied if we were to look at the change over a long period of time.

Another example of PI is in the fertilizer industry. Here melted fertilizer is sprayed into the top of a tall tower, where it falls as droplets countercurrently with rising air, which cools and solidifies the drops. This process is referred to as 'prilling'. Such a tower can be more than 100 m tall. An alternative is granulation, in which the melt is sprayed on to a rotating, cooled surface. This involves a dramatically smaller process unit. Settling and centrifugation are both practised. Obviously the economics do not dictate that the PI solution is superior, but the two options co-exist.

13.3 Process Case Study

It is useful to discuss the concept of PI and its implications in the context of a specific process. The absorption–desorption process for the capture of CO_2 from a gas stream has been chosen for a case study, specifically its application to flue gas treatment. This process is reasonably simple to grasp and an overview is easy to obtain such that the key points are not lost in a multistep process train description (Figure 13.2).

The flue gas is transferred to the capture plant through a long gas channel with a large cross-sectional area. The gas enters the process at a temperature higher than that desired for CO_2 capture. It is first cooled in a direct-contact cooler by a circulating water flow, which also washes out some impurities. A fan is needed to overcome the process pressure drop. The gas is then contacted with aqueous monoethanolamine (MEA) in the absorber, where 85–90% of the CO_2 is typically removed. The gas desorbs a little MEA from the absorbent, which is essentially removed in a closed-loop water wash in a separate section at the top the absorber. The CO_2-rich absorbent is then pumped through a heat exchanger, where heat is recovered from the regenerated absorbent before it enters the desorption column. CO_2 is desorbed in the desorption column, aided by water vapour generated in the reboiler. The water vapour has two functions: one is to provide desorption heat along the column and the other is to lower the partial pressure of CO_2 to aid the desorption. Water vapour is recovered from the overhead stream, rendering moist CO_2. The recovered water is pumped back to the desorber. The CO_2 lean absorbent leaving the reboiler is pumped back to the absorber via the heat recovery exchanger and a cooler.

The CO_2 extracted from the flue gas is compressed and dried for transport and storage (a compressor train is shown in Figure 13.3). A drying plant is typically introduced at a

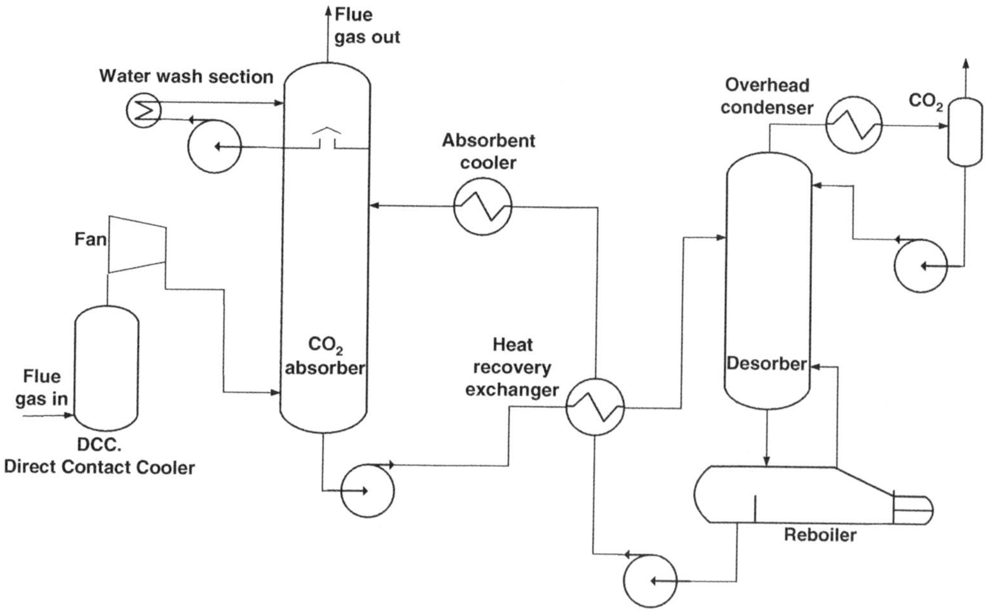

Figure 13.2 Absorption–desorption process for the capture of CO_2 from flue gas. Reclaimer, filters and tanks not shown.

convenient pressure level. It should be obvious that the higher the desorption pressure, the cheaper the compression train and the less energy it consumes.

Chemical engineering does not live a protected life. Costs are always important. Hence it is meaningless to analyse a process without including the cost aspects, even in the idea stage. In the present case, it is assumed that there is a generic 'brown field' site in, for example, the Rotterdam area. Site cost is excluded, and all utilities are assumed to be available at the battery limit at a cost.

This case study is an early phase estimate type, where limited information is available. However, mass and energy balances are needed to size equipment and to calculate operating

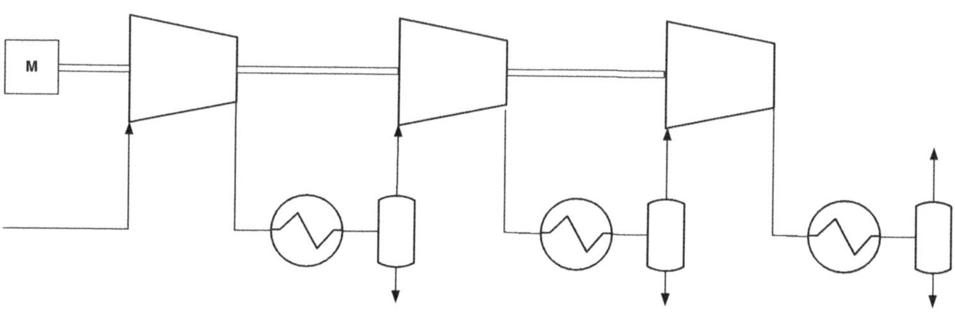

Figure 13.3 Compressor train for CO_2 compression downstream of recovery from flue gas. There will typically be three stages, with interstage cooling and condensate drainage.

costs. Such balances may well be rough and approximate. The idea is to obtain a picture of the process to be evaluated. In the present case, these balances are well documented through many published studies, but that is not the key point here. Equipment estimates can be obtained using a number of sources; in this case they are based on 'questimate'. These estimates are not very interesting as they appear. It is the total plant cost we are after. This is obtained by a factorial estimate using a so-called 'installation factor' (sometimes referred to as the 'Lang factor'), which is normally applied to the sum of the equipment cost and is based on experience from related plants. It should be quite clear that this factor will be much higher for low-cost equipment than for expensive equipment. For equipment costing around 10 000 EUR the factor is 10–15, while for 1 000 000 EUR it is around 4. In the present estimates, installation factors have been applied to equipment item by item and then summed. In this way we obtain a more realistic insight with respect to the cost impact of a selected item. This is turn makes it easy to see which items are worth our while addressing for the radical change implied by PI.

In the present case, this breakdown has been taken one step further by also allocating energy and chemical costs to each process item or group of items, as appropriate. Such costs have been set at a net present-value basis. Armed with this, it is possible to discuss the process in the context of radical improvements as exemplified by PI. It would be silly to run the risk of introducing radical new technology if there were no business driving force involved. Such a driving force does not have to be lower costs as such – it could equally be improved product quality or even a new product.

The key estimate normally asked for in a CO_2 capture process to make comparisons is cost per unit of CO_2 avoided released. ("CO_2 avoided" is explained in Reference [8]). In the present case the cost of CO_2 captured has been used. Next they would ask for kg steam used per kg CO_2 recovered and the estimated cost of the erected plant. This is fine for comparing technologies but is not enough to determine where in the process to look for future research targets. Even just to evaluate the merits of a research proposal, it is necessary to have a finer knowledge grid in order to make educated judgements.

Using a bit of chemical engineering insight in the process illustrated in Figures 13.2 and 13.3, it seems convenient to divide the process into three sections where the equipment in each section is primarily dimensioned by one consideration only:

- *"Absorber section"*: Dimensioned based on how much gas is to be treated. Includes the absorber and related items
- *"Amine Stripping"*: Dimensioned based on how much CO_2 is to be absorbed (Includes the desorber and related items including the absorbent heat exchangers)
- The CO_2 compression train (includes CO_2 drying). It is dimensioned based on how much CO_2 is absorbed.

With reference to Figure 13.4, it can easily be seen that the impact of OPEX (OPerating EXpenditure) is bigger than that of CAPEX (CAPital EXpenditure). This tells us that process improvements must focus on reducing OPEX. There are developments underway in this field in the form of proprietary absorbents, and so far a 30% reduction in stripping energy is claimed. In Figure 13.4 this would reduce the OPEX part of the 'amine stripper'. Other research efforts are ongoing, but so far nothing radical has emerged.

Going deeper into the 'absorber' section, the local costs involved here are depicted in Figure 13.5. Two items stand out. Not surprisingly, the largest is the flue gas absorber itself. The other is the flue gas fan. The CAPEX part of the absorber is to a large extent caused by

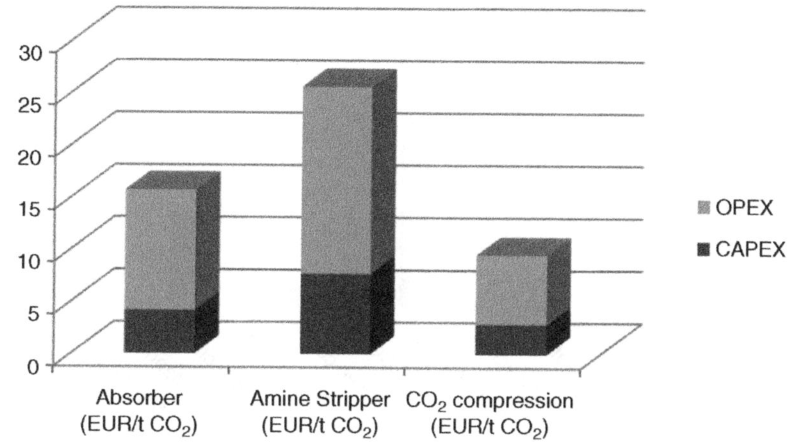

Figure 13.4 Costs in euros/tonne CO_2 for each of the three beakdown sections described. OPEX is operating expenditure on a net present-value basis. CAPEX is capital expenditure.

the use of stainless-steel column packings, supports and liquid distributors. The OPEX is the surprising part, since there is no energy use associated with the column itself. The absorber costs are further broken down in Figure 13.6, where it emerges that the amine replacement and its associated destruction costs are the contributing factors (based on published figures for the use of 30% weight MEA solutions). These replacement costs are to a large extent caused by amine degradation arising from SOx, NOx and oxygen in the flue gas, and possibly some dust particles. Clearly any process alterations that were to reduce the absorption of these components relative to the absorption of CO_2 would have an impact on this issue. The OPEX associated with the fan is associated with the pressure drop in the process. Any reduction in pressure drop could reduce the fan's need for energy.

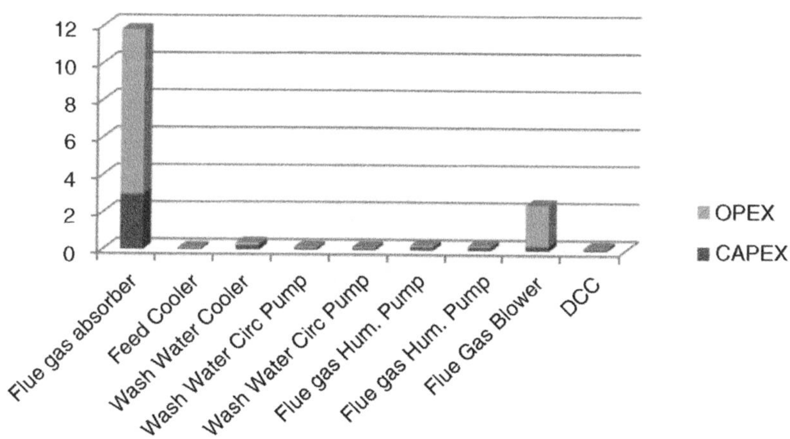

Figure 13.5 CAPEX and OPEX of the absorption section. Values are in euros/tonne CO_2.

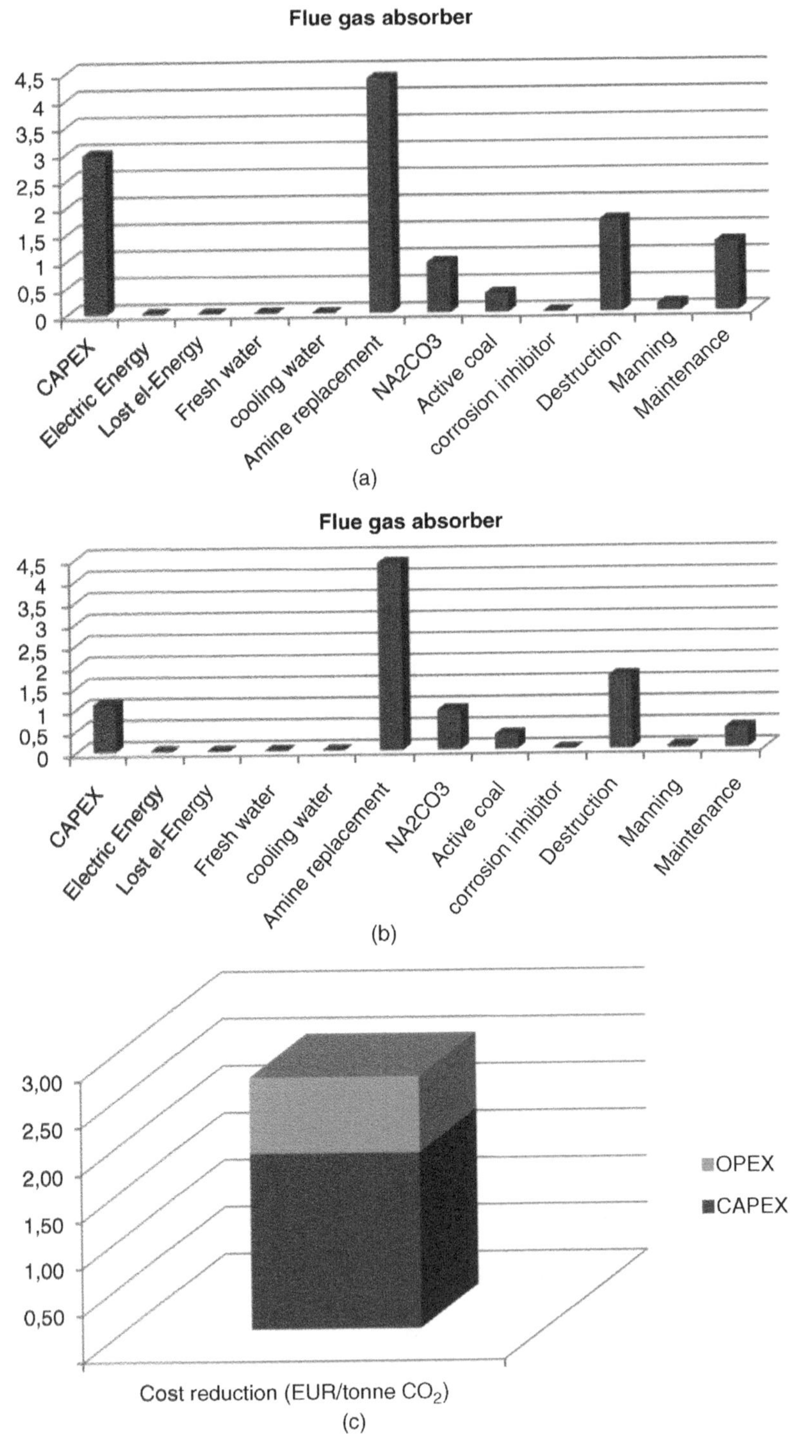

Figure 13.6 *(a) Allocation of costs to the absorber itself. (b) Allocation of costs to the absorber after cost reduction. (c) Contribution from cost reduction. Values are in euros/tonne CO_2.*

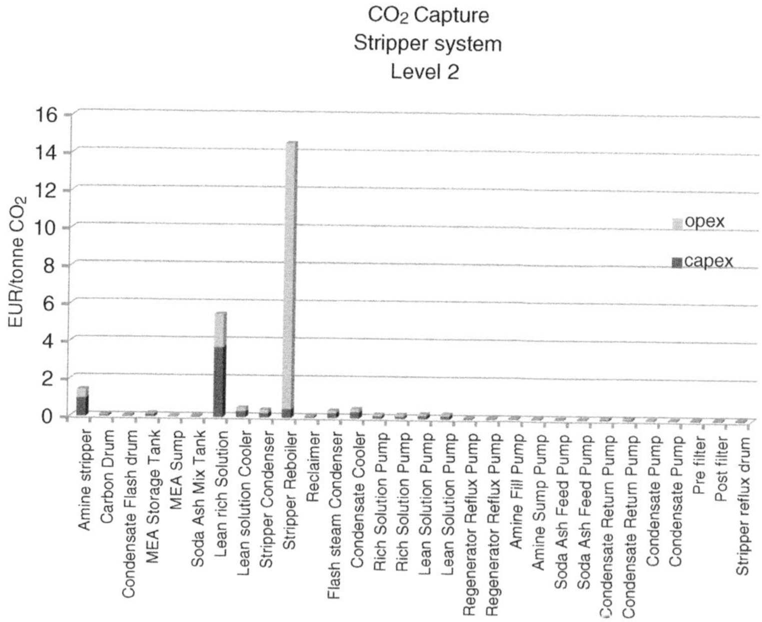

Figure 13.7 *CAPEX and OPEX of the amine stripping section. Values are in euros/tonne CO_2.*

A similar approach may be taken to analyse the 'amine stripper' section. Figure 13.7 shows the results of this analysis. The one cost that really stands out is the OPEX of the reboiler. This reflects the energy used by the process to desorb the CO_2, usually as steam at around 5 bar pressure from the power plant, which could otherwise be used to produce electricity using a condensing steam turbine. The desirability of reducing this energy is obvious. The other big cost is the CAPEX associated with the heat-recovery exchanger. Compact heat exchangers could well make an impact, particularly if the need for parallel units were reduced. The optimization trade-off for this heat exchanger is between heat recovery and energy needed for the reboiler. Its OPEX arises from imperfect heat recovery.

Amine degradation costs are normally associated with the absorption of components that react with the amine in the absorber. Since this is a cocktail of organic compounds, more chemistry would be expected to take place in the warmer amine stripping section. The higher the temperature, the faster the chemistry. A shorter residence time associated with a lower inventory would reduce this chemical 'production'.

The picture would not be complete without a few words about the CO_2 compression section. Taking the same approach as before, Figure 13.8 shows that the cost is dominated by the CO_2 compressor. Figure 13.9 shows that this can be further pinpointed to the electricity needed to drive this compressor. Energy for compression is associated with pressure ratios. It should be clear that an increased suction pressure would reduce the compression energy needed. Hence it would be advantageous if the desorption pressure were increased. This, however, would increase the desorption temperature, and a higher-quality steam would eventually be needed. Then there is the degradation chemistry issue. There is also an

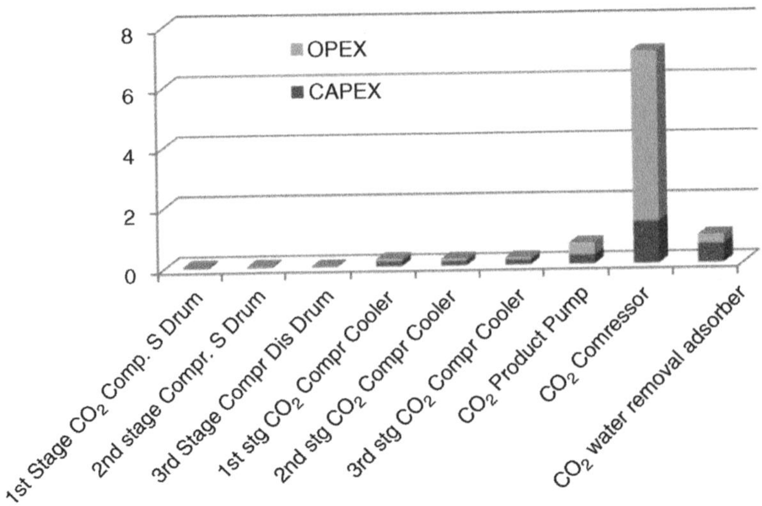

Figure 13.8 Distribution of costs for compressor train items, including CO_2 drying, not shown in Figure 13.3.

investment cost aspect to suction pressure. Gas is at its most voluminous when it enters. The first stage of a compressor thus tends to be expensive. A reduction in suction volume through a higher suction pressure would also reduce the cost of the compressor.

The impact of replacing the absorber with a rotating packed bed is indicated in the Tables 13.1 and 13.2.

13.3.1 Essential Lessons

Why is this case study relevant to a PI discussion? Essentially, it is useful to obtain a map of the problem you are trying to solve. It is an adage that 'if you do not know

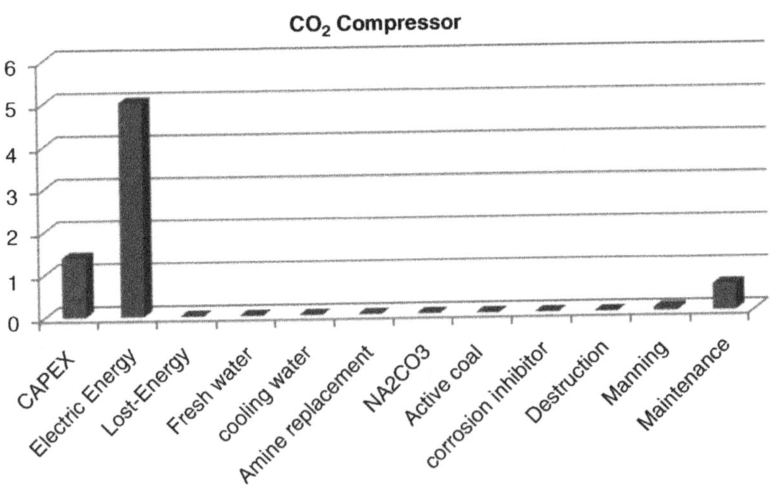

Figure 13.9 Split of compressor costs.

Table 13.1 *Equipment characteristics.*

Feature	Packed column	HiGee
Shell material	Carbon steel	Carbon steel
Shell dimensions	Diameter: 1.37 m Height: 13.7 m	Diameter: 1300 mm Length: 2640 mm
Others particular characteristics relevant to equipment	Random packing: stainless-steel 2″ Pall rings (assumed) Height of packing: 10 m (one bed)	Rotor dimensions: Outer diameter: 660 mm Inner diameter: 280 mm Axial length: 250 mm

Table 13.2 *Equipment costs.*

	Packed column	HiGee
Present estimate	119 000 euro	80 000 euro
Bucklin *et al.*'s estimate [6]	165 000 USD	150 000 euro

where you are going, you are not likely to get there'. This case study is about how early phase analysis of a process, including its economics, becomes a powerful tool in helping define research targets. The approach used here is generally applicable to any project.

The very first step in a creative process is to define a focus: a target on which to concentrate. There are several foci that more or less precipitate from this case study. Care has to be taken in how to formulate them. They must not be so tightly defined as to hamper the creative process, but too wide a definition would lead the process astray. Examples from the case study are not limited to, but include:

- How can the pressure drop be reduced to decrease the energy used by the fan?
- How can the erected cost of the absorber be reduced?
- How can hot and cold absorbent be heat exchanged more cheaply?
- How can the energy needed for desorption be radically reduced?

The role of PI in this is to add the term 'radically' to the last question. Let us go back to the description of PI as a way of reducing the size of equipment by two orders of magnitude. It is clear that such a challenge would help set a very different agenda for creativity than if the goal were simply to 'improve'. A different, but equally radical, angle would be to ask: Do we need this piece of equipment, or could the function be taken care of by another item already in the process? Or slightly differently: Could we combine two or more pieces of equipment? As an example, it has previously been suggested that the reboiler could be integrated with a rotating packed bed for distillation [9].

Many people working in research would argue that the type of analysis performed in this case study would be impossible until considerable work had been done. This is not true. Estimates, including the economics, can be made even if the idea is barely sketched. As is generally taught in chemical engineering design classes, the first thing to check is whether the product is more valuable than the raw materials. When an idea is first formulated and mulled over through a few idea development sessions, it should be possible to sketch a flow sheet and set up a simple mass and energy balance. Next some consideration will have to be paid to equipment sizes. For experienced engineers this would not be all that difficult. A lot of information would be lacking, but by making intelligent assumptions through analogies or on the basis of past experience it is possible to get there. Each assumption made becomes a research target for any next step. All assumptions should be on the optimistic side – that is important. If the analysis eventually shows that the idea is hopeless in spite of positive assumptions being made from A to Z, the idea should be discarded or taken back to the drawing board.

Radical assumptions in the best tradition of PI may be made. A PI based approach should be used to identify further cost effective solutions. The only word of warning here is that if a proposal is dependent on 10 radical ideas to succeed, the probability of carrying it off may be low. There would certainly an uphill struggle to convince investors, but if there is enough driving force, it is not impossible.

13.4 Business Risk and Ideas

There is no denying that introducing radical new technology involves some form of risk. However, history is full of cases where risk has been managed and the industry in question has gone on to new strengths. Risk taking is necessary in any business as no situation remains in a *status quo*. Change is constantly occurring. Those who do not manage change will disappear over time. Development will at some point meet a rate of diminishing returns, and this is where risk aversion really sets in. Unfortunately, there comes a time when nature limits further progress. At that point, completely new roads must be laid out, and visionary leadership is needed to take business to the next step change level. Engineers must respond to this.

If a company is to survive in a regime of changing technology, there has to be risk management. There are many forms this can take and many ways it can be organized. In the context of technology development, this essentially means carrying out the right tests and making sure there are highly qualified people available to specify them – and to foresee any stumbling blocks. The detailed engineering carried out based on a newly developed technology represents a different challenge to 'repeats'. More time should be allowed, and there may be questions which merit further visits to the laboratory. Engineering is the next step in the development process.

At some stage, however, there is a need to sit back and conclude that the technical work is complete, and business evaluation is needed. At this stage the risks involved should be clearly defined. If they are a bit on the 'woolly' side, this should be clearly stated! The questions to be answered are:

- Is the perceived risk justified? Could it be reduced?
- What is the upside?
- What is the downside?

If the conclusion is negative, it may be asked whether more work could rectify the situation. The stumbling points should be well described.

Pursuing radical ideas in an industrial environment requires a commitment from the top. Research efforts must be directed. It must be realized that most of the ideas created will be discarded. This should happen early, such that as few resources as possible are spent on failures. A conscious filtering process such as this will help to build confidence in those ideas that are pursued to completion.

13.5 Conclusion

In summary, PI has a role to play in the chemical processing and handling industry. That role may at times be subtle, but it is there. PI is a way of instilling radical angles in meeting challenges and thus helping facilitate step changes by shaking people out of their set ways.

References

(1) D. Reay, C. Ramshaw and A. Harvey, Process Intensification, Butterworth-Heinemann: Oxford (2008).

(2) E.de Bono, Think! Before it's Too Late, Vermillion: London (2009).

(3) D. Eimer, M. Sjøvoll, N. Eldrup, R. H. Heyn, O. Juliussen, M. McLarney and O. Swang, Creative chemical approaches for carbon dioxide removal from flue gas, in Carbon Dioxide Capture for Storage in Deep Geological Formations – Results from the CO_2 Capture Project, Volume 1, D. C. Thomas (Ed.), Elsevier (2005).

(4) D. Trent and D. Tirtowidjojo, Commercial operation of a rotating packed bed (RPB) and other applications of RPB technology, Proc. 4th Int.Conf on Process Intensification for the Chemical Industry, bHr Group, 11–19, Brügge, 10–12 September 2001.

(5) C. Zheng, Recent progress of high gravity engineering and technology: cases and strategies in China, IChemE Conf PII, Manchester, 21–22 October 1998.

(6) R. W. Bucklin, P. A. Buckingham and S. C. Smelser, The Higee demonstration test of selective H_2S removal with DEA, Laurence Reid Gas Conditioning Conference, University of Oklahoma, Norman, 7–9 March 1988, 227–237.

(7) R. Fowler, K. F. Gerdes and H. F. Nygaard, A commercial-scale demonstration of Higee for bulk CO_2 removal and gas dehydration, 21st Ann. Offshore Technology Conference, Houston, TX, USA, 1–4 May 1989, OTC 6121, 653–662.

(8) T. Melien, Economic and Cost Analysis for CO_2 Capture Costs in the CO_2 Capture Project Scenarios. Chapter 3 in, Carbon Dioxide Capture for Storage in Deep Geologic Formations – Results from the CO_2 Capture Project, volume 1, D.C. Thomas (ed), Elsevier, 2005.

(9) C. Ramshaw, The opportunity for exploiting centrifugal fields. Heat Recovery Systems & CHP, 13(6), 493–513 (1993).

14

Process Economics and Environmental Impacts of Process Intensification in the Petrochemicals, Fine Chemicals and Pharmaceuticals Industries

Jan Harmsen

Harmsen Consultancy BV, Nieuwerkerk aan den Ijssel, The Netherlands

14.1 Introduction

Process intensification (PI) can have a significant impact on greenhouse gas reduction in the process industries [1]. It is also an important new means of innovation in the chemical industry, promoting economic, environmental and social changes [2–6].

The value of these claims is explored in this chapter. The petrochemical (oil refining and bulk chemicals) industry is very different to the fine chemicals and pharmaceuticals industries. It has different drivers, different conventional technologies and different PI technologies. Therefore the two sectors are treated separately in this chapter.

Process Intensification for Green Chemistry: Engineering Solutions for Sustainable Chemical Processing, First Edition. Edited by Kamelia Boodhoo and Adam Harvey.

14.2 Petrochemicals Industry

14.2.1 Drivers for Innovation

14.2.1.1 Economic Drivers

The economic drivers for process innovation in the non-state-owned petrochemicals industries are lowest marginal cost, lowest capital investment and highest return on investment (ROI) [3].

Marginal Cost. Transport fuels and bulk chemicals are produced in very large-scale processes, ranging in capacity from 100 to 30 000 kton/year. Global competition is fierce as the cost of transport over sea is only a few per cent of the total manufacturing cost. This means that often the market price is close to or below the marginal manufacturing cost of most companies. Thus only companies with the lowest marginal cost can make a profit and survive in a recession (and subsidized state-owned companies). These companies also have to have the highest ROI, as only then will shareholders be willing to provide the capital needed for new investments.

Table 14.1 gives an insight into the main elements of the marginal and the capital cost [7]. It is clear that feedstock cost is by far the major element. The feedstock cost per ton of product is dependent on the market price per ton and on the product yield of the feedstock. The market price is generally the same for all companies. This means that those process technologies with the highest yield of product from the feedstock have the lowest feedstock cost. The highest product yield is obtained by the most selective catalyst, the most optimized reactor design and the best reactor conditions.

The second major cost element is the energy requirement. In most cases this is determined by the reactor yield, the separation technology and the amount of heat integration in the whole industrial complex, including the utility production section.

Transport cost is lowest by ship. Therefore, most oil refineries and bulk chemical processes are found at sea ports and often in combination with a river mouth. The latter is used to transport products inland.

Interest on capital, labour and maintenance is dependent on the process capital expenditure, so marginal cost depends on the capital expenditure of the process. All in all it is clear that profit or loss is strongly dependent on the process technology applied.

Table 14.1 *Fuels and bulk chemicals – marginal cost elements and their relative contributions.*

Cost element	Contribution (%)
Feedstock cost	75–90
Energy and utilities	5–10
Transport cost	1–3
Interest on capital	2–6
Labour cost	1–3
Maintenance cost	1–3
Total	**100**

Table 14.2 *Petrochemicals capital cost elements and their relative contributions (data derived from [7], Table 25–50 fluids processing).*

Element	Factor	%
Process equipment delivery (free on board)	1	22–18
Process equipment installation	0.27–0.47	6–9
Piping	0.66–1.20	15–22
Steel and concrete for the structure	0.13	3–2
Electrical and instruments	0.09–0.11	2–2
Building and site preparation	0.92–1.19	21–22
Auxiliaries	0.70	16–13
Contractor fees	0.21	5–4
Contingencies	0.42	10–8
Total	**4.40–5.43**	**100**

Capital Cost. Table 14.2 lists the major capital cost elements of petrochemical processes. The cost estimation method starts from the process equipment cost, obtained from the vendors. All other cost is then calculated using the factor values in Table 14.2. This method is of course only valid for conventional equipment.

If, for instance, a microreactor were to be chosen which had a higher equipment cost than the conventional reactor, but a much lower volume and weight, then this cost calculation method would result in a higher cost for all other elements, but it is clear that the steel and concrete cost would in reality not go up, but rather down, and all other costs would stay the same. To avoid this mistake it is better to take the same value for all cost elements outside the microreactor as for the conventional reactor. The steel and concrete cost is a small fraction of the total, so this simple method is justified.

Table 14.2 implies that reducing the amount of equipment reduces all other capital cost. Therefore, reducing the amount of equipment using PI, as in reactive distillation (RD), is a very effective way of reducing total cost.

Capital Cost and Scale-up. The capital cost of the petrochemical process typically increases with process production capacity according to the power law [7]:

$$\text{Capital cost} = \text{constant} \times \text{capacity}^{0.6} \tag{14.1}$$

This implies that the cost per ton of product decreases with increasing capacity:

$$\text{Capital cost/ton product} = \text{constant} \times \text{capacity}^{-0.4} \tag{14.2}$$

Other costs, such as labour, increase even less with capacity, while transport costs are constant per ton of feedstock and product. For this reason, petrochemical processes are designed for very large capacities.

Return on Investment. Accurate ROI calculations are complicated as they must take into account the capital expenditure over the installation time and the additional costs and low sales of the start-up period. For our purpose – a direct comparison between conventional and innovative technologies – a simple estimate of the ROI is sufficient. This takes the estimated annual profit:

$$\text{Profit} = \text{sales} - \text{cost} \tag{14.3}$$

as a fraction of the capital invested:

$$\text{ROI} = \text{profit}/\text{capital cost} \times 100\% \tag{14.4}$$

There is one problem with this simple method: sales prices vary enormously over time and cannot be predicted. However, as it is meant for a relative comparison between conventional and innovative technologies, it is not sensitive to sales prices. It is recommended that the product sales price and the feedstock price be taken for the same date, as they are generally strongly correlated.

The highest ROI is thus obtained by having the lowest marginal cost and the lowest capital investment. If these cost elements are in conflict, an optimization study must be carried out. In general the optimum is then found near the lowest marginal cost.

14.2.1.2 Environmental Drivers

The main environmental driver for the oil refining and bulk chemicals industries is the reduction of carbon dioxide emissions from their furnaces [3]. In some locations, reducing diffusive emissions of hydrocarbons is also a driver, as these can cause low ozone and smog formation [3]. Such emissions stem mainly from flanges connecting pipes to vessels. Therefore, another driver is reducing the number of flanges, by reducing the amount of equipment. The products of oil refineries must also have less impact on the environment, which means sulfur reduction in diesel and benzene reduction in petrol. This drives the search for energy-efficient processes, deeper hydrodesulfurization in the production of petrol and diesel and increased benzene extraction from refinery streams.

14.2.2 Conventional Technologies Used

Petrochemical processes consist mainly of conventional unit operations. The most frequently used unit operations are distillation for separation, fixed-bed catalytic reactors for conversions, heat exchangers for heating and cooling, pumps for liquid mass movement and compressors for gas mass movement. All processes are operated continuously.

14.2.3 Commercially Applied PI Technologies

The following PI technologies have been implemented many times in the oil refining and bulk chemical industries: RD, more than 150 times [8]; divided wall column distillation, more than 100 times [9]; reverse flow reactors (RFRs), more than 100 times [3]; while other PI technologies have been hardly or not at all applied.

Here we explain why certain PI technologies have been implemented many times and others not at all by using the economic and environmental drivers mentioned in Section 14.2.1.

14.2.3.1 Reactive Distillation

RD combines reaction with distillation in a single column. The heat of reaction is directly used for evaporation, thereby saving on capital cost for exchangers, while primary energy is saved since less unit operations – such as pumps and compressors – are needed, as is shown in Figure 14.1 for the conventional process for hydrodesulfurization (HDS) and the RD process of CDTECH. The capital cost and energy savings are both in the range 20–80% [8]. Because there are less flanges, diffusive emissions will also be decreased.

The rapid commercial-scale implementation is also due to the fact that the individual equipment pieces in the RD – column, heat exchangers, gas/liquid separator – are all conventional; only the catalytic packing is novel. Manufacturing of the pieces is therefore

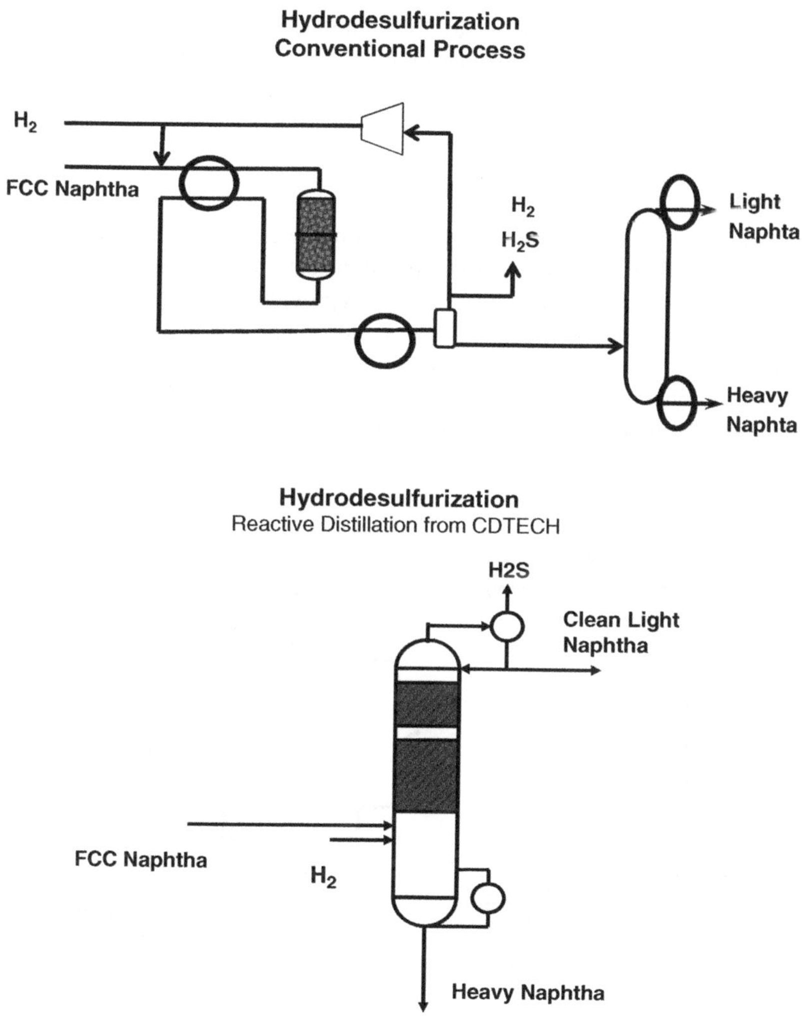

Figure 14.1 Conventional hydrodesulfurization and the RD process.

conventional. The catalytic packing can be obtained from Sulzer Chemtech, CDTECH or others. Scale-up and design methods are also available [8].

14.2.3.2 Divided Wall Column Distillation

Divided wall column distillation combines two distillations in one column. In the majority of applications, the column contains a dividing wall in the middle section, while the top and bottom section are common, as shown in Figure 14.2. In the front middle part a sloppy separation takes place, separating A from C, while component B stays in the top and bottom flows. In the back middle part, B is then separated from A and from C. This technology saves both capital cost and energy in the range 10–30%. [3,10]. Diffusive emissions of hydrocarbons will also be decreased, due to there being fewer flanges.

Divided wall columns have been rapidly commercially implemented in the last 2 decades in over 100 applications [9]. The individual elements are nearly all the same as those in a conventional distillation column. Scale-up, design and control methods are available [10].

14.2.3.3 Reverse Flow Reactors

In RFRs, heat exchange to heat the feed flow is combined with recovery of heat from the product flow and with a fixed-bed catalytic reaction section in a single vessel. Hayes [11] mentions that the technology is in use in industry, but provides no numbers. Mc Namara [12], of Shell, describes the commercial-scale reverse flow catalytic combustion of a waste gas stream from an ethylene oxide plant. A crude estimate using the number of implementations at Shell and the number of ethylene oxide plants and other partial oxidation plants worldwide results in a figure of more than 100 commercial applications [3].

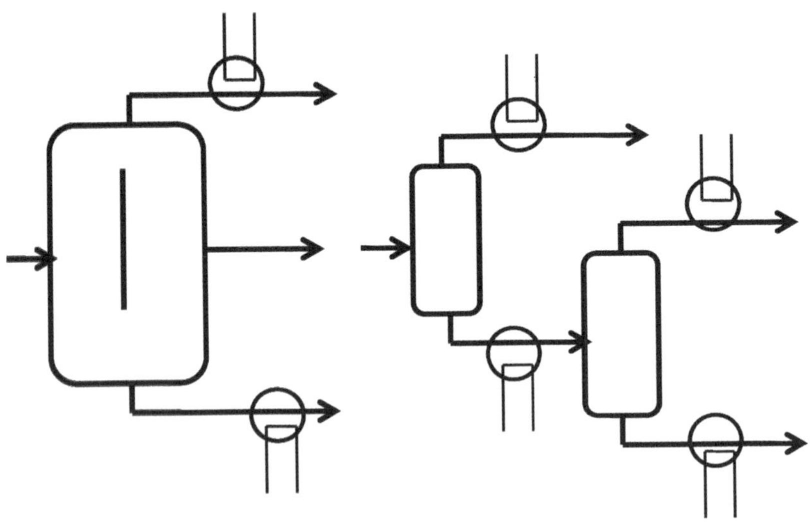

Figure 14.2 Divided wall column versus conventional distillation.

The business drivers for RFR implementation are more stringent emission requirements and capital cost reductions over conventional heat exchangers plus fixed-bed reactors. The hurdles are low as the individual elements of the technology – a vessel with a conventional catalyst and some valves – are all conventional to the industry, and design methods for commercial-scale design are available [13]. Scale-up is straightforward as it is a fixed-bed technology.

14.2.3.4 Microreactors

Microreactors have not yet been implemented at commercial scale in the petrochemicals industry. The reasons for this absence have to do with the use of solid catalysts and the capital expenditure of microreactors. We will treat both items in this section.

Almost all reactions in refineries and bulk chemicals are catalysed by solid catalysts. The use of this type of catalyst in microreactors poses the following challenges for the designer: how to fix the catalyst in the micropores; how to remove spent catalyst from the microchannel; and how to avoid plugging of the catalyst-filled narrow channels. So far, none of these challenges has been addressed.

The capital cost of microreactors for very large-scale applications is also an important item. We will quantify the capital cost of microreactors using capital cost data for microreactors and conventional reactors provided by Tonkovich [14] for the case of hydrogen production by steam methane reforming (SMR).

Tonkovich shows a design case for a hydrogen production capacity of $2\,m^3/s$. The microreactor system consists of five modules connected together via manifolds. Each microreactor has two sets of microchannels separated by heat-exchange walls. In one set, the solid particle catalytic steam reforming takes place. In the other, solid particle catalytic combustion of methane occurs, providing the heat for the endothermic SMR reaction. The installed capital cost of the microreactor system is 7.9 million USD, while the conventional reactor (a furnace fitted with the catalytic SMR reactor as pipes) has an installed capital cost of 11.2 million USD. So at this scale the microreactor has a lower capital cost than the conventional reactor.

Now the effect of capacity on capital cost is explored for this SMR case. According to Tonkovich [14], the capital cost of the microreactor rise linearly with the capacity – hence $n = 1$ – while for the conventional reactor the capital cost exponent is 0.6. To facilitate the use of this information in a more general way, we have plotted the capital costs of the microreactor and the conventional reactor against the heat-exchange duties of the reactors in Figure 14.3. The heat-exchange duty is a key parameter for reactors with high reaction temperatures. For SMR, the reaction enthalpy is 68.7 kJ/mol H_2 (206.1 kJ/mol CH_4). This graph is therefore now of more general use.

Figure 14.3 shows that beyond a heat-exchange duty of 15 MW the capital cost of the microreactor system exceeds that of the conventional reactor. In petrochemical reactor systems, heat-exchange duties in the order of 100 MW are common. Hence, for these large-scale capacities the microreactor is far more expensive than a conventional reactor.

Only in those cases where a microreactor with a highly selective and fast catalyst has a much higher selectivity (lower feedstock cost) than conventional reactors might it become attractive. However, such cases are hard to find in the petrochemicals industry.

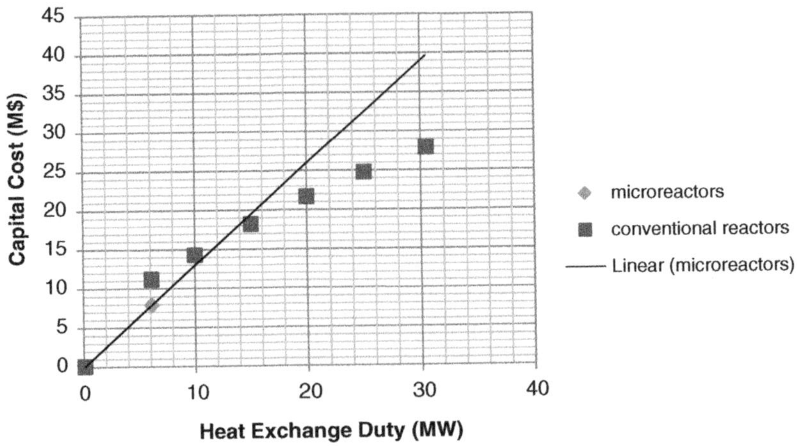

Figure 14.3 Capital costs of SMR reactors (microreactors versus conventional reactors). Data obtained from [14].

14.2.3.5 PI Technologies Using External Fields

The only commercially applied technology using external energy or force fields is the rotating packed bed (RPB) column [15,16]. It is applied by Dow Chemical for a chlorination reaction in which mass transfer affects the selectivity [17] and it is applied commercially in China for the de-aeration of water [16]. For all other technologies using external fields [15], no commercial applications have yet been reported.

The hurdles to implementing these technologies are higher because the scale-up knowledge required is more complex and the reliability of the technologies is unknown [3]. The commercial-scale implementation rate is expected to be low in the petrochemicals industry as reliability is an important item for economic operation [2,3].

14.3 Fine Chemicals and Pharmaceuticals Industries

14.3.1 Drivers for Innovation

The economic drivers for innovation in the fine chemicals and pharmaceuticals industries are speed-up of scale-up from exploratory chemistry to commercial scale, reliable product quality at commercial scale by the removal of scale-up effects and higher product yield on feedstock [18].

The raw material cost accounts for 30–80% of the total operating cost. The operating cost outside the raw material cost, averaged over 22 different commercially produced fine chemicals, is as follows: labour, 27%; plant cost, 10%; quality assurance and control, 12%; waste treatment, 15%; logistics and transport, 15%; and cleaning, 21% [18].

The batch reactor is 15% of the total capital investment of a fine chemicals plant, so capital savings on the reactor alone have a small impact on the total capital cost.

Environmental drivers are not explicitly mentioned by Roberge [18], but waste treatment is 15% of the total operating cost, so the environmental driver here is the same as that for the higher product yield on feedstocks.

Safety is mentioned by DSM [19] as a driver for implementing continuously operated microreactors instead of batch large-vessel reactors. Microreactors are inherently safer due to their very small reactive hold-up volume (factor 1000–10 000 smaller than batch reactors).

14.3.2 Conventional Technologies Used

The production capacity of fine chemicals and pharmaceuticals plants is in the order of 500 tons per annum [18]. The conventional process technology is a batch or semibatch process. A major piece of equipment is the mechanically stirred jacketed reactor with a vapour top cooler, which allows cooling by evaporation of a solvent. The reactions, and often also some separations (crystallization and filtering), take place in this vessel. Because the heat exchange of this reactor is very limited and the reactions are often highly exothermic, the operation is frequently carried out in semibatch, with slow feeding of one of the reactants. The reactions and separations can take a few days [18]; during this time, undesired consecutive reactions of the product may take place.

Note that the reactions of organic molecules with reagents that occur homogeneously in the liquid phase are in most cases very fast, with conversions taking place in seconds.

14.3.3 Commercially Applied PI Technologies

Only a few applications of commercial-scale PI technologies are known in these industries to date: DSM [19] for example has a continuously operated microreactor process in Linz with a capacity of 300 tons in 10 weeks. However, many applications are in the pilot-plant stage at various companies [19–21], and in the near future many commercial-scale microsystem applications are expected in these industries.

The reasons for the increased use of microsystems in these industries are the higher yield of product from feedstock and the increased safety, due to their very low reactive hold-up, much lower cleaning cost and lower waste treatment cost compared to conventional mechanically stirred jacketed vessels operated batchwise. But perhaps the most important factor will be the certainty that the product quality at commercial scale will be the same as that at the small experimental scale, due to the multichannel numbering scale-up design.

References

(1) D. Reay, The role of process intensification in cutting greenhouse gas emissions. Applied Thermal Engineering, 28, 2011–2019 (2008).

(2) S. Becht, R. Franke, A. Geiszelmann and H. Hahn, An industrial view of process intensification. Chemical Engineering and Processing: Process Intensification, 48(1), 329–332 (2009).

(3) J. Harmsen, Process intensification: its drivers and hurdles for commercial implementation. Chemical Engineering and Processing: Process Intensification, 49, 70–73 (2010).

(4) J. Harmsen and J. B. Powell, Sustainable Development in the Process Industries: Cases and Impact, John Wiley & Sons: New York (2010).

(5) G. J. Harmsen, G. Korevaar and S. M. Lemkowitz, Process intensification contributions to sustainable development, in Re-engineering the Chemical Processing Plant, Process, Intensification, A. Stankiewicz and J. A. Moulijn (Eds.), Marcel Dekker: New York (2004).

(6) A. Stankiewicz, J. A. Moulijn, Re-engineering the Chemical Processing Plant, Process Intensification, Marcel Dekker: New York (2004).
(7) R. B. Perry, Perry's Chemical Engineers' Handbook, 6th edition, McGraw-Hill: Singapore (1984).
(8) G. J. Harmsen, Reactive distillation: the frontrunner of industrial process intensification: a full review of commercial applications, research, scale-up, design and operation. Chemical Engineering & Processing: Process Intensification, 46, 774–780 (2007).
(9) G. Parkinson, Dividing-wall columns find greater appeal. CEP, May, 8–11 (2007).
(10) T. Adrian, H. Schoenmakers and M. Boll, Model predictive control of integrated unit operations: control of a divided wall column. Chem Eng. And Processing 43, 347–355 (2004).
(11) R. E. Hayes, Catalytic solutions for fugitive methane emissions in the oil and gas sector, Chem, Eng. Sci. 59, 4073–4080 (2004).
(12) D. Mc Namara *et al.*, Mission impossible. Hydrocarbon Engineering, March, 1–4 (2006).
(13) P. Marín, S. Ordóñez and F. V. Diéz, Simplified design methods of reverse flow catalytic combustors for the treatment of lean hydrocarbon–air mixtures. Chemical Engineering and Processing: Process Intensification, 48, 229–238 (2009).
(14) A. L. Tonkovich *et al.*, Micro channel technology scale-up to commercial capacity, 7th World Congress of Chemical Engineering, Glasgow, Congress Manuscripts, IChemE, 10–14 July 2005.
(15) A. Stankiewicz, Energy matters: alternative sources and forms of energy for intensification of chemical and biochemical processes. Chemical Engineering Research and Design, 84, 511–521 (2006).
(16) J.-F. Chen and L. Shao, Progress of high gravity technology – from fundamentals to nanomaterials, chemicals and pharmaceutical industrial applications, lecture and abstract, WCCE 8th Congress, Montreal, 23–27 August 2009.
(17) D. L. Trent, Chemical processing in high gravity fields, in Re-engineering the Chemical Processing, Plant, A. Stankiewicz and J. A. Moulijn (Eds.), Marcel Dekker: New York, 33–68 (2004).
(18) D. M. Roberge, L. Ducry, N. Bieler, P. Cretton and B. Zimmermann, Microreactor technology: a revolution for the fine chemical and pharmaceutical industries? Chem. Eng. Techn., 28, 318–323 (2005).
(19) DSM Corporate communications: microreactor, www.dsm.com, last accessed 14 September 2012.
(20) GSK Case studies, a radical alternative manufacturing approach, www.gsk.com, last accessed 14 September 2012.
(21) V. Hessel, Micro process technology – microreactors, novel process windows and applications, course presentations, http://ebookbrowse.com/course-hessel-lecture-1-micro-process-technology-2011-pdf-d208637273

15

Opportunities for Energy Saving from Intensified Process Technologies in the Chemical and Processing Industries

Dena Ghiasy and Kamelia Boodhoo

School of Chemical Engineering & Advanced Materials, Newcastle University, UK

15.1 Introduction

In 2010 the UK industrial sector consumed more than 27.5 million tonnes of oil equivalent, accounting for 18% of the total final energy spending [1]. Businesses are motivated to reduce their energy consumption due to the associated economical and environmental benefits, as well as the obligation to comply with legislations currently in place to cut CO_2 emissions. Energy savings in the chemical industry, which is the largest energy-consuming industrial subsector, can be achieved by improving the existing processes, commercializing new processes and novel technologies and making use of renewable energy sources, recycling, waste minimization and so on.

Process intensification (PI) has been identified as a powerful tool for drastically reducing the energy consumption of chemical processes. In the context of potential energy savings, Neelis *et al.* [2] estimate that energy loss from poor selectivity in conventional processes accounts for 500 PJ worldwide each year. Therefore, with increased selectivity being one of the highlighted benefits of PI, technologies derived from intensification principles may be exploited to increase the energy efficiency of processes across a range of sectors.

Process Intensification for Green Chemistry: Engineering Solutions for Sustainable Chemical Processing, First Edition. Edited by Kamelia Boodhoo and Adam Harvey.

On a broader scale, the potential energy saving due to investment in PI is estimated to be 40 PJ per annum for the UK's chemical sector, as outlined in a study carried out by Linnhoff March [3]. In the Netherlands, overall energy savings of 20% have been predicted by 2050 [4] from PI implementation in the petrochemical, fine and pharmaceuticals and food industries, largely due to improved selectivity, reduced energy consumption in separation processes and improved process control. It is even estimated that some sectors, such as the food industry, could save as much as 75% of their present energy consumption over the next 15 years through the application of more efficient water removal technologies [5]. Similar studies have been carried out by the US Department of Energy (DOE), predicting potential savings of more than 70 PJ per annum by 2020 [3].

In this chapter, we first highlight examples of the PI technologies which might be applied to those industrial operations that are particularly energy intensive in an effort to minimize energy consumption in the chemical and processing industries. We then present a case study which draws a comparison between the energy consumptions of a conventional and an intensified process. The process selected is based on the thermally initiated polymerization of styrene and is compared in a conventional agitated polymer reactor and in a spinning disc reactor (SDR).

15.2 Energy-Intensive Processes in UK Chemical and Processing Industries

In a recent survey of UK chemical and processing industries, which included pharmaceuticals, fine/specialty chemicals, oil and petrochemicals and the food industry [6], process heating was unsurprisingly found to be the most common operation, followed by mixing and low-temperature operations (Figure 15.1). These operations are carried out in a range of standard processing equipment, as outlined in Table 15.1.

15.2.1 What Can PI Offer?

As has been highlighted throughout this book, PI can improve the efficiency of common operations by removing heat- and mass-transfer limitations through the development of

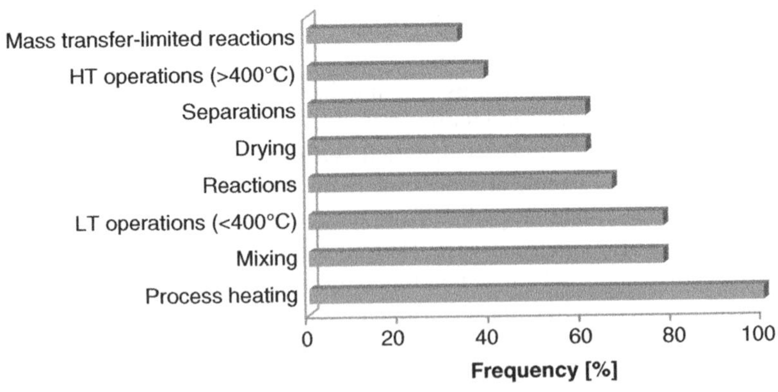

Figure 15.1 Processes commonly used in industry. HT- high temperature; LT, low-temperature.

Table 15.1 *Typical processing equipment used in chemical and processing industries.*

Process	Equipment typically used (frequency)
Process heating	Heat exchanger with hot liquid (78%)
	Heat exchanger with steam (72%)
	Direct steam injection (50%)
	Heat exchanger with hot air (22%)
Mixing	Stirred tank vessel (72%)
	Static in-line mixer (50%)
	Mechanical in-line mixer (17%)
Reactions	Batch stirred tank reactor (39%)
	Continuously stirred tank reactor (CSTR) (22%)
	Plug flow reactor (PFR) (6%)
Drying	Fluidized bed reactor (17%)
	Spray dryer (17%)
	Tray dryer (11%)
	Tunnel dryer (11%)
	Other (33%)

intensified equipment and novel processing methods. The design of novel equipment may involve the development of multifunctional units, such as reactive distillation (RD) columns and heat-exchanger reactors, or the development of systems based on more revolutionary approaches, such as high-gravity-fields technology. Ultimately it is desirable to combine intensification techniques in order to gain synergistic effects in process efficiency. A number of such intensification strategies have been established and a few, such as rotating strippers, rotating packed beds, RD and static mixers, have been fully developed and implemented, with widely demonstrated resulting benefits, while others are still in a premature phase and require more research and development [7].

The study carried out by Linnhoff March [3] estimated the potential energy savings in the UK chemical sector due to investment in a range of PI process unit operations (Table 15.2). According to these estimates, the UK chemical sector could have cut its energy consumption in 2010 by more than 21.5% as a result of overall plant intensification.

We present in this section a few potential technologies for the intensification of the various processes identified in the survey and shown in Table 15.1.

Table 15.2 *Potential energy savings in the UK chemical sector due to investment in PI. Adapted from [ref 3] © 2008, with permission from Elsevier.*

	Energy saving[a] (PJ/annum)	% of energy consumption in the UK chemical sector in 2010[b]
Compact heat exchangers	16	8.6
Separators	6.2	3.3
Reactors	11	5.9
Overall plant intensification	40	21.5

[a]Savings estimated in Linnhoff March report [3].
[b]Based on a total of 4439 thousand tonnes of oil equivalent consumed in 2010 [1].

15.2.1.1 Process Heating

A range of heat-transfer equipment, from compact to micro heat exchangers, has been developed [3] in order to enhance heat transfer efficiency. Active and passive enhancements may also be applied to improve the efficiency of heating processes [8]. The use of extended surfaces in the form of fins is just one of many passive enhancement techniques for the improvement of the rates of heat transfer in straightforward heating/cooling applications.

15.2.1.2 Mixing and Reactions

For a mixing-limited reaction, improving the mixing eliminates the physical barriers, which can lead to faster reaction rates, shorter processing times and higher product yields, all of which can result in lower energy requirements. Therefore, mixing is often an integral part of reactors (Section 15.2.1.3). Static mixers are perhaps one of the most commonly used pieces of PI equipment [9]. They can achieve plug flow, good radial mixing and uniform energy dissipation. As indicated in Table 15.1, static in-line mixers have already been implemented in industry. This is probably not all that surprising, considering that, of all the PI technologies available, they are one of the most well established and have found wide industrial application in many areas, such as polymerization [9].

15.2.1.3 Reactions

The heat/mass transfer and mixing capabilities of stirred tank reactors deteriorate significantly with increasing size, as the specific surface area and mixing efficiency decrease in consequence. This type of reactor configuration tends to impede reactions with inherently fast kinetics and is particularly unsuitable for parallel, competitive reaction schemes. Intensified reactor technologies, such as microreactors, SDRs and oscillatory baffled reactors (OBRs), on the other hand, provide enhanced mixing and heat/mass-transfer performances, as well as plug flow conditions.

As has been highlighted in Chapter 3, SDRs are particularly favoured for inherently fast, highly exothermic reactions, such as neutralization and polymerization [10,11], as well as reactions that require rapid mixing, such as crystallization [12]. SDRs can also offer benefits to processes involving heat-sensitive materials, due to their very short, controllable residence times. Microreactors operate on the basis of reduced diffusion paths in order to enhance the rate of heat and mass transfer and are mainly applicable to the pharmaceuticals and fine chemical industries [13,14], where the target production rate is low. Another potential application of microreactors is in distributed energy systems [15,16]. OBRs are compact units that provide plug flow characteristics with uniform and controllable mixing [17]. One of their main application areas is in biodiesel production [18].

15.2.1.4 Drying

Drying is one of the most energy-demanding operations in the chemical industry. The drying process can be intensified by reducing the drying time, and hence the energy consumption needed to achieve the desired outcome. Electric fields such as microwaves and radio frequency fields can be used to intensify the dying process. One example is the work carried out by Prabhanjan *et al.* [19], which showed that the use of microwaves for the drying of carrots resulted in a substantial reduction of between 25 and 90% in drying time. Although microwave drying in food processing was recognized as a promising technology a few decades ago and there are some current industrial applications, at present the uptake of this enhancement at industrial level is rather limited. There are a number of formerly successful applications – such as pasta drying and finish drying of potatoes, biscuits and crackers – that have been discontinued, partly due to the technical difficulties, which at first were not well understood [20].

15.2.1.5 Separations

Separation processes are important and energy-intensive operations in the chemical sector, commonly carried out in distillation columns. Several techniques for improving the energy efficiency of separation process plants have been identified and implemented in industry. One notable example is the HiGee rotating packed beds. With their small footprints and short residence times, these units offer the prospect of a more efficient separation process. Wang and Xu [21] provide an overview of the development of HiGee technologies over the past 3 decades and their existing industrial applications. However, as the rotation of the column requires input of energy, the net energy savings of such units may not be a strong driving force for their implementation. Rather, it is their much smaller footprint compared to conventional distillation towers which is their distinctive selling point.

Membranes can also be employed to achieve a separation task, offering high energy saving potential. The driving force for the separation may be a difference in pressure, concentration or electrical potential. The energy requirements of some membrane processes can approach the thermodynamic minimum, making them highly efficient, as discussed by Mulder [22].

Heat integration, retrofit and debottlenecking techniques can also be used to improve the energy efficiency of separation processes. Errico *et al.* [23] outline an industrial case study on retrofit of a multicomponent distillation plant, which led to 23% saving on energy with minimal modification to the existing plant.

15.2.1.6 Hybrid Technologies

PI may involve combining a number of unit operations into one multifunctional unit, which commonly involves the integration of reaction with at least one other function, such as heat transfer or separation. RD [24], membrane reactors [25], membrane distillation [26] and exchanger reactors [27] are some examples of this. Hybrid technologies can potentially lead to enhanced reaction rates and increased conversion/selectivity, as has been highlighted in various chapters in this book. Their reduced energy consumption is also an attractive feature.

15.3 Case Study: Assessment of the Energy Saving Potential of SDR Technology

The objective of this case study is to compare the energy consumed in polymerizing styrene via a free-radical mechanism in two types of an industrial-scale process. The first comprises a conventional stirred tank reactor operating in batch mode, while the second involves continuous processing in an SDR subsequent to a batch prepolymerization stage

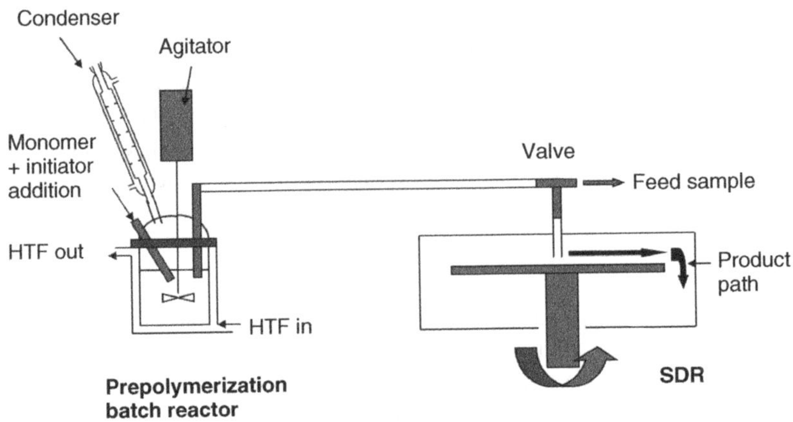

Figure 15.2 Schematic of combined batch and SDR processing (HTF: Heat transfer fluid).

(Figure 15.2) to reach the same final conversion level as in the first. The feasibility of the second process has been established on the basis of extensive laboratory data gathered in a previously reported experimental study [10,28].

15.3.1 Basis for Comparison

A production capacity of 20 tonnes of polystyrene at a conversion of 75% is used in this energy-evaluation study as a basis for comparison between the two processes.

15.3.2 Batch Process Energy Usage

The batch reactor volume is taken to be 35 m^3 (measured at 20 °C) and initially consists of 84% v/v styrene, 16% v/v toluene and 0.05 mole/l benzoyl peroxide initiator. In the lab-scale process, the required conversion level of 75% is achieved in 2 hours following the onset of polymerization [28]. In practice, the rate of polymerization would be expected to be much slower in larger-scale reactors, especially as viscosity increases with conversion. A reaction time scale of 10 hours in the industrial process is therefore a more realistic estimate.

Energy in the batch is consumed in the following ways:

15.3.2.1 Preheating Energy to Reaction Mixture

Assuming that styrene monomer and toluene solvent are heated to the required operating temperature of 88 °C by transfer of heat from heating oil circulating through the hollow agitator shaft, the heat required, ΔH_1, is calculated as:

$$\Delta H_1 = (\rho_s V_s C_{ps} + \rho_t V_t C_{pt}) \cdot \Delta T \tag{15.1}$$

The volumes of styrene and toluene in the initial batch process are 29.4 and 5.6 m^3, respectively. Assuming a temperature rise of 68 K to heat the contents from 20 °C to the operating temperature of 88 °C, ΔH_1 is estimated as 3470 MJ. Assuming a heat-transfer efficiency of 80%, the energy delivered by the heating oil is estimated as 4340 MJ.

15.3.2.2 Pumping Energy of Heating Oil

It is assumed that Dow Corning heating fluid (average kinematic viscosity assumed to be 135 cS at 88°C) is used as the oil in the heating process. Based on an average fluid velocity of 2 m/s in a 40 mm-diameter pipe and 20 m equivalent length of coil, the estimated pressure drop is calculated as 105 kN/m^2, giving a pumping power requirement of 263 W. If the period of operation of the pump is 10 hours, the energy input, with an assumed efficiency of 80%, is 12 MJ.

15.3.2.3 Pumping Energy of Cooling Water in Jacket

Assume cooling water of mass flow rate 1 kg/s flows in a carbon steel pipe of optimum diameter, d(optimum), given (for water) as [29]:

$$d_{optimum} = 293 \cdot \dot{m}^{0.53} \rho^{0.37} \tag{15.2}$$

The optimum diameter is calculated as 22.7 mm, giving a fluid velocity of 2.5 m/s. The pressure drop in an equivalent pipe length of 20 m is estimated to be 56 kN/m^2. The pumping power requirement is calculated as 56 W, giving a total energy input of 2.5 MJ over a period of 10 hours of operation based on a pump efficiency of 80%.

15.3.2.4 Mechanical Agitation Energy

Close-clearance impellers, such as an anchor impeller or a helical impeller, are the most suitable agitators for handling the highly viscous mixtures formed at high conversion levels of the styrene polymerization process. The typical power requirement of a close-clearance type of impeller in a bulk polymerization process is between 1.0 and 6.0 kW/m^3 [30].

For the purpose of this case study, we will assume that the average agitator power requirement for the duration of the batch polymerization is 3 kW/m^3. The zero shear viscosity of the polymerizing mixture at 75% conversion has been measured to be 36 Pa.s at 35 °C, while, at the higher operating temperature of 90 °C, the viscosity has been estimated as 0.028 Pa.s using literature correlations [28]. The middle-of-the-range value of 3 kW/m^3 for the power requirement of the agitator is selected on the basis of these relatively modest viscosities in this polymerization process. For a total volume of reaction mixture of 35 m^3, the agitator power consumed is 105 kW, and over 10 hours of polymerization time the energy input is 3785 MJ.

The total energy requirement for 75% conversion in the batch process is summarized in Table 15.3.

Table 15.3 *Energy consumed in batch polymerization process over 10 hours.*

Source of energy consumption	Energy consumed (MJ)
Preheating energy to reaction mixture	4340
Pumping energy of heating oil	12
Pumping energy of cooling water in jacket	3
Mechanical agitation energy	3785
Total consumption of the process	**8140**

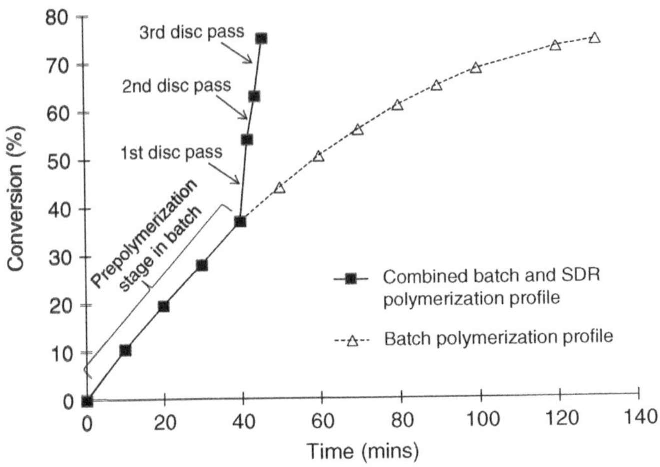

Figure 15.3 *Comparison of laboratory-scale styrene polymerization in batch and in a combination of batch and cascade of multiple disc passes to achieve 75% conversion [28].*

15.3.3 Batch/SDR Combined Energy Usage

Based on the laboratory experimental results for styrene polymerization in the SDR [10,28], it is predicted that a cascade of three successive passes on discs of 360 mm diameter rotating at 850 rpm, with the first disc fed with a 37% converted prepolymer from the batch stirred tank reactor, would be able to achieve the desired increase in conversion to 75%, as illustrated in Figure 15.3.

In the industrial-scale version of the combined batch and SDR process, the prepolymerization would be carried out in the batch vessel over the first 3.3 hours, giving a conversion of 37%. As explained earlier, the polymerization in the large batch reactor would be expected to be slower than that in the laboratory-scale reactor, hence the extended timescale for prepolymerization to above 3 hours. The prepolymer is then fed on to the spinning discs, rotating at 850 rpm. We assume that the cascade arrangement of three rotating discs in series, each of 360 mm diameter, can be replaced by a single disc of diameter 1.2 m to give the required performance. Three such discs would be used in a parallel arrangement (Figure 15.3), with a prepolymer flowrate of 50 cm^3/s to each, in order to handle the volume of material leaving the prepolymer reactor. The required processing time in the spinning disc stage under the selected operating conditions for 35 m^3 of 37% converted prepolymer is estimated to be about 66 hours. For the feed flow of 50 cm^3/s chosen in this case study, the film thicknesses at the periphery of the disc ($r = 0.6$ m), halfway across the disc ($r = 0.3$ m) and at $r = 0.15$ m are calculated from the film-thickness expression given in Equation 15.3 as 110 microns, 170 microns and 270 microns, respectively, with an average fluid viscosity of 0.15 Pa.s. The corresponding residence time (Equation 15.4) is estimated as 3.7 seconds.

$$\delta(r) = \left(\frac{3}{2\pi}\frac{\nu}{\omega^2}\frac{Q}{r^2}\right)^{1/3} \tag{15.3}$$

$$t_{res} = \left(\frac{81\,\pi^2\,\nu}{16\,\omega^2\,Q^2}\right)^{1/3}\left(r_o^{4/3} - r_i^{4/3}\right) \tag{15.4}$$

15.3.3.1 Batch Prepolymerization Energy Usage

The sources of batch energy usage for the period of prepolymer preparation are similar to those in Section 15.3.2, with the exception of the shorter timescale of 3.3 hours for the prepolymer stage reaction. The individual sources of energy consumption during the prepolymer preparation in batch are thus computed as follows:

- heat input to batch reaction mixture to increase temperature from 15 to 88 °C = 4340 MJ;
- pumping energy to heating fluid in coil immersed in reaction mixture = 4 MJ;
- pumping energy for water in cooling jacket = 0.8 MJ;
- agitation energy consumption in batch = 421 MJ (assuming an average power requirement of 1 kW/m^3 of reactor volume for agitation of low-viscosity reaction mixture at 37% conversion [30]);
- total energy requirement for prepolymerization in batch = **4766 MJ**.

15.3.3.2 Spinning Disc Energy Usage

The main sources of energy consumption in the SDR are:

- rotational energy of the dry disc, $E_{rotation}$;
- overall kinetic energy supplied to the fluid due to the change in its angular and radial velocity components between the entry and exit positions on the disc, $E_{K.E.}$;
- frictional energy dissipation of the fluid on the disc, $E_{friction}$;
- heat and pumping energy requirement of heat-transfer fluid in SDR, $E_{h.t.f}$.

These will be considered in turn below.

Rotational Energy Requirement, $E_{rotation}$. Assuming a torque of 10 Nm is required to keep the disc rotating at constant speed against the retarding frictional forces of the bearings, rotary union and surrounding medium (air), the rotational power consumed on a dry disc at 850 rpm is given by:

$$\begin{aligned} P_{dry} &= \tau\omega \\ &= \tau.\frac{2\pi N}{60} \\ &= 10 \cdot 2 \cdot \pi \cdot 850/60 \\ &= 890\,\text{W} \end{aligned} \tag{15.5}$$

With a motor efficiency of 70% and operating time of 66 hours, the total rotational energy delivered to the three disc systems is **904 MJ**.

Kinetic Energy of Liquid Film, $E_{K.E.}$. The change in kinetic energy of the film per unit time has been formulated by Khan [31] as:

$$P_{K.E} = \tfrac{1}{2}Q_f\left\{\rho_o\left(r^2\omega^2 + u_m^2\right)_o - \rho_i\left(r^2\omega^2 + u_m^2\right)_i\right\} \tag{15.6}$$

The mean radial velocity u_m of the film is derived on the basis of the centrifugal model, as described in [11]:

$$u_m = \frac{Q_f}{2\pi r\delta} = \left(\frac{\rho Q_f^2 \omega^2}{12\pi^2 \mu r}\right)^{1/3} \tag{15.7}$$

Taking $r_i = 0.05$ m and $r_o = 0.6$ m, $\rho_o = 978$ kg/m^3 and $\rho_i = 904$ kg/m^3, $\mu_o = 0.1$ Pa.s, $\mu_i = 0.002$ Pa.s, $\omega = 89$ rad/s (corresponding to a disc rotational speed of 850 rpm), the energy supplied by the motor (70% efficiency) due to the increase in kinetic energy for three discs over a total processing time of 66 hours is estimated to be **70 MJ**.

Frictional Energy Dissipation, $E_{friction}$. Another source of energy consumption in the disc system is the energy required to compensate for the power dissipated by the polymer material due to friction with the disc surface. The frictional power dissipation has been derived as [31]:

$$P_{friction} = \frac{1}{2}\rho Q_f \omega^2 \left(r_o^2 - r_i^2\right) \tag{15.8}$$

The power requirement for one disc is calculated as 95 W based on an average polymer density on the disc of 940 kg/m^3. Over the total period of operation of the three discs, the energy to be delivered by a 70% efficient motor is estimated as **68 MJ**.

Energy Requirement of Heat Transfer Fluid, $E_{h.t.f.}$. The heating of the disc surface is achieved by circulation of a heat-transfer fluid (e.g. water) underneath the disc surface through the shaft system. For each of the three disc systems, the heating fluid will be heated in a 30 l-capacity, constant-temperature bath by electric coils with an assumed efficiency of 80%. The pump operates at 80% efficiency and delivers water from the bath at the rate of 20 l/minute. On the basis of the given conditions, the heat and pumping energy for each SDR can be calculated:

Heat input to raise the temperature of 30 l of water in the bath from

$$\begin{aligned} 15 \text{ to } 88\,^\circ\text{C for one disc system} &= m_{water} \cdot C_{pwater} \cdot (T_{out} - T_{in})/\text{efficiency} \\ &= 30 \cdot 4.18 \cdot 10^3 \cdot (88 - 15)/0.8 \\ &= 1.1 \times 10^7 \text{ J} \\ &= 11 \text{ MJ} \end{aligned} \tag{15.9}$$

Assuming that the disc reaches the required temperature of 88 °C in 30 minutes and that the maximum temperature drop in the heating water circulating at a rate of 20 l/minute through the system during the heat-up period is 10 K, additional heat energy is needed to

Table 15.4 *Energy usage for combined batch/SDR process.*

Source of energy consumption	Energy consumed (MJ)
Batch prepolymerization	4766
$E_{rotation}$	904
$E_{K.E.}$	70
$E_{friction}$	68
$E_{h.t.f.}$	156
Total consumption of the process	**5964**

bring the water back to 88 °C:

$$\text{Additional heat supplied by coils operating at an efficiency of } 80\% = (20/60)\cdot 4.18\cdot 10^3\cdot 10\cdot 30\cdot 60/0.8 = 3.1\times 10^7\,\text{J or } 31\,\text{MJ} \quad (15.10)$$

$$\text{Total heat supplied for three disc systems} = 3\cdot(11+31) = \mathbf{126\,MJ}$$

Assuming a total pressure drop of 101.3 kN/m^2 in the internal heat-transfer system of the SDR, the power input to an 80% efficient pump required to compensate for the pressure drop in heating fluid flowing at a rate of 20 l/minute through one disc is 42 W. Based on a processing time of about 66 hours in three SDRs operating in parallel, the total energy consumed in pumping heating fluid is **30 MJ**.

The energy requirement for the combined batch/SDR system is summarized in Table 15.4.

15.3.4 Energy Savings

The percentage energy saved using the combined batch/SDR in comparison to the purely batch process is estimated as 27%.

A more realistic estimation of the percentage saving in energy achieved from the use of the rotating disc system can be obtained by comparing the energy consumed in the batch process and that consumed by the SDR to raise the conversion from 37 to 75%. The energy usage in batch for the stated change in conversion amounts to about 3374 MJ (= 8140 − 4766 MJ), while the SDR uses about 1200 MJ to reach the same conversion. The percentage saving in energy is thus calculated to be about 64%, which highlights the tremendous energy saving potential of the SDR as a polymer reactor.

15.4 Conclusion

PI offers the prospect of considerable energy savings for industrial processes, leading to reduced greenhouse gas emissions and a more sustainable future. This chapter has provided an overview of the various PI technologies that can enable such energy savings to be attained in a variety of common processes employed by the chemical and processing industries. The energy saving capability of intensified systems was illustrated in a theoretical case study involving SDRs for styrene polymerization. It was shown that the production of 20 tonnes of polystyrene at 75% conversion in a semibatch process consisting of three large discs of 1.2 m diameter rotating at 850 rpm and a feed flow of 50 cm^3/s on each disc could be achieved with more than 60% saving in energy compared to polymerization in an industrial-scale batch reactor.

Nomenclature

C_p	Specific heat capacity (J/kg K)
d	Pipe diameter (m)
$E_{friction}$	Frictional energy dissipation on disc (J)
$E_{h.t.f.}$	Energy requirement of heat transfer fluid (J)
$E_{K.E}$	Kinetic energy input into liquid film on disc (J)
$E_{rotation}$	Disc rotational energy (J)
ΔH	Heat input (J)
$\dot{m}$	Mass flow rate (kg/s)
N	Disc rotational speed (rpm)
P_{dry}	Rotational power consumed by dry disc (W)
$P_{friction}$	Frictional power dissipation (W)
ΔP	Pressure drop (N/m^2)
Q	Volumetric flow rate (m^3/s)
r	Radial distance from centre of disc (m)
T	Temperature (°C or K)
t_{res}	Mean disc residence time (s)
u_m	Mean radial velocity of film on disc (m/s)
V	Volume (m^3)

Greek Letters

δ	Film thickness (m)
μ	Dynamic viscosity (Ns/m^2 or Pa.s)
ν	Kinematic viscosity (m^2/s)
ρ	Fluid density (kg/m^3)
τ	Torque (Nm)
ω	Angular velocity ($= 2\pi N/60$) ($rads^{-1}$)

Subscripts

s	styrene
t	toluene
w	water
f	film
i	inner
o	outer

Appendix: Physical Properties of Styrene, Toluene and Cooling/Heating Fluids

$$\rho_s = 906\,\mathrm{kg/m^3} \text{ at } 20\,^\circ\mathrm{C}$$

$$\rho_t = 867\,\mathrm{kg/m^3} \text{ at } 20\,^\circ\mathrm{C}$$

$$\rho_w \approx 1000\,\mathrm{kg/m^3} \text{ at } 20\,^\circ\mathrm{C}$$

$$\rho_{oil} = 970\,\mathrm{kg/m^3} \text{ at } 25\,^\circ\mathrm{C}$$

$$C_{ps} = 1.598\,\mathrm{kJ/kg\ K}$$

$$C_{pt} = 1.738\,\mathrm{kJ/kg\ K}$$

$$\nu_{oil} = 200\,\mathrm{cS} \text{ (at } 25\,^\circ\mathrm{C)}$$

$$\nu_{oil} = 70\,\mathrm{cS} \text{ (at } 150\,^\circ\mathrm{C)}$$

References

(1) Industrial Energy Consumption in the UK. UK Department of Energy & Climate Change (2011).
(2) M. Neelis, M. Patel, P. Bach and K. Blok, Analysis of energy use and carbon losses in the chemical industry. Applied Energy, 84(7–8), 853–862 (2007).
(3) D. A. Reay, The role of process intensification in cutting greenhouse gas emission. Appl. Therm. Eng., 28, 2011–2019 (2008).
(4) Interim Report Action Group Process Intensification. Regiegroep Chemie (2007).
(5) European Roadmap for Process Intensification (2007).
(6) J. Smith, Assessment of energy saving potential of process intensification technologies in the UK Chemical and Processing Industries, MEng Dissertation, Newcastle University (2011).
(7) I. Stankiewicz and J. A. Moulijn, Process intensification: transforming chemical engineering. Chem. Eng. Prog., 96(1), 22–34 (2000).
(8) D. A. Reay, Compact heat exchangers, enhancement and heat pumps. Int. J. Refrig., 25(4), 460–470 (2002).
(9) R. K. Thakur, C. Vial, K. D. P. Nigam, E. B. Nauman and G. Djelveh, Static mixers in the process industries – a review. Chem. Eng. Res. Des., 81(A7), 787–826 (2003).
(10) K. V. K. Boodhoo and R. J. Jachuck, Process intensification: spinning disk reactor for styrene polymerisation. Appl. Therm. Eng., 20(12), 1127–1146 (2000).
(11) K. V. K. Boodhoo and R. J. Jachuck, Process intensification: spinning disc reactor for condensation polymerisation. Green Chem., 2(5), 235–244 (2000).
(12) L. M. Cafiero, G. Baffi, A. Chianese and R. J. J. Jachuck, Process intensification: precipitation of barium sulfate using a spinning disk reactor. Ind. Eng. Chem. Res., 41(21), 5240–5246 (2002).
(13) H. Pennemann, V. Hessel and H. Lowe, Chemical microprocess technology – from laboratory-scale to production. Chem. Eng. Sci., 59(22–23), 4789–4794 (2004).
(14) B. Pieters, G. Andrieux and J. C. Eloy, The impact of microtechnologies on chemical and pharmaceutical production processes. Chem. Eng. Technol., 30(3), 407–409 (2007).
(15) P. Pfeifer, L. Bohn, O. Gorke, K. Haas-Santo and K. Schubert, Microstructured components for hydrogen production from various hydrocarbons. Chem. Eng. Technol., 28(4), 474–476 (2005).
(16) A. Y. Tonkovich, J. L. Zilka, M. J. LaMont, Y. Wang and R. S. Wegeng, Microchannel reactors for fuel processing applications. I. Water gas shift reactor. Chem. Eng. Sci., 54(13–14), 2947–2951 (1999).

(17) X. Ni, M. R. Mackley, A. P. Harvey, P. Stonestreet, M. H. I. Baird and N. V. R. Rao, Mixing through oscillations and pulsations – a guide to achieving process enhancements in the chemical and process industries. Chem. Eng. Res. Des., 81(A3), 373–383 (2003).
(18) A. P. Harvey, M. R. Mackley and T. Seliger, Process intensification of biodiesel production using a continuous oscillatory flow reactor. J. Chem. Technol. Biot., 78(2–3), 338–341 (2003).
(19) D. G. Prabhanjan, H. S. Ramaswamy and G. S. V. Raghavan, Microwave-assisted convective air-drying of thin layer carrots. J. Food Eng., 25(2), 283–293 (1995).
(20) H. Schubert and M. Regier, The Microwave Processing of Foods, Woodhead Publishing: Cambridge (2005).
(21) G. Q. Wang, Z. C. Xu and J. B. Ji, Progress on Higee distillation – introduction to a new device and its industrial applications. Chem. Eng. Res. Des., 89(8A), 1434–1442 (2011).
(22) M. Mulder, Energy requirements in membrane separation processes, in Membrane Processes in Separation and Purification, J. G. Crespo and K. W. Boddeker (Eds.), Kluwer Academic: The Netherlands (1994).
(23) M. Errico, B.-G. Rong, G. Tola and I. Turunen, Process intensification for the retrofit of a multicomponent distillation plant – an industrial case study. Ind. Eng. Chem. Res., 47(6), 1975–1980 (2008).
(24) J. L. Degarmo, V. N. Parulekar and V. Pinjala, Consider reactive distillation. Chem. Eng. Prog., 88(3), 43–50 (1992).
(25) K. K. Sirkar, P. V. Shanbhag and A. S. Kovvali, Membrane in a reactor: a functional perspective. Ind. Eng. Chem. Res., 38(10), 3715–3737 (1999).
(26) E. Curcio and E. Drioli, Membrane distillation and related operations – a review. Sep. Purif. Rev., 34(1), 35–86 (2005).
(27) C. Habchi, T. Lemenand, D. Della Valle and H. Peerhossaini, Turbulent mixing and residence time distribution in novel multifunctional heat exchangers-reactors. Chem. Eng. Process., 49(10), 1066–1075 (2010).
(28) K. V. K. Boodhoo, Process Intensification: spinning disc reactor for the polymerisation of styrene, PhD Thesis, Newcastle University (1999).
(29) R. K. Sinnott, Chapter 5 of Chemical Engineering, volume 6, 4th edition, Elsevier Butterworth-Heineman: Oxford (2005).
(30) J. Y. Oldshue, Chapter 15 of Fluid Mixing Technology, McGraw-Hill (1983).
(31) J. R. Khan, Heat transfer on a rotating disc with and without phase change, PhD Thesis, Newcastle University (1986).

16

Implementation of Process Intensification in Industry

Jan Harmsen

Harmsen Consultancy BV, Nieuwerkerk aan den IJssel, The Netherlands

16.1 Introduction

Implementations of process intensification (PI) technologies in various process industries, notably oil refining, bulk chemicals, fine chemicals, pharmaceuticals and biomass process industries, are described in this chapter. The most important aspects of commercial implementation are given under the headings 'Practical Considerations for Commercial Implementation', 'Scope for Implementation in Various Process Industries' and 'Future Prospects'.

16.2 Practical Considerations for Commercial Implementation

Commercial implementation of a new technology in the process industries in general is preceded by various innovation stages, as shown in Figure 16.1. Essential elements in these stages are:

- *Exploration stage*: Concept design and experimental research at bench scale.
- *Development stage*: Scale-up and pilot plant.

Process Intensification for Green Chemistry: Engineering Solutions for Sustainable Chemical Processing, First Edition. Edited by Kamelia Boodhoo and Adam Harvey.

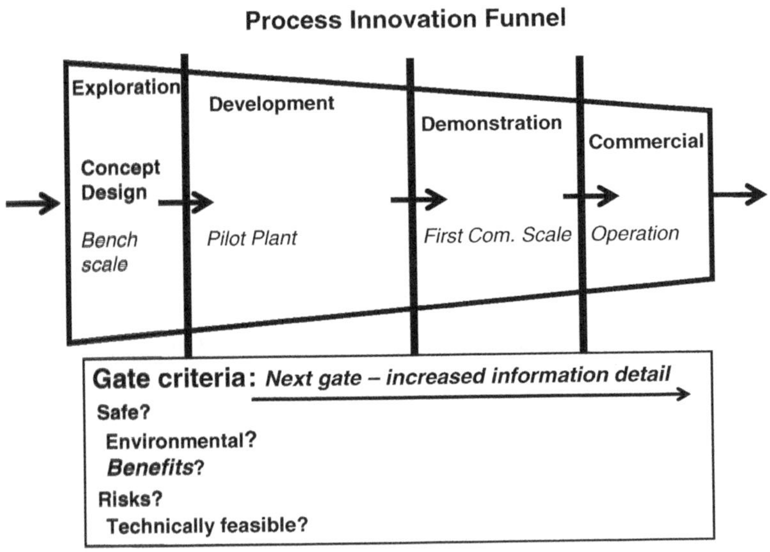

Figure 16.1 *Process innovation funnel.*

- *Demonstration stage*: Detailed design and construction.
- *Commercial stage*: Reliable operation.

The practical considerations of each stage are described in this section for a number of PI technologies.

The term "brute-force scale-up" is used a number of times. This means that all critical elements for scale-up performance are kept the same in both the pilot plant and the commercial-scale plant. In other words, the pilot plant is a perfectly scaled-down version of the commercial-scale plant. In this way the scale-up risk is greatly reduced.

For example, in the development of reactive distillation (RD) processes brute-force scale-up means that all items critical to scale-up are kept the same: the column height, packing, the location of the reactive section and the feed points and all gas and liquid velocities; hence it is a down-scaled version of the commercial design with minimal changes. The feed to the pilot plant is also the same as that in the commercial-scale plant; that is, it is obtained from the same industrial process. The commercial scale has then only a wider diameter and a larger capacity than the pilot plant, as shown in Figure 16.2.

This brute-force scale-up method can also be applied to microsystems, by keeping the micro channel dimensions and conditions on the small scale exactly the same as at the commercial scale, by ensuring that the feed distribution system over the channels at both scales is also kept the same and by applying the same feed quality.

16.2.1 Reactive Distillation

Concept design of RD is a far from trivial act. First, the choice to use it has to be made. Figure 16.3 shows some heuristic rules applied in the process industry [1]. Then the location and sizing of the reactive section, the locations of feeds in the column, the catalytic

Scale-up methods
Reactive Distillation

Commercial Scale	Pilot plant "Brute Force"	Pilot plant + Rate Models
Diameter	Smaller	Smaller
Height	Same	Smaller
Packings	Same	Smaller
Velocities	Same	Smaller
Feed quality	Same	Same
Good G/L distr.	Avoid wall flow	Avoid wall flow

Figure 16.2 *Scale-up methods for RD.*

packing size, the sizing of the distillation sections in top and bottom, the reflux ratio and the column pressure (and related temperature profile) must all be determined, while their effects on the column performance are all highly nonlinear. Information on catalytic packing types is provided by Harmsen [1] and a description of design methods is given by Sundmacher [2]. The software package SYNTHESISER is of particular help with process synthesis services provided by PDC [1].

For a large number of processes, concept designs are available from technology providers such as CDTECH and Sulzer Chemtech [1]. The concept design must be optimized by flow sheet packages. First an equilibrium-stage simulation package is used to give a quick impression of feasibility, then a rate-based package determines the column sizing.

When is Reactive Distillation Suitable?
Some Heuristic rules

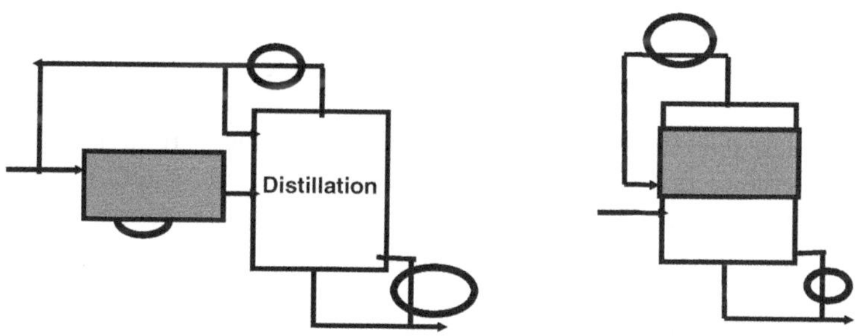

	IF	
Reactor expensive Heat exchange	and	Column same size
Catalyst fouling, deactivation	and	Avoided
Byproduct consecutive, equilibrium	and	Product out reaction phase
Azeotrope separation problem	and	Reaction

Figure 16.3 *When is RD suitable? Some heuristic rules.*

Experimental laboratory-scale work is needed to provide input for the concept design. Notably, the reaction chemistry and kinetics, the physical transport properties and the vapour–liquid equilibrium (VLE) model must be determined.

During development, a pilot plant will be needed to remove uncertainties concerning the selection of construction material, the tendency of fouling, catalyst deterioration and foaming.

The scale-up uncertainties can be removed by the brute-force scale-up method. This is applied in the oil refinery industry, in which an enormous number of chemical components are present that are not individually identified. Hence specific kinetics are not known, only lumped kinetics. The tall pilot plants required are available from CDTECH [1].

Another scale-up method is based on rate-based modelling and validation using a small-scale pilot plant with known mass transfer and residence time distribution (RTD) behaviour. This method is used for chemical applications. Sulzer Chemtech, for instance, provides pilot plant-sized columns [1].

To ensure a reliable detailed design and construction at the commercial scale, technology providers and engineering contractors with experience in RD are recommended. For refinery applications (hydrogenations, hydrodesulfurization), CDTECH is available. For esters and ether productions, Sulzer Chemtech and CDTECH can be used.

16.2.2 Dividing Wall Column Distillation

Concept design for dividing wall columns starts with concept design of a thermally coupled complex column system. Design methods are provided by Doherty [3]. The question of whether to combine two columns into one dividing wall column is in general less important than making the complex column design. The optimal complex design will reduce the energy requirement compared to simple column sequences. The divided column design will only save capital costs compared to separate columns.

Besides the conventional vapour–liquid transfer simulation, model simulations will include the impulse transport (pressure drop) simulation of the whole column, in order to ensure a design with proper vapour flow in the two column sections. Dynamic modelling facilitates the process control system design and optimization [4].

Experimental research will be needed to determine the VLE model if it is not available.

Whether a pilot plant development stage is needed is debatable. A reliable engineering contractor with experience in dividing wall column design and construction is generally sufficient to ensure successful commercial implementation as industrial pilot plant operations are not mentioned in the industry-implemented cases reported in Chapter 14.

16.2.3 Reverse Flow Reactors

The concept design of reverse flow reactors can be carried out simply, using a dynamic fixed-bed reactor model. Experimental laboratory research is needed to determine the reaction kinetics for the whole temperature range employed. A small-scale pilot plant in which the heat exchange via and through the wall is minimized is then needed to determine the catalyst stability under the dynamic conditions [5].

Scale-up can be done using the brute-force scale-up method or the design modelling method. Scale-up experiences are described by Harmsen [6].

A large number of technology providers are available to carry out the detailed design and construction of reverse flow combustors, used to remove small amounts of combustibles in a gas stream.

16.2.4 Microreactors

Concept design of microreactors is generally the same as for normal reactors. Detailed and accurate descriptions of heat transfer, RTDs, mass transfer and pressure drop integrated in a reactor model are however essential for the design of reliable concepts.

Experimental research at laboratory scale of a brute-force scaled-down design of the commercial scale is sufficient to prove the validity of the integrated model. The down-scaled reactor should have several microchannels in order to prove that the fluid distribution over the various channels is sufficiently even. Long-term runs of the micro-reactor will be needed to show the stability of the catalyst system and that no fouling is building up.

For the commercial scale, detailed design and construction dedicated technology providers are available for a variety of microreactor processes. Particular attention should be paid to even fluid distribution over the channels.

16.2.5 Rotating Packed Bed Reactors

Concept design of rotating packed bed (RPB) reactors is similar to that for conventional fixed-bed reactors, except that different expressions for mass transfer and pressure drop must be used [5,7–9].

Experiments at laboratory scale are needed to determine the reaction kinetics and, if not available, the physical transport properties.

Scale-up development will include a pilot plant to validate the reactor model with the special correlations for mass transfer, residence distribution and pressure drop [5,7–9].

Commercial-scale detailed design and construction should be carried out by an engineering contractor with proven experience in reliable design and construction of rotating bed reactors; alternatively, this detailed information can be obtained from Dow Chemical.

16.3 Scope for Implementation in Various Process Industries

16.3.1 Oil Refining and Bulk Chemicals

The business drivers for innovation in oil refining are providing the market with new petrol and diesel products with new specifications on very low sulfur content. In oil refining and bulk chemicals they are increasing the capacities of existing facilities, increasing the reliability of operation, reducing energy requirements and reducing diffusive emissions to the environment [1,6].

The real and perceived hurdles for PI in this industry are [6]:

- complex novel design outside the conventional design practice;
- need for an expensive pilot plant;
- need for a demonstration plant;

- difficult start-up;
- unreliable operation.

For RD, dividing wall columns and reverse flow combustors, the application range in the oil refining and bulk chemical industries is high. Furthermore, they have a good fit with the business drivers and all the hurdles mentioned have been overcome, mainly by technology providers with a robust and easy to understand scale-up method. The number of commercial-scale implementations for each of these technologies exceeds a hundred, and it is ever increasing [6].

For microreactors, the business drivers mentioned are less fitting. In particular, the economy of scale rule works against the application at the required large scale of oil refining. This is probably why so far no commercial-scale applications have been reported for the oil refining and bulk chemicals industries. Hydrogen production from methane in microreactors may be the first candidate for commercial-scale implementation, as the required production scale is relatively small and the economics look favourable [10].

For RPBs the applicability is very limited, the business drivers are not so fitting and the scale-up certainty and reliability of operation are lower (due to the rotating equipment involved) [6]. Dow Chemical has a commercial-scale RPB process in operation [11] and China has several RPB processes that use the same technology as Dow [8].

In general, the scope for PI technology in the oil refining and bulk chemicals industries is large for all technologies in which several functions are combined into a single vessel, such as RD and dividing wall columns, because of the fit with the previously mentioned business drivers and the methods available to reduce the uncertainties in commercial-scale implementation.

16.3.2 Fine Chemicals and Pharmaceuticals Industries

The drivers for these industries are the lowest cost of feedstocks and rapid commercial-scale implementation. Lower feedstock cost per mass of product can be obtained by reducing byproduct formation. This in turn can be done by using short residence times and perfect temperature control. The majority of reactors used in this industry belong to the mechanically stirred type, which has an overhead vapour condenser for cooling, is operated fed-batchwise and has a residence time of hours. This long residence time is generally due to the limited cooling capacity of the reactor, by which slow fed-batch feeding is needed in order to control the reaction temperature. Because of this reactor type there is ample scope for PI.

Notably, there is a large application area for high heat-exchange rate microreactors operated in batch or continuously. The reaction time of hours can then in most cases be reduced to a few seconds, which often also greatly reduces byproduct formation. An example of a microreactor for this type of application is provided by Zuidhof [12]. Moreover, the development time and the risk inherent in scale-up from laboratory-scale experiments to commercial scale can be greatly reduced by these microreactors, as the commercial-scale reactor is essentially the same as the laboratory-scale one, if it is brute-force down-scaled.

There is also scope for a multifunctional PI, such as RD, in these industrial branches, particularly where byproduct formation can be reduced by *in situ* removal of product from the reactive phase, which is feasible using RD [13] or reactive extraction [14].

16.3.3 Biomass Conversion

Biomass conversions to fuels and chemicals are generally more expensive than fossil oil conversions to the same products. This is due to the fact that biomass is produced on land and has to be transported by truck to the conversion plants. Moreover, a large part of the crude biomass is water, which does not end up in the product, further increasing the transport cost per ton of final product. Therefore, biomass conversions are carried out on a relatively small scale near the biomass growth areas. This small scale generally means higher capital cost per ton of product, however, due to the low economy of scale power (see Chapter 14).

The business drivers for biomass conversions are therefore: low transport cost, low-cost separation of water from biomass and low capital cost per ton of biomass at small pants (low plant production capacities) [15]. These drivers then provide ample opportunities for PI technologies in biomass conversion processing.

Biomass processing to fuel and chemical products involves reactions and separations. In most cases the latter means separating oxygenates like alcohols, esters or ethers from water. In this field, multifunctional PIs, such as RD, extractive RD, hybrids of membranes and distillation have a large scope of application.

Microsystems such as microreactors are already applied in the small-scale Fischer–Tropsch synthesis of diesel fuel using syngas from biomass gasification at Guessing biorefinery. The microreactor, with 900 microchannels, catalytically converts the syngas to alkanes [16].

N.B. An overview of biomass conversions to chemicals, using process intensification technologies, has appeared in ref. [17].

16.4 Future Prospects

In future, PI technologies will see increasing implementation due to the good fit with business drivers mentioned in Section 16.3.

In the oil refining and bulk chemical industries, future applications will be in all kinds of innovative multifunctional processes, such as reactive extractive distillation, reactive dividing wall column distillation and hybrid systems of distillation with membranes. All these applications will reduce the energy requirements, the capital cost, the operational cost, the maintenance cost and environmental diffuse emissions, and they will often also reduce the need for precautionary safety measures, since runaway behaviour will be prevented by the low reactive hold-up.

In the fine chemicals and pharmaceuticals industries, microsystems will be increasing applied to the reducing the feedstock cost per mass of product and to improving commercial-scale production, as scale-up effects are avoided by these reactors.

The number of commercial PI applications will be further increased by the introduction of courses on PI in BSc and MSc Chemical Engineering programmes.

References

(1) J. Harmsen, Reactive distillation: the frontrunner of Industrial Process Intensification: a full review of commercial applications, research, scale-up, design and operation. Chemical Engineering & Processing: Process Intensification, 46(9), 774–780 (2007).

(2) K. Sundmacher and A. Kienle, Reactive Distillation Status and Future Directions, Wiley-VCH: Weinheim (2003).
(3) M. F. Doherty and M. F. Malone, Chapter 10 in Conceptual Design of Distillation Systems, McGraw-Hill: Boston (2001).
(4) T. Adrian, H. Schoenmakers and M. Boll, Model predictive control of integrated unit operations: control of a divided wall column. Chem Eng. And Processing 43, 347–355 (2004).
(5) P. Marín, S. Ordóñez and F. V. Diéz, Simplified design methods of reverse flow catalytic combustors for the treatment of lean hydrocarbon–air mixtures. Chemical Engineering and Processing: Process Intensification, 48(1), 229–238 (2009).
(6) J. Harmsen, Process intensification: its drivers and hurdles for commercial implementation. Chemical Engineering and Processing, 49, 70–73 (2010).
(7) J. F. Chen, Characterisation of micromixing efficiency in rotating packed bed reactors. Chemical Eng. J., 121, 147–152 (2006).
(8) J.-F. Chen and L. Shao, Progress of high gravity technology – from fundamentals to nanomaterials, chemicals and pharmaceutical industrial applications, lecture and abstract, WCCE 8th Congress, Montreal, 23–27 August 2009.
(9) Y. Fei, Z. Haikui, C. Guangwen, S. Lei and C. Jianfeng, Modelling and experimental studies on absorption of CO_2 by Benfield solution in rotating packed bed, Chemical Eng. J., 145, 377–384 (2009).
(10) A. L. Tonkovich *et al.*, Micro channel technology scale-up to commercial capacity, 7th World Congress of Chemical Engineering, Glasgow, Congress Manuscripts, IChemE, 10–14 July 2005.
(11) D. L. Trent, Chemical processing in high gravity fields, in Re-engineering the Chemical Processing Plant, A. Stankiewicz and J. A. Moulijn (Eds.), Marcel Dekker (2004).
(12) N. Zuidhof, The Beckmann rearrangement of cyclohexanone oxime to ε-caprolactam in micromixers and micro channels, PhD thesis, TU Eindhoven (2011).
(13) M. C. de Jong, Reactive distillation for cosmetic ingredients: an alternative for the production of isopropyl myristate? PhD thesis, TU Eindhoven (2010).
(14) B. Schuur, Enantioselective liquid-liquid extraction in centrifugal contactor separators, PhD thesis, University Groningen (2008).
(15) J. Harmsen and J. B. Powell, Sustainable Development in the Process Industries: Cases and Impact, John Wiley & Sons: New York (2010).
(16) Velocys, www.velocys.com, last accessed 14 September 2012.
(17) J. P. M. Sanders, *et. al.* (a.o. Jan Harmsen), Process intensification in the future production of base chemicals from biomass, Special Issue of the journal "Chemical Engineering & Processing: Process Intensification", 51, 117–136 (2012).

Index

Note: Figures are indicated by *italic page numbers* and Tables by **emboldened numbers**; abbreviations: OBR = oscillatory baffled reactor; PI = process intensification; SDR = spinning disc reactor

Process Intensification for Green Chemistry: Engineering Solutions for Sustainable Chemical Processing, First Edition. Edited by Kamelia Boodhoo and Adam Harvey.

Printed and bound by CPI Group (UK) Ltd, Croydon, CR0 4YY
16/09/2026
14963473-0005